DIGITAL REVIEW of *AsiaPacific*
2007–2008

Supplementary news, reports and analyses are available for download at:
http://www.digital-review.org

DIGITAL REVIEW of *Asia Pacific*
2007–2008
\<dirAP\>

CHIEF EDITOR: Felix **Librero**
ASSOCIATE EDITOR: Patricia B. **Arinto**

EDITORIAL BOARD:

Danny **Butt**
Claude-Yves **Charron**
Suchit **Nanda**
Maria **Ng** Lee Hoon
Milagros **Rivera**
Rajesh **Sreenivasan**
Krishnamurthy **Sriramesh**
Jian Yan **Wang**

CONTRIBUTING AUTHORS:

Frederick John **Abo**
Musa **Abu Hassan**
Ilyas **Ahmed**
Zorayda Ruth **Andam**
Lkhagvasuren **Ariunaa**
Batpurev **Batchuluun**
Axel **Bruns**
Danny **Butt**
Donny **B.U.**
Elizabeth V. **Cardoza**
Claude-Yves **Charron**
Kapil **Chawla**
Masoud **Davarinejad**
Deng Jianguo
Hj Abd Rahim **Derus**
João Câncio **Freitas**
John **Fung**
Atanu **Garai**
Goh Seow Hiong
Lelia **Green**
Nalaka **Gunawardene**
Shah M. Ahsan **Habib**
Mohd Safar **Hasim**
Sarmad **Hussain**
Jong Sung **Hwang**
Malika **Ibrahim**
Seungkwon **Jang**
Jihyun **Jun**
Keisuke **Kamimura**
Kyungmin **Ko**
Thaweesak **Koanantakool**
Shelah **Lardizabal-Vallarino**
Heejin **Lee**
Lawrence **Liang**
Yu-li **Liu**
Geoff **Long**

Salman **Malik**
Muhammad Aimal **Marjan**
Jamshed **Masood**
Ram **Mohan**
Charles **Mok**
Rapin **Mudiardjo**
Frederick **Noronha**
Thein **Oo**
Sushil **Pandey**
Adam **Peake**
Phonpasit **Phissamay**
Gopi **Pradhan**
Ananya **Raihan**
Naomi **Robinson**
Massood **Saffari**
Lorraine Carlos **Salazar**
George **Sciadas**
Basanta **Shrestha**
Abhishek **Singh**
Rajesh **Sreenivasan**
Krishnamurthy **Sriramesh**
Tan Geok Leng
Suranart **Tanvejsilp**
Myint Myint **Than**
Tran Ba Thai
Tran Ngoc Ca
Kalaya **Udomvitid**
Brian **Unger**
Sajan **Venniyoor**
Eunice Hsiao-hui **Wang**
Sangay **Wangchuk**
Chanuka **Wattegama**
Esther Batiri **Williams**
Andy **Williamson**
Yong Chee Tuan
Zhang Guoliang
Zhang Xinhua

International Development Research Centre
Ottawa • Cairo • Dakar • Montevideo • Nairobi • New Delhi • Singapore

Los Angeles • London • New Delhi • Singapore
www.sagepublications.com

The views expressed in this publication are those of the authors and editor and do not necessarily reflect the views of the publishers. The designations employed and the presentation of material throughout this publication do not imply the expression of any opinion whatsoever on the part of the publishers concerning the legal status of any country, territory, city or area of its authorities, or concerning the delimitations of its frontiers or boundaries. The publishers do not guarantee the accuracy of the data published here and accept no responsibility whatsoever for any consequences of their use.

Copyright © Orbicom and the International Development Research Centre 2008

All rights reserved. No part of this publication may be reproduced, stored in a retrieval system or transmitted in any form or by any means, electronic, photocopying or otherwise, without the prior permission of the publishers.

SAGE Publications India Pvt Ltd
B1/I-1 Mohan Cooperative Industrial Area
Mathura Road, New Delhi 110 044, India
www.sagepub.in

SAGE Publications Inc
2455 Teller Road
Thousand Oaks, California 91320, USA

SAGE Publications Ltd
1 Oliver's Yard, 55 City Road
London EC1Y 1SP, United Kingdom

SAGE Publications Asia-Pacific Pte Ltd
33 Pekin Street
#02-01 Far East Square
Singapore 048763

International Development Research Centre
P.O. Box 8500
Ottawa, ON, K1G 3H9
Canada
www.idrc.ca
ISBN (e-book) 978-1-52550-377-5

International Network of UNESCO Chairs in Communication, Orbicom
Suite J-4351
Université du Québec à Montréal
P.O. Box 8888, Downtown Station
Montréal, QC, H3C 3P8
Canada
www.orbicom.ca

Published by Vivek Mehra for SAGE Publications India Pvt Ltd, typeset in 9.5/13 Times by Star Compugraphics Private Limited, Delhi, and printed at Swan Press, New Delhi.

Library of Congress Cataloging-in-Publication Data Available.

ISBN: 978-0-7619-3674-9 (Pb)   978-81-7829-840-5 (India-Pb)

The SAGE Team: Sugata Ghosh, Maneet Singh and Rajib Chatterjee

The *Digital Review of Asia Pacific* wishes to acknowledge the International Development Research Centre (IDRC) for its technical and financial support, for its commitment and encouragement to regional research and authorship, and for its dissemination of research results within and beyond the Asia Pacific region.

This edition of the *Digital Review of Asia Pacific*
is dedicated to the memory of
PROFESSOR VANNIARACHCHIGE KITHSIRI SAMARANAYAKE
whose contribution to ICT institution building
and human capacity development in Sri Lanka was outstanding,
as was his commitment to and engagement with
regional and international ICT4D platforms.

Contents

Foreword Muhammad Yunus **ix**
Preface Claude-Yves Charron and Maria Ng Lee Hoon **x**
Introduction Felix Librero **xi**
Acronyms **xiv**

Issues for the region

ICT4D in Asia Pacific—An overview of emerging issues Danny Butt, Rajesh Sreenivasan and Abhishek Singh **3**

Mobile and wireless technologies for development in Asia Pacific Tan Geok Leng and Suranart Tanvejsilp **19**

The role of ICTs in risk communication in Asia Pacific Krishnamurthy Sriramesh, Chanuka Wattegama and Frederick John Abo **29**

Localization in Asia Pacific Sarmad Hussain and Ram Mohan **43**

Key policy issues in intellectual property and technology in Asia Pacific Elizabeth V. Cardoza and Lawrence Liang **59**

State and evolution of ICTs: A tale of two Asias George Sciadas **73**

Review of individual economies

.af **Afghanistan:** Muhammad Aimal Marjan **89**
.au **Australia:** Lelia Green and Axel Bruns **92**
.bd **Bangladesh:** Ananya Raihan and Shah M. Ahsan Habib **102**
.bt **Bhutan:** Sangay Wangchuk and Gopi Pradhan **109**
.bn **Brunei Darussalam:** Yong Chee Tuan and Hj Abd Rahim Derus **117**
.kh **Cambodia:** Brian Unger and Naomi Robinson **122**
.cn **China:** Zhang Guoliang, Zhang Xinhua and Deng Jianguo **131**
.hk **Hong Kong:** Geoff Long, John Fung and Charles Mok **142**
.in **India:** Frederick Noronha and Sajan Venniyoor **150**

.id	Indonesia:	Donny B.U. and Rapin Mudiardjo 161
.ir	Iran:	Masoud Davarinejad and Massood Saffari 172
.jp	Japan:	Keisuke Kamimura and Adam Peake 180
.la	Lao PDR:	Phonpasit Phissamay 188
.my	Malaysia:	Musa Abu Hassan and Mohd Safar Hasim 196
.mv	Maldives:	Malika Ibrahim and Ilyas Ahmed 204
.mn	Mongolia:	Lkhagvasuren Ariunaa and Batpurev Batchuluun 211
.mo	Macau:	Geoff Long 217
.mm	Myanmar:	Thein Oo and Myint Myint Than 223
.np	Nepal:	Sushil Pandey and Basanta Shrestha 230
.nz	New Zealand:	Andy Williamson 236
.kp	North Korea:	Kyungmin Ko, Seungkwon Jang and Heejin Lee 244
	Pacific Island Countries:	Esther Batiri Williams 251
.pk	Pakistan:	Jamshed Masood and Salman Malik 263
.ph	Philippines:	Lorraine Carlos Salazar, Shelah Lardizabal-Vallarino and Zorayda Ruth Andam 268
.sg	Singapore:	Goh Seow Hiong 278
.kr	South Korea:	Jong Sung Hwang and Jihyun Jun 289
.lk	Sri Lanka:	Nalaka Gunawardene 296
.tw	Taiwan:	Yu-li Liu and Eunice Hsiao-hui Wang 304
.th	Thailand:	Thaweesak Koanantakool and Kalaya Udomvitid 313
.tl/.tp	Timor-Leste:	João Câncio Freitas 321
.vn	Vietnam:	Tran Ngoc Ca and Tran Ba Thai 325

Review of sub-regional associations

Association of Southeast Asian Nations: Lorraine Carlos Salazar and Shelah Lardizabal-Vallarino 335

South Asian Association for Regional Cooperation: Atanu Garai and Kapil Chawla 345

About the contributing authors 348
Index 362

Foreword

The overview of emerging issues in information and communication technologies (ICT) for development in Asia Pacific in this edition of Digital Review of Asia Pacific (DirAP) takes up the question of whether ICT ranks equally in priority with other sectors of development for investing the scarce resources of poor countries. It then takes the position that ignoring ICT will only lead to further excluding poor countries from the circuits of power and prosperity.

Indeed, this is a small world today and ICT is making it even smaller. The list of Impossibles in this world is shrinking. We should not wait too long to cross off a few more items from this list:

- It is impossible to eliminate poverty from this world.
- It is impossible to provide basic education to all.
- It is impossible to ensure necessary health care to the needy.
- It is impossible to make universal access happen.

ICT is quickly changing the world, creating a distance-less, borderless world of instantaneous communication. Increasingly, ICT is becoming less costly. Thus, ICT has much potential to create opportunities for growth and development in the rural areas of Asia. As ICT begins to create income generation activities in rural areas and as it becomes an instrument of rural economic and social activities, it begins to pay back on our hard-earned investments.

The Grameen Bank in Bangladesh, one of the poorest countries of the world, long ago made the choice to invest the present and future of the poor in ICT. ICT is a new opportunity for grassroots innovation. I saw an opportunity for the poor people to change their lives but only if this technology could be brought to them to meet their needs.

Towards this vision, we created Grameen Phone and we provided loans to poor women to buy phones to sell mobile phone services in the villages where they live. In this endeavour, we see the linkage between microcredit, our established strategy, and ICT, our newer strategy. Today, Grameen Phone is the largest phone company in Bangladesh and serves more than 12 million subscribers.

Social businesses such as Grameen Phone can play a significant role in creating opportunities that will help societies and their members to continue in the path of progress. Social business is a very important concept to me and very close to my heart. I define social businesses as a non-loss, non-dividend company, dedicated to achieving social objectives. Investors can take back their investment money, but they cannot get any dividend beyond that. Promoters of social businesses are the catalyst for positive change in a society.

Today, I would like to challenge our intellectuals, innovators, business leaders, corporations and institutions to help identify ways and means to help create social ICT businesses locally, nationally and internationally. Social business is a promising concept that I would like to bring into the ICT world, for we are applying it in earnest to our work with the poor in the villages of Bangladesh. I would like to emphasize that my challenge to our thought leaders is not only to create social business ideas in the ICT arena, but also to develop replicable designs that will help others in non-ICT sectors to be innovators of social ideas and social businesses.

In the future, I look to DirAP to document the stories of grassroots ICT innovation and learning for the Asia Pacific region, in technology deployment and research, as well as in innovative systems of delivery that bring useable ICT in a sustained manner to the doorstep of the poor.

Muhammad Yunus
Founder, Grameen Bank
Nobel Laureate, 2006

Preface

Information and communication technology for development in the Asia Pacific region: Encountering Rashomon's diversity of perspectives

The *Digital Review of Asia Pacific* (DirAP) has the mission of generating new descriptive, analytical and predictive knowledge about the field of ICT for development in the Asia Pacific region. It attempts to provide in-depth analyses and syntheses of ICT policy, developments and applications, and issues and debates concerning the significance of policy and technology enabling environments for national and regional socio-economic development. DirAP targets both regional and global audiences, especially decision and policymakers and practitioners from both government and NGOs.

From our perspective as publishers, DirAP's key contributions to the state-of-practice and state-of-the-art in ICT and ICT for development in Asia Pacific may be summarized as follows:

1. It adds a major source of research-based data and information to a field that is growing into a discipline with as yet relatively little research literature especially relating to Asia Pacific.
2. It gives ICT stakeholders in Asia Pacific opportunities to develop skills in research methods, research processes and research documentation.
3. It draws together a large number of leading ICT players from both developed and developing countries in Asia to reflect on platforms they identify as important for engagement to influence change.
4. It permits a time series narrative macro view of how total project investments by all parties aggregate into national syntheses on both country-level performance and issues-based performance.
5. It harnesses the intellectual contribution of a sizable community of practitioners and researchers from a multitude of disciplines from most of the developing countries of the region.

The voices of DirAP are independent and if they are ideological at all, they are the voices of these writers who are the key movers and shakers in the ICT for development arena in the region. We believe that this multiplicity of voices, which includes those of policymakers, professionals from the private sector and senior scholars, offers a unique opportunity to access the richness and the complexity of the debates, of the choices being made and to be made, and of the major issues faced in the interface between communication and development. And we strongly believe in the importance of this complementarity and diversity of voices, ensuring that, as in Kurozawa's *Rashomon*, the perspectives of the different actors are represented but also debated through research and statistical evidence.

The previous editions of DirAP were launched at the UN World Summit on the Information Society in Geneva (December 2003) and Tunis (November 2005) in both English and French versions, and they were extremely well received. We hope that this edition will provide you with an important source of perspectives about the major achievements in the midst of constraints as well as the challenges ahead, in your respective working environments within Asia Pacific. And we hope that this edition will provide a well-deserved visibility to the different types of ongoing experiments in the region to stakeholders in other parts of the world, on the different lanes of the Information Highway towards knowledge societies.

Claude-Yves Charron
The Network of UNESCO Chairs in Communications,
ORBICOM
Maria Ng Lee Hoon
International Development Research Centre

Introduction

In his Nobel lecture and in his Foreword to this edition of the *Digital Review of Asia Pacific* (DirAP), 2006 Nobel Laureate Muhammad Yunus, founder of Grameen Bank in Bangladesh, observed that information and communication technology (ICT) is transforming the world into a 'distanceless, borderless world of instantaneous communications' and that poor people can change their lives if they had access to and can use ICT to meet their needs.

For Muhammad Yunus, the first step in bringing ICT to the poor in Bangladesh was the creation of a mobile phone company called the Grameen Telecom of Bangladesh, a part of Grameen Communications, which was undertaken jointly with Telenor of Norway. Grameen Bank provided loans to the poor women of rural Bangladesh to purchase their own mobile phones and to sell mobile phones to villagers. The mobile phone business proved to be a brisk one and today there are almost 300,000 women engaged in the mobile phone business serving 10 million subscribers to Grameen Telecom of Bangladesh.

The operations of Grameen Communications are anchored on the basic principle that 'empowerment of disadvantaged individuals and groups can be accelerated through access to information'. In the Grameen Digital Center website, it is reported that because 'information regarding government, community, health, education, agriculture, environment, etc., is not available to all people, specially to rural people', the rural areas 'suffer considerably from lack of adequate information services' and the rural areas of Bangladesh have 'become distant centres of poverty and hunger due to the lack of communication and other support facilities'. By providing computer, Internet and mobile phone services in the rural areas, therefore, Grameen Communications is 'creating opportunities for addressing poverty and hunger through technological intervention'.

Indeed, there is growing recognition of the role that the new digital technologies can play in fostering human development. In 2002, former UN Secretary General Kofi Annan issued this 'challenge to Silicon Valley':

> The new information and communications technologies are among the driving forces of globalization. They are bringing people together and bringing decision-makers unprecedented new tools for development. At the same time, however, the gap between information 'haves' and 'have-nots' is widening, and there is a real danger that the world's poor will be excluded from the emerging knowledge-based global economy.

Also in 2002, the United Nations adopted the Millennium Declaration, considered a landmark document reflecting the aspirations and concerns of peoples worldwide, setting specific targets to reduce poverty, and calling for 'concerted action to fight injustice and inequality and to protect our common heritage, the earth, for future generations'. One of the many commitments made in that document is to 'ensure that the benefits of new technologies, especially information and communication technologies, are available to all'.

Between 2002 and 2006, several documents on ICTs and their role in people empowerment, poverty alleviation and development have found print. Moreover, a number of conferences have been held by governments and by international organizations, including the World Summit on the Information Society in 2003 and 2005. Two major recent initiatives are the Global Alliance for Information and Communication Technologies in Development (GAID), which was convened the second time in Kuala Lumpur on 19–20 June 2006, and the First Meeting of the Internet Governance Forum convened in Athens from 30 October to 2 November 2006. A key message from the Internet Governance Forum is that the Internet must be 'accessible, usable and safe for all'.

The DirAP seeks to contribute to the ongoing discussion of how best to put ICTs in the service of human development. DirAP is a biennial publication that aims to provide descriptive, analytical and reflective analysis of current initiatives and issues in ICTs for development (ICT4D) in the region. Given the diversity of concerns of more than 30 countries, economies and sub-regional organizations in the region, providing a descriptive analysis is not an easy task. However, the intention is not to focus on a specific direction that growth and development of ICT4D might take in the region, but to highlight some possibilities in light of current developments and the efforts of governments and institutions in the region.

In its 2003/2004 edition, DirAP reported on the status of ICT4D initiatives in 23 countries and economies, and provided an overview of issues on governance, open source and Internet politics in Asia Pacific. In the 2005/2006 edition, the qualitative analysis of the state of ICTs in 29 countries was complemented by a quantitative analysis that also provided a visual representation which revealed the wide gaps in the growth and development of ICTs across countries in the region. The 2005–2006 edition also included reviews of two sub-regional groups, ASEAN and APEC, and thematic chapters on: (a) bridging the digital divide in Asia Pacific, (b) Internet governance, (c) the social, political and cultural aspects of ICT, and (d) appropriate ICT for Asia Pacific.

This edition (2007/2008) continues the tradition of providing an analytical overview of the state of ICT4D in Asia Pacific. It covers 31 countries and economies, including North Korea for the first time. Each country chapter is an attempt to provide a relatively comprehensive coverage of the various aspects of ICT4D in each of the countries at the time that the chapter was written (in 2006). To provide a broad perspective of the issues covered, the chapters are written by a team of authors representing different sectors, such as government, academe, industry and civil society. There are also five thematic chapters providing a synthesis of some of the key issues in ICT4D in Asia Pacific today.

The banner thematic chapter titled 'ICT4D in Asia Pacific—An Overview of Emerging Issues' by Danny Butt, Rajesh Sreenivasan and Abhishek Singh analyzes current and emerging concerns regarding the growth and development of ICT4D in Asia Pacific, including the impact of ICTs on economic inequality, the environment, culture and content, and policy concerns with reference to Internet governance, e-Governance, regulation, and competition and security. The authors note that the dominant approach to ICT4D in Asia Pacific tends to be patterned after the approach adopted by advanced economies, with its focus on new technologies that might make older structures obsolete, limited discussion of potential risks or unexpected consequences, and little attention to cultural and social issues that are critical to project success. Rather than accepting a one-size-fits-all philosophy, governments in the region should 'formulate a strategy of engagement that suits their particular situation' while also 'foster[ing] networks where we can learn from the experiences of others in similar situations'.

In the area of regulation, for example, Butt, Sreenivasan and Singh note that because countries in the region differ greatly in terms of level of development, 'each country needs to develop its own set of culturally sensitive and national priority-consistent policies'. In so doing, countries must consider the need for regular and effective cooperation and coordination among ICT regulators and industry, a holistic view of the national and regional landscape, and focused and coordinated implementation. In general, governments must not forget that the non-ICT components of development are equally important, and that political will is necessary to ensure success.

In 'Mobile and Wireless Technologies for Development in Asia Pacific', Tan Geok Leng and Suranart Tanvejsilp discuss key technological developments—mobile phones, Wi-Fi, WiMax and meshed wireless networks—that bode well for ensuring universal access to the knowledge economy. The chapter highlights some highly innovative applications, especially of mobile phones, in Asia Pacific, including distance education via SMS, small-value transactions and e-governance. The chapter also discusses the barriers to use of mobile and wireless technologies in many parts of Asia Pacific that are caused by language and literacy, as well as some efforts to overcome them. The chapter concludes with 'a discussion of "development-friendly" policies that policymakers can adopt to expedite the rollout of [mobile and wireless] communication infrastructure and spur greater take-up of services and applications'.

An important application of mobile and wireless technologies in the Asia Pacific region is in risk communication and disaster management. In the thematic chapter titled 'The Role of ICTs in Risk Communication in Asia Pacific', Krishnamurthy Sriramesh, Chanuka Wattegama and Frederick John Abo provide a comprehensive analysis of experiences in the use of ICTs in dealing with serious public health emergencies such as the SARS and avian flu outbreaks and natural disasters such as the Asian Tsunami, volcanic eruptions, typhoons and other natural disasters that have recently occurred in the region with dramatic and tragic dimensions. The chapter highlights the importance of effective risk communication in disaster management in particular and in development efforts in general. Key regional and international programmes harnessing ICTs in risk communication during all phases of disaster management are described and a number of recommendations for policymaking in Asia Pacific countries are given. The recommendations include not only establishing the necessary ICT infrastructure that will enable the use of ICTs as tools in disaster management but also promoting 'risk communication in local languages, given that English is spoken by a mere fraction of the close to three billion people who live in the Asia Pacific region', harnessing the power of mass media and promoting coordination among them during disaster situations, including risk communication 'as one of the dimensions of the activities of telecentres, which are found in many rural Asian societies today', and participating in regional efforts.

Taking up the challenge of localization in Asia Pacific is the focus of the thematic chapter by Sarmad Hussain and Ram Mohan. Localization, defined as the 'process of developing, tailoring and/or enhancing the capability of hardware and software to input, process and output information in the language, norms and metaphors used by a community', is of utmost importance in a region where more than half of the world's 6,800 languages is spoken, including 21 of the 30 most spoken languages in the world, but where only about 10 per cent of the population is reached by the Internet. According to Hussain and Mohan, 'Asia Pacific is lagging behind in the use of ICTs not only because of the unavailability of affordable hardware and connectivity, but also because computing is still primarily in non-Asian languages', in particular, English.

Because of the 'great linguistic diversity of the region', localization is a complex undertaking for which there are no easy or short-term solutions. Hussain and Mohan point out that for policymakers, the decision to embark on localization projects involves striking a balance between the requirements of majority and minority languages and between basic and advanced localization, developing the necessary linguistic and technical expertise, generating resources, choosing the appropriate licensing option and computing platforms, and participating in regional and international standardization activities. They conclude that although it is challenging, the task of localization must be understood as 'an opening for Asia Pacific to revitalize its IT industry and to develop its knowledge economy'.

The intersection between intellectual property (IP) and technology is the focus of the thematic chapter written by Elizabeth Cardoza and Lawrence Liang. The chapter discusses key issues in IP policy for countries in Asia Pacific, particularly in light of increasing pressure from developed countries to impose IP regimes that fail to take into account the socio-economic, cultural and technological needs of developing countries. The key issues include copyright and its impact on access to knowledge and technology; emerging practices in IP protection that are likely to impact on technological development and innovation, such as digital rights management; and the implications of IP provisions in bilateral free trade agreements that go beyond the minimum requirements of the World Trade Organization (WTO) Agreement on Trade-Related Aspects of Intellectual Property Rights (TRIPS).

In the light of the possible trade-offs between stronger IP enforcement and technological and economic development in developing countries, Cardoza and Liang highlight the need for policymakers in Asia Pacific to make use of available flexibilities and exceptions under TRIPS and to adopt non-proprietary models of knowledge production and exchange such as free and open source software (OSS) and open content. In negotiating bilateral free trade agreements, countries need to take the initiative of proposing 'novel rules, related incentives and alternative mechanisms on policy areas where there are clear and significant interests' such as proposals relating to data protection and cultural heritage/traditional knowledge. They also need to ensure transparency in negotiations, foster better public awareness of what the stakes are, and weigh the impact of stronger enforcement of IP rights on other development priorities. Cardoza and Liang also recommend a regional or multilateral approach to negotiating trade agreements with powerful countries as this is more 'likely to result in commitments that are of general value and impact' and will thus help them avoid decisions that might jeopardize national socio-economic development goals.

DirAP's mission is to generate new descriptive, analytical and predictive knowledge about the field of ICT for development in the Asia Pacific region. It attempts to provide in-depth syntheses and analyses of ICT-related policies, developments and applications, and issues and debates that meet the needs of policymakers, academics, scholars and practitioners from government, the private sector and civil society. It is hoped that this edition of DirAP fulfills its mission.

Felix Librero
Chief Editor

Acronyms

3G 3rd Generation wireless networks
4G 4th Generation wireless networks
A2K Access to Knowledge
ACE ASCII Compatible Encoding
ADB Asian Development Bank
ADSL Asymmetric Digital Subscriber Line
AiDA Accessible Information on Development Activities
AP Access Point
APT Advanced Packaging Tool
ASCII American Standard Code for Information Interchange
ASEAN Association of Southeast Asian Nations
ASR Automatic Speech Recognition
ATM Automatic Teller Machine
AUSFTA Australia-US Free Trade Agreement
B2B Business-to-Business
B2C Business-to-Consumer
B2G Business-to-Government
BcN Broadband Convergence Network
BPO Business Process Outsourcing
BTS Base Transceiver Station
BWA Broadband Wireless Access
ccTLD Country-Code Top-Level Domain
CDMA Code Division Multiple Access
CD-ROM Compact Disc Read-Only Memory
CIT Cheque Imaging & Truncation
CPP Caller-Party-Pay
CPU Central Processing Unit
DAB Digital Audio Broadcasting
DLT Distance Learning Technology
DMB Digital Multimedia Broadcasting
DNS Domain Name System
DRM Digital Rights Management
DSL Digital Subscriber Line
DTT Digital Terrestrial Television
DVB Digital Video Broadcasting
DWDM Dense Wavelength Division Multiplexing
EDGE Enhanced Data Rates for GSM Evolution
FDI Foreign Direct Investment
FLOSS Free/Libre and Open Source Software
FOSS Free and Open Source Software
FTA Free Trade Agreement
FTTH Fibre-to-the-Home
FWA Fixed Wireless Access

Gbps Gigabits per second
GDP Gross Domestic Product
GPLv-3 General Public License of the Free Software Foundation (FSF)
GPRS General Packet Radio Service
GSM Global System for Mobile Communications
GSMA GSM Association
gTLD Generic Top-Level Domain
HDD Hard Disk Drive
HDSL High bit-rate Digital Subscriber Line
HDTV High Definition TV
HSDPA High-Speed Downlink Packet Access
HTML Hyper Text Mark-up Language
iB3G Integrated Beyond 3rd Generation
IC Integrated Circuit
ICANN Internet Corporation for Assigned Names and Numbers
ICDL International Computer Driving License
ICP Internet Connection Provider
ICT Information and Communication Technology
ICT4D Information and Communication Technology for Development
IDM Interactive and Digital Media
IDN Internationalized Domain Name
IDRC International Development Research Centre
IEEE Institute of Electrical and Electronics Engineers
IETF Internet Engineering Task Force
IOSN International Open Source Network
IP-DC Internet Protocol Datacasting
IPR Intellectual Property Rights
IPTV Internet Protocol TV
IPv4 Internet Protocol version 4
IPv6 Internet Protocol version 6
IR Information Retrieval
ISDN Integrated Services Digital Network
ISM Industrial, Scientific and Medical
ISO International Standards Organization
ISP Internet Service Provider
IT Information Technology
ITES Information Technology Enabled Services
ITU International Telecommunications Union
IVR Interactive Voice Response
IXP Internet Exchange Point

Kbps Kilobits per second
LAN Local Area Network
LCD Liquid Crystal Display
LDC Least Developed Country
LF Low Frequency
Mbps Megabits per second
MHP Multimedia Home Platform
MMS Multimedia Messaging Service
MPEG Moving Picture Expert Group
NAP Network Access Provider
NGN Next Generation Network
NGO Non-Governmental Organization
OA Open Access
OCR Optical Character Recognition
OECD Organization for Economic Co-operation and Development
OPGW Optical Ground Wire
OSS Open Source Software
P2P Peer-to-Peer
PBS Public Broadcasting Service
PC Personal Computer
PDA Personal Digital Assistant
PHS Personal Handy Phone System
PLC Power Line Communications
POP Point of Presence
PPP Public–Private Partnership
PSTN Public Switched Telephone Network
PTT Post, Telegraph and Telephone
PWLAN Public Wireless Local Area Network
R&D Research and Development
RFID Radio Frequency Identification
RIO Reference Interconnect Offers
SAARC South Asian Association for Regional Cooperation
SARS Severe Acute Respiratory Syndrome
SCPC Single Channel Per Carrier
SDH Synchronous Digital Hierarchy
SIM Subscriber Identification Module
SME Small and Medium-sized Enterprises

SMP Significant Market Power
SMS Short Message Service
SSML Speech Synthesis Mark-up Language
S&T Science and Technology
STM Synchronous Transport Module
TDM Time Division Multiplexing
TD-SCDMA Time Division Synchronous Code Division Multiple Access
TLD Top-Level Domain
TPM Technological Protection Measures
TRIPS Agreement on Trade-Related Aspects of Intellectual Property Rights
TTS Text-to-Speech synthesizer
TVRO Television Receive Only
UHF Ultra High Frequency
UNDP United Nations Development Programme
UNESCO United Nations Educational, Scientific and Cultural Organization
UPS Uninterruptible Power Supply
URL Uniform Resource Locator
USP Universal Service Provision
UTF Unicode Transformation Format
VAS Value-Added Service
VoiceML Voice Mark-up Language
VoIP Voice over Internet Protocol
VSAT Very Small Aperture Terminal
W3C World Wide Web Consortium
WAN Wide Area Network
W-CDMA Wideband CDMA
WGIG Working Group on Internet Governance
WiBro Wireless Broadband
Wi-Fi Wireless Fidelity
WiMAX Worldwide Interoperability for Microwave Access
WIPO World Intellectual Property Organization
WLAN Wireless Local Area Network
WLL Wireless Local Loop
WSIS World Summit on the Information Society
WTO World Trade Organization

Issues for the region

ICT4D in Asia Pacific—An overview of emerging issues

Mobile and wireless technologies for development in Asia Pacific

The role of ICTs in risk communication in Asia Pacific

Localization in Asia Pacific

Key policy issues in intellectual property and technology in Asia Pacific

State and evolution of ICTs: A tale of two Asias

ICT4D in Asia Pacific–
An overview of emerging issues

Danny **Butt**, Rajesh **Sreenivasan** and Abhishek **Singh**

Introduction

In 2007, it cannot be denied that information and communication technologies (ICTs) have had a transformative impact on the entire Asia Pacific region. Even in the least developed areas of the region, where ICTs have yet to make a significant mark on everyday life, the processes of lawmaking and the flow of economic goods are in some way influenced by globalization and networked markets enabled by ICTs.

As ICTs become central to the economic structure of countries all over the world, the approach to their role in social and economic development has become more sophisticated. In contrast to earlier policy agendas which sought to increase the use of ICTs as a pathway to achieving development, there is an increasing recognition that ICTs cannot be seen as inherently good or bad, as their effects are dependent upon the particular context of use. For example, access to e-commerce facilities may allow a producer to sell goods to an external market, resulting in higher sales. However, this same channel will allow the importation of goods and services which will pose a threat to local industry development.

Therefore, the decision to enter a global electronic market is a strategic one: the success factors for local businesses and regions will depend to a large degree on access to e-business skills, marketing budgets and distribution capacity. Where these exist, there is the possibility of greatly enhanced economic prospects through e-commerce. Where they do not, it is likely that exposure to global markets will result in overwhelming competition. The early Internet dream that held e-commerce as the saviour of the artisan producer has not come to fruition, even though there have been well-promoted individual success stories. Instead, the rise of ICTs has brought unprecedented consolidation in markets.

Globalization researcher Saskia Sassen explains that this is due to the separation of organizational functions enabled by ICTs: the distinctive way information technologies facilitate dispersal of routine activities and centralization of control activities explains the increasing dominance of cities in global economic activity (Sassen 1991). While there will always be success stories among the poor, there is no doubt that ICTs are, overall, increasing the gap between wealthy and poor businesses, countries and regions.

Little of the policy literature on ICT for Development (ICT4D) explicitly considers these larger overall effects of ICTs. More commonly, ICT4D discussions share uncomfortable similarities with ICT analysis emerging from highly developed economies, such as:

- a focus on new, 'revolutionary' technologies designed to make obsolete older structures and ways of working, with little assessment of the total costs of such a transformation;
- limited discussion of risk or unintended consequences from ICT developments; and
- an abstract theoretical model for economic development through technology that downplays cultural and social issues which are critical to individual project success in the ICT field.

Throughout Asia Pacific, the theme of Universal Access drives ICT4D policy. Policy initiatives carry ambitious titles such as 'Computers for all' or 'One laptop per child'. These are worthy ideals but in the policy setting they become problematic as they are never finally achievable and they provide little guidance for the tough decision-making that is required to support the use of ICTs where basic poverty issues, such as access to food, water and basic health care, remain unsolved. Kerry McNamara (2003, p. 1) points out this significant gap in evaluation:

> Despite a proliferation of reports, initiatives and pilot projects in the past several years, we still have little rigorous knowledge about 'what works'. There are abundant 'success stories', but few of these have yet been subjected to detailed evaluation. There is a growing amount of data about the spread of ICTs in developing countries and the differential rates of that spread, but little hard evidence about the sustained impact of these ICTs on poverty reduction and economic growth in those countries.

Given this unhappy trend, why should developing regions even consider expanding their investments in ICT? Would it not make more sense, as some advocate, to concentrate on traditional industries and means of development?

Our view is that even though ICTs are making economic development more challenging for developing areas, ignoring ICTs will only lead to further exclusion from the circuits of power and economic prosperity which rely on these technologies. But the challenges for a truly inclusive information society remain substantial and there are no magic solutions to follow from increased levels of investment in ICTs. It remains the task of each business, government, NGO and individual to formulate a strategy of engagement that suits their particular situation. More importantly, given the very unstable nature of ICT-enabled markets and relationships, it is imperative to foster networks where we can learn from the experiences of others in similar situations, rather than accepting one-size-fits-all philosophies proposed by those benefiting from the status quo.

Documenting and sharing these experiences is one of the key objectives of the *Digital Review of Asia Pacific* (DirAP). The thematic chapters in this issue, on Mobile and Wireless Technologies, Risk Communication, Localization, and Intellectual Property Regimes, highlight the wide range of options that are available for policymakers and ICT4D practitioners in these critical areas. The chapters on individual economies also highlight the diversity of ICT4D projects being undertaken throughout the region. In the rest of this chapter we outline the major trends in ICT as they affect human development—including technology, the knowledge economy, digital and economic divides, security, environment and e-Government. There is also an overview of the regulatory issues facing policymakers in Asia Pacific. After a brief look at the regulatory focus in developed ICT market countries, the chapter focuses on trends observed in developing ICT market countries and seeks to distil the key elements of regulatory approaches that have seemed to work, that have managed to encourage the growth of a healthy, competitive and innovative culture around ICTs. Our hope is that regulators and authorities in Asia Pacific can assess their own policy framing and implementation mechanisms against these measures and in cases where the measures are already adopted and part of the regulatory approach, there may be scope for a modified approach tailored to a country's culture and business environment.

Technological developments

Technological developments continue to bring about significant changes in cultural and economic life in Asia Pacific. The most significant changes are coming about through three linked areas of innovation: broadband, convergence and wireless. These technological changes reshape the social and economic opportunities that are available in the online environment.

Broadband

Broadband diffusion is continuing at a rapid pace. At the technological level, DSL has established itself as the dominant protocol for broadband delivery among regions with a high investment in fixed-line Plain Standard Telephone Network (PSTN) infrastructure. The growth in available bandwidth via broadband changes the kind of content that is available to users, because of two distinctive characteristics.

First, broadband is 'always-on', which means that broadband Internet networks take on the form of a utility or basic service, in the same way that telephone or broadcast networks are consistently available. This means that Voice over IP (VoIP), for example, can become a viable replacement for the telephone, although the opportunities to significantly decrease telephony costs are somewhat offset by the lack of control and reliability that nations expect from critical infrastructure. Who gets to decide what is 'good enough' for service delivery remains a critical issue.

The second effect of broadband is in expanded bandwidth, which means that the online distribution of audiovisual material increases. This distribution mechanism replaces broadcast networks for many young and affluent consumers who engage in 'series stacking' (downloading many episodes of a series at once) or downloading pre-release music or movies. Further, the growth in processing power of the personal computer now allows users to treat their personal computer as a music and photo library, video player and home video editing suite. Users increasingly send their own audio-visual content among Internet networks, and this has led to the growth of popular user-generated content services such as the video-sharing site YouTube and the photography website Flickr. The expanded possibilities for audio-visual Internet communication among those without high levels of text literacy should not be underestimated.

The Asian region is a world leader in the development of broadband, with countries such as South Korea, Hong Kong and Taiwan having high penetration rates and offering new kinds of applications and services through high-speed data.

Convergence

Convergence is coming to the fore, as previously distinct forms of media (radio, music stores, television, film, telephony) are now both emulated by and reconfigured within Internet networks. These raise challenging business and regulatory issues for communications regulators who have previously relied on regulation of different physical infrastructure for specific media forms. For example, spectrum allocation allowed for a finite number of free-to-air television and radio broadcasters and this also meant that content control via the broadcasters was a relatively simple affair. In the new media environment where content from a particular producer may be hosted offshore, and with an infinite number of content 'channels', maintaining complete control over local content is almost impossible. There are important cultural policy implications: for example, the concept of a 'quota' of local programming may no longer be appropriate in a non-channel based environment.

Wireless

There is a continuing proliferation of mobile devices and wireless networks, facilitating both increased movement and reduced costs for 'last mile' delivery. In their country chapter in this volume, Ananya Raihan and Shah M. Ahsan Habib note that Bangladesh had 12.6 million phone users in April 2006, an increase of more than tenfold from 1.2 million phone users in 2001. During the same period, fixed line subscribers merely doubled. This reflects the central role that wireless and mobile technologies now play in extending the reach of communications networks in developing countries.

Platform choice for cell spectrum is also critical in a developing country with limited resources. In highly developed market economies, there is sufficient investment capability to allow firms to establish competing infrastructure (for example, both GSM and CDMA). For developing countries, a combination of donor and private sector investments from foreign countries will influence the technological platform that is deployed. This platform choice may have far-reaching consequences in the future.

Innovative mobile operators are embracing the changes that are occurring. As Keisuke Kamimura and Adam Peake point out in this volume, in Japan 600,000 'one-seg' telephones capable of receiving digital television broadcasts were sold in the first three months of the service being available. While this is encouraging, customized content still cannot be produced due to licensing regulations and a range of unresolved issues such as royalties, which are disbursed according to a system designed for a standard television environment. As usual, regulators are on the back foot with respect to the new technological developments.

Once again, Asia is playing a leading role in the deployment of wireless. While Japan has long been a leader in the provision of mobile data services such as i-mode, South Korea, Hong Kong, Taiwan and the Philippines are becoming recognized for their role at the forefront of the 'mobile information society'.

An important issue with respect to mobile and wireless is security. In a fixed line environment, it is usually possible to track the source of a particular communication made over a network. With the emergence of pre-pay calling and wireless Internet, such tracking is not always possible. Different policy remedies are available. For example, Australia requires identification to be shown at the point of sale of pre-paid calling services (Australian Communications and Media Authority 2006), although such initiatives also raise privacy issues.

Overall, recent developments reflect the ongoing reality that technological change will continue to occur quickly and regulatory or business propositions that are tightly tied to particular technological solutions are at risk of becoming redundant when these conditions change. A focus on developing the capacity to adapt to change and developing a clear picture of the desired social, economic and cultural objectives, whether in public or private sector bodies, is required.

Electronic waste and environmental impacts

These new technological developments often make previous technologies obsolete. However, the physical items themselves do not disappear. A serious question regarding the ongoing sustainability of ICT4D is electronic waste or e-waste, which is the most rapidly growing waste problem in the world. Toxic elements of ICTs such as lead, beryllium, mercury, cadmium and flame retardants pose both an occupational and environmental health threat. Most of the consumption of information technologies occurs in wealthy economies, while the waste products are often shipped to developing countries (where the raw materials came from in many cases). This makes ICTs dependent upon a process which for many countries results in a conversion of their natural resources into toxic material.

International movement of hazardous waste is controlled by the Basel Convention 1992. There is also a proposed amendment, the 'Basel Ban', which prohibits international trade in waste classified as hazardous. Although this has not been ratified, the European Union is voluntarily abiding by the ban (Terazono et al. 2006, p. 6). Nevertheless, close to 40,000 tons of used electronic equipment find their way to India every month, unnoticed, and they are getting routed to illegal electronics dump grounds, reports Ravi Agarwal, director of the non-profit environmental group Toxics Link. The equipment is then incinerated, contaminating the environment with toxic organic compounds and metals (Basu 2006).

There are positive aspects of the international trade in e-waste. The working lives of products can be extended through reuse, and second-hand goods can still be used in many recipient countries even if the goods are considered obsolete in the exporting countries. This not only increases resource utilization efficiency, but also provides economic benefits for people in the importing countries. However, it is difficult to ignore the vastly asymmetrical risks associated with this trade. Often, related industries in recipient countries, especially in the informal sectors, do not consider externalities such as environmental effects. Market price alone does not reflect the true economic value of a material. There are human costs through illness and long-term spoiling of natural resources that can occur through toxic spillage.

There are initiatives in the region to control the e-waste problem. The EECZ Programme (http://www.eecz.org/index.html), for example, is dedicated to reducing the environmental pollution caused by industrial activities and hazardous waste in the Chinese province of Zhejiang. The programme is focused on establishing a well-regulated hazardous waste management system and supporting eco-efficient production.

Combating e-waste will require a range of strategies. As a net importer of waste, Asia Pacific could negotiate an agreed range of standards that will allow shared monitoring and enforcement. Much of the current trade occurs in the informal sector and it is difficult to determine the long-term effects. Of course, more eco-friendly manufacturing processes and more serious attempts to reduce the generation of e-waste will be critical. Ultimately, the selling price of technology should also reflect its true cost, including environmental and human costs which are largely ignored by ICT producers.

ICTs and economic inequality

ICTs and poverty alleviation

While the entire ICT4D sector would claim to focus on development in favour of the poor, it remains challenging to find simple solutions or agreement on priorities for ICT4D and poverty. The various technology parks invested in by governments as a key part of ICT strategic plans underscore the difference within countries between those able to make use of ICTs (largely in urban centres) and the rural poor. The question of how to foster social mobility among the rural poor is complex and will not be easily resolved.

McNamara (2003, p. 4) points out that this development issue is not confined to ICTs and that 'the enthusiasm for ICTs has also mirrored earlier fads in development thinking in overemphasizing one factor and failing to focus adequately on the complexity and difficulty of fostering pro-poor change, and on the political and structural constraints on that change in a given country'.

This is not to say that ICTs do not have a role to play, but it is one that needs to be integrated into a larger analysis of structures (for example, trade, education) that may not be largely driven by ICTs. For example, a recent UNDP report on telecentres aimed at empowering the poor found that user satisfaction with the centres is 'closely associated with the capability of the staff at the centres and this in turn affects the degree of community acceptance that the centres enjoy' (Harris and Rajora 2006, p. 13). Such findings show how technology by itself is far from sufficient to achieve clear results in the ICT field and that a range of human skills are required.

ICTs can, however, make a difference to many specific factors that exacerbate poverty. In the area of disaster alleviation, as outlined by Krishnamurthy Sriramesh, Chanuka Wattegama and Frederick John Abo in this edition, ICTs have a critical role to play in preparation, warning and response. This comes from the ability of ICTs to duplicate and deliver instantaneous data in many different formats. Krishnamurthy et al. note, however, that ICTs alone will not save lives and that they need to be integrated into a larger, holistic management system.

Another issue in pro-poor development is that transferring critical services to ICTs needs to take place without dismantling services that are relied upon by those who are not online. In this edition, Lelia Green and Axel Bruns discuss the further marginalization experienced by those unable to access online services after cutbacks to face-to-face bank transactions in Australia.

ICT-related economic development

At the regional and nation-state level, ICTs and technological convergence pose significant challenges for economic development. Many countries throughout the region are beginning to focus their economic development policies on ICT industries. These industries can be difficult to develop without an existing base as they often require integration with dominant platforms and standard-setting bodies which are based outside the region (for example, operating systems such as Microsoft or Apple). A common strategy is to undertake clustering of ICT industries to enable learning from each other and develop regional linkages. The Multimedia Super Corridor (MSC) in Malaysia is a prime example. The informal exchanges and learning that take place in technology parks lead to overall skill development. This is why countries such as Iran are providing large subsidies for such initiatives in their five-year development plan. As Masoud Davarinejad and Massood Saffari point out in their country chapter, there are now nine such parks in Iran alone.

Another significant and sometimes controversial strategy is the use of tariffs to protect local ICT manufacturers. Without such tariffs it is very difficult to develop a local hardware industry and without such an industry technological skills that are necessary for future competitiveness may remain undeveloped. On the other hand, the tariffs may also result in ICT hardware remaining unnecessarily expensive and out of reach of many users and thus prevent the development of new markets. For countries with a limited local industry, like Lao PDR, this debate is ongoing as the chapter on this economy by Phonpasit Phissamay makes clear. It is difficult for policymakers to view these issues objectively due to the tremendous pressure applied by industries with economic interests in the outcome of policy decisions.

Business Process Outsourcing (sometimes termed BPO or, more commonly, simply 'outsourcing') continues to be a growing phenomenon that is having a transformational impact on the economies of the region. On one level, outsourcing has become a significant source of income for many Asia Pacific countries as European and US firms make use of ICT to have labour-intensive service tasks such as call centres, animation work and data processing performed in lower wage countries. However, these industries are challenging to forecast, as they are subject to task migration: if one's own city can host a call centre for a firm, there is a strong chance that the firm could shift to outsourcing to somewhere else if it gets a better deal, as the required skills are relatively transferable (May 2000).

The Internet as private infrastructure

One of the newest pressures affecting ICT-enabled trade is the concept of 'Network Neutrality' (Butt 2006a). The theoretical model for the Internet suggests that Internet Service Providers (ISPs) carry any and all Internet traffic equally, rather than being able to prioritize or block certain traffic or charge differential rates for different kinds of data. User groups advocate for legislative measures to maintain this openness, arguing that it is necessary because users do not have true competition in the telecommunications area due to high switching costs and limited choice. These groups are concerned that the increasing attempts to link content with network service provision will result in users being required to sign up to a particular ISP in order to receive certain kinds of content. Competing businesses might bundle exclusive access to particular content packages along with network access in order to extract the maximum revenue per user, because provision of basic network access alone has low-profit margins. This will result in a fragmentation of the network, as users will not generally purchase more than one access technology in order to access all possible content.

Many people assume that the Internet is a public facility because the technical protocol for transferring information (TCP/IP) is public. But the actual physical networks are owned primarily by private entities who interconnect via market transactions. This raises significant new challenges for governments in particular who are dealing not only with a global facility that is relatively impervious to national interests, but also with a private governance structure that has little incentive to consider the needs of poorer potential users not already connected. In the early days of the Internet, the bulk of the bandwidth was owned by academic or research networks and the profit motive was not usually present, even though the networks only served a small elite. As the Internet becomes important infrastructure for communities around the world, serious challenges are developing for those seeking to expand Internet use in the public interest, when the infrastructure is privately owned.

Peering and exchanges

Related to the issues arising from the ISP business model, there are also policy challenges emerging due to interconnection agreements between networks. Because the Internet is not a single network but a network of networks, the market transactions between these networks can have effects that seem unintuitive from a policy perspective. If two networks within a country do not have a 'peering' arrangement, for cost reasons, traffic may end up travelling to quite remote physical destinations before returning to the very same city. This is akin to how smaller economies are served by airlines—for example, travel between small islands in the Pacific can be more expensive than around-the-world tickets with stops at major cities. While many believe that a market economy will naturally lead to the establishment of Internet Exchange Points due to the economic efficiencies involved, there are indications in some highly developed countries that this is not the case. In New Zealand, for example, major ISPs are choosing to de-peer from exchange points (Bertram 2006). In Lao PDR, as Phissamay describes in this volume, the Lao National Internet Committee (a government agency) is investing in an exchange point to link traffic between the five ISPs and academic networks internally to help alleviate the current situation where national data will often be routed via Thailand or Singapore.

Security

Security is an increasingly important concept covering a range of diverse areas with different political and social consequences. Security, dependability and trust are critical factors in stimulating the take-up of new ICT services. From the computer owned by the individual user which must be protected from hackers, to phishing scams, through to the use of ICTs for border control, the breadth of security threats is enormous. Tarimo (2006) characterizes approaches to ICT security as increasingly focused on the concepts of availability, confidentiality and integrity:

- Availability—The prevention of the unauthorized withholding or unscheduled inaccessibility of information or resources.
- Confidentiality—The prevention of the unauthorized disclosure of information.
- Integrity—The protection of information from unauthorized modification.

What makes security so challenging is that these concepts are often in tension with each other. For example, the desire to make information always available entails the potential for compromise at the level of confidentiality. The existence of a potential threat does not always mean that it can be eliminated. As Tarimo (2006, p. 46) notes, 'The way one defines ICT security will influence how one models the same and ultimately the design of the corresponding solution—approach and model. Many security designs are poor because they are based on unrealistic threat models.' Highlighting unlikely but easily fixable security issues is certainly more lucrative for security solution-providers than attending to more likely and difficult-to-solve security problems.

ISO/IEC 17799, the Code of Practice for Information Security Management, is an internationally recognized standard that provides a framework for security in the private and public sectors covering small to medium enterprises as well as large corporations. The 2005 version of the standard (International Organization for Standardization 2005) contains the following 12 main sections which constitute a useful outline of the issues that need to be considered in the area of information security:

1. Risk assessment and treatment
2. Security policy
3. Organization of information security
4. Asset management
5. Human resources security
6. Physical and environmental security
7. Communications and operations management
8. Access control
9. Information systems acquisition, development and maintenance
10. Information security incident management
11. Business continuity management
12. Compliance

Internet governance

The questions of peering and neutrality are currently addressed in forums associated with Internet Governance. The topic of Internet Governance, discussed by Adam Peake in the 2005–2006 edition of DirAP, has lost some of its momentum after the World Summit on the Information Society (WSIS) failed to come to any agreement on the issues in 2005. WSIS has developed the Internet Governance Forum (IGF) to discuss the ongoing coordination of the Internet, but it has no authority to implement changes in the Internet's governance bodies. There can be broad or narrow definitions of Internet Governance. Increasingly, many actors favour a broader definition comprising the traditions, institutions and processes that determine how power is exercised, how stakeholders are given a voice and how decisions are made with respect to the Internet. The UN Working Group on Internet Governance (WGIG) differentiated clusters of issues related to Internet Governance that can be summarized as (a) access issues; (b) issues related to use of the Internet; and (c) issues around coordination of Internet resources.

One of the common myths about the Internet is that it is not governed (Ang 2005). Technically, the Internet is coordinated rather than governed and it is true that there is no single place where the Internet and a coordinating agency can be identified. But in reality, there are a number of different areas where self or state regulation is in place and these are areas where the analysis of governance is useful.

An Internet Governance issue which does seem to be ongoing is that of Internationalized Domain Names or IDNs, and it has particular relevance to the Asia Pacific region (Butt 2006b). As the number of non-English speakers on the Internet grows exponentially, the Domain Name System (DNS) overseen by the Internet Corporation for Assigned Names and Numbers (ICANN), which works on a subset of Roman script, has proven to be incapable of effectively providing Internet navigation services in other languages and scripts. ICANN recently announced a new focus on the issue and it is deploying testbeds for new IDN systems. For Asia Pacific, however, many feel that this is too little, too late, and alternate systems are needed to allow people to use their own languages online. IDN testbeds were established by the Asia Pacific Network Group in 1998. There are also a number of IDNs already established by particular ISPs.[1] Organizations such as the Multilingual Internet Names Consortium (MINC) are attempting to develop a coordination framework to ensure that fragmentation of the Internet does not occur through 'leakage' of these IDNs into different zones.

The debate on IDNs remains polarized: there are those supporting universality, standardization, stability and control, on the one hand; versus those advocating for multiplicity, diversity, loose coordination and accountability to local language groups, on the other. From the perspective of bodies such as ICANN and the IETF, a single system for IDNs should be established which can serve the interests of all stakeholders and multiple systems should be avoided. However, this 'universal' approach to IDNs raises much more complex technical, political and economic issues than developing a viable system for a particular language group. This complexity partially accounts for the slow progress on IDN development within the ICANN system. The major challenge will be to create viable mechanisms for mediating between these philosophies. While all agree that the goal is to ensure that the Internet remains a single, interoperable public facility, many increasingly believe that the right of all people to communicate in their own language must be maintained and expanded within this new medium.

Culture and local content issues

The ready availability of relevant local language content is critical for the development of productive capacity in new media. One of the challenges in the early years of Internet diffusion lay in the dominance of the English language and US-centric content, with little relevance to many Asia Pacific Internet users. Without locally relevant content in local languages, immediate uses of ICT for day-to-day activities may not be apparent.

As Sarmad Hussain and Ram Mohan point out in their chapter on 'Localization' in this edition, there are many aspects to the technical issues in localization, including encoding, keyboard and input method, fonts and rendering, locale and local language interfaces. They state, 'Localization is conventionally defined or understood in a narrow sense—that is, it is usually limited to interface translation and other basic changes in the computing platform. We suggest that localization has a broader scope that includes the entire range of script, speech and language technology to enable access to information for the end-user.' Different language groups have very uneven capacities to deliver all of the components necessary for a fully localized experience, and therefore the priorities for Chinese language initiatives, for example, differ from those for Khmer.

There are a number of notable projects working to build local language computing capacity in Asia Pacific. In Nepal, the Dobhase project is currently building an engine for English-to-Nepali machine translation on the Web. Critically, they are also incorporating Nepali-to-English functionality to ensure that the Nepalese become Web producers and not just consumers of online content. Other projects are happening at the platform level: in Bhutan there has been successful development of Dzongkha Linux, a localized operating system.

Increasingly, regional ICT plans are focusing on content and cultural issues. For example, Lorraine Carlos Salazar and Shelah Lardizabal-Vallarino note in the ASEAN chapter in this volume that the ASEAN ICT Fund has endorsed the Brunei Action Plan which will (a) empower home workers in ASEAN countries, (b) conduct workshops on public domain and content development, (c) conduct e-learning, e-culture and e-heritage training for youth, and (d) develop an ASEAN Skills Standard.

In the area of IPRs and patents, countries are increasingly aware of the ability of overseas companies to patent technologies and materials based on traditional products. ICTs play an important role in allowing this commercialization and publication. However, ICTs can also potentially provide a mechanism to mitigate against such exploitation. The government of India's Traditional Knowledge Digital Library initiative has built a database of over 36,000 Ayurvedic formulations and other traditional medicines (see Noronha and Venniyoor, this volume). It is hoped that this can work as a defensive mechanism against inappropriate patents that may be taken out on such materials, which might otherwise result in their becoming inaccessible for India's citizens.

The question of viable financial models for local content companies remains troublesome in the Asia Pacific region, whether in the public or private sector. It is easier for nations with a high number of cultural producers using common languages, such as English or Chinese, to engage in cultural exports.

But for smaller nations, export audiences may not be as easy to reach, and local audiences are at risk from increased competition from global sources, which contribute to difficulties in language maintenance.

Because media policies are based on control of national borders, the global nature of Internet media poses many questions. Overall, these can be seen in the shift in national interest content policies from control and quotas to growing effective new content producers who are able to thrive in an international market. Such initiatives are increasingly undertaken by economic development agencies pursuing the 'creative industries' rather than by cultural development agencies. This reflects the forum shifting that has occurred in the intergovernmental sphere, where media and entertainment are now more commonly discussed in the WTO as part of intellectual property agreements rather than in agencies such as UNESCO that have traditionally been the agency for the discussion of cultural issues. As Braithwaite and Drahos (2000) point out, Europe and the US have led these moves, more so than countries in the Asia Pacific region.

Even in nations managing to achieve significant control over content, such as China's 'national firewall', the controls are a partial measure. New social media such as blogs and wikis provide a challenge to holistic media policy, as much of this cultural activity takes place informally, outside of organizations easily subject to policy initiatives. However, these independent forms also allow an unparalleled opportunity for local content to emerge. Donny B.U. and Rapin Mudiardjo estimate that there are currently around 70,000 to 90,000 blogs within Indonesia alone (see the Indonesia chapter in this volume). Ultimately, there is much for the region to gain by supporting the user-generated content platforms that reflect the diversity of all countries in the region.

E-government and regulatory issues

Range of regulatory and policy focus in Asia Pacific

Perhaps the most important point about ICT regulation in Asia Pacific is this: Each country needs to develop its own set of culturally sensitive and consistent national priority policies. There is no 'best' approach to policy formulation and neither are there 'best' types of policies for dealing with specific aspects of ICT. Indeed, most developing Asia Pacific countries are striving to evolve their own set of policies and implementation methods, each using their own discernibly unique methods of consultation, policy formulation and execution/implementation. The differences in regulatory approaches in Asia Pacific are based largely on cultural as well as economic issues, such as the level of development of a country's ICT infrastructure, the penetration rates of different forms of ICTs, the emphasis people place on culturally unique content, the willingness to invest, and of course, two factors that greatly influences all of the above— per capita income and the level of education.

In terms of ICT market maturity, countries in the region can be classified as either 'Developed ICT Market Countries' or 'Developing ICT Market Countries'. 'Developed ICT Market Countries' refers to countries such as Japan, South Korea, Singapore and Taiwan, where ICT markets are relatively mature, where current generation infrastructure is in place and there are definite plans (some already in the execution phase) to install next generation infrastructure, where there are multiple competitors in a market and where the market demand for services and products is healthy enough to encourage innovation. 'Developing ICT Market Countries' refers to two types of countries: (a) countries where ICT markets are rapidly growing due to economic prosperity and large populations capable of sustaining such growth, but where this has been a recent phenomenon (for example, China and India) and (b) countries whose economies and ICT take-up is not expanding at a rapid pace, either due to their small population and market sizes, geographic locations or economic size.

In economically developed countries with relatively mature ICT markets, such as Singapore, South Korea and Japan, infrastructure is a given and the build-out of future-proofed cutting-edge infrastructure and upgrading also tends to be a given. This is because the relatively high per capita income and critical mass of educated population able to use and fully benefit from ICTs tend to make new build-outs and massive capital commitments viable over a medium term for private operators. In such countries, market forces and dynamics make it viable for private operators to consider build-outs without too much assistance from government, although public–private partnerships (PPPs) are being seen as a mutually advantageous option for infrastructure build-outs.[2] Here, monopolies in mainstream broadcasting, telecommunications and Web access have ceased and the regulators and policymakers have accepted the technology neutrality argument and have enacted (mostly) technology-neutral laws and regulations to regulate and enable convergence-driven technologies. The focus of policy and regulators in these countries tends to be less on the penetration rates of basic technology or technology accessibility and more on ensuring that the country is 'future-proofed' and that growth and sustainability of a competitive market are assured. This is evident in Singapore, where the IT and telecoms regulator, the Infocomm Development Authority of Singapore, has a number of infrastructure and training programmes already running under an 'Intelligent Nation 2015' masterplan.[3]

In contrast, in many developing countries such as Mongolia, Vietnam, Bhutan and Nepal, regulators and policymakers are tasked with creating the conditions that would enable an active ICT market to come about. That said, developing countries in the nascent stages of ICT adoption, with less mature markets but with access to current technologies, may actually be at an advantage compared to developed countries. This is because they may be able to 'leapfrog' some of the issues faced by the developed countries that can often distort or skew the ICT development process. For instance, developed markets may have had to make a tough political and economic choice when faced with the prospect of having to impose prohibitively high termination payments (using taxpayers' money) as a cost of dismantling tightly controlled or nationalized monopolies or oligopolies. Faced with such a prospect, a government might delay liberalization to ensure that the monopoly period granted runs out first.

Another perhaps more important issue is an outdated regulatory and legal framework. Developing countries may be at a relative advantage in so far as they can start from a 'fresh page' when it comes to developing regulatory and legal frameworks for ICT, especially since they can draw on the collective regulatory experience of developed countries to bolster and build their own expertise. Some developing countries have been able to use World Bank aid to consult professionals from countries with maturing regulatory regimes, such as Singapore, to get advice on structuring ICT regulation. This allows them to gain insight into the issues and challenges involved in successfully regulating for growth—ensuring end-user protection without stifling industry growth by over-regulating—and to put in place a market-conducive legal and regulatory framework. For example, through World Bank funding, both Mongolia and Lesotho in Africa have tapped professionals from developed ICT markets to assist in the formulation of enabling and technology-neutral ICT regulation and legislation designed with the above goals in mind.

The policy and regulation focus in developing countries tends more toward increasing ICT penetration rates especially in rural and semi-urban areas. In nearly all developing countries, there are direct or indirect policies aimed at increasing the usage of ICT. So for instance, Bangladesh did away with import taxes and duties on computers in 1998 (see Raihan and Habib, this volume), while Nepal is constructing a national optical fibre backbone for telecommunications partly with aid from the Indian government (see Pandey and Shrestha, this edition).

Some developing countries are devising novel ways to grow their nascent ICT industries via indirect policies encouraging growth in technology, content and software development, especially to encourage innovation. This allows policy to have an influence in a space that is rarely regulated because it usually does not need to be—the space between infrastructure and end-user. In Pakistan, as described by Jamshed Masood and Salman Malik in this volume, the government is engaged in a study to consider the viability of a government-backed venture capital fund that can provide subsidized funding for Pakistani ICT ventures. A further impetus is tax holidays granted to the income of ICT-focused venture capital funds.

Leading by example—e-government and e-governance

The continuing diffusion of ICTs and the emergence of the Internet as the communications medium of choice among many businesses and citizens has increased the pressure on government departments to provide information and services through electronic means. This year, for example, Australia published *Responsive Government: A New Service Agenda* which recognizes the need to deliver a more coordinated and citizen-focused programme of activities and ensure that the capabilities to support this are present. This reflects many other similar initiatives in the region such as South Korea's *u-Korea Master Plan (2006–2010)*, described by Jong Sung Hwang and Jihyun Jun in this volume.

There is a large variety of e-government projects and obviously, many of these require substantial expertise and experience that may not be available in governments with low IT capacity. Thomas Parks (2005) points out that very few government decision-makers have direct experience with IT. Even if they see the opportunity, they often lack experience in planning and implementation, which leaves them at the mercy of vendors and/or individual consultants. One common problem is the lack of time and budget for software acquisition and implementation, systems integration and training. Training in particular is vital and should thus be given as much budgetary allocation as the computer hardware itself.

Examples of laudable ongoing e-government initiatives can be found even in the remotest Asia Pacific countries like Bhutan (see Wangchuk and Pradhan, this volume) and the Maldives (see Ibrahim and Ahmed, this volume). Bhutan is working on implementing initiatives in border management, passport control and civil registration, while the Maldives is in the process of interlinking government offices via a WAN network as a precursor to e-government services. These initiatives show how policymakers can lead by example to encourage the uptake and penetration of ICT.

Open source

Nearly all of the Asia Pacific countries surveyed in this edition of DirAP encourage the development of open source software. Localized versions of the Linux OS are popular in a number of countries such as Vietnam, Bangladesh, Bhutan and Mongolia. This is not surprising, given the relatively high cost of licensed software. Perhaps the most visible examples of policy favouring open source are in China and India: since 2002 governments in both countries have taken policy stances advocating open source over proprietary solutions.[4] This has been followed more recently by a number of institutions in Southern India successfully deploying Linux-based solutions and state governments advocating their adoption.[5]

From an economic perspective, this makes absolute sense: one should use something that can do the job for $0. The chink in Linux's armour, however, is usability. Especially in countries with a relatively low literacy rate, it is much harder to learn on Linux than on any other platform. Policymakers should encourage the development of easy-to-use localized user interfaces in open source platforms.

Open content

Regulators and policymakers in Asia Pacific need to grapple with the issue of how best to get local content online and make it accessible to the widest audience. This can have a direct bearing on penetration rates for ICT as more and more people may choose to come online or take up ICTs if the services are available in local languages and a range of local content is accessible.

As for a licensing model for open content, there is a discernible interest in Asia Pacific in the more flexible licensing models offered by the likes of the Creative Commons[6] and the Free Software Foundation (FSF). For example, India has been involved in the updating of the FSF's General Public License (GPL) from version 2 to 3 (see Noronha and Venniyoor, this volume).

General IPR and ICT laws

Legislation on IT and digital signatures is now in place in a number of countries in the region. When enacting such laws, attention must be given to the principle of technology neutrality: the laws should be applicable regardless of the technology.

With respect to approaches to applying and enforcing laws, the trend in developing ICT countries has been to follow the lead of the developed countries, which is to use a 'light touch' in ICT regulation—that is, regulators step in only when needed to correct imbalances in the market which the market cannot correct by itself (for example, monopolistic or cartelized competition).

Building and implementing sound policy and regulation

We have found that most of countries in the region that have enjoyed productive and sustained ICT growth share certain features in terms of regulatory approaches. These are:

- Regular and effective cooperation and coordination among ICT regulators and industry;
- A holistic view of the national and regional landscape when seeking to determine and implement ICT policy; and
- Focused and coordinated implementation.

The regulatory and policy environment of some countries has all of these features, while others have only some of the features and they may be present in varying degrees. The point is that their presence appears to encourage holistic and forward-looking policy that is capable of being both visionary and realistic in terms of implementation resource requirements. A regulatory tool set containing some or all of these features (depending on national circumstances) can help to realistically address two key considerations facing the majority of Asia Pacific countries, namely:

1. Managing the digital divide: 'Digital divide' refers to the gap that in many cases has opened up between citizens who are versed in ICTs and able to derive maximal benefit from using them and those not versed in ICTs due to lack of opportunity or inadequate infrastructure.
2. Managing the convergence process: 'Convergence' in this sense denotes the convergence of content from disparate analogue formats provided over separate analogue infrastructure, to a single digital format (digital broadcast, digital sound, digital data) capable of delivery to multiple I/O, interactive devices via a single fibre optic 'fatpipe' infrastructure.

These tasks require regulators and policymakers to engage with ICT industry and users when formulating policy and laws. Coordination and consultation among the responsible government ministries and regulators, including those not directly responsible for ICT, are likewise essential. Singapore provides a good illustration of this model in practice. For nearly all major policy decisions, the Singapore IDA (www.ida.gov.sg) holds a public and industry consultation, and makes a decision only after considering the responses given.[7] Thus, for example, the new IDA-administered *Code of Practice for Competition in the Provision of Telecommunication Services 2005* was released with significant revisions most of which were partly or wholly attributable to concerns raised via consultations. Such coordination can inform the industry and users of the directions policymakers intend for ICT to take and it can give them a chance to adapt to it or, even better, to constructively influence it. Used properly, this collective decision-making process can make it easier to implement and get results from policies, as users feel the policies are cognizant of their interests.

A holistic view of the national and regional landscape

Policymakers need to think through the objectives of a policy, law or regulation and assess whether its expectations are realistic in a national and regional context. If it is unclear how the expectations and goals can be reached, then steps to make those clear should be identified and implemented.

This may seem simple, but there are ICT policies drafted as vision statements, without guidelines as to who needs to do what to achieve certain goals. For example, a government might come up with a 10-point plan for putting a country on the ICT growth path but does not provide clear instructions to the concerned ministries and governmental bodies on what coordination is expected or should take place.

The regional angle can also come into play, especially if it can enhance a country's position on ICT issues, such as open source software development and content licensing. With respect to open source software, given the marked encouragement it has received in Asia Pacific, it is conceivable to imagine policy encouraging cross-border collaborations (whether between governments/regulators or commercial entities or both) on open source development, especially between countries with the same languages. This can accelerate the development process for both locally developed software and/or the localization process for all software.

As for content licensing, as countries consider more flexible licensing models for open content, they may at the same time hesitate to have stringent standards thrust upon them, complete with obligations to adhere to digital rights management. An example is Australia, where a number of critics have cried foul over the government's decision to bring its copyright legislation in line with that of the United States, as part of its obligations under a free trade agreement which came into force in 2005. Australia's modification of its copyright and intellectual property laws included adopting provisions rendering illegal any measure to circumvent technology protection measures (TPMs), which in turn could put Australian copyright law at odds with the hitherto unassailable fair use access rights available to users of copyright materials (see Green and Bruns, this volume).

In such cases, it may be possible for developing country regulators to collaborate across borders to agree on principles for flexible licensing and copyright protection and to then use this common ground to seek better terms when negotiating with countries with traditional licensing terms (see Cardoza and Liang, this volume). This common stance could be used collectively when negotiating as a trade bloc or it could be used to strengthen an individual country's bargaining position when negotiating individually. Such a policy stance could also be used to advantage by private entities negotiating content licensing terms even in the absence of a free trade or IPR-specific agreement.

Forming a regional position also involves determining all of the governmental and non-governmental agencies that need to be co-opted to work together to bring high-level ICT policy to practical fruition. First, the practical goals of the policy must be worked out. Second, who is empowered to do what is necessary to bring it to fruition should be identified. And third, the responsibilities towards achieving the policy goals should be delegated and a clear review process and schedule to monitor progress should be agreed upon. In short, political will is necessary to achieve ICT growth. That said, it must be clear that the process of guiding and implementing focused and coordinated policy implementation is the ICT policy goal that requires perhaps the greatest amount of political will.

Overall impact of ICT on governance and policy

It is frequently assumed that the introduction of more advanced ICT reduces opportunities for corruption. However, as Wescott (2002) notes, the reality is more complex. While ICT sometimes helps in combating corruption, it can also have no effect, or even provide for new corruption opportunities, including fraud. Indeed, unintended consequences are common in e-governance projects, even those that have positive outcomes. Consider the following related but contrasting examples.

The government of Andhra Pradesh developed a land registration system where the land owner enters some details of his/her property, such as location, dimensions and other factors that affect the value of the land, and then calculates the value. Prior to the system, land valuation was performed in an entirely non-transparent system by assessors and agents and often required weeks and some additional payments. According to Subash Bhatnagar of the Indian Institute of Management, Ahmedabad, 'Land registration can be completed in a few hours (with the new system), whereas earlier it took 7–15 days' (Parks 2005, p. 6).

However, researcher Solomon Benjamin (2005) has found that new land regimes can have very uneven effects. He notes that in Bangalore, the reduction of complexity in titles and centralization has made land much more open to larger purchasers. 'This has allowed very large real estate companies catering to the IT industry to access land in Bangalore, resulting in dramatic changes in land markets' (Benajmin 2005, p. 8). Gentrification becomes an issue and the rights of the poor are made more tenuous when ICT enables companies and politicians to collaborate on larger 'real estate development projects', which may be good for a region's overall economy but result in the transfer of security away from the poor to the benefit of the wealthy. After all, it is unrealistic to think that the poor will be trading on the ICT-enabled property market.

Education for the information society

While the discussion of ICTs is often posed in relationship to a 'knowledge economy' or an 'information society', the latter is usually taken to mean an advanced industrialized economy where service industries account for a very large proportion of the national economy. While this economic structure is identifiable in some nation states, there are very few roadmaps for how such an economic state can be achieved by developing countries and in this respect Asia Pacific is no different from the rest of the world.

There are five novel processes that characterize the Information Society in published literature. Identifying these processes provides an insight to the kind of policy responses required to prepare a population for this emerging society:

1. Global networks of finance capital are rapidly expanding, with economic advancement often based on the ability to shift informationalized capital between markets. This suggests the need for an international perspective and experience and a certain level of cosmopolitanism and outwardly-focused thinking among business owners.
2. Information itself is increasingly commodified. That is, there has been significant growth in the sales of 'information products' and media as a proportion of economic activity. This intensifies the need for literacy and for a critical capacity to assess information.
3. Lifestyle and consumption choices increasingly define diverse social structures, requiring businesses to have a more sophisticated understanding of cultural issues and empathy with their chosen markets.
4. Services are becoming an increasingly important economic category, suggesting a shift in the traditional policy focus on science and technology and increased emphasis on human disciplines such as the arts, humanities and social sciences.
5. There is the recognized emergence of significant 'informal' economies, which direct the flow of money outside of the formal market mechanisms where more open relationships take place. For example, most areas of the creative industries are heavily dependent on personal relationships. Skills in relationship development are necessary in more informal environments.

As a representative and influential example, Manuel Castells' view is that there is a new mode of development, which he calls *informationalism*, that is driven by changes in the mode of capitalist production. According to Castells (1996, p. 27), this trend is linked to the rise of the service industry and the informational economy, where the workplace is focused on the generation, manipulation and interpretation of text, images and other symbolic information. The result of these changes is the emergence of a 'skill bias' in changing employment opportunities under information-intensive economies, where jobs requiring manual labour are disappearing and new jobs require higher levels of information literacy and knowledge. Economists have put forward this notion of 'skill-biased technological change' to explain the growing overrepresentation of the least skilled workers in unemployment figures in many countries over the last two decades. Greenan et al. (2002, p. 10) note that this change 'induces an upward drift in the relative efficiency of skilled workers and a downward drift in the cost share of unskilled workers', leading to increased wage inequality without affecting aggregate wage and employment statistics. This is important in ICT4D because regional and national aggregate statistics are usually used as evidence for ICT-induced economic gains but may in fact be coextensive with decreased economic well-being for a majority of people.

While many jobs have been lost through ICT development, many new ones have been created. However, aggregate economic statistics such as job growth, usually used as evidence to support IT-supported economic gains, shed little light on the kinds of jobs that are created and lost in this transformation. Where are the contemporary skills shortages that must be addressed? These are often difficult to map and little is known about changes in on-the-job training. Certification may often be simply for convenience, providing barriers to entry rather than reflecting true business need. For example, the Microsoft Certified Software Engineer (MCSE) qualification became increasingly popular because it provided the type of ICT skills that were/are thought to be relevant in the labour market, but it rapidly produced over-supply (South Africa Human Resource Development Data Warehouse 2004).

There is widespread agreement that education is one of the most important issues in preparing people for the information society and in particular in adapting to technology-enabled networks. European researchers have highlighted three kinds of ICT skills that are important: ICT User Skills, ICT Practitioner Skills and e-Business Skills (European Committee for Standardization 2006):

- **ICT user skills**, such as using the Internet or desktop processing programmes, generally lead to productivity improvements for organizations and allow individual users to get higher paying work.
- **ICT practitioner skills**, such as software development, enable the development of new industries. These generally require a higher level of training and result in an individual who has a career in the ICT sector.
- **e-Business skills** (or what we might call e-organizational skills) are related to an understanding of changes in markets, policy and organizational structures which are occurring due to the rise of the Internet and ICTs. These skills are gained through education but also often by the experience of the user working in a particular field.

Chennells and van Reenen (2002, p. 199) suggest that while there is considerable agreement about skill-biased change, there is little research analyzing the means by which technological change translates into higher demand for skills. Their work points to organizational changes made possible through ICT, such as 'delayering, decentralization and giving greater autonomy to workers' as the link between technology and higher skill requirements or professionalization. While computer interfaces have not changed the knowledge that manufacturers must have about their production process, they do provide far more second-by-second information about the process that needs to be interpreted (Shaw 2002, p. 232). Consequently, firms develop highly skilled job designs that reflect this need for interpretation and cognitive skills.

In ICT4D in Asia Pacific, we can see this reflected most clearly in the changes in the telecentre movement. As Raihan and Habib point out in the Bangladesh chapter, a content-based approach gives a new direction to the global telecentre movement. Previously, a telecentre was essentially a technology learning centre and communication centre (with an Internet connection and telephones) which largely relied on the information processing capabilities of the end users. Now, telecentres are able to provide the core information and knowledge service for things such as market opportunities or information on the visits of aid organizations. These functions are unthinkable without ICT. However, ICT use is not the goal but simply a means for the telecentre to become an effective social and informational hub.

Conclusion

As this overview and the volume as a whole makes clear, the field of ICT4D is extremely diverse and it intersects with a wide range of other issues. There are indications that ultimately, specialization and focus on the non-ICT parts of society remain critical for successful ICT4D projects. Nepal is positioning itself as a communications gateway between China and India, making use of its unique physical location between these two emerging superpowers. In the Maldives, the few software developers who have been successful have been largely focused on the hospitality industry, and they have managed to market point-of-sale and hospitality management software to other countries as well.

Policymakers have to take stock of their true situation and the resources available to them rather than following one-size-fits-all blueprints for ICT-enabled development. In developed ICT market countries, policy and regulation are concerned more with future-proofing. On the other hand, in developing ICT market countries, policy seems to be more concerned with getting up to speed than with future-proofing. It is heartening to see a number of developing ICT market countries in Asia Pacific grappling with how to increase penetration rates and technology access, not just at the infrastructure level but also at the user level via targeted education, while at the same time trying not to over-regulate and maintaining a 'light touch'.

The main challenges ahead are likely to be managing the digital divide and convergence. While we have noted features that effective regulatory and policy frameworks appear to possess, it is still the regulators and the people of a country with ground-level experience who would know best what works there. That said, appropriate policy options and advice on what to do may be available but the political will could be lacking. Malaysia's well known leadership position in ICTs must be partially a function of the high-level support required to transform ideas into action. In 1994, Prime Minister Mahathir Mohamad chaired the newly established National IT Council. Other countries in the region have prime ministerial-level champions, such as Prime Minister Samdech Hun Sen, Chair of the National Information Communications Technology Development Authority in Cambodia, and President Gloria Macapagal-Arroyo in the Philippines, who initiated a National ICT Month in 2005. It is clear that many other countries in the region do not have the same levels of leadership in their government administrations.

ICTs are rapidly transforming many fields and their successful implementation requires high-level champions who can influence the strategic direction of governments, business and community organizations in response to the changing environment. However, such advocacy must remain critical, as ICTs are not simply a good, but offer a powerful capacity to disrupt long-standing markets, communities and technologies. Sometimes such disruption is necessary to enable communities to respond adequately to a rapidly changing environment, although as the current consensus on climate change suggests, we may not be able to evaluate the success or failure of our development efforts until many years into the future. For ICT4D decision-makers whose own fortunes rise and fall quickly, the only option in such uncertainty is to evaluate the situation as rigorously as possible and attempt to chart a course for the right side of history.

Notes

1. Providers of IDN systems include China Internet Network Information Centre (CNNIC) in China; i-DNS.net, a plugin-based architecture; Japan Network Information Centre (JPNIC) which has registered over 60,000 domain names in Japanese; and Korean Network Information Centre (KRNIC), which has registered over 50,000 domain names in Hangul.
2. An example is IDA Singapore's recent Pre-Qualifying RFP for the build-out of Singapore's Next Generation Network on a PPP model basis. For further information see: http://www.ida.gov.sg/Infrastructure/20060919190208.aspx
3. See: http://www.in2015.sg/
4. For a more recent update on Linux in India see: http://www.businessweek.com/magazine/content/06_40/b4003069.htm. The comments appended to this story offer an interesting view on the Indian public's view of Microsoft v Linux in India.
5. See: http://www.financialexpress.com/fe_full_story.php?content_id=138464
6. See: www.creativecommons.org
7. A list of current consultations and results of past consultations can be viewed at: http://www.ida.gov.sg/Policies%20and%20Regulation/20060418215526.aspx

References

Ang, P.H. (2005). The legacy of the working group on internet governance. In D. Butt (ed.), *Internet governance: Asia-Pacific perspectives* (pp. 9–17). Bangkok: UNDP-APDIP, Elsevier.

Australian Communications and Media Authority. (2006). Identity checks for pre-paid mobile services. Retrieved 12 January 2007 from http://www.acma.gov.au/WEB/STANDARD//pc=PC_100565

Basu, I. (2006). India the digital dumping ground [Electronic Version]. *Asia Times Online*. Retrieved 12 February 2007 from http://www.atimes.com/atimes/South_Asia/HH03Df01.html

Benjamin, S. (20 November 2005). Analogue to digital: Re-living big business's nightmare in new hydras. *World-Information City*. Retrieved 6 December 2006 from http://static.world-information.org/infopaper/wi_ipcityedition.pdf

Bertram, N. (2006). *New Zealand's Internet landscape: An analysis of peering, content, and scalability*. Retrieved 2 February 2007 from http://www.webbedfeet.net.nz/t3site/fileadmin/stuff/Neil%20Bertram%20-%20BITT489%20-%20NZ%20Internet%20Landscape.pdf

Braithwaite, J. and Drahos, P. (2000). *Global business regulation*. Cambridge; Melbourne: Cambridge University Press.

Butt, D. (2006a). Net neutrality: No easy answers. *Media International Australia Incorporating Culture & Policy*, 120.

Butt, D. (2006b). *Internationalized domain names* (No. 9). Bangkok: United Nations Development Programme—Asia Pacific Development Information Programme.

Castells, M. (1996). *The rise of the network society*. Cambridge, MA: Blackwell Publishers.

Chennells, L. and van Reenen, J. (2002). Technical change and the structure of employment and wages. In N. Greenan, Y. L'Horty and J. Mairesse (eds), *Productivity, inequality, and the digital economy: A transatlantic perspective* (pp. 175–223). Cambridge, Mass.: The MIT Press.

European Committee for Standardization (CES). (2006). European ICT skills meta-framework: State-of-the-art review, clarification of the realities, and recommendations for next steps. Retrieved from ftp://ftp.cenorm.be/PUBLIC/CWAs/e-Europe/ICT-Skill/CWA15515-00-2006-Feb.pdf

Greenan, N., L'Horty, Y., and Mairesse, J. (2002). *Productivity, inequality, and the digital economy: A transatlantic perspective*. Cambridge, Mass.: The MIT Press.

Harris, R. and Rajora, R. (2006). *Empowering the poor: Information and communications technology for governance and poverty reduction—A study of rural development projects in India*. UNDP-APDIP: Elsevier.

International Organization for Standardization. (2005). Information technology—Security techniques—Code of practice for information security management. *Widely used standards* ISO/IEC 17799:2005. Retrieved 11 December 2006 from http://www.iso.org/iso/en/prods-services/popstds/informationsecurity.html

McNamara, K. (2003). Information and communication technologies, poverty and development: Learning from experience. Paper presented at the infoDev Annual Symposium. Retrieved from http://infodev.org/en/Document.17.aspx

Parks, T. (2005). A few misconceptions about eGovernment. Retrieved 2 February 2007 from http://www.asiafoundation.org/pdf/ICT_eGov.pdf

Sassen, S. (1991). *The Global City: New York, London and Tokyo* (Updated Edition 2000). Princeton: Princeton University.

Shaw, K. (2002). By what means does IT affect employment and wages? In N. Greenan, Y. L'Horty and J. Mairesse (eds), *Productivity, inequality, and the digital economy: A transatlantic perspective* (pp. 229–67). Cambridge, Mass.: The MIT Press.

South Africa Human Resource Development Data Warehouse. (25 February 2004). Intermediate skills in ICT as important as high-level skills. *Fact Sheet 7*. Retrieved 15 November 2006 from http://www.hsrc.ac.za/media/2004/2/HRDFactSheet7.html

Tarimo, C.N. (2006). *ICT security readiness checklist for developing countries: A social-technical approach*. Unpublished doctoral thesis, Stockholm University, Stockholm.

Terazono, A., Murakami, S., Abe, N., Inanc, B., Moriguchi, Y., and Sakai, S. (2006). Current status and research on e-waste issues in Asia. *Journal of Material Cycles and Waste Management*, 8, 1–12.

Wescott, C. (2002). *E-government in the Asia Pacific Region*. Retrieved from http://www.adb.org/Documents/Papers/E_Government/egovernment.pdf

Mobile and wireless technologies for development in Asia Pacific

Tan Geok Leng and Suranart Tanvejsilp

Introduction

Development agencies have long recognized the role of telecommunications in the economic and social development of communities and nations. The provision of universal telecommunication services is now built into every nation's development plans. The traditional approach was to construct a wired, copper-based infrastructure and extend that to places where there are pockets of population. Unfortunately, this turned out to be an expensive and lengthy process that has denied large segments of the population access to telecommunications.

Recent developments in mobile and wireless[1] technologies have opened up new possibilities to reach the population at a faster rate and at a lower cost than with the traditional copper-based approach. In this chapter, we describe the capabilities of several types of mobile and wireless technologies that have made a big impact on nations developing their communications infrastructure. These technologies include mobile phones, Wi-Fi, WiMAX and meshed wireless networks. We then discuss the barriers to use of mobile and wireless technologies in many parts of Asia Pacific that are caused by language and literacy, as well as some efforts to overcome them. We conclude with a discussion of 'development-friendly' policies that policymakers can adopt to expedite the rollout of communication infrastructure and spur greater take-up of services and applications.

There are many different forms of mobile and wireless technologies in use today. These include Near Field Communications (NFC), Bluetooth, DECT, mobile phones, Wi-Fi, WiMAX, High Altitude Platforms and satellite telephone networks. Each of these have been designed and optimized to serve a certain market segment. There is no one single technology that is able to meet all of the needs of all of the people, all the time. The greatest development impact is obtained by deploying the most appropriate technology for the situation at hand. For example, Bluetooth is designed as a cable-replacement and it is very inexpensive, but it is capable of supporting only very short range communications of moderate data rates. In contrast, a satellite system is very expensive to set up but it is perfectly suited to cover vast terrains; on the other hand, it works only when there is a direct line of sight between the end-user unit and the satellite.

In this chapter, we focus only on those technologies capable of delivering voice and Internet services to a large population spread across a vast geographical area. Two very important and widely-deployed wireless technologies that meet these criteria are the mobile phone and Wi-Fi LAN systems. We cover these in detail and show how they serve different segments of the community. We also discuss the technological evolutions of these systems and show that the differences between them may diminish in the future.

Mobile phones

Mobile phones[2] have become the most common mode of communication in the world. In 2005, there were over 2.20 billion mobile phone users compared to only 1.26 billion fixed-line telephone users globally (ITU 2005a). Even more interesting are their rates of growth: the Compound Annual Growth Rates (CAGR) for mobile phones and fixed-line phones were 24 per cent and 5.2 per cent, respectively, from 2000 to 2005. If this trend continues, mobile phone penetration will far exceed fixed line phone penetration in the next decade. Mobile phones are not just for developed countries. With over 690 mobile phone network operators operating in 213 countries, virtually every country around the world has access to the technology. And with mobile phone penetration in many developed and developing countries reaching saturation levels, mobile phone providers are developing more cost effective solutions that are targeted at less developed countries. More than half of the world's population resides in these countries and this is where future growth of mobile phones is expected to come from.

A lot of thought has gone into the design of mobile phone systems. It can seamlessly provide voice communications to a large population spread across large geographical areas. Each user is identified with a unique subscriber identification number for authentication and billing purposes. The system is capable of supporting roaming across national boundaries. It is these features that make the mobile phone personal to the user, enabling him or her to be contacted at any time and any place where there is radio coverage. Mobile phone systems are very reliable and communication across the system is secure against eavesdropping. Spam, or unsolicited calls or messages of an advertising nature, is almost non-existent in the mobile phone system. Because of these attributes, users have made the mobile phone their trusted personal device, which has contributed to the dramatic growth in mobile phone use around the world.

While a mobile phone may seem to be very easy and convenient to use, it only works because it is supported by an extensive network of base stations and a complex layer of management software working in the background. A mobile phone system consists of a cellular arrangement of base stations spread across the entire area of coverage. Each base station can provide cover to an area several kilometres in radius. A mobile phone needs only to communicate with the base station closest to it and through it reach any other mobile or landline telephone it wants to connect to. The cellular arrangement of base stations enables the technique of frequency reuse to be exploited, thus enabling a very large increase in the traffic carrying capacity of the system. It would not have been possible for policymakers to allocate sufficient radio spectrum for the general population if the technique of frequency reuse had not been invented.

When mobile phones were first introduced, only businesses and the very rich could afford them. Today they have become much more affordable and they have penetrated all segments of society. Furthermore, we are starting to see a trend of mobile phones displacing fixed-line telephones. In developed countries, it is the convenience of being reachable anytime and anywhere that drives the trend. The phone number is tied to the person and not to a specific location. Whereas residents of a household are likely to share one fixed-line telephone, it is highly likely that each individual within a household will have his or her own personal mobile phone. In the case of developing countries, it is the speed of deployment and lower investment costs that drive the take-up of mobile phones. A strategically placed radio base station can almost immediately bring communications to everyone within a 5–10 km radius compared to the months it may take to plan, obtain all the necessary permits and deploy a copper infrastructure to cover the same area.

The first generation mobile phone system was introduced by AT&T Bell Labs in Chicago in 1978 and it achieved success with the business community. The second generation mobile phone used digital technology to improve its capabilities. In this way, mobile phones began to make an impact on the general population. Besides voice, the second generation mobile phone system also has some basic data capability, such as the General Packet Radio Service (GPRS) and the Short Message Service (SMS). While GPRS saw only limited success, SMS turned out to be a big hit. GPRS provides a data connection capability between the handset and the phone network and from there onto the Internet. The lukewarm reception towards GPRS may be due to the fact that it could only support a relatively low data rate of about 20 to 40 kbps and users have found it to be an expensive service. SMS allows the user to send short alphanumeric messages of up to 160 characters to other mobile phone users. SMS is extremely popular with youngsters and even businesses have found innovative ways of using SMS to take advantage of its low cost and the ubiquity of handsets. Last year, over one trillion SMSs were sent globally and it now accounts for a significant portion of an operator's revenue. For example, 16 per cent of China Mobile's 2004 revenue came from data, and SMS sent in China grew from a mere 0.44 million in 2000 to 300 million in 2005 (ITU 2006).

The third generation mobile phone exploits wideband digital communications to further improve on the second generation phone system. One key improvement is the capability to support multimedia applications. Another key benefit is the fact that third generation mobile phone systems can deliver voice services at a fraction of the cost of second generation systems, potentially making mobile phone service even more affordable than before.

The success of mobile phones is due to the work of many players. These include the European Telecommunications Standards Institute (ETSI), the body which created the technical standards for GSM, the dominant mobile phone system in use today, and the International Telecommunications Union (ITU), an intergovernmental organization of public and private entities. The ITU, working together with National Administrations, endorses standards and regulations and harmonizes the radio frequency spectrum to be used by mobile phone operators around the world. Standardization of equipment enables production volumes to be efficiently scaled up, thus bringing down per unit costs. Through market competition, these savings can then be passed on to the consumer. A harmonized radio spectrum makes international roaming possible with just one handset. An extensive network of inter-country, inter-operator relationships have sprung up to enable seamless service across countries around the world. This is made possible by the fact that each mobile phone subscriber has a direct billing relationship with his or her home operator and can be uniquely identified through the personal subscriber identification module (SIM) associated with his or her phone subscription.

A country's telecommunications regulator has a strong influence on whether there will be a high take-up of mobile phone services in that country. The regulator influences the degree of competition and vibrancy of the market by the number of players it permits to play in that market. If there is insufficient competition, subscribers may end up paying more because operators need not fight too hard for the market share. However, it has also been argued that too much competition will create a market where the operators may not make enough money for re-investment to keep the infrastructure up to date.

Regulators also control the quantum and pricing of the radio frequency spectrum. It can be shown that there is an inverse relationship between the quanta of spectrum allocated and the intensity of infrastructure (related to capital expenditure) needed to service a particular distribution of subscribers. The cost of the radio spectrum can vary dramatically; some governments see it as a means to raise as much revenue as possible, while others may see it as a tool for telecommunications development and thus charge only a nominal fee. However, the trend is for regulators to use the auction mechanism and let the market establish a price for the radio spectrum.

Regulators must also play the role of referee to ensure a level playing field for the market mechanism to work effectively. The regulator needs to establish clear rules for interconnection between mobile operators and between the mobile and landline operators, and ensure access to critical infrastructure such as telecommunications ducts and pipes and access to telecommunications towers and exchanges. The regulator must also ensure that operators do not adopt anti-competitive practices such as locking of handsets and imposition of excessively long service contract periods. The regulator should also ensure that there is a mechanism for mobile number portability so that users can freely switch operators without worrying about losing the telephone number that has become personal to them.

Another very important player is the GSM Association (GSMA), a global trade association representing more than 690 mobile phone operators and more than 180 equipment manufacturers and suppliers. Their primary goal is to ensure that mobile phones and services work globally, creating business opportunities for their members. Ground-breaking initiatives launched by the GSMA include the 'Emerging Market Handset Programme' (EMH) designed to bridge the digital divide by catalyzing the development of affordable handsets. The programme seeks to achieve an ex-factory price of below USD 30 per handset. This is much lower than the price of handsets in the market today. EMH handsets have been supplied through 10 participating operators in countries such as India, South Africa, Nigeria, Democratic Republic of Congo (DRC), Egypt, Algeria, Tunisia, Bangladesh, Turkey, Thailand, Philippines, Malaysia, Indonesia, Pakistan, Yemen, Sri Lanka and Kenya. Other important programmes launched by GSMA include the GSMA Development Fund and studies on the impact of government taxes on the affordability of mobile phone services.

Mobile phone operators play a very important role in the mobile phone ecosystem. A savvy operator can extract maximum revenues from businesses and yet still provide affordable services to the masses with a well structured service offering. Some of the most important tools for developing the mass market include the availability of handset subsidies, prepaid service plans, Caller-Party-Pays (CPP) plans and discounted or free buddy call plans.

Studies by GSMA have shown that handset cost is one of the major impediments to mobile phone penetration in developing countries. This problem can be mitigated by an operator bundling a subsidized handset with a service plan and amortizing the cost of the handset over a service period of 24 or more months. Another very successful tactic is for the authorities to encourage the development of an orderly and active market for second-hand handsets. In many developed markets, handsets have become fashion items and they may be traded in for new models after only one or two years of use. These used handsets can be purchased at a fraction of their original price and they can be redeployed in the price-sensitive market segment.

The availability of prepaid or pay-as-you-go price plans is particularly important to stimulate growth of the price-sensitive market segment. This type of plan appeals to people with a limited or uncertain income stream and who may not be able to commit to any fixed monthly subscription plan. Prepaid plans also permit users to moderate their usage based on what they can currently afford. Prepaid plans have been so successful that in Africa, prepaid subscribers accounted for 87.4 per cent of all subscribers in the year 2004 and the global average for that year was 46.3 per cent (ITU 2005b).

Another very important tool for growing the price-sensitive market segment is a CPP plan. Under this settlement regime, it is the originating party that pays for the call while the terminating party is not charged at all. Operators have found that CPP appeals to the price-sensitive market segment, especially students, teenagers and blue-collar workers even though it may cost more on a per-minute basis compared to the Receiving-Party-Pays plan. This is because CPP subscribers can manage their telephone bills by controlling the amount of calls they make while freely accepting calls to their mobile number. A good example of how the introduction of CPP boosted user uptake is the case of Mexico: the number of mobile subscribers in the country grew by over 100 per cent to more than four million within a year from the introduction of CPP plans in 1998 (Petrazzini 2000).

Wireless local area networks

The wireless local area network (WLAN) is a relatively new technology. Work to develop the WLAN standard (IEEE 802.11) was started in 1990 by the Institute of Electrical and Electronics Engineers Inc., a US-based professional engineering society, to exploit the Industrial, Scientific and Medical (ISM) frequency band for high-speed networking. Unlike most other radio frequencies which are tightly controlled and accessible only to specific licensed personnel, anyone with the appropriate equipment can use the ISM band. The IEEE completed the initial 802.11 standard in 1997 and this was followed in quick succession by an upgrade (IEEE 802.11b) which exploits new technology innovations to improve performance from two to 11 Mbps. Another upgrade (IEEE 802.11a) exploits the newly open ISM bands in the 5.3 to 5.8 GHz bands. To be released in 2007 is yet another variant (IEEE 802.11n) that exploits advanced digital signal processing technologies such as Multiple Input Multiple Output (MIMO) algorithms to bring WLAN data rates up from 54 Mbps to 110 Mbps and beyond. Fortunately, when developing these upgrades, the IEEE standards workgroups have ensured that there is backward compatibility between the new and old. This means that infrastructure installed before the release of these new standards can still work with equipment made to the newer WLAN standards.

Unlike mobile phones where the equipment supplier base is relatively small, the number of manufacturers supplying WLAN equipment is much larger. Many of these manufacturers originate from the Far East and they have a reputation for bringing out very cost competitive products. Competition to supply WLAN equipment is very intense and prices have come down to a point where they have become quite affordable to the masses. For example, currently it is possible to purchase a WLAN PC card for under USD 25 and the price is still dropping. One of the main reasons why WLAN has taken off is that Intel, the US-based computer chip giant, strongly endorsed the technology with its Centrino campaign to put WLAN capability into every notebook computer that is manufactured. The trend to WLAN-enable portable devices is spreading to Personal Digital Assistants (PDAs) and mobile phones. This trend points to a situation where sometime in the not too distant future, almost everyone will carry a WLAN-enabled end-user device everywhere he or she goes.

Recognizing the need to ensure that WLAN equipment from various WLAN manufacturers can work with each other, the leading players from the WLAN community set up a global non-profit organization called the Wi-Fi Alliance to drive the adoption of a single worldwide accepted standard for high-speed wireless local area networking. They created a logo (WiFi) which is affixed to products that have successfully gone through a rigorous certification programme designed to ensure inter-operability between products from different Wi-Fi manufacturers. Users have come to see the logo as a mark of quality and assurance that the product can work with Wi-Fi devices from other manufacturers that may be found in the home, at the office or in public hotspots.

The simplest Wi-Fi system is just a Wi-Fi card plugged or pre-built into a computer, establishing radio contact with a nearby Access Point (AP), which then provides onward connectivity to the global Internet. Because Wi-Fi operates in the unlicensed ISM band, regulators have imposed strict limits on the maximum permissible transmitter power[3] for the Wi-Fi card and AP. This is so that the ISM band can be shared by as many people as possible and it is not monopolized by any one single user. The transmit power limitation in turn causes the maximum communication range between a Wi-Fi card and AP to be limited to about 100 metre. If there is no direct line of sight, say if there is a wall between the Wi-Fi card and the AP, the range may drop to below 10 metre. By default, all Wi-Fi cards and APs use omni-directional antennas because a user may approach an AP from any direction. In some special cases where it is desired to link two known fixed points together using Wi-Fi, it is possible to replace the omni-directional antennas with highly directional antennas and extend the range to tens of kilometres. In such situations care must be taken to align the two antennas for maximum effect and the user should note that the mobility capability of Wi-Fi is lost.

Wi-Fi has made major in-roads into the Office, Home and Public Hotspot markets. In the Office market, some technology-savvy companies have Wi-Fi enabled their offices so that their employees can be more productive. Their employees can access corporate resources such as databases, e-mail and the Internet and conduct business transactions even when they are away from their desks. Wi-Fi is also taking root in the Home market. In places where broadband penetration is high, it has become quite widespread to use Wi-Fi to create a wireless home network so that members of a household can share one cable or DSL connection to the Internet. The third market segment is the Public Hotspot market. Public Hotspots are places where Wi-Fi APs have been put up and made available to the general public, sometimes for a fee, or for free, or in exchange for patronizing a business. These Hotspots can typically be found at coffee outlets, fast food chains, shopping malls, libraries and community centres. The public can use their personal Wi-Fi devices to access the Internet at these Public Hotspots.

Because Public Hotspots are relatively cheap and easy to set up, many community leaders see the potential of using Public Hotspots to bridge the digital divide. The city of Philadelphia, working with Earthlink Inc., a commercial Internet Service Provider (ISP), is one noteworthy example of such an effort. Philadelphia is deploying over 4,000 APs on lamp posts all over the city. The capital costs of the infrastructure will be shouldered by Earthlink while the city contributes access to its lamp posts and takes responsibility for obtaining approvals from relevant city departments for the service. Through this public-private collaboration, Philadelphia will be able to offer Wi-Fi Internet access to its needy residents for half the cost of normal commercial Internet access using cable or DSL technologies. The service is also putting pressure on the traditional ISPs to either improve their service offerings or drop their prices to compete. Alarmed by this impending threat to their business, incumbent broadband providers in Philadelphia have responded by pushing for State legislation to prevent City administrations from using public funds to deploy such networks because they supposedly represent unfair competition. We can expect a repeat of this legal battle in many of the cities trying to deploy cheap and affordable community Wi-Fi systems in the United States and elsewhere around the world. Regulators should monitor the case closely for clues on how they may handle a similar situation within their jurisdiction.

Following on the heels of Philadelphia is the city of San Francisco where they approved a joint bid by Google and Earthlink to offer free and paid Wi-Fi Internet access to city residents. The service will be funded through location-aware advertisements based on the context of the users. In Asia Pacific, Singapore announced its Wireless@Sg programme in December 2006, which will see the number of Public Hotspots in Singapore climb from 600 to 5,000 within one year. Through this service, residents and visitors will get to enjoy free basic rate Internet access for a period of three years.

Wi-Fi Public Hotspots are not just meant for cities in developed countries. First Mile Solutions (2003), a Cambridge-based company with close links to the Massachusetts Institute of Technology, has developed and launched several innovative mobile Wi-Fi Hotspot solutions for developing countries. One solution uses a store-and-forward technique for emails to be dropped and picked up by Wi-Fi equipped motorcycles (called 'Motomen') or trucks as they move in a circuit around a central hub that has an external Internet connection. Villages along the circuit can send and receive e-mail through these roving Wi-Fi carriers, albeit with some delay. Villagers can also access the Internet through content that has been previously downloaded and stored locally. This technology has brought some form of basic communications to rural towns and villages in Cambodia and India for a fraction of what a traditional communication system would have cost. Recently, Goswami and Purbo (2006) published a paper reviewing the uses of Wi-Fi in less developed nations and they highlighted certain 'innovative' uses of Wi-Fi to overcome the hostile market and regulatory environment in Indonesia. These innovations included the use of Wi-Fi to provide last mile access and as a low capacity backhaul technique to overcome infrastructure limitations in Indonesia. Many users also resorted to becoming unlicensed re-seller ISPs to recover their own costs.

Although Wi-Fi came into existence some 20 years after the mobile phone and it does not yet enjoy the widespread adoption of mobile phones, it is nevertheless a technology with tremendous potential for development. As mentioned earlier, a mobile phone requires a complex supporting infrastructure in the background for it to work. Creating that infrastructure requires government licensing, extensive capital expenditure, and detailed radio and network planning. Only large companies with significant capital resources are able to undertake these tasks. In contrast, a Wi-Fi system is much easier to set up. Many countries have created regulations to permit systems to operate in a relatively unconstrained manner in the ISM band. Such ISM systems, which include Wi-Fi, can operate without a specific radio frequency allocation or an operating license.[4] Wi-Fi would never have taken off if each and every user had to apply for a license from their home regulator! It is also much easier to set up a Wi-Fi radio network because Wi-Fi uses a much simpler though admittedly less efficient radio channel access method to manage multiple users attempting to access the base station. This makes radio planning for Wi-Fi systems much simpler at the expense of certain performance parameters. Because of these features, Wi-Fi systems can be set up in a much shorter time using relatively less skilled human resources compared to mobile phone systems.

Mobile phone base stations can cost up to USD 100,000 per unit, whereas a high-grade Wi-Fi AP can be purchased for only about USD 5,000 per unit. This is because the technology used in Wi-Fi is less complex and there is much greater competition in the Wi-Fi market. The severe competition has resulted in price wars and breakthrough innovations that bring better value to Wi-Fi users. In the 10 years since its introduction, Wi-Fi's data rate improved from 1 Mbps to 54 Mbps. During that same period, the mobile phone's data rate improved only from 40 Kbps to 384 Kbps.

Possibly because of these factors and the fact that there is much more competition for the supply of Wi-Fi services, Wi-Fi tends to offer much better value than mobile phones. With Wi-Fi, data rates tend to be much higher and prices much lower. For example, under Singapore's Wireless@Sg Wi-Fi service, users can connect at 512 Kbps for free, whereas a mobile phone call would cost SGD 0.20 per minute and a 40 Kbps GPRS data connection would cost SGD 0.0037 per kilobyte transferred. While it has become quite common for Wi-Fi operators to offer unlimited data plans, mobile phone operators tend to charge for the amount of data carried or by connection time. Both of these charging methods can expose the user to a very large bill at the end of the month.

On top of the endorsement by Intel and the emergence of Public Hotspots, several new developments have given Wi-Fi a further boost. These include the launch of Skype, a Voice over Internet Protocol (VoIP) technology for making phone calls over the Internet. Skype managed to attract over 130 million registered users since its launch in 2003 because it offers free on-net telephone calls and very affordable off-net calls. On-net calls are calls between two people connected via the Internet using their Skype IDs. Off-net calls are calls between a Skype user and someone with a normal fixed-line or mobile telephone number. Equipment manufacturers are rushing to put Wi-Fi capabilities into devices such as notebook computers, tablet personal computers, game machines, cordless phones, PDAs and even mobile phones, to exploit use of Hotspots and services such as Skype. With the number of Public Hotspots growing, interesting new Wi-Fi devices being introduced and with the availability of services such as Skype, more and more people are acquiring Wi-Fi-enabled devices to take advantage of this trend. In time, Wi-Fi may come to challenge mobile phones as the dominant means of communication in the world.

However, some of the weaknesses of Wi-Fi systems should be noted. First, Wi-Fi is primarily a short-range system. Thus, building an infrastructure that can provide seamless coverage similar to that of a mobile phone system in a reasonably-sized city would require the deployment of tens, if not hundreds of thousands, of Wi-Fi Hotspots. The backhaul connectivity costs and maintenance requirements of such a system are not trivial and may come to dominate the operating expenses of the Wi-Fi service provider.

Second, Wi-Fi is not able to provide the seamless connectivity that mobile phones seem to do so well; it is able to provide only nomadic access capability. Once a Wi-Fi end-user device moves out of range of the AP it is associated with, connectivity is lost and the link is broken. It has to search for and establish a new connection with another AP whose signals it can detect. There is no automatic handover or mobility capability.

Third, unlike with mobile phones where there are only a limited number of operators per geographical region, there are usually many more smaller and independent Wi-Fi operators in a geographical region. Each of these may have a different sign-on mechanism and they are likely to issue a log-in identity that is unique to their system. Unlike with mobile phones where a single identity suffices for travel around the world, a traveller using Wi-Fi may pass through several of these Wi-Fi operators and thus end up with a whole bunch of different log-in identities and an assortment of bills at the end of his or her journey.

Fourth, many Wi-Fi operators have adopted simplified log-in processes that may have security weaknesses. In these implementations, data exchanges between the end-user device and the AP are unencrypted and unscrupulous people could eavesdrop on the data conversation and extract information that is private to the user.

Future evolutions of Wi-Fi and mobile phone systems

The proponents of Wi-Fi are well aware of its limitations and have been taking steps to address them. One such effort led by Intel with the support of many leading players from the Wi-Fi space is the Wireless Interoperability for Microwave Access (WiMAX) Forum. The technical standards being developed by the WiMAX Forum include the ability to cover large distances of up to 30 miles, provide seamless handover from one base station to another and assure end-to-end security. In short, they are trying to eliminate the shortcomings of Wi-Fi by replicating many of the features of mobile phone systems while trying to deliver them at a much lower cost. In August 2006, the WiMAX camp got a big boost when one of the leading cellular phone operators in the US, Sprint Nextel, announced that it will be using the WiMAX platform to construct a fourth generation nationwide broadband network.

Wi-Fi equipment vendors are starting to build mesh networking capabilities into their products. Mesh networking connects several individual APs into an interconnected network that can share a backhaul link into the Internet. The sharing of a backhaul connection can reduce the operating costs of large Wi-Fi systems very significantly and make a big difference to the business case of city-wide Wi-Fi systems. Mesh networking also gives the system the ability to respond to isolated equipment failures by re-routing traffic to other paths unaffected by the failure. The WiMAX Forum is also looking into building mesh networking technologies to its standards.

The players in the mobile phone ecosystem are not taking the challenge from Wi-Fi and WiMAX lightly. Where previously they were able to introduce new improvements at their own pace, which is about one refresh cycle per decade, they are now deploying improvements at a much faster rate. These improvements include the launch of mini and micro-sized mobile phone base stations which are much cheaper, lower cost handsets that are more suited for developing markets and the expedited deployment of new technologies such as HSDPA (High-Speed Download Packet Access), the mobile phone's response to criticism that its data rates cannot compete with Wi-Fi.

Services and applications—The next frontier

Historically, development efforts have concentrated on providing basic voice communications to as many of the population as possible. However, given the state of communications technology today, one may be able to build a communication infrastructure that is capable of supporting both voice and data for the same cost as for a voice-only infrastructure. Data capability provides access to the world of the Internet and multimedia, making for a much richer experience for users. Technologies such as Wi-Fi are inherently data communications channels that can also support voice, whereas mobile phones are primarily voice communication channels with some limited data capabilities.

Beyond the realm of voice communications, the issue of language and literacy needs to be addressed. Over half of the world's population resides in the Asia Pacific region where over 2,000 languages are being used and English is understood by only about 20 per cent of the population. A related issue is the fact that most electronic devices are designed to support Romanized characters only, which means that it is not possible to input or display without modification pictographic languages such as Japanese, Chinese, Indian and Thai. Yet another difficulty is the fact that a large number of people in underdeveloped countries are illiterate. They not only lack the capability to understand a foreign language such as English, they also cannot read even in their own language, which means they cannot communicate using text-based applications such as email and SMS.

Several countries have shown that it is not necessary to use English to achieve success in getting their population to enjoy and benefit from advanced communication services. Countries like Japan, Korea, China and Thailand have shown that it is possible to develop comprehensive and innovative communications services using their native languages. For example, China Mobile, the world's largest mobile phone operator with about 300 million subscribers, uses the Chinese language, and Japan and Korea have built two of the world's most sophisticated mobile phone services using the Japanese and Korean languages, respectively. These facts point to the need to develop communications services and devices that support languages native to the target population.

Some noteworthy attempts to develop native language capabilities in Asia Pacific include Malaysia's Murasu Communications (M) Sdn. Bhd., which is developing software to enable sending and receiving text messages in the Tamil language. In Cambodia, a private company, iWOW Communications Pte. Ltd., has developed a system for keying in and reading Khmer characters for sending and receiving SMS. There are cross-country efforts, such as some Chinese companies developing Korean language (Hangul) capabilities for sending text messages from China to Korea. Similarly, Pock Translate (http://www.pocktranslate.com/) runs a service for translating English SMS text into the Thai language to help visitors to Thailand communicate with the local people. Finally, Microimage, a Sri Lankan company, won the prestigious GSMA Asia Mobile Innovation Award 2006 for their multilingual (Sinhala, Tamil, Telugu, Malayalam and Thaana) SMS software for mobile phones.

The Internet is a treasure trove of information that can be very useful for development efforts in Asia Pacific. Unfortunately, most of the information is in English and beyond the reach of the majority of people in the region. To bridge the language barrier, research organizations are developing automatic machine language translators to convert the information into a form that the local population can understand.

NECTEC (the National Electronics and Computer Technology Center) in Thailand has developed such a suite of software applications called Parsit, an English-to-Thai language translation engine that has an accuracy of over 90 per cent. Parsit automatically translates Internet content in English into the Thai language, making it comprehensible to most people in Thailand. Parsit can work in conjunction with Vaja, a text-to-speech conversion software that takes in Thai-language characters and generates human voice of that text. Using Vaja, one can overcome the literacy barrier because the information from the website can be 'read out' to interested parties who may not be able to read but who can comprehend the spoken word. The third component of the software suite is iSpeech, a speech recognition program using an isolated word recognition technique. It is designed to convert speech inputs into text—that is, the user can just talk into a microphone connected to his or her personal computer and iSpeech will convert the voice data into text. Parsit, Vaja and iSpeech can be combined to create a 'speech-to-speech' translation application with wide ranging potential for development, especially in the tourism industry. First, a spoken English sentence is fed through iSpeech which then generates text in English. The English text is then passed on to Parsit for translation into Thai language text. Finally, the Thai text is input into Vaja for conversion into audio sounds in the Thai language.

While NECTEC's efforts are focused on Thailand, the PAN Localization Project, which is supported by IDRC's Pan Asia Networking Program, the National University of Computer and Emerging Science, Pakistan and its Centre for Research in Urdu Language Processing, is a multi-agency, multi-country effort to solve language localization issues across countries in Asia Pacific, including Afghanistan (Pashto, Dari), Bangladesh (Bangla), Bhutan (Dzongkha), Cambodia (Khmer), Laos (Lao), Nepal (Nepali) and Sri Lanka (Sinhala, Tamil). It aims to develop character sets, fonts, lexica, spelling and grammar checkers, search and replace utilities, speech recognition systems, text-to-speech synthesis and machine translation for these languages. Different aspects of localization technology that will be addressed include linguistic standardization, computing applications, development platforms, content publishing and access, effective marketing and dissemination and intellectual property right strategies for the output products.

Other regional language translation activities include the English-to-Malay translator being developed by the University Sains Malaysia and the collaboration between NECTEC, Thailand and a private Japanese company on machine language translation. Globally, there are also Internet-based language translation tools such as Yahoo's Babel Fish (http://babelfish.yahoo.com/) and Google's language translation service (http://www.google.com/translate_t) that can be brought into service.

The Asia Pacific region is the home of some really innovative mobile applications that have captured the attention of the world. The Philippines is recognized as one of the heaviest and most innovative users of SMS in the world. Paul Budde, an Australian telecommunications analyst, noted that in 2006, some 250 million SMSs were sent in the Philippines every day accounting for about 10 per cent of the world's SMS. One reason for the popularity of the text service is the fact that it is much more affordable than the voice service. It costs only 1 peso per SMS while a normal voice call can cost up to 20 pesos per minute. One of the most innovative uses of SMS in the Philippines is for small-value financial transactions. It is possible to pay for goods and services, transfer money from person to person and even donate money using SMS. For example, many Filipinos working overseas use SMS to transfer money to loved ones at home. The SMS channel is also being used as an information network by fishermen and flower farmers in outlying areas to get the latest prices in the city so that they can match supply to demand and obtain the best prices for their products. The Philippine government is another heavy user of SMS: it uses the channel to broadcast information to certain targeted groups and to obtain feedback from citizens. Specific channels for people to lodge complaints on air pollution and corruption, for example, and to give feedback to key decision-makers, including the President of the country, have been created.

In Indonesia and Malaysia, an innovative mobile phone application is called 'Solat Times' or 'Prayer Times' to help the Muslim population get their timing for prayers accurately. To use Solat Times, a person sends an SMS to request the Muslim prayer time and he or she will be informed by a return SMS the officially sanctioned prayer times appropriate for his or her particular location. This unique application which makes use of SMS and location information has now been exported to the United Kingdom and South Africa.

Development-friendly regulatory policies

The stance taken by a nation's telecommunications regulator has a big impact on the vibrancy of the communications ecosystem of the country. The policies the regulator adopts will have a strong bearing on the types of systems deployed, their affordability, and the take-up rate of services in that country. Some policymakers are more concerned with protecting the incumbent telecom provider, perhaps because the incumbent is a sister organization and is part of the government. Another reason could be that the incumbent is providing jobs to a large number of people and there could be political unrest if competition causes a large number of people to lose their jobs. In such cases, the policymaker might adopt policies that limit competition to the incumbent such as restricting the number of players and imposing terms that disadvantage other players with respect to the incumbent. With globalization, such practices have come under increasing scrutiny by world bodies such as the World Trade Organization (WTO). But irrespective of WTO, policymakers should have seen by now enough examples of how nations moving away from a 'pro-incumbent approach' and taking the path of market competition, have met with tremendous success.

The telecommunications regulator is in a position to determine the degree of competition in the market through the number of licenses issued and the license conditions, such as the minimum quality of service for coverage, network up-time, fault response times, and any Universal Service Provision (USP) that may be imposed. All of these conditions affect the quantum of investment needed by the operator and ultimately the profitability of the service. However, given that mobile operators are among the most profitable companies around the world, there seems to be little danger that regulators are being too strict with their licensing requirements today even though mobile operators have been required to come out with very large sums of up-front money to roll out their networks.

For market competition to work there must be mechanisms to limit the market power of the incumbent. These include clear rules for local loop unbundling, establishing and enforcing Reference Interconnect Offers (RIO), transparent rules for radio spectrum allocation, and vigilance against practices that can unfairly lock in consumers such as leveraging on handset subsidies to lock down handsets or impose an unreasonably long contractual period. Other measures include introducing rules to permit number portability when a subscriber moves from one provider to another. In certain markets taxes can be a drag on development. By removing handset taxes and subscriber registration taxes, India and China, for example, have made the service much more affordable, which has given their markets a tremendous boost.

Besides creating an environment for market forces to work, policymakers need to consider issues that affect society in general. Asian societies tend to be more conservative than Western societies, which means that applications such as adult content, chat line services and gambling, which are very popular in the West, cannot be so readily deployed in Asia Pacific. If these applications were permitted, they could give rise to certain social ills and there could be vigorous objections from certain segments of society. Many regulators in Asian countries have chosen to prohibit such services.

The telecommunications regulator should also ensure that subscriber privacy is protected at all times through strict rules against unlawful tapping of phone conversations and disclosure of a subscriber's location information and personal information such as his or her home address, calling habits and circle of contacts.

Some forward-looking policymakers see their role to be larger than simply regulating for market competition: they take it upon themselves to stimulate the development of the telecommunications industry in their country. Some of the actions they have taken include the development of human resources through education and training, nurturing indigenous equipment and service providers, encouraging research and development of communication technologies to reduce dependency on external providers, and facilitating infrastructure-based competition in their country.

Conclusion

The last decade has been a period of rapid technological developments in the wireless and mobile space. Unlike the networks built a decade ago, these technologies are now sufficiently advanced and versatile that they are capable of supporting multiple applications such as voice, data and broadcasting simultaneously on one single network. Less developed nations have the opportunity to use these modern technologies to leapfrog and not be bogged down by legacy equipment installed earlier. Equipment standardization has also helped create a critical mass of users and providers, which in turn drives down the cost of equipment, making communications much more affordable than before.

While the provision of basic voice services was the initial driver, future communication systems should also support access to the Internet to maximize their development potential. In such cases, care must be taken to overcome any language and literacy challenges so that the people who most need the tools can exploit them meaningfully.

Finally, telecommunications policymakers have a very important role to play. They should not expect the rules that served them well in the past decade to be still appropriate for the new era. New technologies call for new rules, in particular forward-looking rules that permit rather than hinder these new technologies and services to flourish for the betterment of their people.

Notes

1. Taken literally, the terms 'mobile' and 'wireless' apply to products that are capable of functioning without a tethered connection. This definition would include MP3 music players, digital cameras, as well as communication devices such as mobile phones, Wi-Fi-enabled notebook computers and personal digital assistants (PDAs) with a built-in mobile and/or Wi-Fi connectivity. In this chapter, we shall take mobile and wireless to mean only those products designed for voice and/or data communications and which use radio frequency to connect back to a base station or network.
2. Mobile phones are also known as cellular phones.
3. The maximum transmit powers (EIRP) for WLAN at 2.4 GHz for point to multi-point applications are +30 dBm and +23 dBm for the US and Singapore, respectively.

4. Unfortunately, not all developing countries have opened up the ISM frequency band for general use.

References

Budde, P. (2006). *Philippines—mobile communications—market overview*. Retrieved from http://www.budde.com.au/Reports/Contents/Philippines-Mobile-Communications-Market-Overview-1530.html

First Mile Solutions. (2003). *Internet village motoman project in Ratanakiri, Cambodia*. Retrieved 2 June 2007 from http://www.firstmilesolutions.com/projects.php?p=ratanakiri.

Goswami, D. and Purbo, O. (2006). *Wi-Fi 'innovation' in Indonesia: Working around hostile regulatory and market conditions*. WDR discussion paper 0611. Available online: http://www.regulateonline.org/content/view/737/31

ITU. (2005a). *Global main line telephone and mobile cellular subscriber data*. Retrieved 2 June 2007 from http://www.itu.int/ITU-D/icteye/Indicators/Indicators.aspx#

ITU. (2005b). *ITU world telecommunication indicators database*. 9th Edition.

ITU. (2006). *World information society report 2006*. Retrieved 2 June 2007 from http://www.itu.int/osg/spu/publications/worldinformationsociety/2006/index.html

Petrazzini, B. (2000). *Fixed-mobile interconnection case studies, slide 18/29 (Impact of CPP on cellphone subscribers)*. ITU Workshop, 20–22 September 2000.

The role of ICTs in risk communication in Asia Pacific

Krishnamurthy **Sriramesh**, Chanuka **Wattegama** and Frederick John **Abo**[1]

Introduction

In recent years, the Asia Pacific region has experienced a spate of crises that have brought untold misery to large sections of the populace. Two types of crises have received the most attention in the region in the past few years because of their adverse impact—natural disasters (the Asian Tsunami and earthquake in Kashmir) and public health emergencies (SARS and Avian Flu).[2]

Although natural disasters cannot be prevented, their adverse impact can be mitigated through effective risk communication. Likewise, effective risk communication can help avert or mitigate the adverse impact of public health crises. This chapter focuses on the use of ICTs in risk communication to help avert or mitigate the adverse impact of natural disasters and public health emergencies in the Asia Pacific region.

Risk communication

Lerbinger (1997) referred to *risk* as the 'probability that death, injury, illness, property damage... will stem from a hazard' (p. 267). In this chapter, the term *crisis* is used as a synonym for *hazard*. Seymour and Moore (2000) defined risk as 'the sensitive task of dealing with a latent or slowly advancing crisis *before* it breaks in full force' (p. 17). Fearn-Banks (1996) saw risk communication as 'an ongoing program of informing and educating various publics about issues that can affect... [them]' (p. 13).

Simply put, risks often are precursors to crises. A lack of, or inadequate, risk management may lead to a crisis with grave consequences. It is wiser to avoid a crisis, which requires that we pay attention to risk management. Effective risk communication underpins robust risk management. In essence, effective risk communication is often the best way to avert a crisis.

The importance of risk communication has received international recognition at the highest levels. *Coordinating risk communication* was one of three 'key policy areas for immediate attention' identified by the international meeting of health ministers held in Ottawa in October 2005 (Ottawa 2005). The ministers' communiqué was reinforced a couple of months later at the East Asia Summit when the Heads of State of ASEAN, Australia, China, India, Japan, South Korea and New Zealand included in their declaration a statement (East Asia Summit 2005) calling for '[E]nhancing capacity building in coping with pandemic influenza, including establishing information sharing protocols among countries and multilateral organizations to ensure effective, timely and meaningful communication before or during a pandemic influenza outbreak.' The World Health Organization (WHO) recently published its third complementary strategy for Avian Flu aimed at 'rapidly detecting, and potentially stopping—or containing—an emerging pandemic virus near the start of the pandemic' (WHO 2006a).

Different types of media have been used for communicating risks to small and large audiences. However, there is a paucity of published literature on the use of new media for risk communication. In this chapter, we discuss the use of ICTs for effective risk communication vis-à-vis natural disasters and public health emergencies. We discuss risk communication vis-à-vis long-term planning before a crisis strikes, shorter term planning much closer to a predicted crisis, and during and in the immediate aftermath of a crisis. We then conclude by offering some recommendations on how ICTs can be better harnessed for more effective risk communication in the Asia Pacific region.

Long-term programmes for ICT use in risk management

Long-term planning is crucial to effective management of public health and natural disaster mitigation efforts. In the preparedness phase, emergency managers develop plans of action to be carried out when the disaster strikes. Common preparedness measures include proper maintenance and training of emergency services, development and exercise of emergency population warning methods combined with emergency shelters and evacuation plans, stockpiling of supplies and equipment, and development and practice of multi-agency coordination. Traditional media such as television and radio, as well as modern ICTs like Internet and e-mail, can be used in these disaster preparedness activities.

Mathew (2005) highlighted the strategic role of ICT in managing disasters and public health emergencies, suggesting that effective health response to disasters will depend on three important prongs of action: (a) long-term disaster preparedness, (b) emergency relief and (c) management of disasters. He proposed that work in these three critical areas may be facilitated by communication and space technologies, especially the Internet and remote sensing satellites. Matthew presented a model that manages disasters through the Health and Disaster Information Network that operates in coordination with Internet community centres. 'The Model for Public Health Management of Disasters for South Asia' deploys ICT in a strategic manner to meet the unique and urgent requirements of personnel involved in disaster management. The infrastructure proposed is intended to serve governments, non-governmental organizations and institutions working in the areas of disaster and emergency medicine. The creation of such an infrastructure is intended to provide connectivity to support the rapid transfer of data, information and knowledge from senior government officials to grassroots organizations, as well as across national borders to other concerned organizations in the South Asian countries.

Mathew (2005) also presented a case study of the statewide ICT network for disaster management established by the Government of Maharashtra in April 2000, as part of the Maharashtra Emergency Earthquake Rehabilitation Programme following the Latur earthquake that left more than 10,000 people dead and 200,000 homes destroyed. The main role of the network is to facilitate and provide a rapid flow of information, online connectivity, response planning, and control and monitoring of the situation in disaster areas. The network comprises an emergency operations centre, a central control room, an alternative central control room, control rooms at each of the six divisional headquarters of the state and district control rooms at each of the 32 districts. The network is connected via VSAT telecommunication links. It is designed to provide data, voice and video teleconferencing facilities. All sub-districts in the state are linked through a VHF wireless network with nodes located at the district control rooms. The case shows that ICT has an important role in disaster preparedness, response, and coordination and control.

SOPAC: Another example of ICT use in long-term risk management is the South Pacific Applied Geoscience Commission (SOPAC) project which seeks to ensure vulnerability reduction in the Pacific Island Countries through the development of an integrated planning and management system. A key component of the project is GeoCMS, a Content Management System which facilitates the collection and sharing of geographical data among the stakeholders in the project. This was developed using two existing Free and Open Source Software (FOSS) based applications, MapServer and Tikiwiki. This system has made it possible for the Pacific Island states to publish their geographical data on the Internet and to receive contributions from all over the world. Countries involved in this project include Cook Islands, Federated States of Micronesia, Fiji, Kiribati, Marshall Islands, Nauru, Niue, Palau, Papua New Guinea, Samoa, Solomon Islands, Tonga, Tuvalu and Vanuatu (Nah 2005).

AlertNet: Reuter's AlertNet is another example of the use of ICTs for long-term disaster preparedness. It was started in 1997 by the Reuters Foundation, an educational and humanitarian trust, as a humanitarian news network based around a popular website. It aims to keep relief professionals and the wider public up-to-date on humanitarian crises around the globe. Emergencies are categorized into four types: health-related, sudden onset, food-related and conflict. Some emergencies do not fit neatly into these categories, frequently overlapping in a complex manner that makes it difficult to separate cause and effect. AlertNet's presentation of emergency material aims to make clear these areas of overlap. Reuters is the main, though not the sole, source of information for AlertNet. It operates with a surprisingly limited full-time staff (Gidley 2005).

AlertNet tracks all emergencies for which it is possible to find reliable information, including those that receive only sporadic coverage elsewhere in the media, or the so-called 'forgotten' or 'hidden' emergencies. For example, the north-eastern Indian state of Assam has had massive floods several times, with thousands of people displaced and made homeless, but good warning and evacuation procedures have kept the death toll low. AlertNet highlights these types of disasters and attracts upwards of three million users a year. It has a network of 400 contributing humanitarian organizations and its weekly e-mail digest is received by more than 17,000 readers (AlertNet 2006).

ASEAN-Disease-Surveillance.Net: The ASEAN-Surveillance Net (ADSNet) is an Internet-based network established to improve ASEAN's long-term capability in detecting and responding to infectious disease outbreaks. It is sponsored by the Indonesian Ministry of Health and the ASEAN Secretariat and operated by the ASEAN Disease Surveillance Secretariat. The US Naval Medical Research Unit No. 2 (US NAMRU-2) and the South East Asia Foundation for Outbreak Regional Cooperation jointly support ADSNet. The network was established on the basis of a framework approved at the Regional Action Conference for Surveillance and Response: Infectious Disease Outbreaks in Southeast Asia held in Bali, Indonesia in September 2000. Consensus was reached at the conference, which was attended by 150 participants with Ministerial representation from 17 countries, to build systems to facilitate the exchange of outbreak information within Southeast Asia and the WHO-SEARO Grouping.

ADSNet was established to achieve several objectives: rapid dissemination of outbreak information within the region, establishment of a mechanism for sharing important information on epidemic disease transmission without compromising national sensitivity and confidentiality concerns, and provision of directory assistance in identifying regional expertise, including laboratory diagnostic capabilities that can support outbreak investigative activities. It also seeks to provide directory assistance in identifying outbreak causation, as well as training and educational opportunities within the region. ADSNet aims to accomplish its goals by undertaking the following activities which can be classified as examples of risk aversion and risk communication:

- Conducting outbreak response training workshops
- Developing laboratory diagnostic capabilities in identifying causative outbreak etiologies
- Facilitating outbreak investigations or interventions
- Establishing early warning outbreak recognition systems

ADSNet is a part of the Early Warning Outbreak Recognition System (EWORS) that currently maintains 18 surveillance sites in Cambodia, Indonesia, Lao PDR, Singapore, Thailand and Vietnam. Nepal and South Korea are expected to join EWORS in the near future. Local clinics in the participating countries make use of simple EWORS technology to upload syndromic data on disease outbreaks to hubs located at hospital computers. Algorithms running on these hub computers assess whether the uploaded data indicate any disease outbreaks. The system alerts local public health personnel if it determines that an outbreak has occurred. EWORS applications are designed to run on desktop computers, laptops and PDAs. The network's website is hosted in Singapore and jointly maintained by the Singapore Ministry of Environment and NAMRU-2. Plans are underway to link ADSNet with the Pacific Disease Surveillance Network (PACNET), so as to effectively cover 16 time zones, and eventually to integrate it into WHO's global network of networks.

In epidemic intelligence and systematic case detection, WHO Global Alert gathers information from formal sources as well as rumours of suspected outbreaks in order to get a clear picture of the epidemic threat to global health. With the advent of modern communication technologies, many initial outbreak reports now originate from informal sources in the form of electronic media and electronic discussion groups. More than 60 per cent of the initial outbreak reports come from unofficial informal sources, including sources other than the electronic media, which require verification. The information thus gathered is categorized and disseminated to public health professionals. Regular updates are issued to a network of electronically-interconnected WHO member countries (192), disease experts, institutions, agencies and laboratories through an Outbreak Verification List. WHO also reports verified outbreaks on its website, Disease Outbreak News.

Some private schools in Singapore are teaming up with Apple Macintosh Asia to develop ways of delivering education over the Internet to provide continuity of education during a pandemic. A technology is being tested by which both video and audio content as well as Powerpoint presentations can be broadcast through the Internet using Apple's iTunes.

Various capacity building efforts have been implemented at both international and national levels to help prepare countries to respond to and manage a pandemic incident such as Avian Flu. To support member states in their efforts to build the capacity of health care facilities to manage communicable disease emergencies of different orders of magnitude from a small number of patients to a widespread pandemic, the WHO worked with the Asian Disaster Preparedness Center (ADPC) to conceptualize the Health Care Facility Emergency Preparedness and Response to Epidemics and Pandemics (HCF-EPREP) Programme. The first phase of this programme culminated in the first training workshop in September 2006 in Bangkok attracting 39 participants from Bhutan, the Philippines, Thailand and Vietnam with facilitators from WHO, ADPC, Infection Control Plus (Australia) and the University of Texas School of Public Health. Associates from Brazil, India, Indonesia, Nepal, the Philippines, Syria, Thailand and the USA also participated. Most of the participants were directors of health care facilities; senior level specialists in infectious diseases, emergency medicine and infection control; and public health officials from national Ministries of Health and sub-national health authorities. The workshop paid special attention to using ICTs to disseminate information to both the public and health care professionals during the inter-pandemic phase (when the disease is spreading but has not yet become widespread) and the pandemic alert phase.[3] Participants were also taught how to use official websites to educate the public about the pandemic.

Strategic Health Operations Centre: In the global arena, the WHO Department of Epidemic and Pandemic Alert and Response (EPR) is continuously monitoring and tracking evolving infectious disease situations, sharing expertise, sounding the alarm and initiating an appropriate response to protect the population from the effects of a communicable disease emergency. To provide a well-coordinated and systematic response, which is crucial in any public health emergency, the Strategic Health Operations Centre (SHOC) was established. It utilizes the latest ICT to support member states in managing health emergencies.

The SHOC was used effectively during the SARS epidemic when the WHO responded rapidly and effectively in collaboration with many key partners. The response was widely praised by the public health and infectious disease research community, especially the role of ICT in early detection and in fostering global collaboration and information exchange during the SARS epidemic. On 17 March 2003, the WHO called upon 11 laboratories in nine countries to join a collaborative multi-centre research project on SARS diagnosis. The network took advantage of e-mail and a secure WHO website to share outcomes of investigations of clinical samples, electron-microscope pictures of viruses, sequences of genetic material for virus identification and characterization, and post mortem tissues from SARS cases. Individual departments of affected hospitals also used websites and e-mail to rapidly disseminate clinical findings to health professionals. However, a review of the system revealed that it was not used to its full potential. This assessment led to the formation of the Global Public Health Intelligence Network (GPHIN)[4] by Health Canada and the revised International Health Regulation or IHR (2005)[5] that requires governments to report public health threats—in particular, public health emergencies of international concern (PHEIC) or disease outbreaks and natural disasters that could have an international dimension.

Animal Health Management and Biosecurity: Since August 2004, Australia has been working with ASEAN on a project aimed at Strengthening Animal Health Management and Biosecurity in ASEAN (SAHMBA). Funded through the ASEAN Australia Development Cooperation Program (AADCP), the project seeks to enhance the capability of ASEAN member countries to manage risks to the biosecurity of livestock industries particularly those related to trade and impacting on the poor. The project focuses on strengthening capabilities in risk analysis, disease surveillance and animal health information management at the ASEAN regional level. It has four components. Component 1 aims to develop a consistent approach to risk analysis based on international best practices. Component 2 seeks to develop a consistent approach to wide-area disease surveys including the collection and analysis of information to substantiate reduction of disease in livestock populations. Component 3 expects to facilitate the development of a regional animal health information database that will (a) promote the trade of animals and animal products throughout the region by clearly establishing a country's disease status, (b) facilitate the sharing of animal health information within the region and (c) streamline production of disease reports required to meet international obligations such as those of the World Organization for Animal Health (OIE). Finally, Component 4 responds to project management and monitoring requirements for successfully implementing the technical components and keeping stakeholders advised of the project's progress and performance (AADCP 2005).

In public health, measures to control and prevent disease often rely upon timely and accurate information transfer. Thus, the Internet is increasingly becoming the preferred platform for agencies designing systems to protect public health. The WHO's Communicable Disease Cluster is spearheading a global partnership called the Global Atlas of Infectious Diseases (WHO 2006b), a Web-based system that brings together an interactive information system, surveillance data, reports and documents on the major diseases affecting developing countries, including malaria, HIV/AIDS, tuberculosis, as well as diseases that are close to being eradicated, such as guinea worm, leprosy and lymphatic filariasis. The global atlas also covers epidemic-prone and emerging infections like meningitis, cholera and yellow fever. It also provides information on essential support services such as the network of collaborating centres and the Global Outbreak Alert and Response Network.

Use of ICTs in risk communication about impending disasters

We next focus on the use of ICTs for shorter term risk communication—early warnings in the face of an impending disaster. In the last few decades, national level early warning systems have been used with varying degrees of efficacy during many natural disasters that have affected the Asia Pacific region. The powerful tropical cyclone that hit the Chittagong district of

Bangladesh in 1991 is an example. Though not as devastating as the infamous 1970 cyclone that killed an estimated 300,000 in that country, the cyclone of 1991 recorded a death toll of 140,000 and left more than 10 million homeless. Although early warnings about the impending storm had been issued to some extent, the number of casualties was high because of a lack of adequate cyclone shelter facilities (ADRC 2005). The Kobe earthquake of January 1995 (7.2 on the Richter Scale) caused 5,000 casualties and displaced over 300,000. The damage caused to roads, houses, factories and infrastructure was estimated at about USD 120 billion (Geo Resources 2002). Apparently, there had been little or no warning, causing many Japanese citizens to lose faith in the technology behind forecasting. Forecasting crises is a key ingredient of risk communication.

Undoubtedly, it was the fateful Asia Tsunami in 2004 that caused a spurt in interest in early warning systems for natural disasters. The death toll was as high as 200,000 while a disaster of comparable magnitude the following year—Hurricane Katrina—recorded less than 2,000 casualties. The high death rate in the Tsunami was attributed to the absence of timely and effective warning systems. At the 2005 World Conference on Disaster Reduction in Kobe, Japan, it was found that most national early warning systems were rather simple alert systems with limited capacities to collect, analyze and distribute information. This prompted UNESCO to coordinate the activities to launch an early warning system for the Indian Ocean (UN/ISDR 2005). The Ministerial Meeting on Regional Cooperation on Tsunami Early Warning Arrangements held in January 2005 in Phuket, recognized the ADPC's readiness to serve as a regional centre or focal point for a multi-nodal Tsunami early warning arrangement in the region (ADPC 2006).

Tsunami warning systems

The Indian Ocean Tsunami warning system coordinated by UNESCO is currently being implemented. The Intergovernmental Coordination Group for the Indian Ocean Tsunami Warning and Mitigation System (ICG/IOTWS) serves as the regional body to plan and coordinate its design and implementation. For the interim period, the Pacific Tsunami Warning Centre (PTWC) and Japan Meteorological Agency (JMA) are expected to provide Tsunami warnings to Asian nations. Australia, Bangladesh, Timor Leste, India, Indonesia, Iran, Malaysia, Maldives, Myanmar, Pakistan, Singapore, Sri Lanka and Thailand are among the 26 countries that have established official Tsunami Warning Focal Points (TWFP) to receive interim advisory information based on seismological and sea-level information from the operational centres serving the Pacific in Hawaii and Tokyo (UNESCO 2006).

Meanwhile, the regional Tsunami and multi-hazard observation and monitoring network for Southeast Asia is also being implemented by ADPC as a joint venture with the national governments of Cambodia, Lao PDR, Myanmar, Thailand, Vietnam, China, the Philippines, Singapore, Bangladesh, Sri Lanka and the Maldives. This system will be implemented in two phases. The first phase, covering the most vulnerable areas of Southeast Asia, involves the installation of five sea-level gauges and five seismic stations across Cambodia, Lao PDR, Myanmar, the Philippines, Thailand and Vietnam. Phase 2 will see the proliferation of these technical components, in addition to several deep-sea buoys, across the region in order to provide a comprehensive network (ADPC 2005).

An earthquake near Indonesia in July 2006 tested the interim measure of early warning systems. Unfortunately the technology failed on that occasion too. More than 5,000 casualties were reported when two-metre high waves triggered by an undersea earthquake hit the island of Java. The 20 minutes scientists had to analyze data from 30 seismological stations and send out a warning were not sufficient. Furthermore, even if they were in a position to issue a warning, the scientists did not have the media needed to disseminate it to the isolated communities in the islands (BBC 2006).

The communication subsystem is an essential part of any early warning programme. In its most basic form, an early warning system is a communication channel between those who monitor the prospective disaster and the community for whom the messages collected from monitoring are intended. The disaster warning message should be reliable, authoritative, timely and clear to the target communities especially about what is expected of them. The anticipated response of a community differs according to the nature of the possible disaster. Incorrect responses can create additional problems.

'Last mile' communication in disaster warning

In the Asia Pacific context, crossing the 'last mile' in the communication chain is a major challenge because ICT penetration in most countries in the region is still far below satisfactory levels. For example, the teledensities of Cambodia, Laos and Bangladesh are 38, 32 and 15 per 1,000 of the population (UNDP 2005). In rural areas, many households are not connected to any medium. Multimedia communication is sometimes needed for Early Warning Systems to be effective in this situation.

Of the available media channels, radio and television are most likely to be used for warning about impending disasters. However, television penetration is not high in many Asian societies, especially in rural areas. Radio is the preferred medium for the dissemination of disaster warnings. Bangladesh, for example, relies heavily on radio to issue warnings about impending disasters, effectively using it in several flood and cyclone related incidents since the early 1970s (UNEP 2001). However, neither radio nor television is interactive and they are of limited use if a warning is to be disseminated late at night when most stations in the region are not in operation.

Telephones, both fixed and mobile, overcome lack of interactivity and limited use at night, but they have their own limitations. One obvious disadvantage is potential congestion in the period immediately prior to a crisis. This limitation is applicable for both voice and SMS messages. To overcome this, mobile phone manufacturers have introduced a feature called 'cellular broadcasting' which helps disseminate a warning message to a selected group in a short period of time using a different band to avoid congestion. There are no additional costs as this feature is already available in most network infrastructures and phones. This combined geo-scalability and geo-specificity of mobile phones helps disaster managers to avoid panic and traffic jamming (MobileIN.com 2005).

There are many examples of how simple phone warnings helped save many lives in South Asian countries during the Tsunami in 2004. Perhaps the most famous example occurred in the small coastal village of Nallavadu in Pondicherry (India) where a timely telephone call about the impending Tsunami is said to have saved the village's 3,600 inhabitants as well as the people of three neighbouring villages. Nallavadu was part of the very successful M.S. Swaminathan Research Foundation's Information Village Research Project. A former project volunteer who was working in Singapore heard the Tsunami alert issued there and immediately phoned the research centre in Nallavadu and asked that its early warning alert system be used to warn the villagers. His quick thinking and the swift and coordinated action in Nallavadu led to the evacuation of the four villages before the Tsunami hit the coast (Subramanian 2005).

First conceptualized in 1992, the Information Village Research Project under the aegis of the M.S. Swaminathan Foundation, is now being implemented in seven villages in Pondicherry. The objective is to test whether information technology can become an ally in poverty alleviation and whether it can be used as a tool in empowering the rural poor. Seven Village Knowledge Centres have been set up, each with a computer, a modem and a wireless system, backed by solar power since there is irregular power supply. The services provided by these centres include gathering and transmission of information such as commodity prices, weather, government announcements and the daily news. The centres also help in the generation of data (for example, surveys, library references, discussions, bulletins) and assist in the creation and maintenance of locality-specific databases on local hospitals/doctors, training programmes, high school/college course guidance, government welfare programmes/entitlements and soil agronomy/weather/cropping patterns (M.S. Swaminathan Research Foundation 2006).

Amateur radio and community radio are two channels that can be used effectively for disaster management purposes. However, there is little evidence so far that these media have been used effectively except in rare instances. The Indian NGO, National Institute of Amateur Radio (NIAR) that promotes amateur radio or ham radio in the country as a scientific and socially useful activity used amateur radio for communicating with its team in Port Blair, the capital of the Andaman Islands, in the aftermath of the Asian Tsunami. Though this channel had not been used prior to the disaster, it served as a key communication

Bridging the 'last mile' in warning about natural disasters

When the Tsunami struck the costal areas and took nearly 40,000 lives, or one in every 500 Sri Lankans, the obvious question was why it was not forewarned. Arthur C. Clarke, author and long-time resident of Sri Lanka, remarked: 'The Asian Tsunami's death toll could have been drastically reduced if the warning was disseminated quickly and effectively to millions of coastal dwellers on the Indian Ocean rim. It is appalling that our sophisticated global communications systems simply failed us that fateful day' (IDRC 2006).

Indeed there had been a warning. Scientists at the Pacific Tsunami Warning Centre (PTWC) in Hawaii who had detected the extraordinary seismic activity issued a local Tsunami warning one hour after the undersea quake. It was received in Sri Lanka but unfortunately not effectively communicated to the communities. History repeated itself on 17 July 2006 when a Tsunami caught Java island by surprise, in spite of PTWC disseminating a warning 17 minutes after the 7.7 magnitude undersea earthquake. A timely public warning could have saved many of the nearly 600 people who died.

The 'weakest link' on both these occasions was the so called 'last mile'. The warning did not cross the last mile. And even if it had done so, there is no guarantee that the communities would have taken the correct action because they had not been given any training on how they should behave during such events.

> The Last Mile Hazard Information Dissemination Project is an attempt to address this critical issue. It is multi-stakeholder initiative to complement other actions being taken at national and regional levels. It involves four Sri Lanka-based entities that value the role of ICTs and community mobilization in disaster preparedness: Sarvodaya, an NGO with a presence in all Sri Lankan villages; LIRNE*asia*, a regional ICT research and capacity building organization; TVE Asia Pacific, a media organization specializing in communicating development; and Dialog Telekom, a leading telecommunications service provider. Financial support comes from the International Development Research Centre (IDRC) of Canada.
>
> The action research project aims to study which ICTs and community mobilization methods will work most effectively in disseminating information on hazards faced by Sri Lanka's coastal communities. The research is not confined to Tsunamis alone; coastal erosion, cyclones, drought and floods are among the other hazards covered (LIRNEasia 2006). Focusing on the crucial 'last mile' dissemination, the project will test different ICTs in delivering timely warnings to the local people immediately at risk and build community capacity to respond to such warnings rapidly and systematically.
>
> In the first phase, the project involves 32 villages from the eastern, western, northern and southern coastal areas of Sri Lanka. The project will evaluate several factors that contribute to the design of an effective last mile hazard information dissemination system, such as the reliability and effectiveness of various ICTs as warning technologies, how community training influences effective warning responses, how the level of organizational development of a village contributes to an effective warning response, and gender-specific response to hazard mitigating action. Among the first and most important activities was training 30 youth leaders attached to Sarvodaya. The training, delivered by TVE Asia Pacific, covered such topics as understanding vulnerability and hazards, community-based hazard identification using Participatory Rural Appraisal (PRA) techniques, communicating risks and hazards, understanding and responding to early warnings and community response planning (TVE AP 2006).
>
> Different combinations of ICTs and community mobilization will be tested in the participating villages. These include fixed telephones, Sinhala/Tamil SMS (text messaging) with alarm for Java-compatible mobile phones, Very Small Aperture Terminals (VSATs), Disaster Warning Response and Recovery (DWRR) units based on addressable satellite radio developed by the WorldSpace Corporation under the WHO, and an Early Warning Network Remote Alarm Device developed by Dialog Telekom with the assistance of the University of Moratuwa, Sri Lanka. While some ICTs have been in public use for years or decades, others are recent innovations whose utility in disaster warning communication is being tested for the first time through this initiative. The plan is to identify the optimum combinations of training, community mobilization, and technology tools that could help Sri Lankan communities to receive hazard warnings and disseminate them locally.
>
> Sources: IDRC. (2006). *Bridging the 'Last Mile': Building grassroots capacity for disaster warning and preparedness in Sri Lanka*. Retrieved 10 November 2006 from http://www.idrc.ca/uploads/user-S/11465104711General_intro.pdf
>
> LIRNEasia. (2006). *Evaluating last-mile hazard information dissemination: A research proposal project document*. Retrieved 10 November 2006 from http://www.lirneasia.net/wp-content/uploads/2006/05/HazInfo%20Proposal.pdf#search=%22Evaluating%20Last-Mile%20Hazard%20Information%20Dissemination%20%22
>
> TVE Asia Pacific. (2006). *Taking hazard warnings to the grassroots: New project to mix technologies and training*. Retrieved 10 November 2006 from http://www.tveap.org/news/0605tra.htm

channel between the mainland and the islands in managing aid for the displaced (i4d 2005).

Electronic health initiatives

Noting the potential impact that advances in ICT could have on health care and health-related activities, Resolution WHA58.28 urges WHO Member States to plan for appropriate eHealth services in their countries. eHealth activities at the global level fall into two broad categories:

1. access to reliable, high quality health information for professionals and for the general public; and
2. use of ICTs to strengthen various aspects of country health systems, such as eLearning for human resources development and support for delivery of care services.

The WHO launched the Health Inter Network Access to Research Initiative (HINARI) in 2002, in partnership with leading biomedical publishers, academic institutions and organizations of the United Nations system to provide free or very low-cost online access to 2,900 major journals in biomedical and related social sciences to local, non-profit institutions in developing countries. It is one of the world's largest collections of biomedical and health literature. At present, 1,400 institutions in 104 countries are participating in the network. In 2004, users downloaded over 1.7 million articles.

To cater to the needs of the general public, the WHO began the Health Academy in December 2003. This innovative approach to improving public health provides the general public with health knowledge through eLearning packages designed to help people make the right decisions for preventing disease and leading healthier lives. The initiative draws on the WHO's information resources and expertise in health and its access to health information worldwide.

Use of ICTs in risk communication during crises

Whereas risks are often precursors to crises and risk communication is important in averting or mitigating the effects of crises, different types of risks arise even after a crisis strikes especially in the aftermath a natural disaster. The most difficult period of a disaster is the immediate aftermath, when prompt and swift action is essential. Disasters cause significant numbers of deaths and injuries and displace even larger numbers of survivors. There are physical as well as emotional injuries such as witnessing the loss of loved ones. Essential items such as food and other supplies need to be delivered, temporary shelters need to be put up and medical attention needs to be provided. All these need to be simultaneously addressed and ICTs can play a critical role in connecting the diverse groups needed to manage effective resource collection and distribution during these critical situations.

Think positive: The Asian face of HIV/AIDS

ICTs are being used to create HIV/AIDS awareness in many Asia Pacific countries. Recently, the UNDP Regional HIV and Development Programme for Asia, UNDP Asia-Pacific Development Information Programme (UNDP-APDIP), UNAIDS-Asia Pacific Leadership Forum, UNICEF, Asia-Pacific Broadcasting Union (ABU), MTV International and the Kaiser Family Foundation joined hands for one such initiative—the production of a series of 'made for television' programmes to raise awareness of the global HIV/AIDS pandemic. *Think Positive: The Asian Face of HIV/AIDS*, as the programme is called, focuses on the impact of HIV/AIDS on the contributing producer's home country, with emphasis on the human or social dimension. Completed productions are available for exchange between the participating broadcasters and are made available rights-free to all ABU member broadcasters.

Participating television producers from Bangladesh Television; China Central Television; PT Surya Citra Televisi, Indonesia; PT Indosiar Visual Mandiri Tbk, Indonesia; Sistem Televisyen Malaysia Berhad (TV3); Nepal Television; Media Niugini, Papua New Guinea; ABS-CBN Broadcasting Corporation, the Philippines; MediaCorp News, Singapore; National Broadcasting Services of Thailand (Channel 11); and Vietnam Television, each created segments for use by all participating broadcasters as individual short-form programmes. 'This was a first co-production initiative arranged by the ABU for its member broadcasters and in association with the Global Media AIDS Initiative', said Craig Hobbs of the ABU. 'It has resulted in strong interest and participation by our broadcasters, who moved very quickly to complete this project in time for World AIDS Day, and is one that is already stimulating many additional broadcast activities relating to the increasing awareness and changing behaviour for fewer HIV infections.'

MTV International supported the co-production project with the contribution of an executive producer who provided technical and creative direction to the participating producers while drawing on the achievement of MTV's long-running Staying Alive campaign. The Kaiser Family Foundation, UNDP, UNAIDS and UNICEF lent substantive expertise based on their work in HIV/AIDS communication, while the ABU played a coordinating role in the production of the content by soliciting applications from its member broadcasters.

Source: Asia Pacific Development Information Programme. (2006). Launch of HIV/AIDS TV programme—'Think Positive: The Asian Face of HIV/AIDS'. Retrieved from http://www.apdip.net/news/hivaidslaunch on 10 November 2006.

Role of the Internet

In spite of its relatively low penetration in some societies, the Internet is an ideal tool for risk communication in the post-disaster period for many reasons: the interactivity of the medium, its ability to reach a large group of people within a short period of time, its multimedia character (making it possible to present information in different formats—text, image, audio, video) and its universality.

Perhaps one of the first instances where the Internet was fully utilized for post-disaster response, was the 1999 earthquake that devastated western Turkey. The earthquake caused extensive damage to the telecommunication infrastructure, rendering fixed-line telephones useless. Although some of the mobile phone networks were still working, they were operating with reduced bandwidth. Many of the microwave repeaters mounted on apartment buildings had been damaged. The Internet was the only medium connecting the affected areas to the outside world. The Internet was used primarily for collecting aid and finding information about missing people in order to link them with families and relatives. Many organizations formed 'Message Lines', which acted as a database of people found, their condition, or the degree of damage to the region in which relatives lived (Zincir-Heywood and Heywood 2000). In addition, NGOs used discussion lists to coordinate donations so that donors could identify where help was needed the most as well as the nature of help needed.

Kalemoglu et al. (2005) studied the consequence of the absence of an effective communication and information system in the aftermath of the Turkey earthquake. They found that hospitals were overwhelmed during the critical six hours immediately after the earthquake, due to the exceptionally large number of injured requiring medical care and 'failure of communication with the disaster area'. The communication problem was eventually solved through the deployment of 'wireless and military communication systems'. Kalemoglu et al. (2005) also identified lack of forward planning and preparedness as a contributing factor to the failure of emergency services immediately after the earthquake. They concluded that lessons learned from the earthquake suggest that emergency response teams should be established as part of a larger contingency plan. These teams should then make advance preparations 'that include general precautions, work schedule, hospital care, equipment, transport, registration [of patients admitted to hospitals for treatment], communication and security'.

The ICT deployed need not be sophisticated to make a difference in managing an emergency, according to Ochi et al. (1999) who studied the aftermath of the Cambodian flood of 1997. E-mail, a basic ICT service, helped the WHO Cambodian field office to respond to a medical emergency that occurred after extensive flooding in Cambodia in August 1997. Unusually large numbers of people were being bitten by venomous snakes that had been washed into populated areas by the rising floodwaters. The Cambodian health authorities made an urgent request to the national field office of the WHO for 100 doses of polyvalent type snake antiserum. The WHO had only a few types of monovalent antiserum on stock; in addition, the WHO field office lacked essential taxonomic information about the snakes in Cambodia which the organizations that could provide the required antiserum needed. The coordinator at the WHO field office subsequently sent an e-mail to a member of the Global Health Disaster Network (GHDNet) to seek help in obtaining the required taxonomic information. The GHDNet member in turn forwarded the email to three mailing lists at the network. Members of the mailing lists recommended that the WHO contact several specialists and institutes in the region, including the Japan Snake Institute and the Serum Institute of India. This information-sharing led to the speedy procurement of the required snake antiserum that was then airlifted to Cambodia, saving hundreds of lives.

Alternative communication channels in the aftermath

In the immediate aftermath of a disaster, the traditional communication system is often overloaded, if not destroyed. This makes coordination of emergency response difficult especially in remote areas. To address the need for communication support in times of disaster especially in the Asia Pacific Region, Télécoms Sans Frontières (TSF), which specializes in emergency telecommunications, has established an Asian base at the ADPC. As soon as a catastrophe or conflict occurs, joint teams from ADPC and TSF are able to arrive anywhere in the world in less than 48 hours and install within minutes an operational telecommunications centre to provide communication support to enable NGOs, UN agencies and the affected population to connect with the outside world.

State-of-the-art satellite mobile telecommunication equipment, the miniaturization of components and the development of satellite networks make possible the rapid assembly of the telecommunication centre whatever the type of terrain. The group uses a network of four geostationary satellites whose 'spot beams' cover 98 per cent of the Earth's land surface. TSF has a number of Mini-M devices (Capsat Phone TT-3060A) which, in addition to digital phone facilities at 4,800 bps, allow fax, data transfer and e-mail at 2,400 bps. The main advantage of these devices is that they are small and light. The emergency telecommunications centres enable live reports, pictures and videos to be transmitted via broadband Internet connections.

Setting up a fixed centre with a permanent satellite Internet connection can prove to be a particularly effective tool for the organization of humanitarian work. Recently, TSF formalized an agreement with the UN Office for the Coordination of Humanitarian Affairs (UNOCHA) to provide telecommunications support for the UN Disaster Assessment and Coordination (UNDAC) teams. The team worked with the UNDAC team in operating the Disaster Operations Centre coordinating both local and international response during the landslide in the province of Leyte in the Philippines in February 2006. The team also operated in Thailand, Indonesia and Sri Lanka in response to the Asian Tsunami of 2004.

During the SARS outbreak in 2003, Singapore relied on ICTs as monitoring devices to enforce the quarantine law. For example, RFID (Radio Frequency Identification) was used to trace individuals who may have come into contact with SARS victims. Hospital workers, visitors and other patients with the potential of coming into contact with SARS victims were given a card containing an RFID transponder that tracked their movements between different zones in the hospital, making it easy to detect who may have come into contact with a patient later confirmed to have contacted the disease, which had different incubation periods extending to a maximum of 10 days. The government also passed a law permitting the installation of surveillance equipment such as electronic picture (ePIC) cameras in the homes of quarantined individuals (the ePIC cameras were monitored by a private company contracted for the purpose). In addition, video facilities were often the only tools available to families to communicate with family members who were gravely ill with SARS in the hospital.

Use of new media such as blogging

Blogging, although still a novelty in most areas of the Asia Pacific region, helped in many ways in the critical period following recent disasters. Sarvodaya, Sri Lanka's largest and most broadly embedded people's organization with a network of 15,000 villages, used blogs successfully in the immediate aftermath of the Asian Tsunami for fund-raising purposes. Two young volunteers started a blog on behalf of Sarvodaya that was later referred to by portals such as Google, Nortel and Apple, which helped to raise USD 1 million in a few weeks from donors around the world. The NGO used this money to provide much needed relief to the victims long before government agencies could react.

Hundreds of blogs emerged in the Asia Pacific region in the first few days following the Tsunami. These were used for information-sharing, locating missing persons, fund-raising and making donations to the needy. Some of these blogs are: http://tsunamihelp.blogspot.com (regional), http://tsunami-penang.blogspot.com (focusing on Malaysia and Thailand), http://indonesiahelp.blogspot.com (Indonesia), http://news.bbc.co.uk/2/hi/asia-pacific/4129521.stm (a journalists' blog hosted by the BBC), http://sltsunami.blogspot.com (Sri Lanka), http://consciouscitizens.blogspot.com (post-tsunami rehabilitation), http://phukettsunami.blogspot.com (Thailand), http://tsunamimissing.blogspot.com (regional) and http://tsunamihelpindia.blogspot.com (India). Many of these sites have remained active, helping in subsequent disasters. There were also many discussions about post-disaster help among e-groups, such as Bytes-for-all.

Long-term recovery

Indonesia effectively used radio to help reduce the trauma of the Tsunami victims. A weekly one-hour programme assisted by UNDP was launched after the Tsunami struck for the 13,000 internally displaced victims in Meulaboh, Aceh. The radio programme covered topics derived from interactions with the community, such as how to control emotions, family relations, worries about employment and income, housing conditions and establishing a community support network. A counsellor and a psychologist provided advice on how to cope with various forms of stress (UNDP 2006).

Wireless LAN technology is another ICT tool that has proven useful in post disaster periods. Ericsson implemented a WLAN solution in Pakistan during the recovery stage of the Kashmiri earthquake at the request of UNOCHA. The Ericsson response team was hosted in the Swedish Rescue Services Agency's camp in Muzaffarabad, Pakistan. As the affected population was scattered in remote and inaccessible locations, the major concern was to ensure that help reached those in need. Relief personnel were connected to an intranet through which information transfer, both within and across the relief organizations' own networks, could take place. The camp was connected via a VSAT connection to the UN Children's Fund (UNICEF) in New York, where a connection to the Internet was provided. The benefit of this system is that all relief workers have access to a common network and can share the same local information (Ericsson 2006).

Recommendations and conclusion

The importance of risk communication in averting many public health crises and natural disasters cannot be overstated. Yet risk communication has received attention only in recent years. Moreover, ICTs have not been harnessed to their full potential to mitigate risks from natural disasters and public health emergencies. For example, a review of risk communication case studies in the Asia Pacific region on issues such as earthquakes, flood, fire threats and disaster mitigation revealed that ICTs were rarely, if ever, used or even considered as a tool.

In order to be effective, risk communication, of which ICT should be an integral component, should be a continuous process that is integrated into the overall developmental scheme of every country. Objectives such as integrating the principles of sustainable development into country policies, building a healthy society and preventing loss of environmental resources cannot be achieved without placing due emphasis on effective risk communication strategies. Current development practices do not necessarily reduce the vulnerability of communities to disasters. Indeed, ill-advised and misdirected development practices may actually increase a society's vulnerability to the risks of disasters. A considerable challenge remains in raising awareness and capacity building so that communities can confidently meet any crisis situation.

We also recommend that national governments promote risk communication in local languages, given that English is spoken by a mere fraction of the close to three billion people who live in the Asia Pacific region. Localization of information about risks will make such content more useful to the population.

Political commitment by governmental and organizational policymakers and community leaders, based on an understanding of risks and disaster reduction concepts as well as the role of ICTs, is fundamental to giving risk communication its rightful place in disaster management. Some national governments fail to implement ICT-friendly policies despite the many benefits of ICTs, particularly in developing countries with poor infrastructures as noted by some cases in this chapter. ICTs can play a significant role in communication of risks arising from natural disasters and public health emergencies. They are not just commercial tools whose sole purpose is to increase corporate profits in the 'emerging economies' of Asia Pacific. ICTs can also contribute to social development.

Telecommunications regulators have a special role in promoting the media that are used in disaster warning systems. Whereas under normal circumstances, mass media such as television, radio, telephones and the Internet may compete with each other as commercial entities, regulation can encourage them to work in harmony during disaster situations for the public good. Thus, making disaster management a part of telecommunication regulation ought to be considered.

Ongoing ICT4D programmes should be made more comprehensive through the inclusion of risk communication. After all, developmental progress is severely curtailed when crises befall a community. Risk communication should be included as one of the dimensions of the activities of telecentres which are found in many rural Asian societies today. The use of an established channel such as telecentres in risk communication may be more economical and more reliable than using a system specially meant for the purpose. Similarly, risk communication can also be made a part of community radio programmes.

Finally, it is essential to give priority to regional efforts since natural disasters and health risks such as pandemics often cross national boundaries. It is prudent for national governments to invest in regional efforts while also focusing on national priorities. Regional and international organizations serve as critical allies in this task by sharing knowledge and creating a common platform that national governments can harness for the benefit of the populace.

Notes

1. The authors would like to acknowledge the contributions of Mr Chin Saik Yoon to this chapter.
2. There are many types of crises: natural disasters ('acts of God'), malevolence (product tampering, kidnapping, rumors), technical breakdowns (software failures, industrial accidents, product recalls), human breakdowns (industrial accidents due to human error), activist challenges (boycotts, strikes, lawsuits), accidents (oil spills, radioactive contamination), pandemics (such as AIDS) and terrorism, which is relatively more recent in origin.
3. The WHO has developed different alert phases for pandemics. Each phase provides the status of the pandemic from a circulating strain, possible infection to humans and the full blown epidemic event.
4. The Global Public Health Intelligence Network (GPHIN) developed by Health Canada in collaboration with the WHO is a secure Internet-based multilingual early-warning tool that continuously searches global media sources such as news wires and websites to identify information about disease outbreaks and other events of potential international public health concern.
5. The broadened purpose and scope of the IHR (2005) are to 'prevent, protect against, control and provide a public health response to the international spread of disease and which avoid unnecessary interference with international traffic and trade.'

References

Academy for Educational Development (AED). (2006). *Avian influenza: A critical challenge in changing behavior*. Retrieved 12 October 2006 from http://www.aed.org/avianflu/birdflu.cfm

ADPC. (2005). *Regional early warning in South East Asia*. Retrieved 12 October 2006 from http://www.adpc.net/general/TEWS-Orawan.rtf

ADPC. (2006). *Early Warning System*. Retrieved 23 July 2007 from http://203.159.16.18/v2007/Programs/EWS/Default.asp

ADRC. (2005). *Total disaster risk management—good practices*. Retrieved 12 October 2006 from http://web.adrc.or.jp/publications/TDRM2005/TDRM_Good_Practices

AlertNet Website. (2006). Available at http://www.alertnet.org

ASEAN-Disease-Surveillance.Net Website. (2006). Available at http://www.ads-net.org/Default.asp

ASEAN-Australia Development Cooperation Programme. (2005). Strengthening Animal Health Management and Biosecurity in ASEAN. In *AADCP Newsletter* Number 8, September. Retrieved 23 July 2007 from http://www.aadcp.org/document/September%202005%20No.%208.pdf

BBC. (2005). *Tsunami aid: Coordinating relief*. Retrieved 12 November 2006 from http://news.bbc.co.uk/2/hi/health/4154295.stm

BBC. (2006). *Indonesia tsunami system 'not ready'*. 19 July 2006. Retrieved 12 October 2006 from http://news.bbc.co.uk/2/hi/asia-pacific/5191190.stm

Brahmbhatt, M. (2006). *Economic impacts of avian influenza propagation*. Speech at the First International Conference on Avian Influenza in Humans, 29 June, Institut Pasteur, Paris, France. Retrieved 12 October 2006 from http://web.worldbank.org/WBSITE/EXTERNAL/NEWS/0,contentMDK:20978927~menuPK:34472~pagePK:34370~piPK:34424~theSitePK:4607,00.html

Cambodian flood in 1997. *Japanese Journal of Disaster Medicine*, 4, 47–50. Retrieved 12 October 2006 from http://plaza.umin.ac.jp/~GHDNet/99/k7snake.htm

Center for Disease Control and Prevention. (2006). *PHIN: Overview*. Retrieved 12 October 2006 from http://www.cdc.gov/phin/index.html

East Asia Summit. (2005). *East Asia Summit declaration on avian influenza prevention, control and response*. Kuala Lumpur, 14 December.

Ericsson. (2006). *Pakistan earthquake: When there is human need to communicate, Ericsson is there*. Retrieved 16 October 2006 from www.ericsson.com/ericsson/corporate_responsibility/ericssonresponse/documents/Pakistan_earthquake.doc

FAO. (2004). *FAO recommendations on the prevention, control and eradication of highly pathogenic avian influenza (HPAI) in Asia*. Retrieved 16 October 2006 from http://www.fao.org/AG/AGAInfo/subjects/en/health/diseases-cards/27septrecomm.pdf

Fearn-Banks, K. (1996). *Crisis communications: A casebook approach*. Mahwah, NJ: Lawrence Erlbaum.

Geo Resources. (2002). *Kobe earthquake*. Retrieved 12 October 2006 from http://www.georesources.co.uk/kobehigh.htm

Gidley, R. (2005). *Covering conflicts and disasters: The media mandate*. Proceedings of the AMIC Annual Conference, Beijing, July.

i4d. (2005). Ham radio connects tsunami survivors in no time. *i4d*, January. Centre for Science, Development and Media Studies (CSDMS), New Delhi

International Federation of Red Cross and Red Crescent Societies. (2005). East and Southeast Asia: Avian influenza. *Information Bulletin*, 22 December. Retrieved 16 October 2006 from http://www.ifrc.org/docs/appeals/rpts05/ESEAai22120503a.pdf

International Monetary Fund (IMF) (2006). *The global economic and financial impact of an avian flu pandemic and the role of the IMF*. Retrieved 16 October 2006 from http://www.imf.org/external/pubs/ft/afp/2006/eng/022806.pdf

Kalemoglu M., Keskin Ö., and Ersanli D. (2005). Analysis of an emergency department's experience. *The Internet Journal of Rescue and Disaster Medicine*, 4(2). Retrieved 16 October 2006 from http://www.ispub.com/ostia/index.php?xmlFilePath=journals/ijrdm/vol4n2/emergency.xml

Lerbinger, O. (1997). *The crisis manager: Facing risk and responsibility*. Mahwah, NJ: Lawrence Erlbaum Associates.

M.S. Swaminathan Research Foundation. (2006). *Ongoing today: 1998 to 2004*. Retrieved 7 November 2006 from http://www.mssrf.org/about_us/history/ongoing.htm

Mathew, D. (2005). Information technology and public health management of disasters: A model for South Asian countries. *Prehospital Disaster Medicine*, 20(1), 54–60. Retrieved 7 November 2006 from http://pdm.medicine.wisc.edu/20-1%20PDFs/Mathew.pdf

MobileIN.com. (2005). *History and importance of cell broadcast*. Retrieved 12 October 2006 from http://www.mobilein.com/Perspectives/Authors/CB_History_Importance.htm

Nah, S.H. (2005). *Breaking barriers: The potential of free and open source software for sustainable human development. A compilation of case studies from across the world*. UNDP/APDIP, Elsevier, New Delhi.

Ochi, G., Shirakawa, Y., Asahi, S., and Toriba, M. (1999). Information transmission through the Internet for preparedness against venomous snakes as the aftermath of Cambodian flood in 1997. *Japanese Journal of Disaster Medicine*, 4, 47–50. Retrieved 7 November 2006 from http://plaza.umin.ac.jp/~GHDNet/99/k7snake.htm

Ottawa. (2005). *Global Pandemic Influenza Readiness: An International Meeting of Health Ministers Communiqué*. 25 October.

Otte, M.J., Nugent, R., and McLeod, A. (2004). *Transboundary animal diseases: Assessment of socio-economic impacts and institutional responses*. Livestock Policy Discussion Paper No. 9, FAO, Rome. Retrieved 7 November 2007 from http://www.fao.org/ag/AGAinfo/resources/en/publications/sector_discuss/PP_Nr9_Final.pdf

Public Health Agency of Canada. (2006). *Canadian Integrated Public Health Surveillance (CIPHS)*. Retrieved 7 November 2006 from http://www.phac-aspc.gc.ca/php-psp/ciphs_e.html#whatis

Seymour, M. and Moore, S. (2000). *Effective crisis management: Worldwide principles and practice*. London: Continuum.

Subramanian, T.S. (2005). Their own warning systems. *Frontline*, 22(02), 15–28 January 2005. Retrieved 12 October 2006 from http://www.hinduonnet.com/fline/fl2202/stories/20050128006701600.htm, accessed 15 April 2006.

Telecoms Sans Frontieres Website. Available at http://www.tsfi.org/html_e/equipment_of_tsf.html

Tiensin, T., Chaitaweesub, P., Songserm, T., Chaisingh, A., Hoonsuwan, W., and Buranathai, C. (2005). Highly pathogenic avian influenza H5N1, Thailand. *Emerging infectious diseases.* Retrieved 18 November 2006 from http://www.cdc.gov/ncidod/EID/vol11no11/05-0608.htm

UN/ISDR. (2005). Summary of national information on the current status of disaster reduction, as background for the World Conference on Disaster Reduction (Kobe, Hyogo, Japan, 18–22 January 2005). Retrieved 12 October 2006 from http://www.unisdr.org/wcdr/preparatory-process/national-reports/summary-national-reports.pdf

UNDP. (2005). *Human development report.* New York.

UNDP. (2006). *UNDP supports radio program to reduce tsunami trauma.* Retrieved 16 October 2006 from www.undp.or.id/tsunami/view.asp?Cat=st&FileID=20060711-1

UNEP. (2001). *State of the environment, Bangladesh.* Retrieved 12 October 2006 from http://www.rrcap.unep.org/reports/soe/bangladesh_disasters.pdf#search=%22radio%20bangladesh%20disaster%22

UNESCO. (2006). Indian Ocean Tsunami Warning and Mitigation System IOTWS. Implementation Plan, Third Session of the Intergovernmental Coordination Group for the Indian Ocean Tsunami, Warning and Mitigation System (ICG/IOTWS-III), Bali, Indonesia, 31 July–2 August 2006, IOC Technical Series No. 71.

UNICEF. (2006a). *Progress report on avian influenza and human influenza pandemic preparedness, January–June.* Retrieved 16 October 2006 from http://www.unicef.org/avianflu/files/Final_Progress_Report_AI_HI_19_June.pdf

UNICEF. (2006b). *The CREATE! Framework: A communication strategy for avian flu response and pandemic flu preparedness.* Retrieved 16 October 2006 from http://www.unicef.org/avianflu/index_31607.html

WHO. (2003). *WHO collaborative multi-centre research project on Severe Acute Respiratory Syndrome (SARS) diagnosis.* Retrieved 12 November 2006 from http://www.who.int/csr/sars/project/en/

WHO. (2005a). International Health Regulations (2005). Adopted by the Fifty-Eighth World Health Assembly through WHA 58.3 Revision of the International Health Regulations. Retrieved from http://www.who.int/gb/ebwha/pdf_files/WHA58/WHA58_3-en.pdf

WHO. (2005b). Resolution and Decision on eHealth, Fifty-Eighth World Health Assembly, Ninth plenary meeting, 25 May.

WHO. (2005c). *WHO global influenza preparedness plan: The role of WHO and recommendations for national measures before and during pandemics.* Geneva, Switzerland: Department of Communicable Disease Surveillance and Response.

WHO. (2006a). WHO pandemic influenza draft protocol for rapid response and containment, 26 January, Geneva.

WHO. (2006b). *About the global health atlas.* Retrieved 12 November 2006 from http://globalatlas.who.int/

WHO. (2007). *Epidemic and pandemic alert and response global outbreak and response network.* Retrieved 12 November 2006 from http://www.who.int/csr/outbreaknetwork/en/

WHO and ADPC. (2006). *International training programme on health care facility emergency preparedness and response to epidemics and pandemics (HCF-EPREP).*

Zincir-Heywood, A. and Heywood, M.I. (2000). *In the wake of the Turkish earthquake: Turkish Internet.* Proceedings of the Internet Society's iNet 2000 Conference. Retrieved 16 October 2006 from http://www.isoc.org/inet2000/cdproceedings/8l/8l_2.htm

Localization in Asia Pacific

Sarmad **Hussain** and Ram **Mohan**

Introduction

The world Internet population crossed one billion users in 2005 (Computer Industry Almanac 2006). However, Asia Pacific continues to lag behind North America and Europe in diffusion of information and communication technologies (ICTs). Compared to 69 per cent of the North American population and 38 per cent of the European population, only about 10 per cent of the Asia Pacific population accesses the Internet (Internet World Stats 2006a), even though China, Japan and South Korea have comparatively high Internet penetration (Internet World Stats 2006b). Low Internet penetration in Asia Pacific's developing nations limits their potential to exploit the benefits of ICT.

Asia Pacific is lagging behind in the use of ICTs not only because of the unavailability of affordable hardware and connectivity, but also because computing is still primarily in non-Asian languages. The Asia Pacific region is home to about half of the world's spoken languages: more than 3,500 languages are spoken in Asia Pacific out of about 6,800[1] languages spoken in the entire world (UNESCO 2004). These languages employ a variety of writing systems (Omniglot.com 2006). Twenty-one of the 30 most spoken languages in the world are also from this region (Katsiavriades and Qureshi 2006). Therefore, enabling ICTs in the local languages is necessary for effective access to information in Asia Pacific (Pimienta 2005). To cite one example, a recent study published in *Korea Times* reports that due to insufficient adaptation to local needs, Google serves only 1.5 per cent of the Korean Internet search market (Wagers 2006). Most Koreans use the Korean search engines which meet their requirements better.

The adaptation of ICT to local needs is called *localization*. Localization can be defined as *the process of developing, tailoring and/or enhancing the capability of hardware and software to input, process and output information in the language, norms and metaphors used by a community*. The localization process must also capture the variances in the use of a language. For example, English speakers in the United States spell words differently from English speakers in the United Kingdom, and Punjabi speakers use Gurmukhi script in India and Arabic script in Pakistan to write the same language. Even more challenging is enabling ICT for oral or unwritten languages like Jatapu and Koya in India (Daswani 1998), as it would be completely dependent on a localized speech interface.

The terms *internationalization* and *globalization* are also used in the context of local language computing but with subtle differences from localization. Internationalizing ICTs requires designing the technology in a generic fashion so that it has the ability to support multiple languages. However, internationalization does not enable any particular language. Enabling technology for *a* particular language is called localization. Globalization of ICTs in this context refers to first internationalizing and then localizing technology to support *multiple* languages.[2]

Localization is conventionally defined or understood in a narrow sense—that is, it is usually limited to interface translation and other basic changes in the computing platform. We suggest that localization has a broader scope that includes the entire range of script, speech and language technology to enable access to information for the end-user.

This chapter provides a brief overview of localization and the process required for enabling it. In addition, the role of regional and international organizations in localization is discussed and the level of localization achieved across different countries within the region is summarized. The chapter concludes with a discussion of policy and planning considerations to achieve wider localization in Asia Pacific, highlighting some of the issues and choices in making localization policy decisions.

The process of localization

Localization requires three steps: linguistic analysis, basic localization and advanced application development. Linguistic analysis is required to unambiguously define the language conventions and norms that are to be modelled by technology. As implied, basic localization caters only to the rudimentary needs of end-users, including input and output of text in a local language. However, to give comprehensive access to novice users and illiterate populations, or to assist in content development in a local language, more advanced applications need to be developed. Further details are given in this section.

Linguistic analysis

Successful language computing is largely dependent on good linguistic analysis based on cultural conventions. Very precise definitions are required for all relevant linguistic phenomena. However, for many languages in Asia Pacific, linguistic details are either incomplete or unavailable. Moreover, relevant cultural conventions are rarely documented. This poses a significant obstacle to localization and requires the involvement of indigenous expertise.

The initial linguistic details, which have to be agreed and standardized for basic localization, include (but are not limited to) the following: the writing system[3] and character set used by the language for its publishing needs; the ordering of these characters; cultural conventions for representing numbers, time and the calendar; and translation of common terms used in the software interface. This has to be done by the appropriate language or cultural authorities at the national level. Experience shows that debate[4] is inevitable in this process of standardization. It is important that the discussions and solutions be based on linguistic merit and not be driven by technology constraints, although all discussions must involve both linguists and technologists, the latter to challenge any ambiguities in the proposals from a technical perspective.

In addition, a detailed linguistic analysis of the script, speech and grammar of the language is required for advanced application development. The analysis encompasses the sound system of the language and its acoustic details, word and phrase structures, and the representation of meaning in the language. These details need to be clearly documented for eventual implementation, as further explained in this section.

Standardization and basic localization support

Once the discussions on the writing system and basic language details are finalized at the national level, the next step is to derive the relevant standards for computing and subsequently develop computer software and hardware to enable local language input and output based on these standards. At the minimum, encoding, keyboards (and input methods), fonts (and rendering engines), definition of cultural conventions (for time, calendar and numbers) and interface translation must to be enabled. Once defined, the keyboard, font and locale support must be incorporated in the operating systems (for example, Linux, Sun Unix, Microsoft Windows, IBM AIX, Apple Mac OS and others) and at least the basic applications, including word processors (for example, Emacs, GEdit, KEdit, Open Office, Word), e-mail clients (for example, Thunderbird, Outlook), Web browsers (for example, Firefox, Internet Explorer), chatting software, and the like, according to end-user requirements. These steps are briefly discussed below.

Encoding

As computers can only manipulate numbers and not characters, to process a language each character in it has to be assigned a unique number.[5] This process is called encoding. The process can be done in a non-standard way by arbitrarily assigning numbers to different letters in the language.[6] However, non-standard encoding inhibits data sharing across multi-user applications, including Web access, e-mailing and chatting. Therefore, the encoding should be done through the international standard ISO 10646 or Unicode.[7] If the Unicode standard does not support a language, or only partially supports it, this standard should be enhanced by submitting a proposal, channelled through appropriate national bodies, to add new characters.[8]

Even if standardization is achieved, there still remains a large repository of information based on arbitrary encodings. Thus, in addition to standardization, additional file-mapping applications need to be developed that will allow the legacy or concurrent content in other encodings to be converted to the standardized encoding.

Keyboard and input method

Once the character set is standardized, keyboard mapping—that is, the placement of characters on the keyboard—needs to be defined. This mapping can be facilitated by extending existing keyboard layouts or doing character frequency analysis.[9] Some languages require complex input methods. For example, because it is not possible to put the thousands of Chinese symbols on the keyboard, different methods based on strokes, Latin character transliteration, and handwriting recognition are used to input text in Chinese (Wikipedia 2006; Hussain et al. 2005). Input methods must be openly and consistently defined and openly standardized to allow users to type in the same way across all computing systems.

Fonts and rendering

Fonts for languages are required for on-screen display and printing. Simpler writing systems like Latin and Cyrillic scripts (which are used for most languages spoken in the Americas and Europe) have been modelled by earlier font formats, such as the True Type Fonts (TTF). However, most scripts used for languages in Asia Pacific are more complex due to their cursive nature and context-sensitive character shaping and positioning (Hussain 2004) and therefore require the enhanced font formats, such as the Open Type Fonts (OTF).[10] Once fonts are created for a language, computer software is used to display the fonts on-screen, in a process called rendering. Complex writing systems, such as the Nastalique writing style for the Urdu language (Hussain 2004), require a sophisticated rendering engine capable of displaying the font.

Locale

The locale for a language contains information about the local language and the cultural representation of time, calendar, numbers and other related information normally visible on computer screens—for example, in English the date stamp '4/1/2005' is usually included with e-mail messages. This represents '1st of April 2005' in the USA but '4th of January 2005' in the UK. Thus, to define and to interpret this information, the language and region of the locale must be clearly declared. In addition to time, date and digit conventions, the locale also defines the sequence in which the words in a language are ordered, which is very important for many applications, for example, to develop a voter list or to make a telephone directory.

The locale may be defined by filling in a given template and submitting it to the Common Locale Data Repository (CLDR) managed by the Unicode Consortium. The locale for each language for each country is defined separately to capture cultural variations, such as bn-BG and bn-IN for the Bengali language (bn) spoken in Bangladesh (BG) and India (IN), respectively. Many Asia Pacific countries have not developed or registered their language locales with CLDR.

Local language interface

Imagine giving a Nepali speaker a computer that is configured for use in the Japanese language. Such a computer would be impossible for the Nepali speaker to operate because he or she cannot comprehend the words and phrases displayed on the screen. For majority of users in Asia Pacific who do not understand a foreign language, words and phrases like 'save', 'print', 'edit', 'file' and the like, need to be translated and displayed in the local language on the computer screen. About 5,000 words comprise the basic glossary to represent menu items for operating systems and basic applications. However, to completely localize all help files and error messages, careful translation of more than 300,000 phases may be required.

Translating a glossary is challenging because there are many words that do not have local language equivalents, such as the word 'cursor'. Either such words are transliterated or new senses of the existing local language words need to be formulated. This creative exercise requires language experts who are proficient in the use of computers, a rare combination of skills in the developing Asia Pacific region.

Once translated, the basic glossary should be verified by language authorities and published as a national standard (for example, DzongkaLinux Team 2007) and supplied to vendors and organizations (for example, Debian, Red Hat, Microsoft, IBM Apple) for incorporation within their platforms.

Basic application localization

Once the basic linguistic analysis is completed and localization support is developed, this support will need to be integrated at two levels. First, the support must be included in the basic operating system being used, for example, Linux, Microsoft Windows, Apple MacOS, IBM AIX, Sun Unix. The operating system would enable the encoding, allow the locale of the language to be defined, and allow the input and output methods to be used effectively. Interface translation in the operating system must also be enabled. Second, once the operating system is enabled, basic applications must be localized. These applications include word processors, e-mail clients, Web browsers, chat clients and other general and customized applications. However, this only provides basic access to trained users. For wider, more effective access for general users, advanced local language computing applications will also need to be developed.

Advanced language computing applications

Basic localization should not be the final goal because it does not completely meet the objective of giving end-users meaningful access to computing. Advanced language technology is required to further facilitate access for end-users and enable them to generate local language content. Advanced language computing requires in-depth speech and linguistic analysis as well as complex programming for implementation, drawing from the fields of phonetics, phonology, morphology, syntax, semantics, signal and speech processing, image processing, language processing, artificial intelligence and statistics. Moreover, a significant amount of local language resources is needed to develop these applications, as further explained in the following sections.

Language resources

Language resources are required by advanced applications to create language models. These resources include first, a list of words in the language tagged with minimal linguistic information (for example, part-of-speech [POS],[11] gender, number).[12] These word lists (or lexicons) are needed to develop applications like spelling checkers. Many applications would also require a large amount of typed text in the language, called the language corpus. This is used to extract word frequencies, word collocations and other grammatical information for statistical language processing. A corpus of 10–100 million words from different text genres is required for different kinds of statistical modelling.

Part of the corpus must also be manually tagged with POS and other linguistic information to infer automatic models for processing text through machine learning[13] techniques. For example, a text corpus manually tagged with POS is used to develop an automatic POS tagger. The POS tagger is used in almost all advanced applications, for example, to decide whether to stress the first or second syllable of a word like 'address'[14] for a text-to-speech system. The Urdu language shows a similar variation, for example, for the word الٹا (*ulta*; 'upside down' vs. 'to turn upside down'). Another such critical system is for word segmentation, since in Asia Pacific, many languages like Chinese, Dzongkha, Khmer, Lao, Thai, Urdu and Burmese do not use spaces between words, which makes it difficult to determine word boundaries in typed text. The word boundaries have to be guessed based on advanced linguistic and statistical techniques. Solving this problem is fundamental for any further processing of these languages through machines, for example, doing line-wrapping in word processing or performing spell checking. In addition to tagging text corpora, the computational grammars of these languages need to be developed and documented.

Speech corpora are required for developing speech applications. These must be recorded for narrative and conventional speech over different channels, including microphone, telephone, mobile phone, and so on, for a variety of speakers and dialects, for the development of the speech applications. Finally, script corpora need to be developed for script processing applications. The corpora must include large samples of different types and handwritings and the corpora must be manually tagged for various linguistic dimensions.

Once the language resources are available, they can lead to the development of advanced applications, which can be broadly categorized into two sets: those which provide access to existing content and others which assist in generating new content in local languages.

Applications to provide access to information

As discussed, basic applications like word processors, e-mail clients, Web browsers and chat clients provide basic access to trained users, once they are localized. However, there are additional applications that can be used to further enrich the computing experience. Most of the population in developing Asia is illiterate and enabling computing in local languages still does not provide this population access to online information that is otherwise available to literate individuals. They need a speech interface, which reads out online text to users (text-to-speech systems or TTS), as well as technology to 'listen' to users (robust automatic speech recognition systems or ASR) and to interpret their requests (language understanding systems). Also needed are search engines and advanced information retrieval (IR) systems that can sift through existing online data and seek out and display requested information. All these must be possible in local languages. While there are generic software programs with open licenses which are already available, these programs have to be trained (and sometimes enhanced) for Asian languages.

Once core technology like TTS, ASR and IR is enabled, it has to be integrated into Interactive Voice Response (IVR) and other dialogue-based systems to 'communicate' with end-users. As the core technologies are developed by a variety of vendors, standardized ways of integrating these technologies will need to be developed. There are ongoing standardization efforts: for example, the World Wide Web Consortium (W3C) is developing Speech Synthesis Mark-up Language (SSML) and Voice Mark-up Language (VoiceML) to allow voice browsing, in addition to the widely used text browsing standard called Hyper Text Mark-up Language (HTML). Voice browsing allows users to interact with a website to access content using speech interface. This can greatly enhance Web use in developing Asia, especially within the illiterate and visually impaired community.

Applications for content generation

Even if access is possible, it is still necessary to have relevant content available in local languages for end-users. At present very limited online content is available in the languages of Asia Pacific. There are three general ways to generate online content: (a) develop original content, (b) copy content from printed sources in local languages and (c) translate existing content in a foreign language. The localized common applications used for access, such as word processors, email clients, Web development tools and chatting software, may be used for content generation as well.

Although online content development is a slow process, script and language technology can accelerate it. And although there is little online content in Asia Pacific languages, there is a lot of printed content. Using Optical Character Recognition (OCR) systems, which scan printed documents and books and automatically convert the images to editable text, this printed material can be quickly transformed into searchable online content.

In addition to content in the local languages, there is also a large amount of universally useful content available in foreign languages, including English (35.2 per cent), Chinese (13.7 per cent) and Japanese (9 per cent) (Global Reach 2004). This content can also be translated to local languages quickly by developing automatic Machine Translation (MT) systems. Automatic translation, although not very accurate, provides access to content that is otherwise completely inaccessible. Automatic translation can be made more accurate with human assistance (where required) at a significantly lower cost compared to a completely manual translation.

TTS, ASR, OCR and MT are advanced applications that require considerable language resources and linguistic and computational analysis. These applications also require dedicated input from specialized human resources over a considerable period. An MT application could take a team of 10 linguists and computational linguists five years to develop.[15] Usable TTS, ASR and OCR systems could take a team of 10 linguists, engineers and computational linguists three years each to develop. To mature and perfect these applications would require continuous focus for an even longer period.

Licensing is an additional problem with online content, even where technology may be available for accelerated online publishing. Much of the content available is normally copyright, which makes it difficult to disseminate. Newer regimes that allow much more open use of content, such as Creative Commons, are emerging. Wikipedia, which allows free-for-all information and is available in many languages, is an excellent outcome of these movements towards open content.

Regional and international organizations

Development of local language computing applications and content requires a sustained effort. Many regional and international organizations have been contributing to this development across Asia Pacific. These organizations are involved in: (a) standards development and (b) technology development. Moreover, there are many funding agencies in the region that are supporting local language computing development, notably the International Development Research Centre (IDRC) of Canada, Center of the International Cooperation for Computerization (CICC) of Japan, National Institute of Information and Communications Technology (NICT) of Japan, United Nations (through UNESCO and the UNDP-Asia Pacific Development Information Programme or APDIP) and Asia IT&C Grants by the European Union.

This section lists some of the major regional standards and technology development organizations supporting local language computing in Asia Pacific and explains the role they play in this context. National and regional initiatives need to develop liaisons with these organizations, for example by subscribing to the multiple online discussion forums that they maintain or by attending the regular meetings, conferences and special workshops organized by them. Where funds are required, the funding organizations listed provide such support.

Unicode consortium

The Unicode consortium develops the Unicode standard, which is the standard encoding scheme for the multilingual Internet and is the same as ISO 10646. The consortium aims to provide standard encoding schemes for all characters and symbols used in different scripts for all languages of the world (Unicode 2006). In addition, it provides guidelines for collation, bidirectionality, reordering and line-breaking, which are fundamental to text processing for many Asian languages based on the Unicode standard. Even though conventional national and proprietary encodings are still being used, most nations across Asia Pacific are now switching to Unicode. In addition to encoding, the Unicode consortium has recently collected and is now maintaining locales for all languages through the CLDR project.

World Wide Web Consortium (W3C)

W3C develops guidelines, standards and software to publish multilingual online content. Its Internationalization Working Group is tasked with keeping these specifications multilingual. W3C maintains the HTML standard which is used for creating multilingual Web pages. In addition, it is developing SSML and VoiceML standards which are used for voice browsing, that is, accessing the Internet through speech. This organization is also developing multimodal content publishing standards for more effective Web accessibility, including access by people with disabilities.

Internet Corporation for Assigned Names and Numbers (ICANN)

Currently Web access requires typing a Web address (also called domain name or URL) in English. For populations who do not understand English, this is one of the significant hurdles in accessing online content. Web addresses, which are the key to entering the multilingual World Wide Web, should also be in local languages. ICANN is responsible for the global coordination of Web addresses[16] and it recently introduced Internationalized Domain Names (IDNs) through reports RFC 3454, 3490, 3491 and 3492, collectively called the IDN Standards (ICANN 2006). IDN would allow Web addresses in local languages. However, due to the seven-bit ASCII-based domain name system, Unicode cannot be used and multi-lingual IDNs are converted to ASCII Compatible Encoding (ACE) before the address is resolved. Still being debated is how to enable Top-Level Domains (TLDs) in local languages and who will control them (Butt 2006; Huston 2006). Due to this continuing controversy, independent systems have also been developed, for example by the Chinese Internet Network Information Center (CNNIC). ICANN and IDNs are bound to play a critical role in making the multi-lingual Internet accessible.

Development of Internationalized Domain Names (IDNs) for India's .IN domain

India's .IN domain first opened to the public in 1992. It was managed by the National Centre for Software Technology (NCST) until 2004, and then by the Centre for Development of Advanced Computing (C-DAC), both research and development institutions run by the Government of India. Until 2004, about 6,600 names existed in the .IN domain database. In late 2004, the Indian government liberalized policies surrounding the .IN domain. This included making available second level domains (example.in) on an unlimited basis, as well as third level domains <co.in>, <net.in> and <org.in> to all registrants. Furthermore, the .IN ccTLD registry separated Registry and Registrar (retail) functions, resulting in the creation of a domain name industry that had until then been dormant. The results of this opening and liberalization have been quite dramatic—100,000 registrations in the first 100 days and over 250,000 new registrations since 1 January 2005.

However, domain name registrations were in English (ASCII script) only, a significant limitation in a nation with 22 official languages, including 400 million speakers of Hindi, 200 million speakers of Bengali, 60 million speakers of Tamil and 70 million speakers of Telugu. This nation of more than a billion has schools that teach in 58 different languages, newspapers publishing in 87 languages, radio programmes broadcast in 71 languages and movies released in 15 languages. To support this diverse, multilingual population, the .IN registry embarked on a programme to internationalize the .IN domain and support the various scripts that are used to represent the 22 official languages in India.

The task of internationalizing the .IN domain is the most complex domain name internationalization project in the world because the 22 languages may be represented completely by merely 11 scripts, leading to significant overlaps and the presence of visually confusable character sequences that are equally valid in multiple languages but which may be represented on a computer by unique encodings. Such visually confusable characters are called *variants*, and one of the most important tasks in localization is the creation of *variant tables* that prescribe which characters are visually confusable between different languages. In addition, some Indian languages support bidirectional text, multiple diacentric positioning and word breaking, and non-empty spaces that are not normally supported in a standard, left-to-right ASCII-based Domain Name System (DNS).

The plan to internationalize .IN may be summarized as follows: build language tables; develop language policies; consider issues brought about by variants; ensure standards compliance and enhance dispute resolution policy to cover IDNs.

To introduce .IN in local languages such as Hindi and Tamil, language and variant tables must first be developed. Homographic variant issues must be determined, which will ensure that characters that look identical are marked clearly and registration of one character in one script automatically reserves the similar looking character in the other script(s). Linguistic experts are needed to ratify the choices of variant and language tables. Finally, steps need to be taken to ensure that the launch of Hindi and Tamil does not disadvantage the later launch of other languages that use similar characters—for example, the Tamil character வ(U+0BB5) is very similar to the Malayalam character വ(U+0D16).

International technical standards exist for IDNs, and .IN has carefully planned to conform to these standards while simultaneously working with the standards community to extend these standards where they are deficient or insufficient. At a minimum, conformance to the IETF RFCs 3490, 3491, 3492 and 3454 are required, as well as general conformance to the ICANN IDN Guidelines (ICANN 2006).

The launch of domain names in local languages requires the development of a robust dispute resolution policy that considers additions for IDNs and has the ability to handle disputes for domain names in either ASCII or the native language representation evenly and equally. Moreover, because variants of one name may conflict with other names, a clear policy has to be developed to resolve such conflicts in a manner that is consistent and conformant to local laws.

In December 2006, the Indian government, in partnership with .IN's technical partner Afilias, completed the first-ever launch of .IN in the Tamil language, implementing the Dravidian script that represents Tamil. Tamil, one of the world's classical languages, will be available for wide use. There are plans to soon thereafter introduce .IN in Malayalam (a related Dravidian script-based language). Language table development for the Devanāgarī script, which is the basis for many northern Indian languages including Hindi, is well underway, although this is a large-scale project whose end-date is yet to be determined.

A new development is the interest in the creation of IDN Top Level Domains (IDN TLDs). This allows the entire domain name to be represented in a local language character set. Technical tests are being conducted to study and ensure feasibility of the following practical issues: (a) Will they work everywhere? (b) Are they backwards compatible? (c) Do they not break application software? (d) Do they support languages appropriately? Certain principles apply towards the roll-out of IDN TLDs, including:

1. retaining the global uniqueness of the TLD system—that is, domain names should remain unique and unambiguous;
2. maintaining the interoperability of the TLD system, that is, 'dot हिन्दी' ('dot Hindi' written in Devanāgarī script) needs to point applications and users to the same place regardless of whether they are accessing the domain from India, the UK or Greece;
3. promoting 'future-proof' solutions that allow seamless introduction of new languages and character sets in the future;
4. avoiding user confusion; and
5. promoting multi-stakeholder involvement.

When implementing IDNs in Asia Pacific, with its large list of languages, character sets and scripts, and relative paucity of experts, important preliminary issues such as language table and variant table development often cannot get off the ground. Government involvement is critical in coordinating and bringing together the right set of individual experts in technology, language and policy to create a model for the implementation of IDNs. The development of IDNs will benefit Internet users who are not literate in English and whose computers do not use ASCII or English character sets by default, provide a good user experience on the Internet and create a multilingual Internet that can be used by all populations worldwide.

International Standards Organization (ISO)

ISO jointly develops the ISO 10646 or Unicode standard with the Unicode Consortium. The technical committee TC37 develops standards for 'Terminology and Other Language and Content Resources', including specifications for lexica, corpora and other language content. The language resource standards are still being discussed and finalized and they are not currently in wide use. Some other related standards include ISO 3166 for country codes and ISO 639 for language codes, which are used for locale definitions by Unicode within CLDR and by other organizations including W3C and ICANN. For example, ur_PK represents the Urdu language locale as used in Pakistan.

Free and Open Source Software (FOSS) initiatives

Notable within software development initiatives for multilingual computing is the FOSS community which provides internationalized software applications that allow rapid localization covered under an open license.[17] Most FOSS operating systems are based on Linux, are internationalized, and are being localized by different groups (for example, Debian, Red Hat and Ubuntu). Debian is currently being localized in more than 150 languages. Open Office, which provides a complete suite of document productivity software, is being localized into 70 languages. The Mozilla project distributes Firefox Web browser and Thunderbird email client. There are many more FOSS initiatives available online, including software for chatting, multimedia, Web development and database.

Asian Federation on Natural Language Processing (AFNLP)

Academic research forums in linguistics and language processing have long existed in many countries in Asia. However, there have been limited regional discussions on Asian languages. The American Association of Computational Linguists (AACL) and European Association of Computational Linguistics (EACL) have been providing a common platform for the Americas and Europe. A similar platform in Asia was created recently by bringing existing national organizations and conferences under a single regional umbrella called AFNLP. The federation is helping organize language computing research and development across Asia by providing a collaborative platform to share academic research and exchange innovative solutions for Asian languages. AFNLP holds a regular conference called International Joint Conference on Natural Language Processing (IJCNLP). Two such conferences have been held so far.

Language resources and vendor initiatives

Many organizations collect and distribute language resources that are essential to perform linguistic and computational research and to develop local language computing. The Linguistic Data Consortium (LDC) at the University of Pennsylvania distributes text and speech corpora, lexica and additional data for many languages, including Chinese, Arabic, Japanese, Hindi, Vietnamese, Tamil, Korean and other languages. The European Language Resource Association (ELRA) distributes similar resources for many Asian languages. Similarly, the Global Wordnet Association is developing lexical-semantic resources for many languages and the South Asian Language Resource Center (SALRC) at the University of Chicago is developing a repository of lexical resources for South Asian languages. No formal centre for the collection and distribution of the language resources of Asia Pacific has been established. However, discussions for establishing an Asian Language Resource Network, similar to LDC and ELRA, are underway. Another language resource organization is the Summer Institute of Linguistics (SIL), an organization of volunteers that has been documenting languages and populations for more than 50 years (see www.ethonologue.com).

The University of California at Berkeley has started the Script Encoding Initiative which is assisting individuals and groups to identify the missing characters, for example from lesser known languages, and helping them get these characters encoded in the Unicode standard.

Some corporations have also been involved in localization. IBM has developed a large repository of C++ and Java code which is called IBM International Components for Unicode (ICU). This library of code is available at http://icu.sourceforge.net/. Microsoft has restructured its localization policy and has started developing local language interfaces, called Language Interface Packs (LIP), which are currently available for seven Asian languages. These efforts will help develop basic localization at least in the languages that have official status in Asian countries or are otherwise commercially viable (for example, languages spoken by large populations).

There is growing interest in localizing the mobile platform, but the effort has mostly been taken up by the manufacturers themselves, for example, Nokia, Samsung, Sony and others. Text-based messaging is now increasingly becoming available through these systems for many Asian languages based on the Unicode standard. However, the localization is driven mostly by commercial interests focused on languages that promise revenues. It is not possible for independent developers to localize these platforms in other languages due to proprietary platforms and lack of open standards.[18]

Status of language technology

Many of the basic standards and applications have already been developed for most of the national languages in Asia Pacific. Many of these standards have been reviewed over time and now align with international standards. However, language computing has matured to different levels in these countries. This section summarizes the status of localization of national languages in different countries in Asia Pacific. There are five levels of maturity that are at best qualitative as it is difficult to make a quantitative assessment (because each country is confronted with its own unique socio-economic, political and linguistic challenges, for example). The comparison is based on the level of work on the national language and research and development capacity in the areas of script, speech and language processing. A checklist of these applications for many national languages from the region is also provided in Table 1. For more

Table 1
Extent of localization for the national language of each listed country of Asia Pacific*

	Encoding	Collation	Keyboard	Fonts	Locale	Interface	Lexicon	Spell-checker	OCR	TTS	ASR	MT
Afghanistan	xxx	x	x	xx	x	x						
Bangladesh	xxx	xx	xx	xxx	x	x	xx	x	x			
Bhutan	xxx	xx	xx	xx	xxx	xxx	x	x				
Cambodia	xxx	xx	xxx	xx	xx	xx	x	x				
China	xxx	xxx	xxx	xxx	xxx	xxx	xxx	xxx	xxx	xxx	xxx	xxx
India	xxx	xxx	xxx	xxx	xxx	xxx	xxx	xxx	xx	xx	xx	xx
Indonesia	xxx	xxx	xxx	xxx	xx	x	xx	x	xx	x		xx
Japan	xxx	xxx	xxx	xxx	xxx	xxx	xxx	xxx	xxx	xxx	xxx	xxx
Korea	xxx	xxx	xxx	xxx	xxx	xxx	xxx	xxx	xxx	xxx	xxx	xxx
Laos	xxx	xx	xx	xx	x	xx	x	xx	x			
Malaysia	xxx	xxx	xxx	xxx	xx	xxx	xx	xx	xx	xx	x	xx
Maldives	xxx	x	xx	xx	xx							
Mongolia	xxx	xxx	xxx	xxx	x	xx				x		
Myanmar	xxx	xx	xx	xxx	xxx	x		x				
Nepal	xxx	xxx	xxx	xxx	xxx	xxx	xx	xx				
Pakistan	xxx	xxx	xx	xxx	x	xxx	xxx	xxx	xx	xxx	x	xx
Philippines	xxx	xxx	xxx	xxx	xx	x		xx	xx			
Sri Lanka	xxx	xx	xxx	xxx	xx	x	xx	xx	xx	xx	x	x
Thailand	xxx	xxx	xxx	xxx	xxx	xxx	xxx	xxx	xxx	xxx	xx	xx
Vietnam	xxx	xx	xxx	xxx	xxx	xxx	x	xx	xx	xx	xx	x

Note: The table lists a comparison for some of the applications. The comparison is qualitative, not quantitative, and is based on the current information available to the authors through the Internet and other sources (for example, Sonlertlamvanich 2002; Tsujii 2005; Hussain et al. 2005). The information has not been independently verified and therefore has some margin of error.
(blank—minimal work; x—initial work started; xx—some work completed; xxx—much work completed; for Year 2006)

information, see Sonlertlamvanich (2002), Tsujii (2005) and Hussain et al. (2005).

Highly localized languages

Leading the development and implementation of local language computing are the more developed countries in the region, including China, Japan and Korea. These countries are very active in international standardization efforts and participate in relevant platforms and discussions. Most software is already localized in Mandarin Chinese, Japanese and Korean. Current research and development is focused on cutting-edge technology, including speech-to-speech translation, as basic localization and advanced applications, including TTS, ASR, OCR and MT, are already developed and available through the commercial sector. These countries have active academic bodies collaborating with the commercial sector, backed by governmental policy and support. Some of the organizations involved are the University of Peking, City University of Hong Kong, Academia Sinica in Taiwan, NICT and Advanced Telecommunications Research Institute International (ATR) in Japan, and Korean Advanced Institute of Science and Technology (KAIST) and Electronics and Telecommunications Research Institute (ETRI) of Korea. Significant research and development is being performed by the commercial sector as well, including Sony, NEC, IBM, Nokia, Microsoft, Hewlett-Packard, Systrans and so on.

Very localized languages

Thailand and India are also very active in local language computing. The National Electronics and Computer Technology Center (NECTEC) of the National Science Technology Development Agency (NSTDA), along with Thai industry and academia, is leading the full localization of the Thai language. A Thai OCR, text-to-speech system, and English-Thai MT are now available. The Thai Language Environment (TLE) project develops and maintains the Open Source Thai Linux distribution.

India also has a thriving and vibrant language computing development sector. The Ministry of Science and Technology has created the Technology Development for Indian Languages (TDIL) department which supports and coordinates active research on Hindi and many other constitutionally recognized languages through research centres at Indian universities and the Centre for Development of Advanced Computing (CDAC). In addition, the IndLinux group localizes Linux distributions in many languages (MIT 2006) and has released the Hindi version. However, commercial-grade applications for end-users are not fully developed and not in wide use due to the complexity and language diversity (currently 22 official languages). Nevertheless, working models of TTS, MT, ASR and OCR for a few languages, including Hindi, Tamil and Marathi, are available. Other language resources, including lexica and corpora, are also available. Government focus and a dynamic language policy are providing the correct impetus and India is seeing an emerging localization and language computing industry.

Moderately localized languages

Indonesia, Malaysia, Pakistan, Sri Lanka and Vietnam have fairly active academic research and development programmes and fairly mature standards and basic language applications, with reasonable work in advanced applications.

Research and development in Indonesia is being carried out by both the public and academic sectors. Basic resources and advanced applications are all being developed with advanced prototypes already released. Badan Pengkajian dan Penerapan Teknologi (BPPT) and the University of Indonesia are two organizations actively involved in this process. Most of the work is on Bahasa Indonesia.

Research in Malaysia started in 1987 through the KANTA project by CICC which developed an MT system for Japanese, Malay, Chinese, Thai and Bahasa Indonesia. Universities, including Universiti Teknologi Malaysia and Universiti Sains Malaysia, are actively involved in research and development.

Localization in Sri Lanka is being led by the University of Colombo School of Computing for Sinhala and Tamil, with support and guidance from the ICT Agency of Sri Lanka. The open source community is also reasonably active through Sri Lanka's Linux User Group (LkLUG), which has made some progress on the development of a Sinhala Linux distribution.

In Vietnam, localization is being led by the Ministry of IT and is also being carried out in some universities. VietKey is an open source office productivity software available in Vietnamese. Work is also underway on advanced applications, like ASR.

Pakistan has shown a promising focus on language computing (see the boxed case study below).

However, very limited development work is being carried out by the commercial sector in these countries, especially for advanced applications.

Language computing development in Pakistan

Pakistan is home to more than 160 million people who speak more than 60 languages. Urdu is the national language and the lingua franca. The official language is English, a legacy of the country's colonial past and a language understood by less than 10 per cent of the population. Punjabi, Seraiki, Sindhi, Pashto, Balochi and Kashmiri are the most spoken languages. Many of the other languages, with small populations, are found in northern Pakistan, where these linguistic communities live in valley 'islands' surrounded by tall Himalayan peaks. Pakistan is a country that has recently re-awakened to the need for local language software and where all stakeholders are coming together in a synergized approach to language computing development. However, Pakistan is still struggling to balance policy, human resource and technology challenges and it is only starting to look at the social challenges and solutions for dissemination of this technology.

Pakistan experienced a boom in language computing in the early 1980s, when the indigenous software industry started developing word processors and fonts for Urdu. Multiple word processing products and fonts were made available. Although the Nastalique script used to write Urdu is very challenging to model, especially with the technology available in the 1980s, numerous solutions were developed, including Inpage, PagePro, Shahkar, Raakim, and the like. Unfortunately, by the late 1980s and early 1990s, most of this industry had vanished because copyright violations made such ventures totally unprofitable.

Language computing has emerged after a decade of stagnation, with the revived interest coming from academia and the public sector. The Center for Research in Urdu Language Processing (CRULP) at the National University of Computer and Emerging Sciences and smaller informal groups led by individual faculty members at various universities in the private and public sectors are at the forefront of this effort. Work has been ongoing in all aspects of localization technology, including MT, TTS, ASR, OCR and handwriting recognition. Universities are also offering specialized courses at master and doctorate level in these areas, thereby developing the essential human resource for this work. Most of the current efforts are focused on technology. However, significant investment also needs to be made in developing specialized linguistic and computational linguistics programmes.

With the emergence of e-governance, the public sector has realized the need for local language computing and incorporated it in the IT policy for the first time in early 2000. Since then, the government has been contributing in multiple ways. The e-government initiatives taken up by federal and state governments now require local language interfaces for many of the software services being developed. The major initiative has been that of the National Database and Registration Authority (NADRA) which is now issuing National ID cards in Urdu to all Pakistanis. NADRA's national database is in Unicode. Other large initiatives, including work on land revenue records and software for recording proceedings of the National Assembly and Senate, all require Urdu components, making localization a viable commercial option again. There are also plans for telecentre projects which will have a significant local language component. The increased demand created by the public sector is now drawing the software industry to invest in local language computing.

However, the industry remains focused on basic localization and is still not developing advanced applications due to the significant level of financial investment required by the latter. The Ministry of IT (MoIT) realizes the requirement for advanced applications and has been funding research and development in this area since early 2000. The first national encoding standard was approved by the President of Pakistan in 2002, through the efforts of a specialized committee (called Urdu and Regional Languages Software Development Forum, URLSDF) formed by MoIT in collaboration with the National Language Authority. This was soon followed by a proposal to update the Unicode standard for complete support of the Urdu language. Since then MoIT has funded a major development project to create Urdu lexical resources, Urdu TTS and English-to-Urdu MT at CRULP. The first phase of this three-year project was completed in 2007 and the content and software is to be released with open licensing to trigger further research and development in the academic and commercial sectors. The project has helped create the necessary linguistic resources, trained a team of more than 50 personnel in speech and languages processing and is bound to have far-reaching effects on language computing in Pakistan. Smaller projects have also been funded by the PTCL R&D fund (now the National ICT R&D fund), including work on developing Web guidelines for local language content publishing, localization of open source software and developing other language processing applications.

Growing awareness in the government sector, along with significant funding allocation for local language computing programmes and requiring local language computing for e-government projects, is creating excitement in the academic and commercial sectors. However, the work is currently limited to Urdu, the national language. It is hoped that other languages will receive the same attention. A more proactive approach by public organizations, civil society and academia to the localization of the languages of smaller populations, is required.

Somewhat localized languages

The national languages of countries like Bangladesh, Myanmar and Nepal belong to this category. In these countries there is an emerging realization of the importance of local language computing and focused public policy is starting to develop, integrate and align existing private initiatives. However, there is only limited work on advanced language computing applications.

Countries like Afghanistan, Lao PDR, Cambodia, Mongolia and Bhutan are also starting to develop basic localization standards and applications in their national languages.

Non-localized languages

Of the approximately 3,500 languages spoken in Asia Pacific, only about 30–40 languages are being localized. Small and developing language communities are left out due to very limited capacity to perform indigenous localization and lack of commercial incentives. This problem is especially severe for countries with exceptionally high linguistic diversity, such as Papua New Guinea (820 languages) and Indonesia (737 languages). Localizing these languages will only be possible through long-term policy initiatives and collaborative effort between national, regional and international organizations.

Policy considerations for localization in Asia Pacific

The goal of localization is to enable communities to share and exchange information through ICTs. Achieving this goal would require planning and executing a strategy that can address the entire spectrum of associated issues. This section presents the considerations and recommendations for national, regional and international organizations to plan the development of language technology, especially in the context of Asia Pacific.

Majority vs. minority languages

National localization planning must strike a balance between the requirements of the majority and the minority. If the policy prioritizes localization based on the speaking population alone, minority languages may not be addressed. More rigorous criteria based on additional demographic and social factors need to be evolved to include minority languages in localization, as these languages present little incentive for commercial interests. Effective planning might even help preserve the linguistic diversity of the region and help protect endangered languages.

Breadth vs. depth of localization

Due to multiple languages spoken in most Asia Pacific countries, resource allocation is a tricky task. Should multiple languages be taken up for basic localization or should fewer languages be taken up for more in-depth advanced application development? If focus remains only on basic localization due to the numerous languages, advanced applications might never be addressed even though it is necessary to provide access to information to a large part of the population in the region. On the other hand, if only advanced applications are considered, only a limited number of languages may be localized because advanced applications take a much longer time to develop.

Again, a complex socio-economic balance must be struck to determine the right formula for each national context.

Human resource training

In most Asia Pacific countries, there is very limited linguistic and technical capacity to develop standards, perform linguistic analysis and create language technology. Training and human resource planning is critical. Depending on the choice of applications and languages, expertise may be required in various branches of linguistics (phonetics, phonology, morphology, syntax, semantics and pragmatics), signal and speech processing, image processing, statistics, computational linguistics and advanced computing. Training for basic localization work could take about six months. To develop advanced applications, experienced linguists and computational linguists are required and dedicated training over many years is necessary. To address national needs and to keep the training process sustainable, diploma and degree programmes in speech, script and language processing should be developed at the universities, through collaboration of the linguistics, computer science and engineering departments. Scholarships dedicated to these areas for study abroad can also help accelerate the process. Regional and international cooperation can play a significant role in these efforts.

The best way to build capacity is to involve the technical development staff in actual hands-on localization work. This can be achieved by national and regional organizations funding language computing projects (*see the case study on the PAN Localization Project below*). Momentum for localization can also be triggered by governments if they create awareness of local language computing and generate market demand by requiring public information to be localized through e-governance initiatives. Regional organizations can organize national and regional training and seminars. Two recent initiatives are the Summer School in Asian Language Processing in 2006 organized by the PAN Localization project and Asian Applied Natural Language Processing for Linguistics Diversity and Language Resource Development (ADD) organized by the Thai Computational Laboratory.

PAN Localization: A regional initiative to develop local language computing capacity in Asia

The PAN Localization Project is a concrete example of a cohesive regional cooperative project to develop and disseminate local language computing technology in Asia Pacific. In the first phase, from 2004 to 2007, the project focused on developing (a) human resource, (b) technology and (c) policy related to language computing across Asia Pacific. In the second phase, from 2007 to 2010, the project will look into social models for enabling local language content access and generation by training rural communities to use local language computing technology. Thus, the project addresses the immediate need for localization in developing Asia.

The project is a collaboration among 11 countries: Afghanistan, Bangladesh, Bhutan, Cambodia, China, Laos, Mongolia, Nepal, Pakistan, the Philippines and Sri Lanka. It is coordinated by the Center for Research in Urdu Language

Processing (CRULP, www.crulp.org) at the National University of Computer and Emerging Sciences (NUCES, www.nu.edu.pk) in Pakistan and funded by the Pan Asia Networking (PAN) programme of the International Development Research Centre (IDRC, www.idrc.ca). The project has also developed formal and informal collaboration with other countries, including India, Iran, Japan, Korea, Myanmar, Indonesia and Thailand.

The project supports a development team of about 100 people across the participating countries who are being trained and who are actively developing local language computing solutions in 15 different Asian languages. The project maintains a team at each collaborating country. The country teams decide the scope of work and the platform to localize based on level of localization and the capacity of the available human resources. Development targets help the teams focus their capacity building efforts. In most cases, the country components are hosted at universities and public sector organizations to ensure sustainability. Sustainability is also addressed by contributing towards the development of formal research groups on localization. The project has already helped establish the Center for Reseach in Bangla Language Computing at BRAC University in Bangladesh, the Research Division at the Department of IT in Bhutan, the Language Technology Research Lab at the University of Colombo School of Computing in Sri Lanka, the Nepali Language Technology Group at the Madan Puraskar Pustakalaya and the Language Technology Lab at the University of Kathmandu in Nepal, the Speech Lab at the Institute of Technology of Cambodia and language and speech technology labs at the National University of Mongolia and Mongolian University of Science and Technology, respectively.

The project has arranged short and long-term national and regional training for its staff. For example, a mentor placement programme has allowed experienced personnel from Pakistan, India and Sri Lanka to be placed in Bhutan, Cambodia and Laos for two to six months. This has been noted as one of the most significant capacity building methods by the partner countries. A two-and-a-half month long Summer School in Asian Language Processing at NUCES, in 2006, addressed training in advanced language computing and helped build capacity in script, speech and language processing for 40 participants from 12 countries. Other training and workshops organized by the project are listed at the project website (see Activities link at www.PANL10n.net). The project has also been training end-users in local language computing applications, for example in Bhutan, Cambodia, Laos, Nepal, the Philippines and Sri Lanka. These have been on multiple platforms—for example, on Open Source platforms in Bhutan, Cambodia, Nepal and Sri Lanka, and on proprietary platforms in Laos, Cambodia and Sri Lanka. These efforts are being extended to all participating countries in the second phase of the project.

In its first phase, the project also developed a variety of local language computing solutions, including Pashto script, keyboard and collation standards; Bangla collation, lexicon, morphological analyzer and OCR; DzongkhaLinux distribution, including Dzongkha fonts, collation, keyboard and localized applications for word processing, e-mailing, Web browsing, chatting and multimedia; Khmer collation, lexicon, word segmentation, spell checker and tagged corpus; Lao fonts, collation, keyboard, lexicon and corpus; Nepali Linux distribution including Nepali collation, keyboard, spell checker and localized applications for word processing, e-mailing, Web browsing, chatting, accounting and multimedia; and Sinhala TTS and OCR, lexicon, collation and corpus (*see the project website for a detailed list of current outputs*). The project has also developed training materials for these and other applications in the local languages. Open licensing allows these outputs to be shared between the partner countries. For example, the OCR software developed for Sinhala by Sri Lanka has been used by the Laos team to retrain it for Lao.

Equally significant is the development of a network of researchers in the region through the project. Experts, practitioners and policymakers have been brought together to interact and guide development teams in the participating countries. The project has also developed a repository of training materials and links to local language resources. It disseminates research outputs with open software and content licenses. Aside from local language software for nine languages, the outputs include research reports specific to the target languages and general guides, such as the *Survey of Local Language Computing in Asia 2005* (Hussain et al. 2005) and *A Guide to Linux Localization*.

The project is helping research the challenges and solutions for creating localization awareness in the region; building sustainable human resource capacity; developing standards and basic and advanced localization technology; and forming a regional network of researchers. It has institutionalized localization in many of its partner countries and is directly and indirectly influencing relevant ICT policy. Thus, the project is addressing local language computing in a holistic fashion across Asia Pacific.

Partnerships and resource sharing

It is redundant and usually expensive to localize independently for all languages. A better model is to reuse the same basic technology for different languages. Most open source software work on this principle. Innovative mechanisms must be put in place to share content, training and other localization work. Regional and international organizations must play a significant role in this context, funding avenues through which research, training, resources and best practices may be shared across nations. Many such initiatives are developing in the region, such as the AFNLP, International Open Source Network (IOSN), Asia Open Source Software (AOSS) and Asia Commons, which are non-governmental organizations. Many other technology frameworks are also available and being developed in universities and other organizations across the world.

Licensing regimes

As discussed, many different licensing regimes are possible both for the software and content being produced. As much as possible, open licensing must be adopted to propagate the work in local language computing. Liberal licenses, such as GPL, MIT and BSD, can allow open source distribution of software for non-profit as well as commercial purposes (cf. Chen 2006). Content must also be made available with liberal licensing for convenient access (for example, Creative Commons). In addition, effective channels are needed to share content and training curricula, perhaps using models similar to the Wikipedia and Sourceforge initiatives.

Because effective coordination cannot be achieved only through virtual communities, there is also a need for face-to-face networking. Regional and international organizations dedicated to social development through ICTs need to play an active leadership role in this regard. For example, the Free and Open Source Software in Asia Pacific (FOSSAP) forum by IOSN has been discussing software licensing and Asia Commons has started addressing content licensing.

Computing platforms

A very important aspect of localization is the choice of computing platform. Both proprietary and open source platforms exist and are currently being used. For end-users in Asia Pacific, the prevalent platforms include Microsoft Windows, Java Virtual Machine (JVM or Java) and varieties of Linux (for example, Red Hat and Debian). Windows is a proprietary software platform which is not free and has some security concerns.[19] Java is a virtual platform and requires a physical platform like Microsoft Windows or Linux on which it can be installed. Linux is open source and free of cost.[20]

However, the choice is not as apparent as it seems. Though Windows is proprietary, closed and vulnerable to security threats, it is still the most widely used software with convenient plug-and-play hardware installation features, making it very convenient for end-users. The Linux platform requires more expertise to use and is more difficult to manage and maintain given the limited administrative and management capacity currently available. Deciding which platform to target for localization is a complex issue. For some languages which are already supported by Microsoft products, Windows may present a more viable short-term solution. For these languages, Linux may present a solution in the longer term, as there is a need to train more human resources to maintain Linux-based systems. For other languages that are not currently supported by Windows, open source platforms may be the only solution, as the localization plans of Microsoft may not align with national priorities.

Participatory standardization

With the growing need and demand for multilingual computing, there is increased standardization activity. Owing to the urgency and multiplicity of the tasks, there are very frequent meetings among the participating organizations across the world, as well as public requests for comments on the developing standards. However, due to lack of expertise and resources, it is difficult for many developing countries in Asia Pacific to participate in these discussions. Unfortunately, lack of participation is always considered to be tacit approval by these standards organizations.

From an academic point of view, assuming approval when there is lack of comment is not always the best strategy for the development of standards despite the operational ease of this process. When multilingual standards are finalized without indigenous feedback, there are bound to be problems (for example, as reported for the Khmer Unicode page) especially once many of these languages catch up to the newer standards. The process of standardization must be proactive from both ends. National bodies must try to actively participate in the process and the standards development organizations should have programmes to train participants from different countries and to proactively seek their feedback before proceeding to finalize multilingual standards. This requires significant financial investment which has to be raised in a sustainable way. For example, the Asian Forum for Standardization of Information Technology (AFSIT) and associated programmes by CICC have contributed significantly in the areas of multilingual computing and related standardization training. Such efforts must continue in the future.

Translation of policy into projects

National policy alone will not ensure the development of local language computing. The policy must be translated into action plans, which in turn must be realized into projects with explicit funding allocation. The first step would be to develop a national committee of experts to discuss and finalize basic standards. Once standards are developed, basic localization for a language is possible for as little as USD 200,000 within one to two years. Developing a complete set of advanced applications would require considerably more effort and time—about three to five years to develop functional models and about a decade to mature—even when using existing software toolkits.[21] Building a complete suite of language technology for a single language could cost more than USD 5 million.[22] Basic localization may be undertaken by the private sector. However, because there are few commercial incentives for advanced applications in developing countries, these would only be developed with explicit support and funding by the government and other organizations.

Concluding remarks

The great linguistic diversity in the Asia Pacific region presents a significant social barrier to widespread use of ICTs. If communities in the region are to cross over into the information age, ICTs must be enabled in their languages. Localization is necessary to give these communities the opportunity to use and benefit from the ICT revolution. However, most of these communities neither have capacity nor currently present the financial incentives for private investment in localization. There is no easy or short-term solution to this problem and a considerable and coordinated national, regional and international effort is required. The initial focus must be on sustainable human resource and technology development within these countries. In addition, a two-tier policy must be adopted—first to support localization through public funding and second to concurrently create enough demand for local language computing, for example through e-government initiatives, to trigger private sector interest.

In conclusion, localization should not be looked at as an obstacle, but as an opening for Asia Pacific to revitalize its IT industry and to develop its knowledge economy. Proper national and regional policy planning and execution can turn the challenges into opportunities.

Notes

1. The Summer Institute of Linguistics reports a total count of 6,912 languages at www.ethnologue.com
2. These terms are normally abbreviated by their first and last letter, infixed by the count of the remaining letters, as I18n, L10n and G11n.
3. For example, the Mongolian government recently decided to adopt Cyrillic script for writing the Mongolian language, abolishing the traditional Mongolian script.
4. The debate is normally between at least two groups: those who would like the language to remain 'pure' and those who would want to adapt the language to 'simplify' its use.
5. For example, 'A' is assigned a code of 65, 'B' 66, and so on, in ASCII and Unicode encoding.
6. Arbitrary assignment has been the traditional way of encoding languages across Asia. Multiple encodings exist for languages across Asia Pacific because each vendor has developed its own assignment (*see Hussain et al. 2005*).
7. The Unicode Standard is the same as the ISO 10646 standard and is co-managed by the Unicode Consortium and a dedicated Working Group of a Sub-Committee of the Joint Technical Committee of the International Standards Organization (ISO JTC1/SC2/WG2).
8. This process can take more than a year. It would normally take about six months for relevant ISO and Unicode committees to evaluate a proposal and another six months for vendors to provide support for these characters within technology, if the characters are approved and included in the standard.
9. Most operating systems allow users to define their own variation of the keyboard layout. 'Phonetic Keyboard Layouts', which map the [p] sounding character on the key with 'P' etc. on the regular QWERTY keyboard layout, are also popular among regular computer users.
10. OTF is an open standard jointly developed by Adobe and Microsoft. There are also other formalisms, including Apple's Advanced Typography (AAT), Postscript, etc. OTF is still an evolving standard, although it can now support the variety in Asian scripts fairly well.
11. POS indicates whether the word is a noun, verb, adjective, adverb, etc.
12. Advanced applications may require as many as 10 of such tags for each word. The following illustrates a sample entry: '*Boy: Common_Noun, Singular, Masculine, Human, Animate.*'
13. Machine Learning is a branch of Artificial Intelligence in which a large amount of data is used to automatically train models to predict certain properties of unseen/new data.
14. The first syllable is stressed if it is a noun and the second syllable is stressed if it is a verb.
15. The estimates vary depending on the source and target language pair, the expertise of available linguists and computational linguists, and

the techniques used. This estimate assumes availability of trained linguists and computational linguists. Systems may be developed within a shorter duration using statistical techniques.
16. Administered through the support of IANA and Regional Registries (RIRs), for example, APNIC for Asia Pacific.
17. Usage depends on the licensing schemes. Most software is available through a standard or limited version of the GNU Public License (GPL).
18. Further discussion of mobile applications is available in the chapter on Mobile and Wireless Technologies in this volume.
19. There is a backdoor to security of Microsoft Windows through _NSAKey. Refer to http://en.wikipedia.org/wiki/NSAKEY, http://www.techweb.com/wire/story /TWB19990906S0003 and http://www.cnn.com/TECH/computing/9909/03/windows.nsa.02/ for related discussions.
20. Although the Linux platform is freely available, there is a cost for installation and maintenance.
21. Toolkits like Festival and MBROLA for TTS, HTK and Sphinx for ASR and XLE for MT are being developed and made available by academic and other organizations across the world, especially for non-commercial use.
22. These estimates are based on availability of human resources with a reasonable level of experience in localization work. If such a pool of human resources is not available, more time and/or funds may be required.
23. The table lists a comparison for some of the applications. The comparison is qualitative, not quantitative, and is based on the current information available to the authors through the Internet and other sources (e.g. Sonlertlamvanich 2002; Tsujii 2005; Hussain et al. 2005). The information has not been independently verified and therefore has some margin of error.

References

Butt, D. (2006). *Internationalized domain names*. Retrieved 15 December 2006 from http://www.apdip.net/apdipenote/9.pdf

Chen, S. (2006). *Free/Open source software: Licensing*. UNDP-APDIP. New Delhi, India: Elsevier

Computer Industry Almanac. (2006). *Worldwide Internet users top 1 billion in 2005*. Retrieved 10 October 2006 from http://www.c-i-a.com/pr0106.htm

Daswani, C.J. (1998). *Language diversity and literacy in India*. In Proceedings of the Second Asia Regional Literacy Forum-Innovation and Professionalization in Adult Literacy: A Focus on Diversity. New Delhi, India. Retrieved 10 October 2006 from http://www.literacyonline.org/products/ili/webdocs/daswani.html

DzongkhaLinux Team. (2007). *Dzongkha computer terms*. Department of IT and Dzongkha Development Authority, Royal Government of Bhutan. Lahore, Pakistan: PAN Localization Project Regional Secretariat.

Global Reach. (2004). *Global Internet statistics by language*. Retrieved 26 December 2006 from http://global-reach.biz/globstats/index.php3

Hussain, S. (2004). *Complexity of Asian scripts: A case study of Nafees Nastalique*. In Proceedings of SCALLA. Kathmandu.

Hussain, S., Durrani, D., and Gul, S. (2005). *PAN Localization survey of local language computing in Asia 2005*. Ottowa, Canada: IDRC.

Huston, G. (2006). *Internationalizing the Internet*. Retrieved 11 December 2006 from http://www.circleid.com/posts/print/internationalizing_the_Internet/

ICANN. (2006). *Guidelines for implementation of internationalized domain names version 2.1*. Retrieved 26 December 2006 from http://www.icann.org/general/idn-guidelines-22feb06.htm

Internet World Stats. (2006a). *World Internet users and population stats*. Retrieved 10 October 2006 from http://www.internetworldstats.com/stats.htm

Internet World Stats. (2006b). *Internet world users by language*. Retrieved 10 October 2006 from http://www.internetworldstats.com/index.html

Katsiavriades, K. and Qureshi, T. (2006). *The 30 most spoken languages of the world*. Retrieved 10 October 2006 from http://www.krysstal.com/spoken.html

MIT, India. (2006). *Annual report 2005–2006*. Retrieved 10 October 2006 from http://mit.gov.in/download/annualreport2005-06.pdf

NEC. (2006). *NEC develops world's first Japanese<->Chinese automatic speech translation software operable on PDA*. Retrieved 10 October 2006 from http://www.nec.co.jp/press/en/0601/0401.html

Omniglot.com Homepage. (2006). Available at www.omniglot.com

Pimienta, D. (2005). Linguistic diversity in cyberspace—Models for development and measurement. In *Measuring linguistic diversity on the Internet*. Paris: UNESCO.

Sonlertlamvanich, V. (ed.) (2002). *Proceedings of the workshop on survey on research and development of machine translation in Asian countries*. Thailand: NECTEC.

Tsujii, J. (ed.). (2005). *AAMT journal* (Special Issue). Thailand: Asia Pacific Association for Machine Translation.

UNESCO. (2004). Retrieved 10 October 2006 from http://www.unesco.org/education/languages_2004/languagesdistribution.pdf#search=%22unwritten%20languages%20of%20Asia%22

Unicode. (2006). *Unicode 5.0*, 5th ed. Addison-Wesley Professional

Wagers, L. (2006). Localization matters: Ask Nokia, Google, Carrefour, Domino's. In *Multi-lingual #80*, 17(4).

Wikipedia. (2006). *Chinese input methods for computers*. Retrieved 20 October 2006 from http://en.wikipedia.org/wiki/Chinese_input_methods_for_computers

Key policy issues in intellectual property and technology in Asia Pacific

Elizabeth V. **Cardoza** and Lawrence **Liang**

Background

Intellectual property (IP) is acknowledged to be a key component of businesses, including those related to or based on information and communication technologies (ICTs) which today constitute a key growth area in Asia Pacific economies. The key forms of IP that impact ICT industries are usually copyright and patents. But increasingly, trademarks, industrial designs and integrated circuit designs are becoming significant focal points. Other key issues are indigenous knowledge, data protection and privacy, and competition policy issues. Thus, when legislating IP laws, policymakers will need to take account of ICT infrastructure, have a basic understanding of the nature of the new technologies, and review consumer protection, licensing and competition policy developments. They need to ensure that IP policies and laws address and reflect national, social and cultural requirements.

IP has had a chequered and contested history in most countries in Asia Pacific. In the early days of the World Trade Organization (WTO) Agreement on Trade-Related Aspects of Intellectual Property Rights (TRIPS), a number of developing countries felt that the linking of IP to trade and the standardization of IP laws was an agenda that developed countries in the north were attempting to impose on them (Correa 1997). TRIPS, which was negotiated in 1994 at the end of the Uruguay Round of the General Agreement on Tariffs and Trade (GATT), lays down the minimum global standard that has to be met by national laws on IP rights, such as copyright, patents and trademarks. The obligations under TRIPS apply equally to all WTO member states. However, developing countries were allowed extra time to implement the applicable changes to their national laws. The transition period for developing countries expired in 2005 while the transition period for least developed countries has been extended to 2016.

Many countries in the global south, including Brazil, India and Thailand, resisted many aspects of TRIPS on the grounds that they would benefit developed countries in the north far more than those in the global south as a result of the economic and technological imbalance between them (Drahos and Braithwaite 2002). Because developing countries are net importers of IP rights, there were serious concerns that the new IP regime would impose a heavy cost with respect to transfer of technology and the development of indigenous technological capabilities. Almost every region in Asia Pacific has at some point or other been accused of not providing adequate protection to IP rights. It is also a fact that most countries in Asia Pacific that have developed strong technological capabilities, including Korea, Taiwan, China and India, have built their capabilities on the basis of poor IP rights enforcement (Kumar 2003). As a recent anthology on IP in Asia (Thomas and Servaes 2006, p. 15) points out:

> The Asian region is a site of numerous contestations over IP precisely because it is home to a variety of countries at different stages of economic development and levels of openness to IP reform despite pressures from multilateral trading system, and the USTR to standardize and harmonize national IP legislations with global requirements.

However, with most Asian countries having signed on to TRIPS and become members of the WTO, the current situation is slightly different. Although TRIPS has created a disconnect between the IP laws of non-industrial, developing countries and their social and economic conditions, since TRIPS did not emanate from the willingness or determination of these countries for forms of IP protection adequate for their needs (Endeshaw 2005), these countries are now obliged to ensure that their national legislations are in conformity with the minimum standards set by TRIPS. They are also required to enforce IP laws in accordance with global standards.

IP thus looms as a concern that poses serious dilemmas to policymakers in Asia Pacific who are attempting to balance their obligations in international law with their commitment to economic and social development. Economic development in many parts of Asia has been extremely lopsided. While a few countries, particularly in South Asia and East Asia, have been able to transform themselves into significant players in the information economy, there are many others that remain very much in the periphery of the knowledge economy. Even in countries like Malaysia, India and China, the digital revolution has reached only a very small percentage of the population. Thus, Asia Pacific countries cannot be treated as a homogenous set—an often-ignored fact. There are clearly marked differences, as well as inequalities, among these countries in terms of scientific and technical capacities, social and economic structures and distribution of wealth.

Even within the World Intellectual Property Organization (WIPO), there is now recognition of the importance of harmonizing IP laws with national developmental goals. The WIPO General Assembly has established the Agenda for Development, a long overdue and much needed first step toward a new WIPO mission and work programme. The WIPO Agenda for Development declares that the WIPO Convention should formally recognize the need to take into account the 'development needs of its Member States, particularly developing countries and least-developed countries'. According to the Geneva Declaration on the Future of WIPO, WIPO's functions should not only be to promote efficient protection and harmonization of IP laws, but also to formally embrace balance, appropriateness and the stimulation of both competitive and collaborative models of creative activity within national, regional and transnational systems of innovation.[1]

This chapter maps out some of the key IP-related issues that policymakers in Asia Pacific will have to address in the coming years, namely:

1. Copyright and its impact on access to knowledge and technology
2. Exceptions and limitations within TRIPS
3. Non-proprietary models
4. Stronger enforcement of IP
5. The impact of IP provisions in bilateral agreements
6. Participation in regional forums

Copyright and its impact on access to knowledge and technology

One of the justifications for a strong IP regime emerges from the argument of economic development. Economists argue that IP is needed for economic growth which is needed to reduce poverty. By ensuring innovation, creativity and productivity through IP development, countries can increase their agricultural and industrial production as well as financial investment.

The argument assumes that the system that has worked for developed countries will work similarly for developing countries. A counter-argument is that IP rights do very little to promote economic development in developing countries and in fact may end up hindering it where the necessary economic and technical capabilities are absent. For instance, the Commission on IPR maintains that IP regimes are ineffective at stimulating research that will benefit poor people because they will not be able to afford the products even if these are developed. Moreover, IP rules limit the option of technological learning through imitation and allow foreign firms to drive out domestic competition by obtaining patent protection and to service the market through imports rather than through domestic manufacture (CIPR 2002).

It is estimated that in 1999 nearly 1.2 billion people lived on less than USD 1 a day, and nearly 2.8 billion people lived on less than USD 2 per day. About 65 per cent of these people are in South and East Asia alone (World Bank 2001). Thus a key issue for policymakers is who to focus on when they consider IP and technology policies. Unfortunately, most IP policies focus on IP owners and producers and not the users. It is important to bear in mind that almost all countries in Asia Pacific, with the exception of Japan, remain net importers of IP. Even countries like India that produce a lot of IP rarely own the legal rights to the products developed locally, since these are created for companies in the northern hemisphere.

The asymmetry between developed and developing countries in relation to technology is further illustrated by the fact that low and middle income developing countries account for about 21 per cent of world GDP (World Bank) but less than 10 per cent of worldwide research and development (R&D) expenditure.[2] The OECD countries spend far more on R&D than India's total national income.[3] Tables 1 and 2 contrast the level of investment and activities with respect to R&D expenditure and patents in developed and developing countries. They provide an insight into the sharp inequalities in the knowledge economy. Table 1 shows that the R&D budgets of developed countries far exceed those of the developing countries. Table 2 shows that the patent share of developing countries is miniscule in comparison to that of the developed countries.

Table 1
Major source countries of technologies in the World, 2000

Country	R&D expentiture[#] $billion ppp $	R&D expentiture[#] Percentage of total	US patents taken, 1977–2000 '000	US patents taken, 1977–2000 Percentage of total	Technology fees received[#] $billion	Technology fees received[#] Percentage of total	FDI outflows $billion	FDI outflows Percentage of total
US	212.8	40.8	1,337.0	57	33.8	42.2	139.3	12.1
Japan	90.1	17.3	429.4	18	6.9	8.9	32.9	2.9
Germany	42.0	8.0	173.8	7	11.9	14.9	48.6	4.2
France	28.1	5.4	68.2	3	2.2	2.7	172.5	15.0
UK	22.6	4.3	67.4	3	5.8	7.2	249.8	21.7
Italy	12.1	2.3	29.0	1	1.6	2.0	12.1	1.1
Canada	11.4	2.2	48.4	2	1.3	1.6	44.0	3.8
Netherlands	7.5	1.4	22.0	1	6.2	7.7	73.1	6.4
Sweden	7.1	1.4	22.9	1	0.4	0.5	39.5	3.4
Switzerland	4.8	0.9	31.0	1	2.8	3.5	39.6	3.4
Subtotal 10	438.5	84.0	2,229.1	94	72.8	91.0	851.3	74.0
World	552.0	100.0	2,364.9	100	80.1	100.0	1,149.9	100.0

Source: Kumar (2003).
Note: [#]Belongs to 1997.

Table 2
Emerging sources of technology in terms of ownership of US patents, 1977–2000

Country	1977–87 Numbers	1977–87 Per cent	1987–90 Numbers	1987–90 Per cent	1991–95 Numbers	1991–95 Per cent	1996–2000 Numbers	1996–2000 Per cent
Taiwan	1,039	0.15	2,496	0.66	7,760	1.41	19,153	2.54
South Korea	236	0.03	704	0.19	4,113	0.75	14,045	1.86
Israel	1,302	0.19	1,156	0.31	1,849	0.34	3,550	0.47
Hong Kong	577	0.08	480	0.13	1,018	0.18	1,842	0.24
South Africa	827	0.12	485	0.13	549	0.10	614	0.08
Mexico	393	0.06	174	0.05	234	0.04	374	0.05
Brazil	245	0.04	156	0.04	299	0.05	435	0.06
China Pub. Rep.	25	0.00	171	0.05	257	0.05	464	0.06
Argentina	206	0.03	78	0.02	136	0.02	225	0.03
Singapore	40	0.01	58	0.02	224	0.04	727	0.10
Venezuela	105	0.02	88	0.02	142	0.03	156	0.02
India	111	0.02	64	0.02	144	0.03	424	0.06
East and Central Europe	4,684	0.69	1,207	0.32	994	0.18	1,143	0.15
Subtotal	9,790	1.43	7,317	1.95	17,719	3.21	43,152	5.72
Others	1,473	0.22	652	0.17	902	0.16	1,731	0.23
Total	682,639	100.00	375,946	100.00	551,902	100.00	754,391	100.00

Source: Kumar (2003).
Note: Based on data presented in US Patents and Trademarks office (2001), *TAF Special Report: All Patents, All Types—January 1, 1977–December 31, 2000*, Washington, DC.

Given this asymmetry, policymakers in Asia Pacific will have to consider the impact of IP on the following four factors: (a) the costs of acquiring technology, (b) the opportunity costs in terms of developmental funding for key areas such as education, health and infrastructure, (c) the need to consider alternatives to paying high royalty costs and (d) the need to focus on developing indigenous technology instead of relying on importing foreign technology. By shifting focus away from protecting IP producers and owners, towards viewing IP through the prism of human rights and development, policymakers will be able to determine what would constitute the best model of IP laws within their economic and cultural context, keeping in mind their obligations under TRIPS.

The ICT revolution of this era promises a radical shift in the paradigm of how information, knowledge and culture are produced, disseminated and accessed (Rifkin 2000). Yet this promise must overcome the challenges posed by severe restrictions that make access to knowledge and culture more difficult for people, especially the poor and underprivileged. Stricter IP laws that raise information costs constitute grave impediments to a more democratized information environment. This is illustrated in the case of partially sighted and blind people whose already limited access to digital content is further curtailed by traditional as well as new copy protocols (*see boxed article*).

The case highlights the broader issue of the relationship between copyright and access to knowledge. Copyright was intended as a system of balances to provide incentives to creators while also ensuring free circulation of copyright works in the public domain for all other creators to build on. This balance has shifted aggressively and it has expanded drastically in favour of

ICT, visual disability and copyright

It is estimated that there are about 180 million people in the world who are blind and partially-sighted and who thus are disadvantaged in their ability to access content. Developments in text recognition software have improved their situation somewhat, although proprietary versions of such software, such as Jaws, still cost up to USD 1,000 per licensed copy. In India, many people with visual disabilities have started using what would be illegally obtained versions of Jaws. The only reason there is no enforcement of IP laws in this case would be the very bad press that a copyright infringement claim against an association for the blind would get if it were pursued.

People who are blind or partially-sighted can only access the written word, whether originally displayed on paper or on computer screen, if its presentation is adapted in some way. Adaptations include enlarging, altering features such as colour or font, and transferring to a tactile code or into an audio format. The result may be hard copy Braille, large print, tape or CD or a temporary output from computer peripherals such as synthetic speech or enlarged screen display. Thus, providing access to content for those who are blind or partially-sighted would include granting them the rights of reproduction, adaptation and perhaps communication, which in turn would mean granting this set of users an exception to copyright.

However, even if exceptions are provided, there could be additional restrictions, such as in the form of digital rights management (DRM). Kerscher and Fruchterman (2002) describe the impact of DRM on the ability of people who are blind to access digital content thus:

> The personal computer is the information access tool of choice for many persons who are blind. The computer is made accessible through a screen reader program. Screen readers use a text-to-speech synthesizer (TTS) to speak aloud the information that a sighted person would visually read on the computer screen. These screen readers intercept the text being written to the display and keep track of it, so that it can be vocalized in response to the user's control. For example, pressing certain keys will cause the screen reader to read the current word, line or paragraph. Screen readers also permit the use of dynamic Braille displays instead of, or in addition to, the TTS.
>
> The screen readers are external applications to the PC-based eBook reading software. The DRM wrappers are designed to work with reading applications that present the text visually without allowing the text to be copied, to prevent the illegal distribution of the book. Unfortunately, these anti-copying provisions also prevent the screen reader from providing access with TTS or Braille. The secure reading application views these external applications as security threats and blocks their access. As a result, people persons who try to use their screen reader with eBook reading systems find that their screen reader is not allowed to do its job… [which] leaves the person who is blind with no access to the ePublication, unless the reading application builds access directly into the user interface.

content owners such as large publishing houses and media conglomerates. It is imperative for policymakers to consider what kind of exceptions or compulsory license mechanisms can be devised to enable greater access to content and information for all of their citizens. In spite of their relatively weak bargaining power in the new global order, policymakers from developing countries in Asia Pacific must consider the best options available to them under the current paradigm.

Flexibilities under TRIPS

TRIPS provides for exceptions and limitations that may be included in national IP legislation. These exceptions may take two forms: (a) fair use or fair dealing and (b) statutory or compulsory licenses. Fair use refers to use of material without the copyright owner's permission that would not amount to an infringement—for example, using extracts of a book in teaching materials. The statutory or compulsory license approach envisages a specific scenario where a work may be freely reproduced after payment of a fixed royalty—for example, publishing a textbook not otherwise available in a country.

The Berne Convention, TRIPS and WIPO Copyright Treaty (WCT) stipulate that limitations or exceptions to copyright shall be confined to certain special cases; that they shall not come in conflict with normal exploitation of the copyright work and that they shall not unreasonably prejudice the legitimate interests of the copyright holder. This is known as the 'three-step test'. Each step makes it more difficult to grant limitations or exceptions to copyright. Historically, the three-step test was inserted into the Berne Convention only in relation to reproduction rights. However, TRIPS widened it to be applicable to all exclusive rights granted by the Berne Convention and TRIPS (Ryan 1999).

TRIPS provides only general parameters for exceptions. For example, Article 10(2) states:

> It shall be a matter for legislation in the countries of the Union, and for special agreements existing or to be concluded between them, to permit the utilization, to the extent justified by the purpose, of literary or artistic works by way of illustration in publications, broadcasts or sound or visual recordings for teaching, provided that such utilization is compatible with fair practice.

Thus, the question of what constitutes 'utilization... [of works] for teaching' is to be determined by national legislation, or by bilateral agreements between Union members. Article 10(2) sets the outer limits without stipulating quantitative limitations for instance. It is up to each country to interpret the provision to enable it to formulate essential exceptions for educational uses of material.

Since Article 9(2) of the Berne Convention and Article 13 of TRIPS allow nation states to determine the extent of the exceptions and limitations to copyright, policymakers must make optimal use of this available flexibility, keeping in mind the wider public policy consideration of making information and technology available to the public. This is particularly useful for educational materials which, with recent and ongoing revolutionary ICT changes, can be produced and accessed in a variety of modes. In many Asia Pacific countries where availability of educational infrastructure and educational materials of a high standard is a problem, distance learning and digital content are useful alternatives. However, as a recent study indicates, 'among the most important obstacles to realizing the potential of digital technology in education are provisions of copyright law concerning the educational use of content, as well as the business and institutional structures shaped by that law' (Fischer 2006). Librarians and educationists argue that governments should use the greatest flexibilities available within the TRIPS agreement to ensure that national copyright laws make adequate provisions for educational use of information (Wong 2004).

In determining optimal use of exceptions and limitations for greater access, policymakers should undertake exhaustive surveys of the best global practices on copyright exceptions and limitations for use as models in the drafting of national copyright policies and laws. With respect to software for example, one good practice is to reserve the right to allow reverse engineering for the purposes of studying the software's functionality and for research and development. Moreover, policymakers should be well versed in the debates surrounding emerging practices in IP protection that have a huge impact on technological development and innovation, such as software patents and digital rights management (DRM) (*see boxed article, next page*).

Non-proprietary models

Along with the current trend of aggressive expansion of IP[4] regimes is a parallel movement rearticulating the importance of the commons of knowledge and cultural production. The idea of a knowledge and cultural commons borrows from the environmental movement and is based on the belief that a vibrant public domain of freely available knowledge and culture is vital for future innovation and creativity (Boyle 2002). Thus, even as copyright, patent and trademark systems are being promoted as the primary mode of understanding the production of knowledge and culture, another paradigm has emerged as a response to this regime of proprietary knowledge—a paradigm that proposes

Copyright, fair use, and digital rights management

DRM refers to technologies that define and enforce parameters of access to digital media or software. There are extensive arguments both for and against the use of software patents and DRM and it is beyond the scope of this article to examine either position in significant detail. However, it makes strategic sense for policymakers to understand the scope and implications of both software patents and DRM.

Even Bill Gates, the most fervent supporter of copyright, has recognized the adverse impact software patents can have on the development of software. In 1991, Gates argued: 'If people had understood how patents would be granted when most of today's ideas were invented and had taken out patents, the industry would be at a complete standstill today. A future start-up with no patents of its own will be forced to pay whatever price the giants choose to impose. That price might be high. Established companies have an interest in excluding future competitors' (qtd. in Stallman 2005).

The ostensible reason for the deployment of DRM is to 'enforce' the copyright of the manufacturer or the copyright holder, as the case may be. It should be noted, however, that DRMs are not envisaged under TRIPS and they are included only in an additional treaty, the WCT. DRM effectively grants to the copyright owner protection that is not available to him or her under traditional copyright law. Take for example a publisher who compiles a database of materials legally in the public domain, such as Supreme Court cases, and then locks the CD under a DRM. Under traditional copyright law, users can access these cases for free. However, with DRM, any attempt to break the technology lock to access the database, even if it does not infringe any copyright, may render users liable under an 'anti-circumvention' provision. Effectively, DRM allows the copyright holder to restrict access to content simply because it is in digital format, even if that same content would be easily accessible under traditional copyright law. Another example is a person wishing to make a copy of a legally purchased media file for personal use or for backup, utilizing the flexibility sanctioned under a fair use provision. This person would not be able to make a copy if anti-circumvention laws under DRM exist. Such laws could also prevent private screenings of digital media, which would otherwise be perfectly legal.

We envisage DRM to have a significant impact on innovation. This is particularly significant for countries where the fruits of innovation need to be accessible to both the innovator and the consumer. An example is the Simputer, a low-cost handheld computer developed in India that would have been more difficult to invent if DRM laws existed in India. With the introduction of DRM and the criminalization of its circumvention, low-cost, locally relevant and contextually appropriate computer hardware and software may never become available to those who can least afford them.

DRM exceeds TRIPS minimum standards and amounts to a TRIPS-plus provision that is neither a necessity nor an obligation. Since TRIPS does not mandate anti-circumvention provisions, there is *no legal obligation* to enact them as law. As for the WCT, a country that is not a signatory to it is not obliged to enact such provisions into the national law. Nevertheless, it is imperative for developing countries to consider all of the implications of failure to resist pressures from the US and Europe to become signatories to the WCT, to introduce DRM or anti-circumvention laws into their IP law, or to take on equivalent commitments under a bilateral agreement.

For the reasons cited above, it seems premature for developing countries to be required to go beyond TRIPS standards and endorse the WCT. Developing countries should decide for themselves the level of protection their laws should afford to technological locks on copyright work and they should adopt anti-circumvention measures that are sensitive to their domestic situations (Garlick 2004). They should retain the freedom to legislate on the regulation of technological measures, in the interest of safeguarding access to knowledge and information and achieving broad socio-economic development, among others.

Should any Asia Pacific government decide to introduce DRM into its copyright law, it must introduce safeguards to protect users from corporate abuse of the anti-circumvention provision, that is, safeguards to allow users to exercise all of the fair dealing clauses specified within their existing law. This is particularly important given that DRM can affect legitimate research, such as use of copyrighted technical journals, educational materials and software by researchers and students in developing countries. Introducing DRM without adequate safeguards could seriously undermine the developmental goals of a country.

'openness', 'collaboration' and 'freedom' with respect to information goods, cultural production and participation in the information economy. This new paradigm has been enabled to a large extent by the success of the free open source software (FOSS) movement and the GNU Linux operating system that has been hailed as a viable alternative to traditional copyright (Wong and Sayo 2004).

FOSS

Free and Open Source Software is an alternative to proprietary software. FOSS grants users the right to use, distribute and modify source code freely.

Open access

Open access (OA) literature is digital, online, free of charge and free of most copyright and licensing restrictions.

Open content

Similar to FOSS, but in the domain of non-software content, such as learning materials, literary works, music, film and the like.

The Copysouth Group argues: 'For the purposes of access to computer technology throughout the global south, both open source software and free software can offer substantial advantages over the proprietary model. Furthermore, these movements offer an alternative to the proprietary model that is important in staking out an independent future for countries in the global south' (Story, Darch and Halbert 2006).

For developing countries, using FOSS significantly reduces the costs of acquisition of technology. As Ghosh (2003) points out:

> ...in developing countries, even after software price discounts, the price tag for proprietary software is enormous in purchasing power terms. The price of a typical, basic proprietary toolset required for any ICT infrastructure, Windows XP together with Office XP, is USD 560 in the US. This is over 2.5 months of GDP/capita in South Africa and over 16 months of GDP/capita in Vietnam. This is the equivalent of charging a single-user license fee in the US of USD 7,541 and USD 48,011 respectively, which is clearly unaffordable. Moreover, no likely discount would significantly reduce this cost, and in any case the simple fact that a single vendor controls any single proprietary software application means that there can never be a guarantee that any discount offered is intended to be sustained for the long term, rather than as a temporary measure used to tempt consumers into a lock-in situation....

Developing countries can customize open source software to suit their needs, and thereby also develop local skills. According to Ghosh (2003) in a 2002 study of FLOSS (Free/Libre/Open Source Software) developers and users, 'the most important reason for developers to participate in open source communities was to learn new skills—'for free'. These skills include programming as well as 'skills rarely taught in formal computer science courses, such as copyright law and licenses', teamwork and team management—skills which 'help developers get jobs and can help create and sustain small businesses'.

Because governments are one of the largest consumers of software, it is critical that they start weighing the costs of using proprietary software for example in comparison with the funding requirements of other developmental priorities. However, most Asia Pacific countries have no official policy with respect to FOSS, open content or open access. One reason is that most of these models work primarily within the domain of private contracts and are completely voluntary. Also, many governments claim vendor neutrality as the reason for not having a policy on FOSS. But given the kind of advantages FOSS can bring to governments, it is time to rethink the idea of vendor neutrality. According to the UK Commission on Intellectual Property Rights,

> Given the considerable needs which developing countries have for information and communication technologies and the limited funds which are available, it would seem sensible that governments and donors should certainly consider supporting programmes to raise awareness about low-cost options, including open source software, in developing countries. Developing countries and their donor partners should review policies for procurement of computer software, with a view to ensuring that options for using low-cost and/or open source software products are properly considered and their costs and benefits carefully evaluated.

Besides the FOSS and open knowledge movements, there are also processes like the proposed Access to Knowledge (A2K) Treaty[5] which is tied with the WIPO Development Agenda. Policymakers need to evaluate how they can integrate the promotion of open models as part of the larger framework of IP and development (Hahn 2002).

With reference to open content, one challenge that policymakers must address is how to deal with two policy questions within the open content movement, namely (a) existing content under copyright and (b) content that may be produced in the future using or with the support of public funding. On the first question there may be little that can be done within an open content framework and some questions are best addressed through a combined strategy of copyright reform and perhaps the use of national right to information laws, wherever they exist. On the second question, however, there may be some interesting possibilities. The demand that IP created using public money should remain within public control is not novel and it can be combined with the normative goals of the open content movement. Furthermore, the success of the open content movement in particular areas can become the basis for strengthening the claim of a direct linkage between open content and greater access to information and knowledge.

Open content has many synergies with existing campaigns and policy reform efforts, including the open access movement. The demand for open policies that would facilitate greater access could be advanced towards public universities, towards publicly funded research and also partially towards privately-owned content for specific uses, including access for visually disabled people. For example, traditional publications can be required to convert their material to open content after a few months of enjoying exclusive publication rights.

Strong vs. weak IPR

Demands for more stringent enforcement of IPR are coming in from all quarters, particularly the US content industries, as piracy has increasingly come to be associated with Asia Pacific economies. However, while TRIPS signatories must provide adequate enforcement of IP, it should be acknowledged that many Asia Pacific countries that have reached an admirable stage of economic and technological development have done so even with relatively weak levels of IP enforcement and through a judicious use of imitation. For example, Taiwan and Korea, which experienced massive transformations from the 1960s to the 1980s, used imitation and reverse engineering to overcome the technological divide and create a strong national capacity in ICTs. Similarly, the Indian pharmaceutical industries benefited for many years from the absence of a pharmaceutical product patent and India is at present one the most significant exporters of generic low costs drugs to many parts of the world.

In 1947, 80–90 per cent of the pharmaceutical patents in India were held by multinational companies and more than 90 per cent of these drugs were not even being produced in India. India changed its patent laws to allow only for 'process patents' and not patents for the end product itself. This essentially meant that an Indian pharmaceutical company could make an existing drug through the process of reverse engineering. During this period, Indian pharmaceutical companies were able to reproduce existing drugs rapidly and at a low cost, thereby making them competitive in both foreign and domestic markets. By 1991, Indian firms accounted for 70 per cent of the bulk drugs and 80 per cent of formulations produced in the country. In 1996, six of the top 10 firms by pharmaceutical sales were Indian firms rather than the subsidiaries of foreign multinationals. Domestic firms (Indian-owned firms based in India) produce about 350 of the 500 bulk drugs consumed in the country. There are over 250 large pharmaceutical firms and about 9,000 registered small-scale units while the Indian Drug Manufacturers' Association (IDMA) estimates about 7,000 unregistered small-scale units producing drugs. The generic drug industry has been vital in ensuring that drugs are available at an affordable price.

Thus a 'weak' IP regime may actually promote local industries and help develop self-reliance in the field of technology (Thomas 2006). As the Copy South Group argues, lower levels of copyright enforcement enable greater circulation of knowledge, culture and technology throughout the developing world, while 'stronger protection and enforcement of copyright rules may well reduce access to knowledge required by developing [countries] to support education and research, and access to copyrighted products such as software'. This in turn could have potentially negative consequences on their ability to develop their human resources and technological capacity.

Policymakers must rise to the challenge of striking a balance between copyright and the need to facilitate affordable technological transfer to enable the emergence of a sustainable indigenous technological sector. To do otherwise would mean perpetual dependence on imported technology and know-how.

Bilateral agreements and TRIPS-plus standards

Over the last decade, developed countries such as the US and the European Union have hotly pursued bilateral and regional free trade agreements (FTAs). US trade policy has been to promote IP rules that reflect a standard of protection similar to that found in US law. Asia Pacific trade partners considering and committing to these more stringent IP rules allegedly do so in exchange for other concessions, such as preferential access to US markets for manufactured and agricultural products. However, such FTAs go beyond TRIPS in terms of protection of patents and pharmaceutical test data, ICT-related areas, copyright protection and enforcement of IP rights. It is also important to note the manner in which FTAs are used by the US as part of a strategy to push for TRIPS-plus standards at a multilateral level. When the US has a whole range of FTAs with small developing countries, it can then claim that these standards are accepted worldwide and should become a global standard.

What follows is a brief case study for the Asia Pacific region—the Singapore approach to FTAs. The discussion highlights some key implications of the IP provisions Singapore committed to under its FTA with the US (hereafter USSFTA).

Singapore is a trade-dependent nation with a cumulative trade volume presently accounting for about thrice its Gross Domestic Product (GDP). While maintaining its commitment to the WTO as a route to global trade liberalization, Singapore actively pursues trade liberalization through regional platforms, such as the Association of Southeast Asian Nations (ASEAN) and the Asia Pacific Economic Community (APEC). The East Asian crisis of 1997–98 and the consequent slowdown in trade and investment liberalization are allegedly the impetus for Singapore's foray into bilateral FTAs with key and strategic trading partners. This strategy aims to increase economic ties and garner 'first-mover' advantage with key and strategic trading partners, enhancing market access opportunities beyond the region and in emerging market economies similarly committed to trade and investment liberalisation in the goods and services sectors.

The USSFTA is ranked as the most comprehensive that the US has achieved with an ASEAN economy and is considered a 'landmark agreement' for its WTO-plus and NAFTA-plus [North American FTA] commitments' (IES 2007). The provisions on IP are particularly remarkable. As described in Singapore's FTA Network,[6] 'stronger IPR protection set[s] [the] ground for knowledge-based industries' and is a means for securing and maintaining competitive advantage on innovation and capability development in fields such as the creative industries, information technology (IT), pharmaceuticals, science and other high-technology industries.

In the trademarks arena, the key provisions are Singapore's commitment to enhance its trademarks regime to register 'unconventional' or non-visually perceptible marks, such as sound or scent marks, and to accord stronger protection for well-known marks to prevent dilution.[7] On patents, commitments were secured to strengthen existing patent regimes of both countries to protect bio-inventions. Singapore was to accede to the International Convention for the Protection of New Varieties of Plants (UPOV) to provide a system for better protection of new plant varieties,[8] and both countries agreed to maintain the current regimes that would allow all inventions to be patentable,

subject to the condition that they are not contrary to morality or public order. Also crucial was the commitment to limit the use of compulsory licences to safeguard against anti-competitive practices, public non-commercial use, national emergencies and other circumstances of extreme urgency. All commitments relating to trademarks and patents were implemented by Singapore on 1 July 2004.

On copyright, Singapore and the US agreed to align their terms of protection for copyrighted works, performances and phonograms. Thus the term of protection in Singapore, effective 1 July 2004, is the period of the 'life of the author plus seventy years' or, as applicable, 'seventy years after first publication, broadcast or performance'.[9] A further commitment was the adoption of additional protection standards in relation to the digital environment and the World Wide Web. Effective 17 January 2005, Singapore acceded to the WCT and the WIPO Performances and Phonograms Treaty (WPPT). Singapore incorporated substantive anti-circumvention provisions to prohibit tampering with technological protection measures and to prevent piracy of copyrighted works over the Internet in view of activities such as online distribution of software, music and publications. Thus, these unlawful acts now carry both civil and criminal liability, independent of any liability for copyright infringement. Provision was made for immunity for Internet (Network) service providers that comply with notification and take-down procedures when suspected infringing material is hosted, stored or transmitted on or through their servers or networks.

To complement these commitments on enhanced IP standards, specific obligations on IP enforcement were entrenched, including anti-piracy enforcement aligned with closer industry consultation and collaboration. Effective 1 January 2005, Singapore incorporated enhanced criminal sanctions to penalize any entity (whether a business or an individual) not only for copyright-infringing activities wilfully carried out for profit or in a commercial setting, but also where the impact of the activities are significant. Effectively therefore, certain forms of end-user infringement or piracy now amount to criminal offences. Another provision worth mentioning is that Singapore agreed to stronger rights for copyright owners in providing, for example, that in trademark and copyright infringement actions, the owners could, where they find it difficult to calculate the actual damages suffered, opt instead, as provided to owners under the US regime, for the remedy of statutory damages (compensation based on a preset range).

Both Singapore and the US resolved on measures for prevention of and enforcement against illegal manufacture, import and export of counterfeit and pirated goods, and in regard to optical disc manufacturing activities, Singapore committed to formalize its regime of regulating these activities through the imprint of

Source Identification Code on optical discs (unless specifically exempted by the rights' owners) and to criminalize businesses that make pirated copies from legitimately purchased products.[10] Overall, the Singapore perspective on the IPR measures incorporated in the USSFTA is that they are relevant and necessary to 'encourage more R&D and knowledge-intensive activities to be located in Singapore' and raise IP 'protection levels to US standards' (Sen 2004).

It is not easy to assess the substantive social and economic implications of these IP commitments. The underlying principle seems to be that a strong IP regime 'provides an incentive for research and development and a ladder on which industry can climb up the value-chain'. Efforts are ongoing to effectively monitor the social and economic effects of the implementation of these provisions through consultations with industry and all relevant stakeholders. According to commentators, the objective of the monitoring efforts is to secure 'a balance between granting exclusivity and allowing for the free flow of ideas and knowledge sharing' (Koh and Lin 2004, p. 133).

Singapore authorities went through public consultation exercises in formulating legislative amendments to its IP laws. One example is obtaining industry feedback[11] to determine what copyright material should be excluded or exempted by the Minister for Law from the prohibition against circumvention of technological access control measures.[12] This highlights the measures that sophisticated Asia Pacific governments must consider for implementation after committing to far-reaching IP provisions in its FTAs.

There are some general advantages and costs that come hand-in-glove with the new IP standards. Generally, the empirical studies do not indicate that countries that strengthen their IP regimes are likely to 'experience a sudden boost in inflows of foreign investment' although it is generally believed this could stimulate cross-border licensing activity and technology transfers. Policymakers must address the growing concerns that rules relating to copyright term extension, technological protection measures, liability of Internet service providers, end-user criminal liability for copyright piracy and the shift of the burden of proof for copyright infringement cases will endanger the rights of local consumers.

The overarching question has been to what extent these rigorous commitments were imperative to Singapore, the extent to which reviews and benchmark studies had been carried out and/or evaluated on necessity and scope and the extent of local and non-governmental involvement in available reviews during the negotiations. In addition, neighbouring Asia Pacific countries[13] have been confronted by the rigour of the Singapore commitments in their own ongoing negotiations with the US (Endeshaw 2005). Thus, it must be acknowledged that the USSFTA could have a disempowering effect on other policymakers in developing countries in Asia Pacific who are dealing with powers such as the US and the EU which are slow to and effectively need not take it on themselves to evaluate the full impact of rigorous IP provisions.

Asia Pacific countries can take away several lessons from the USSFTA TRIPS-plus obligations specifically in relation to ICT-related areas, copyright protection, software protection, indigenous knowledge and enforcement of intellectual property rights. First, in negotiating IP issues in bilateral agreements, each Asia Pacific country should strive to anticipate provisions that their trading partner will propose and be the one to take the bold stance to table novel rules, related incentives and alternative mechanisms on policy areas where there are clear and significant interests. Some important examples are proposals that relate to data protection and cultural heritage or traditional knowledge.

Second, cognizant of the political stakes in these negotiations, policymakers should take measures to secure decisions that are clearly both open and transparent. This will mean that prospective options on policies and negotiating positions should be made accessible to the public to obtain feedback from relevant key local corporations and industry sectors and reconcile differing viewpoints or concerns via transparent processes.

Third, to secure maximum local understanding and participation on issues, policymakers need to foster public awareness. This may require tasking and collaborating with the local media to send clear and consistent messages on these issues to the general public.

Fourth, policymakers of developing countries in Asia Pacific should diligently consider the fact that enhanced enforcement will translate to additional costs in terms of expenses for budget outlays and training of enforcement officers. They should question whether stronger enforcement of IP rights will take away resources from other development priorities and the extent to which this is acceptable.

Finally, policymakers should assess, (a) the need to promote non-proprietary IP, (b) the need to provide a commons to promote innovation and economic development and (c) the need to provide public subsidies for development of free and open source software.

Multilateralism as the way ahead for developing countries

Member countries of APEC agree that the ongoing increase in the number of FTAs[14] adds impetus to their efforts to liberalize trade and investment throughout the region. However, the APEC Business Advisory Council (ABAC) has rung a warning bell that

Asia Pacific countries need to ensure that these agreements do not compromise the regional trading environment for governments and commercial entities. APEC's specific response has been to propose the development of a range of trade and capacity building moves for the region. We see some of these as laudable, such as developing a best practices guide and an FTA/RTA (Regional Trading Agreement) database, and advocating sharing of negotiating approaches and measures. However, these measures must include a review of possibly inconsistent provisions in multiple FTAs that could impact businesses. Some APEC initiatives worthy of mention that policymakers must ensure result in substantive available material are the APEC Intellectual Property Experts Group (IPEG) projects on Public Education and Awareness, the implementation of the APEC Model Guidelines on Anti-counterfeiting and Piracy and the proposal for sharing experiences in negotiation to promote 'High Quality' FTAs and RTAs.

As APEC works on ICT cooperation to increase the capacity of member countries to reap the benefits of the digital era,[15] policymakers particularly from developing countries should closely monitor action on these proposals to ensure close collaboration on both the IP and the ICT front, to avert the danger of 'all sound and fury, signifying nothing'. We propose that a clear message be sent to business and governments in the region on the significance of the role and reach of IP intertwined with ICT, by incorporating IPR activities under the APEC IPEG in APEC's ICT agenda. In the same vein, the APEC IPR Service Centres set up or to be set up in each country and the IPR Education and Awareness programmes should be merged or closely affiliated with APEC Digital Opportunity Center activities.[16]

The ASEAN IPR Action Plan 2004–10 identifies the key objectives of increasing IP asset creation and commercialization in research, science and technology; harmonizing IPR registration, protection and enforcement in the region; promoting public awareness; and empowering national IP offices to collaborate on development of services to business. Progress appears far from swift, one example being that copyright was included for discussion as a specific form of IPR only as recently as 2003. Thus, a review of issues related to the digital environment and ICT and how cultural copyright may better be protected is at an early stage. For a start, policymakers could address a fundamental interplay between IP and ICT by closely reviewing the ICT infrastructure and facilities within each national IP office.[17] Taking steps to enhance what is available within local IP registration regimes would boost ongoing promises to the public that the country will work towards harmonization of IP laws, regional IP registration and business development services to allow for genuine benefits for investors within the region.

Over the last few years, ASEAN has taken on the task of negotiating with other countries, including those in Asia Pacific such as India and China. Policymakers should seriously consider the value of throwing in their lot wholeheartedly with regional groups such as ASEAN or APEC, on the premise that the cumulative negotiation process in any trade agreement with giants such as the US and EU is likely to result in commitments that are of general value and impact, unlike the often overly rigorous IP provisions of bilateral FTAs.

Conclusion

This chapter has sought to highlight some of the key issues with respect to IP and ICT that policymakers in Asia Pacific should bear in mind. The primary policy consideration should emanate from a public interest approach to IP. Treating IP merely as a matter of private property or private interest rights can be seriously detrimental to access to knowledge, culture and technology.

Moreover, policymakers in Asia Pacific must take a close and hard look at moves in their direction that seek to bait them into taking on TRIPS-plus commitments under any guise, whether via a bilateral agreement or a regional multilateral agreement. The overall interests of their public and national social, economic and developmental goals must be carefully guarded and decisions made only after detailed reviews and analytical studies that would enable a reasonable assessment of the necessity and scope of any potential commitment in the fields of IP and ICT.

Notes

1. For more information on the WIPO Development agenda, see http://www.cptech.org/ip/wipo/da.html
2. In 1994, China accounted for 4.9 per cent of global R&D expenditure, India and Central Asia for 2.2 per cent, Latin America for 1.9 per cent, the Pacific and Southeast Asia 0.9 per cent (excluding newly industrialised countries) and sub-Saharan Africa 0.5 per cent (UNESCO 1998).
3. OECD R&D Expenditure in 1999 was USD 553 billion (OECD 2001) while India's national income was USD 440 billion (World Bank Data).
4. A number of activists and scholars have argued that we need to avoid using the phrase 'intellectual property' since it conceals more than it reveals. The phrase covers a range of property claims—trademarks, copyright, patents, geographical indications, etc.—all of which belong to distinct domains. We acknowledge this to be a serious question, and use the phrase in reference to its global usage but with a certain degree of agnosticism.

5. The A2K treaty is a multilateral treaty initiated by Consumer Project on Technology (Cptech) and it attempts to carve out global exceptions to copyright for education and other uses. For a full text of the draft treaty, see http://www.cptech.org/a2k/a2k_treaty_may9.pdf
6. See the FTA Network website at http://www.iesingapore.gov.sg/wps/portal
7. Section 55(4)(b)(ii), *Trade Marks Act, Cap 332, 1999 Rev. Ed.* (available at http://statutes.agc.gov.sg). This provision provides a remedy to proprietors of well-known marks against persons who have business identifiers that are either identical or have an essential part that is identical to the well-known mark. The remedy is available where there has been dilution in an unfair manner of the distinctive character of the well-known mark, or where the said business identifier would take unfair advantage of the distinctive character of the well-known mark.
8. Singapore's obligations as a signatory of the UPOV 1991 Convention have been enacted into law as the *Plant Variety Protection Act 2004, Act 22 of 2004* (available at http://statutes.agc.gov.sg).
9. For the duration of copyright where the copyright subsists in a literary, dramatic or musical work, or in an artistic work other than a photograph, see Section 28(2), *Copyright Act, Cap 63, 1999 Rev. Ed.* For the duration of copyright in other works like sound recordings, cinematograph films, television broadcasts and sound broadcasts, cable programmes and published editions of works, see Sections 92 to 96, *Copyright Act* (available at http://statutes.agc.gov.sg).
10. For provisions on optical disc manufacturing, please see *Manufacture of Optical Discs Act 2004, Act 25 of 2004* (available at http://statutes.agc.gov.sg).
11. Feedback was obtained from copyright owners, educational institutions, archives, scholars, researchers and the general public.
12. For example, the Copyright (Excluded Works) Order 2005, which excludes any literary work in eBook format and for which a technological access control measure was applied to all editions including digital text editions made available by an institution assisting handicapped readers.
13. For example, Thailand and Malaysia.
14. APEC trade agreements include those between Singapore and New Zealand, Singapore and Japan, Singapore and the United States, Singapore and Australia, Chile and five other APEC Member Economies, and the European Free Trade Association and China and Hong Kong. Thailand has signed FTAs with Laos and Australia, while ASEAN signed an agreement with India.
15. See the APEC Action Agenda for the New Economy (2000) and the e-APEC Strategy (2001) for the region to have community-based access to the Internet by 2010 and to increase learning and employment opportunities, improve public services and promote universal for ICT and information services.
16. Set up by Chinese Taipei in 2000, it has led camps and workshops on technology.
17. Examples of IP offices that are state-of-art are those of Singapore, Japan, Korea, Chinese Taipei, Thailand and Malaysia. Such an IP office is also being created in Brunei Darussalam.

References

Boyle, J. (2002) Fencing off ideas. *DAEDALUS*, Spring, p. 13.

CIPR. (2002). *Integrating intellectual property rights and development policy*. Report of the Commission on Intellectual Property Rights (CIPR). Retrieved from http://www.iprcommission.org/graphic/documents/final_report.htm

Copy South Research Group. (2006). *The Copy/South Dossier: Issues in the economics, politics, and ideology of copyright in the global South*. Retrieved from www.copysouth.org

Correa, C. (1997). *Intellectual property rights, the WTO and developing countries: The TRIPS Agreement and policy options*. London: Zed Books.

Drahos, P. and Braithwaite, J. (2002). *Information feudalism*. New York: Norton & Company.

Endeshaw, A. (2005). IP enforcement in Asia: A reality check. *International Journal of Law and Information Technology*, 13(3), 378–412.

Fischer, W. and McGeveran, W. (2006). *The digital learning challenge: Obstacles to educational uses of copyrighted material in the digital age*. The Berkman Center for Internet & Society Research Publication Series. Retrieved from http://cyber.law.harvard.edu/home/2006-09

Garlick, M.K. (2004). Locking up the bridge on the digital divide—A consideration of the global impact of the U.S. anti-circumvention measures for the participation of developing countries in the digital economy. *Santa Clara Computer and High Technology Journal*, May.

Ghosh, R.A. (2003). Licence fees and GDP per capita: The case for open source in developing countries. *First Monday*, 8(12). Retrieved from http://www.firstmonday.org/issues/issue8_12/ghosh/index.html

Hahn, R. (ed.). (2002). *Government policy toward open source software*. AEI-Brookings Joint Center for Regulatory Studies. Retrieved from http://www.aei.brookings.org/publications/abstract.php?pid=296

International Enterprise Singapore (IES). (2007). *USSFTA*. Retrieved from http://www.iesingapore.gov.sg/wps/portal

Kerscher, G. and Fruchterman, J. (2002). The soundproof book: Exploration of rights conflict and access to commercial ebooks for people with disabilities. *First Monday*, 7(6). Retrieved from http://firstmonday.org/issues/issue7_6/kerscher/index.html

Koh, T. and Lin, C.L. (eds.). (2004). *The United States Singapore Free Trade Agreements Highlights and Insights*. Institute of Policy Studies, Singapore and World Scientific Publishing Co. Pte Ltd.

Kumar, N. (2003). Intellectual property rights, technology and economic development: Experiences of Asian countries. *Economic and Political Weekly*, 18 January.

OECD. (2001). *OECD science, technology and industry scoreboard 2001—Towards a knowledge-based economy.* Paris. Retrieved from http://www1.oecd.org/publications/e-book/92-2001-04-1-2987/A.2.htm

Rifkin, J. (2000). *The age of access*. London: Putnam Publishing Group.

Ryan, M. (1999). Fair use and academic expression: Rhetoric, reality, and restriction on academic freedom. *The Cornell Journal on Law Public Policy*.

Sen, R. (2004). *Free Trade Agreements in Southeast Asia*. ISEAS Publications.

Stallman, R. (2005). *Bill Gates and other Communists*. C|Net news.com. Retrieved from http://news.com.com/Bill+Gates+and+other+communists/2010-1071_3-5576230.html

Thomas, P.N. (2006). Uncommon futures: Interpreting IP conflicts in India. In Thomas, P. N. and Servaes, J. (eds.), *Intellectual property rights and communications in Asia: Conflicting traditions*. New Delhi: Sage Publications.

Thomas, P.N. and Servaes, J. (2006). *Intellectual property rights and communications in Asia: Conflicting traditions*. New Delhi: Sage Publications.

UNESCO. (1998). *World science report 1998*. Geneva. Retrieved from http://unesdoc.unesco.org/images/0011/001126/112616eb.pdf

Wong, K. (2004). Free/Open Source Software: Government and policy. Kuala Lumpur: UNDP-APDIP. Retrieved from http://www.iosn.net/government/foss-government-primer/foss_gov_primer_v0_2.pdf

Wong, K. and Sayo, P. (2004). FOSS: A General Introduction. Kuala Lumpur: UNDP-APDIP. Retrieved from http://www.iosn.net/foss/foss-general-primer/foss_primer_print_covers.pdf

World Bank. (2001). *Global economic prospects and the developing countries 2002: Making trade work for the world's poor*. Washington, DC. Retrieved from http://www.worldbank.org/data/databytopic/GDP.pdf

World Bank Data. Retrieved from http://www.developmentgoals.org/Data.htm

State and evolution of ICTs: A tale of two Asias

George **Sciadas**

This section provides a quantitative overview of the state and relative progress of information and communication technologies (ICTs) in the Asia Pacific region, and it is intended to complement the work in this volume, including the individual country chapters. It must be borne in mind that the figures on ICT diffusion and use are affected significantly, if not determined outright, by the multitude of developments concerning the new technologies or their applications, government policies or their absence, regulation and business initiatives described in this publication. Therefore, the analysis that follows aspires to offer a realistic perspective against which to assess the combined reach and effects of these developments as they are reflected in the figures at any given time.

Following the merger of the work of Orbicom and the International Telecommunications Union (ITU) in the area of international benchmarking for the information society,

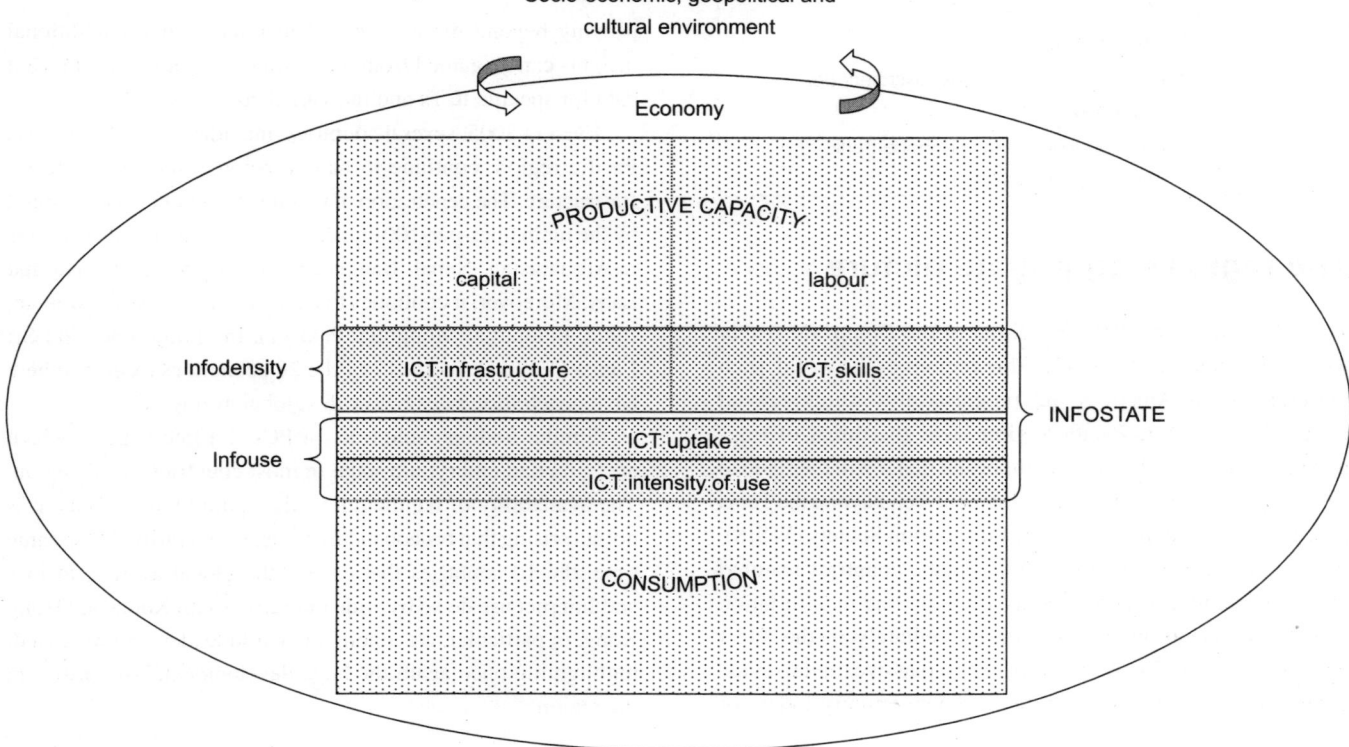

this quantitative analysis is based on ITU's *ICT Opportunity Index* (ITU 2007) and it uses Orbicom's Infostate conceptual framework, which is based on the quantifiable notions of Infodensity and Infouse. This framework enables analysis both across and within countries over time, as well as the monitoring of progress with regard to specific ICTs (Orbicom 2003, 2005).

The ICT Opportunity Index is the aggregation of the following components and indicators:

ICT Opportunity Index	
	Indicators
Infodensity	**Networks**
	Main telephone lines per 100 inhabitants
	Mobile cellular subscribers per 100 inhabitants
	International Internet bandwidth (kbs per inhabitant)
	Skills
	Adult literacy rates
	Gross enrolment rates
	primary
	secondary
	tertiary
Infouse	**ICT uptake**
	Internet users per 100 inhabitants
	Proportion of households with a TV
	Computers per 100 inhabitants
	Intensity of use
	Total broadband Internet subscribers per 100 inhabitants
	International outgoing telephone traffic (minutes) per capita

The region's aggregate picture

The Foreword to the 2005/06 edition of the *Digital Review of Asia Pacific* notes that '[w]hile other regions of the world, such as Europe and the Americas, shift progressively towards regional integration, the Asia-Pacific region faces the threat of fragmentation. This challenge is so important that it will continue to be present in the dynamics of development well beyond the Tunis phase of the World Summit on the Information Society (WSIS)' (p. ix). Indeed, if the group of Asia Pacific countries featured in this publication is perceived as a region, one manifestation of the fragmentation can be seen immediately in the latest published figures. There continues to be a massive digital divide *within* the region as shown by the ICT Opportunity Index for 2005 (Figure 1). Economies such as Hong Kong, Singapore, Australia, Taiwan, Macau and South Korea are not only at the top of the scale for the region, but also among the top countries worldwide—together with Scandinavian, North American and Western European nations. They help pull the regional average higher than the global. Some countries in Asia Pacific form a second tier, with Brunei and Malaysia above the global average and China, Thailand and the Maldives somewhat behind. At the other extreme, Afghanistan, Myanmar, Nepal, Cambodia, Bangladesh, Laos and Pakistan are at the bottom, both in the region and internationally, together with many African states. Afghanistan has been facing extraordinary circumstances and challenges for some time now, but all the other countries have their own unique stories as well. In any case, the digital gaps in these countries are among the largest in the world.

The magnitude of the gaps among the Asia Pacific economies becomes even more pronounced when we focus on the 'networks' component of the overall ICT Opportunity Index (Figure 2). The divide clearly intensifies, with the top countries achieving higher values and the countries at the bottom assuming lower values. Only minor differences are observed in the composition compared to the overall index, such that Malaysia is now below the global average. This underscores the close relationship between the available ICT infrastructure in the country and the uptake and use of ICTs.

Specific ICTs

Moving beyond the aggregate benchmarks, many additional insights can be gained from the examination of the most recent data for specific ICTs and individual countries.

Even in 2005, several countries, including some in the Asia Pacific region, continue to have precious little fixed-line infrastructure. The penetration of wireline telephones, as measured by the main lines-per-100-inhabitants indicator, barely registers in Afghanistan, Nepal, Bangladesh, Cambodia, Myanmar and Laos. Moreover, it continues to be at very low levels in Bhutan, India, Indonesia, Mongolia, Pakistan, the Philippines and Sri Lanka (Table 1). In all, 17 of the 28 economies examined here have penetration rates below the global average.

Similar findings hold true for PCs and Internet use, which are still at an embryonic stage in most countries of the region. The availability of bandwidth is also quite limited. Thus, it is not surprising that broadband Internet use is a rarity. At the same time, the 10 countries that exceed the global average in PCs and Internet use do so by wide margins. South Korea and Hong Kong, in particular, are among the world leaders in broadband, highlighting once again the huge developmental disparities in the region.

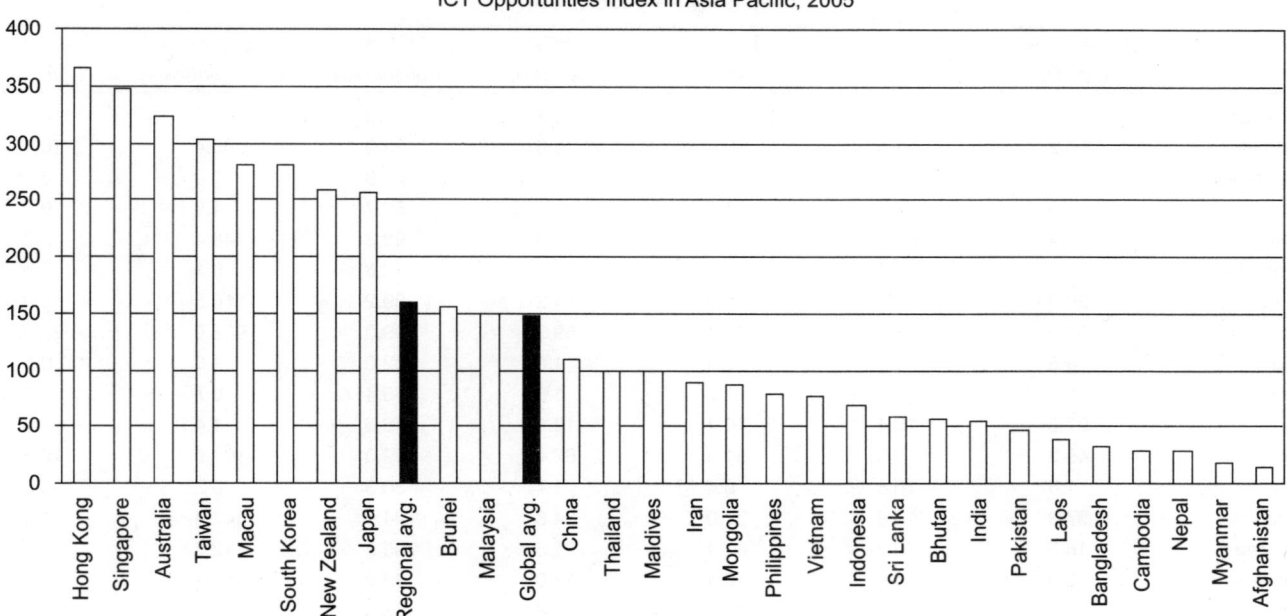

Figure 1
ICT Opportunties Index in Asia Pacific, 2005

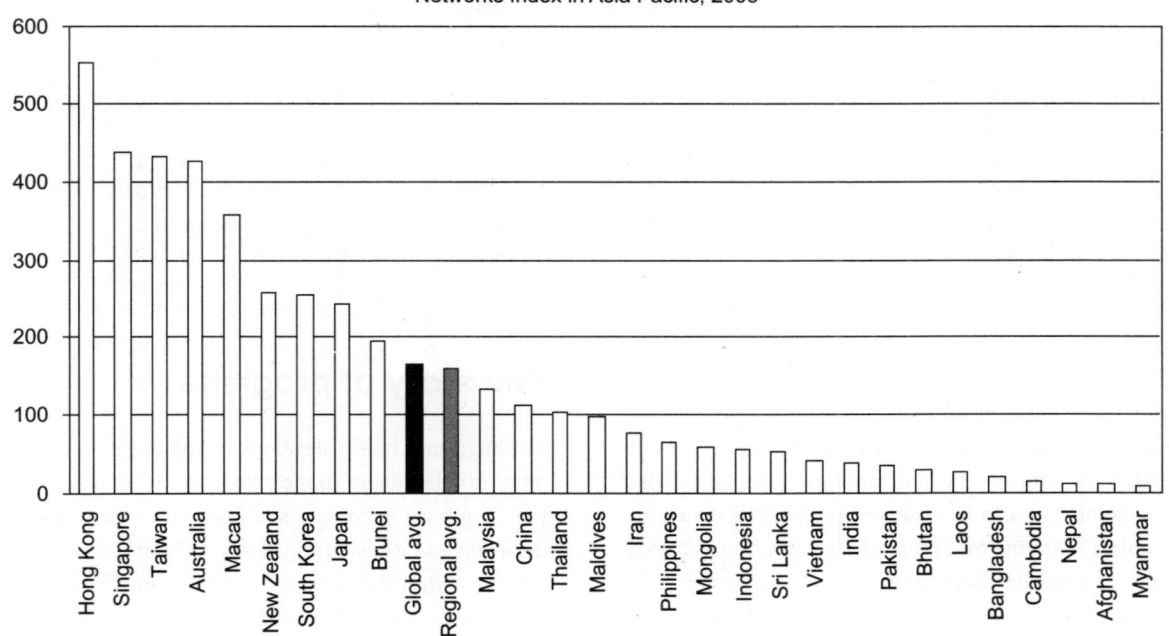

Figure 2
Networks Index in Asia Pacific, 2005

In a few countries, even the diffusion of TVs among households is quite low, both by regional and global standards. (On average, about two in three households in the world have a TV set.) Myanmar and Afghanistan have particularly low TV penetrations among households, at 3 per cent and 6.3 per cent, respectively, while Nepal fares only somewhat better at 13.2 per cent. In Bangladesh, fewer than one in four households have a TV, while the corresponding figure in India and Sri Lanka is less than one TV in three households.

The situation is comparatively much better and more promising when it comes to cellphones. It is by now well documented that the cellphone represents the bright spot in the

Table 1
ICTs in Asia Pacific, 2005

	Main lines (per 100)	Cell phones (per 100)	Internet (per 100)	PCs (per 100)	TVs (% of households)	Bandwidth (kbs/inhabitant)	Broadband (per 100)
Afghanistan	0.3	4.0	0.0	0.1	6.3	0.0	0.0
Australia	50.2	91.4	70.4	76.6	99.0	595.4	10.4
Bangladesh	0.8	6.3	0.3	1.6	22.9	0.0	0.0
Bhutan	3.9	4.6	3.0	1.6	57.7	1.2	0.0
Brunei	22.4	62.3	36.1	8.8	98.3	148.4	2.2
Cambodia	0.2	7.5	0.3	0.3	42.8	0.1	0.0
China	26.6	29.9	8.6	4.2	89.2	10.3	2.9
Hong Kong	53.9	123.5	50.1	59.3	99.0	932.0	23.6
India	4.5	8.2	5.4	1.5	32.0	1.8	0.1
Indonesia	5.7	21.1	7.2	1.5	65.4	0.7	0.0
Iran	27.3	10.4	10.1	12.5	76.6	1.4	0.0
Japan	45.3	75.3	51.5	67.4	99.0	103.5	17.5
Laos	1.3	10.8	0.4	1.7	30.3	0.3	0.0
Macau	37.9	115.8	37.0	34.8	94.0	347.8	14.8
Malaysia	16.8	75.2	42.4	21.5	95.2	12.3	1.9
Maldives	8.6	53.9	5.4	12.0	92.0	14.1	0.9
Mongolia	5.9	21.1	10.1	12.8	63.0	1.5	0.1
Myanmar	0.9	0.3	0.1	0.7	3.0	0.2	0.0
Nepal	1.8	0.8	0.8	0.5	13.2	0.2	0.0
New Zealand	42.9	87.6	68.4	51.6	98.0	113.6	8.2
Pakistan	3.4	8.3	6.8	0.5	46.5	0.5	0.0
Philippines	4.0	41.3	5.5	5.4	63.1	3.8	0.1
Singapore	42.4	100.8	40.2	93.3	98.6	703.7	15.3
South Korea	49.2	79.4	68.4	53.2	99.0	103.0	24.8
Sri Lanka	6.0	16.2	1.3	3.5	31.6	2.4	0.1
Taiwan	59.8	97.4	58.0	57.5	99.0	478.5	19.1
Thailand	11.0	53.3	11.0	6.9	91.9	10.6	0.2
Vietnam	18.8	11.4	12.7	1.4	82.8	4.3	0.2
Global avg.	19.6	45.4	18.2	16.0	65.5	133.2	3.7
Regional avg.	19.7	43.5	21.8	21.2	67.5	128.3	5.0

ICTs-for-development scene. It has made the most inroads among poor populations compared to other ICTs and its penetration has surpassed the availability of fixed telephone lines in most countries. Moreover, this is true both among developed and developing economies, as can be observed in the 2005 data contained in Table 1, with the exception of Iran, Myanmar, Nepal and Vietnam. In some developing countries in particular, cellphones are the only significant telephony option (Bangladesh, Cambodia and Laos), while in others they exceed fixed-lines by a sizeable factor.[1] This is the case in Brunei, Indonesia, Malaysia, the Maldives, Mongolia, Thailand and Sri Lanka, and perhaps nowhere more dramatic than in the Philippines. However, in Myanmar and Nepal, even the cellphone still did not amount to much in 2005 (at a time when in Hong Kong, Macau and Singapore there were more cellphones than people).

The story of progress

The huge gaps in ICT development among countries accentuate other existing gaps. To the extent that ICTs represent powerful tools for development, it is of policy interest to know how their diffusion and use is progressing. The data allow us to examine growth within the region, as well as in a comparative sense vis-à-vis the rest of the world. Aggregate growth, as captured by the ICT Opportunity Index, is shown in Figure 3.

Clearly, much of the progress made in recent years worldwide is due to cellphones and the Internet. In the 2001–05 period, cellphones more than doubled and Internet use nearly doubled. The penetration of PCs increased by 60 per cent, while that of TV increased only marginally, which is not surprising given that in most countries TV reached saturation levels some time ago. On the aggregate, penetration of main lines was stagnant. These trends can be seen in the evolution of the global average (Figure 4). Moreover, bandwidth increased and significant progress was made in the deployment of broadband (virtually non-existent in 2001), as the focus has shifted there for value-added applications.

While similar movements to those encountered globally characterize Asia Pacific economies, the interplay between

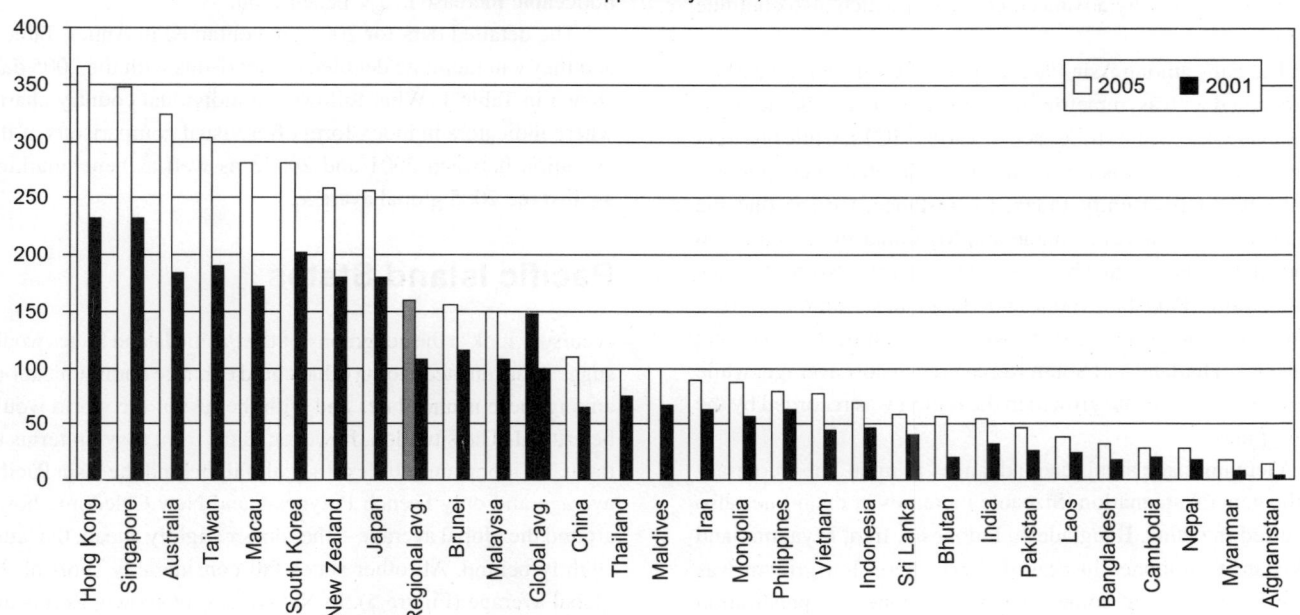

Figure 3
Evolution of ICT Opportunity Index, Asia Pacific

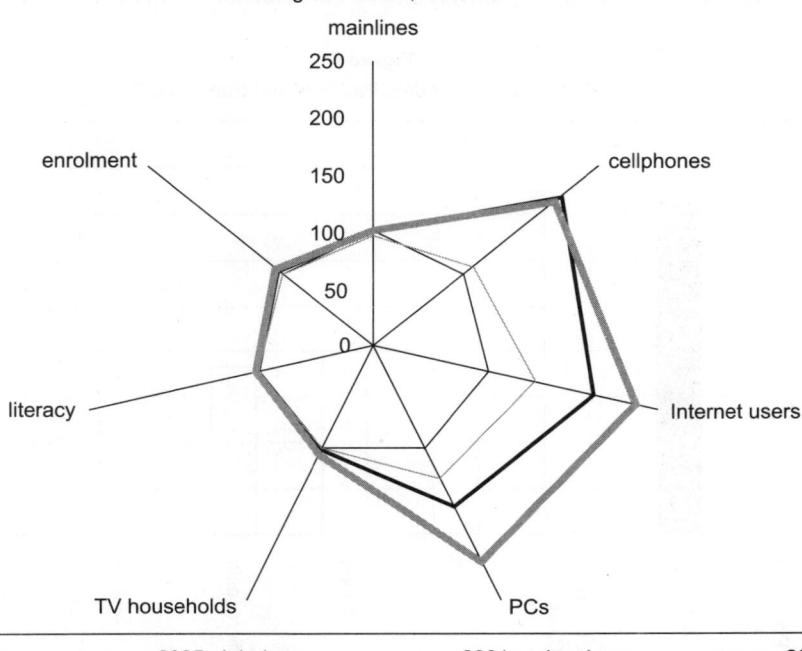

Figure 4
Average Evolution, 2001–05

levels and growth of ICTs must always be placed in its proper perspective. As explained in previous studies (Orbicom 2003, 2005), typically developing countries have much higher growth rates than advanced countries. This is largely the result of insignificant initial levels of ICTs and is not necessarily indicative of catching up—and therefore of a closing divide—as the bar is set higher and higher by advanced countries, which also continue to progress.

The gaps among Asia Pacific economies, discussed earlier, are coupled with asymmetries in the progress made. Some countries made significant strides across various ICTs, while progress in other countries was for the most part limited to cellphones. In the ICT Opportunity Index, for instance, we see that the values for Afghanistan, Bhutan and Myanmar increased much more than others, but above-average growth also took place in Australia, Pakistan, India and Bangladesh, among others. Cambodia and Nepal had growth approaching the average; Malaysia, Thailand and South Korea grew below average; while the smallest aggregate growth in the region was recorded by the Philippines.

Vietnam made significant advances in main lines, contrary to the general stagnation. Sizeable increases in main lines also occurred in China, Bangladesh, Indonesia, Iran, Myanmar and Pakistan. Cellphones increased everywhere, but growth was more spectacular in some countries. Indonesia's penetration rate for instance jumped from 3.1 per cent to 21.1 per cent between 2001 and 2005, while marked increases were also noted in Laos, India and Pakistan (from around 0.5 per cent to 10.8 per cent, 8.2 per cent and 8.3 per cent, respectively).

Cellphones increased the least among countries with very high penetration levels (for example, Taiwan, Japan, South Korea and Singapore). Vietnam experienced impressive growth in Internet use (from 1.3 per cent to 11.7 per cent), while considerable gains were also made by Pakistan, Mongolia and India. Bangladesh and Mongolia led the growth in PCs, whereas Sri Lanka had a noticeable increase in TV penetration.

The detailed data for 2001 are contained in Annex Table 1 and they can facilitate detailed comparisons with the 2005 data shown in Table 1. What follows are individual country charts, where indicators in index form offer visual comparisons of the evolution between 2001 and 2005,[2] as well as benchmarking against the 2005 global average.

Pacific Island States

A cursory look at the geography of the Pacific Island States would suggest that any technology that could enhance communications among these island states and with the rest of the world would be critical. The situation, however, is far from rosy. In terms of the ICT Opportunity Index, they all fall behind the Asia Pacific average and only French Polynesia and New Caledonia hover around the global average—the former slightly ahead, the latter slightly behind. All other states fall considerably short of the global average (Figure 5). In Networks (not shown, as it is not fundamentally different), all countries are below the regional and global averages.

Among the island states in this group, New Caledonia leads in main lines, cellphones (with a much higher penetration than

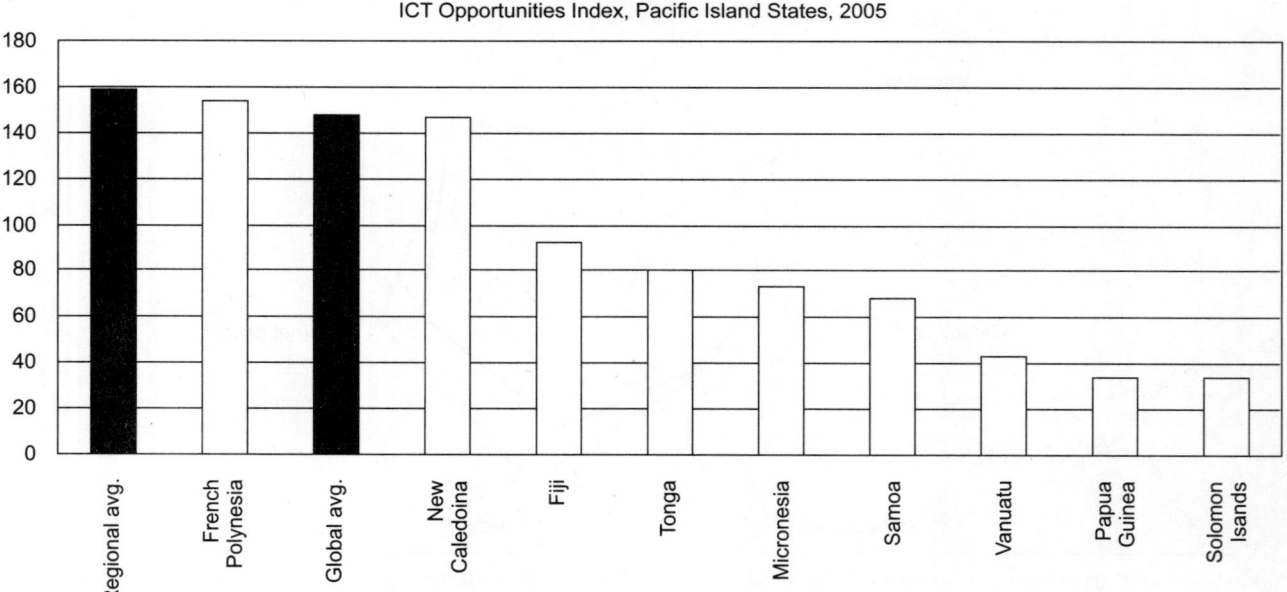

Figure 5
ICT Opportunities Index, Pacific Island States, 2005

the others at 56.7 per cent—up from 31 per cent in 2001) and Internet penetration (Table 2). In terms of levels, main lines are relatively high also in French Polynesia, Tonga and Fiji. Cellphone penetration is once again the success story as it has taken off more than any other ICT, most dramatically in Tonga, which from near zero in 2001 achieved a penetration rate approaching 30 per cent by 2005. Big increases were also recorded in Samoa and Micronesia from near-zero levels. French Polynesia leads in PCs, albeit with a modest 10.9 per cent penetration—significantly lower than the global average. American Samoa and French Polynesia have very high penetration of TVs. French Polynesia and New Caledonia are also the only countries with some broadband—although bandwidth in the islands is very low compared to the global average.

Table 2
ICTs in Pacific Island States, 2005

	Main lines (per 100)	Cellphones (per 100)	Internet (per 100)	PCs (per 100)	TVs (% of households)	Broadband (per 100)	Bandwidth (kbs/inhabitant)
Fiji	13.3	24.2	8.3	5.9	60.0	8.5	0.8
French Polynesia	20.9	34.0	21.5	10.9	92.1	51.6	4.3
Micronesia	11.2	12.7	12.6	5.4	15.1	5.4	0.0
New Caledonia	23.3	56.7	32.1	2.5	78.0	55.7	4.1
Papua New Guinea	1.1	1.3	2.3	6.6	10.0	0.1	0.0
Samoa	10.5	13.0	3.2	1.9	98.0	4.9	0.0
Solomon Islands	1.5	1.3	0.8	4.6	4.1	1.7	0.1
Tonga	13.7	29.8	3.0	6.0	26.3	2.0	0.6
Vanuatu	3.2	5.8	3.5	1.4	6.0	2.3	0.0
Regional avg.	19.7	43.5	21.8	21.2	67.5	5.0	128.3
Global avg.	19.6	45.4	18.2	16.0	65.5	3.7	133.2
American Samoa	18.2						
Kiribati	5.1						
Marshall Islands	8.3						

Notes

1. It is not only that main lines generally do not increase much anymore, but also that in many cases they are decreasing in number even in advanced countries, mainly due to substitution of cellphones for fixed-lines.
2. The indicators used are those of the ICT Opportunity Index, with the exception of bandwidth, broadband and international outgoing traffic which, methodologically, are subject to a monotonic transformation and are not conducive to indices.

References

ITU. (2007). *Measuring the information society 2007*. ICT Opportunity Index and World Telecommunications/ICT Indicators, Geneva.

Orbicom. (2003). *Monitoring the digital divide...and beyond*. National Research Council of Canada.

Orbicom. (2005). *From the digital divide to digital opportunities: Measuring Infostates for development*. National Research Council of Canada.

Saik Yoon, C. (ed.) (2005). *Digital review of Asia Pacific 2005/2006*.

Annex Table 1
ICTs in Asia Pacific, 2001

	Main lines (per 100)	Cell phones (per 100)	Internet (per 100)	PCs (per 100)	TVs (% of households)	Bandwidth (kbs/inhabitant)	Broadband (per 100)
Afghanistan	0.1	0.0	0.0	0.0	6.3	0.0	0.0
Australia	51.8	57.3	39.7	51.5	96.3	36.2	0.6
Bangladesh	0.4	0.4	0.1	0.2	17.2	0.0	0.0
Bhutan	2.6	0.0	0.7	1.0	57.7	0.3	0.0
Brunei	25.9	41.8	12.9	7.3	98.3	17.5	0.6
Cambodia	0.2	1.7	0.1	0.1	40.0	0.0	0.0
China	14.1	11.3	2.6	1.9	87.3	0.6	0.0
Hong Kong	58.0	85.9	38.7	38.7	99.0	106.7	4.2
India	3.7	0.6	0.7	0.6	31.6	0.1	0.0
Indonesia	3.5	3.1	2.0	1.1	54.5	0.2	0.0
Iran	16.9	3.2	1.6	7.0	67.4	0.2	0.0
Japan	48.2	58.8	38.4	35.8	99.0	17.8	3.0
Laos	1.0	0.5	0.2	0.3	29.9	0.0	0.0
Macau	40.4	44.5	23.1	18.3	93.0	27.5	2.2
Malaysia	19.7	30.9	26.6	12.6	85.9	3.1	0.0
Maldives	9.9	6.9	3.6	5.4	61.5	1.6	0.0
Mongolia	5.2	8.1	1.7	1.7	49.0	0.4	0.0
Myanmar	0.6	0.0	0.0	0.3	3.0	0.0	0.0
Nepal	1.3	0.1	0.3	0.4	13.2	0.0	0.0
New Zealand	47.0	59.0	45.4	38.7	98.1	49.0	0.4
Pakistan	2.3	0.5	0.4	0.4	38.5	0.2	0.0
Philippines	4.2	15.5	2.6	2.2	61.5	0.3	0.0
Singapore	47.1	72.4	41.2	50.8	99.4	63.9	3.7
South Korea	54.4	61.4	51.5	47.5	99.0	20.7	4.2
Sri Lanka	4.4	3.6	0.8	0.9	20.6	0.1	0.0
Taiwan	57.3	97.2	34.9	36.4	97.8	32.3	4.2
Thailand	9.8	12.2	5.7	3.2	90.6	1.0	0.0
Vietnam	3.8	1.6	1.3	0.9	79.6	0.0	0.0
Global avg.	19.6	21.8	9.5	10.0	63.0	30.4	0.4
Regional avg.	19.1	24.2	13.4	13.0	63.4	13.6	0.8

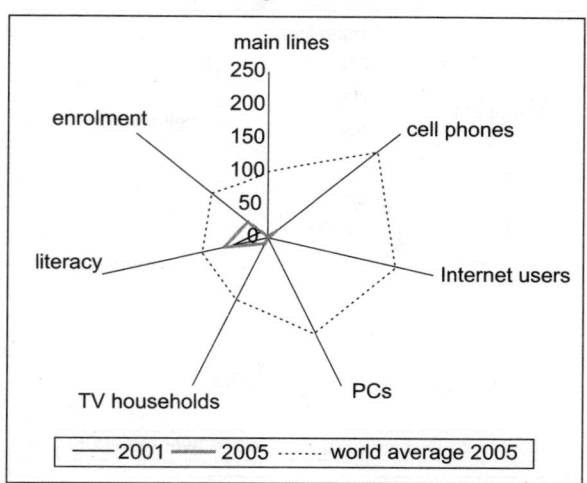

Afghanistan

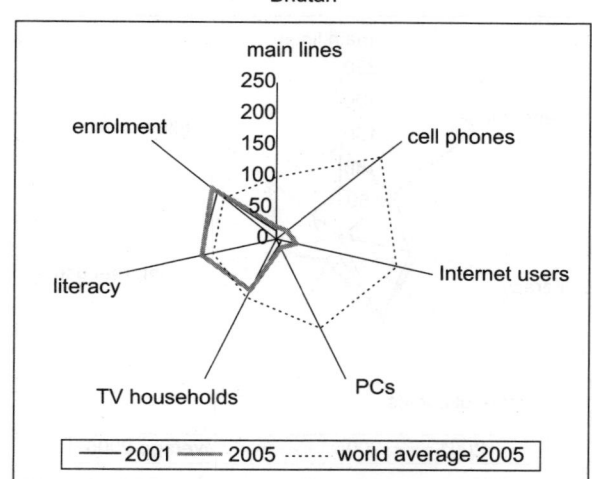

Bhutan

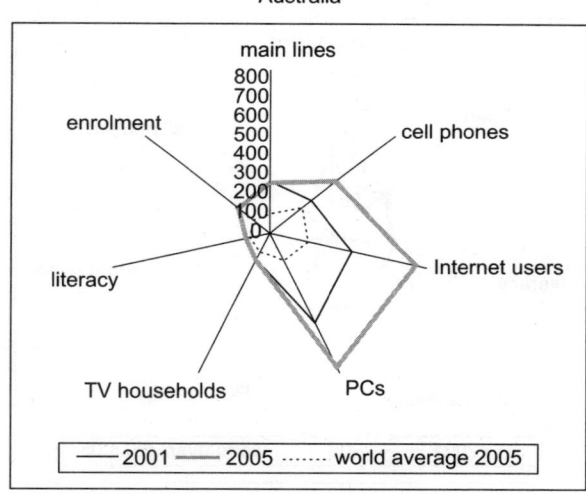

Australia

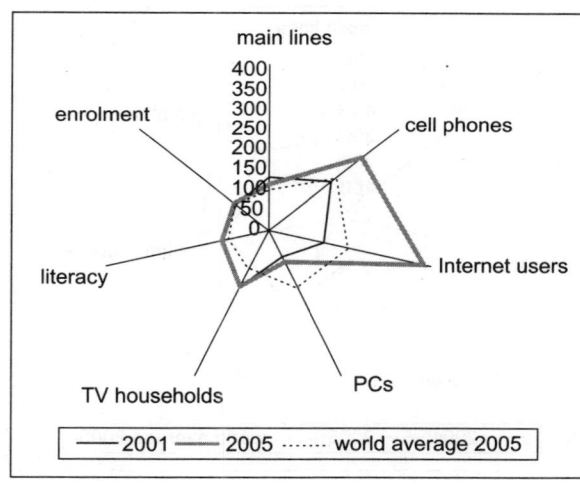

Brunei

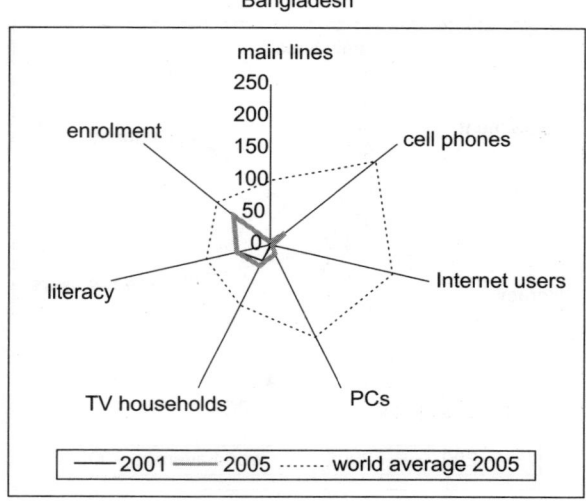

Bangladesh

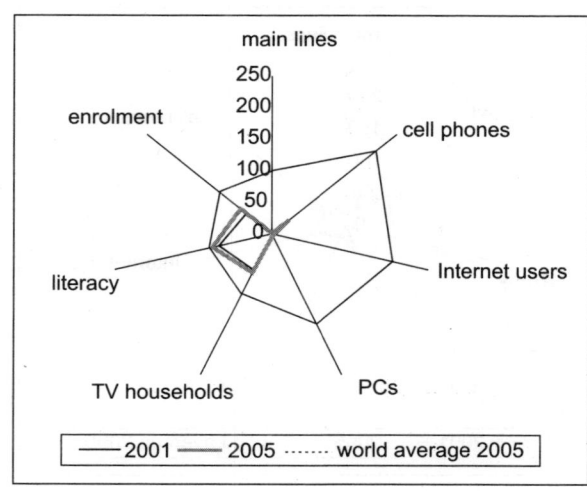

Cambodia

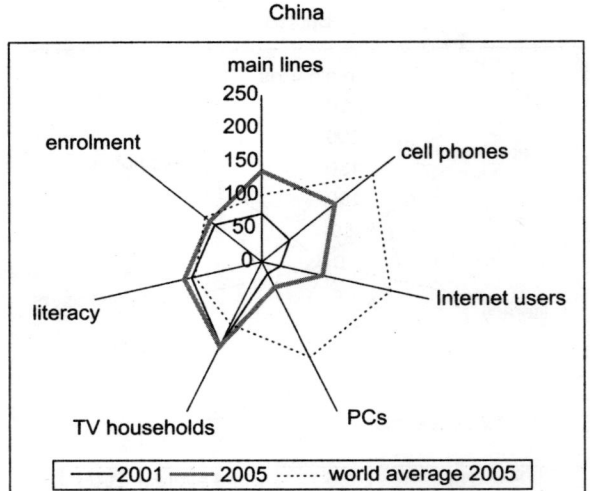
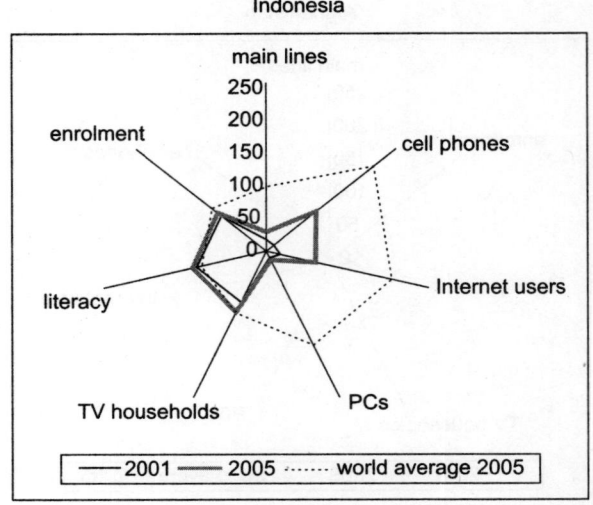
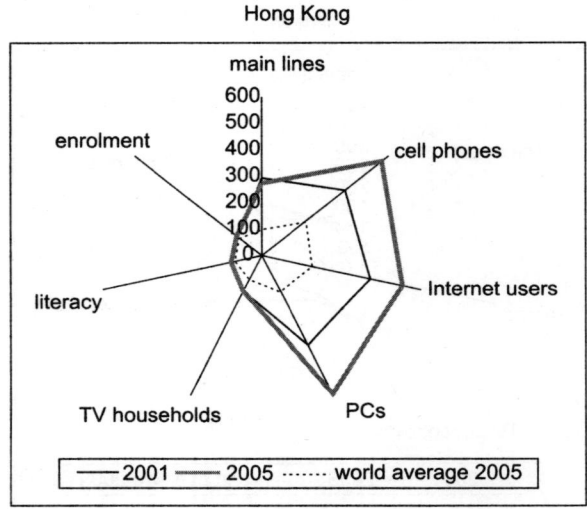
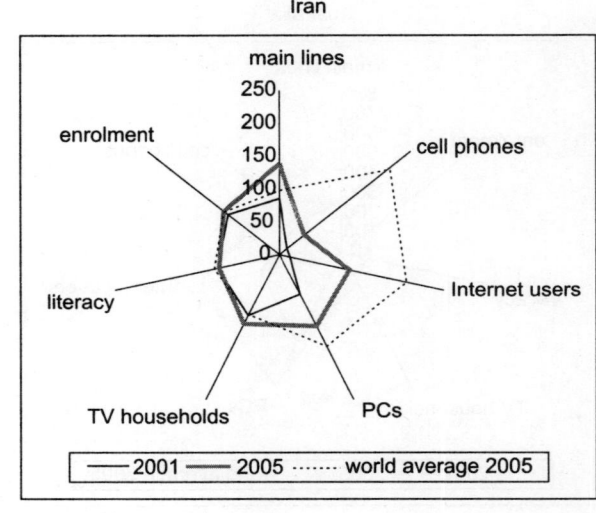
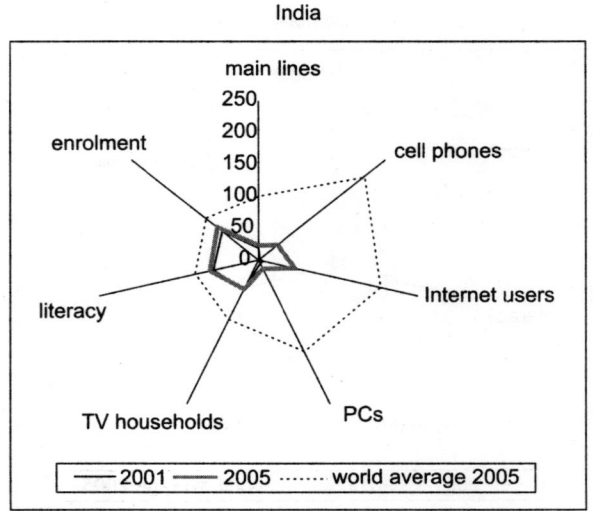
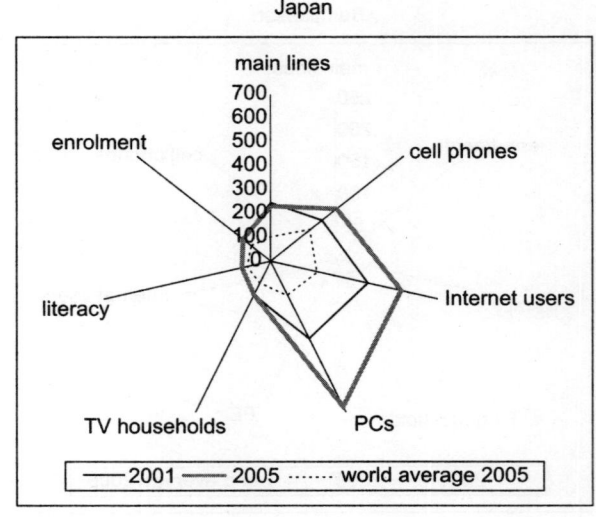

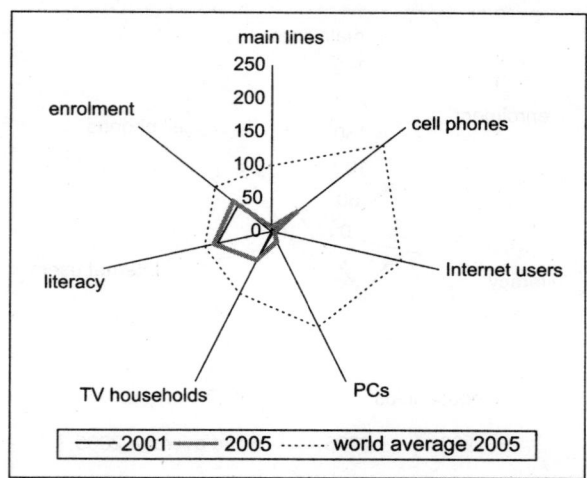
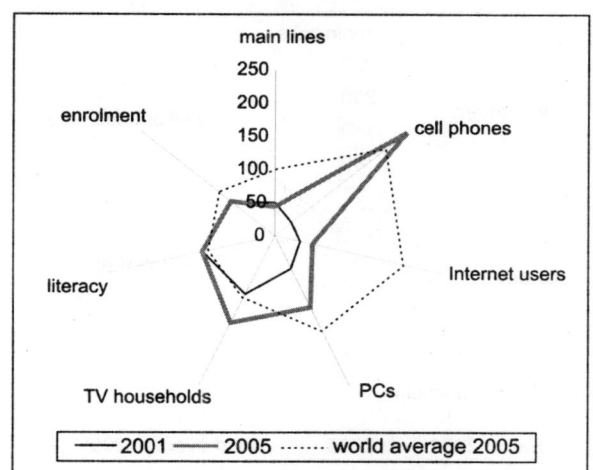
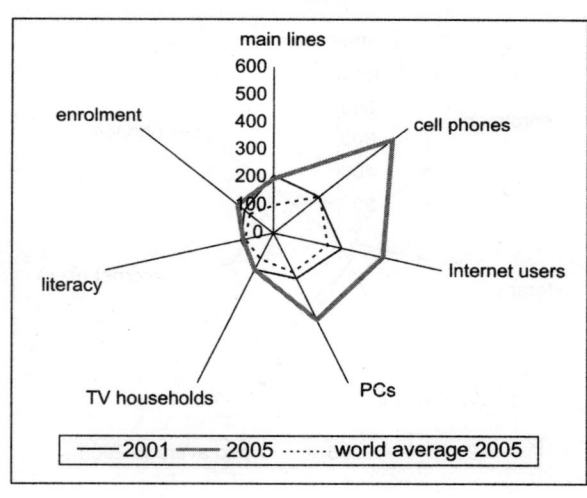
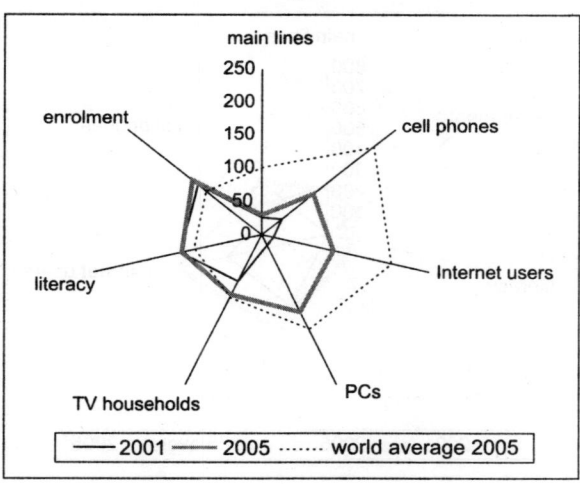
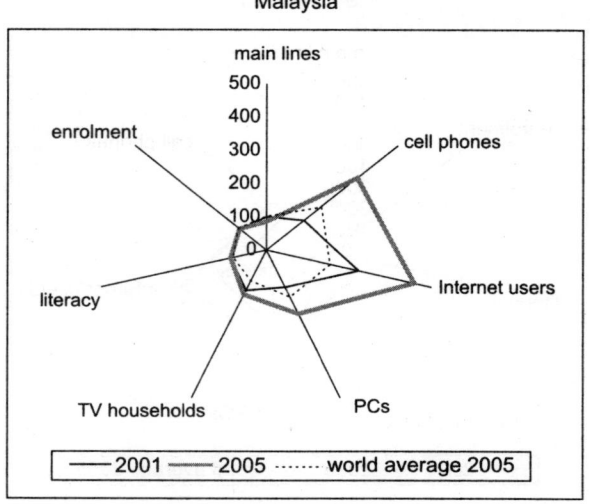
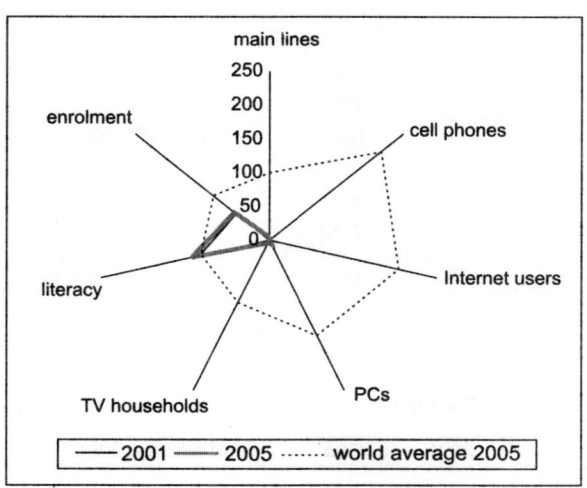

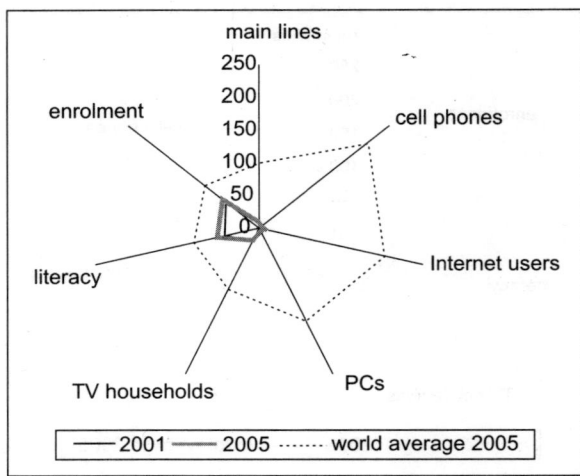
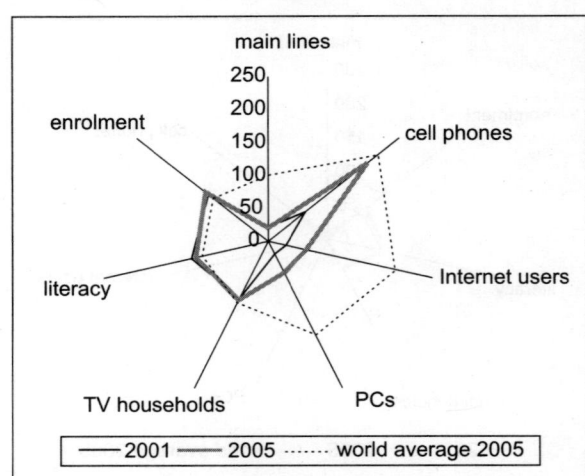
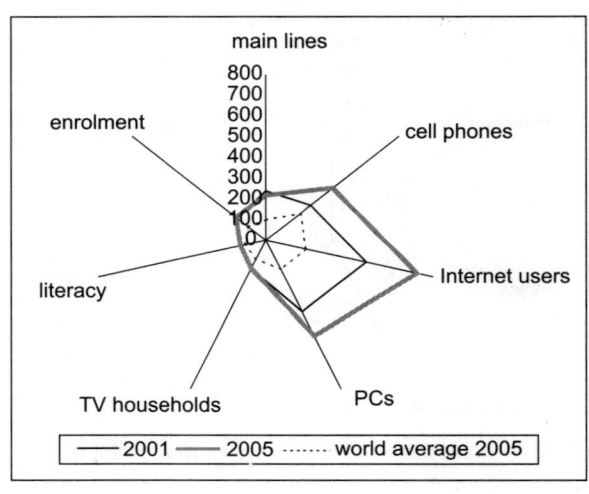
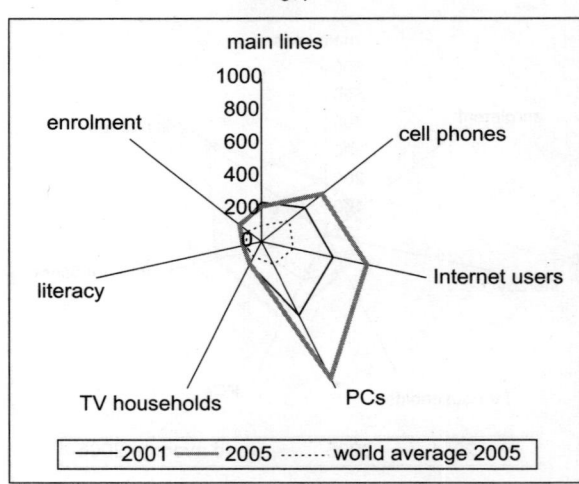
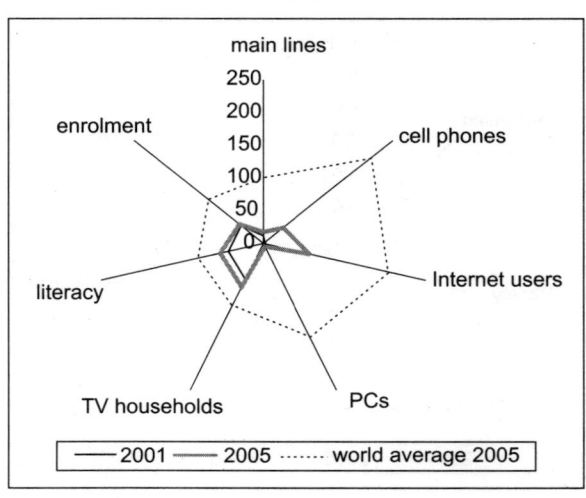
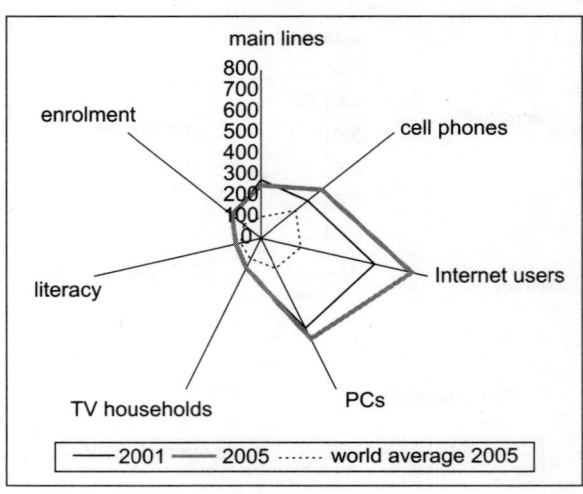

Digital Review of Asia Pacific 2007–2008 **State and evolution of ICTs**

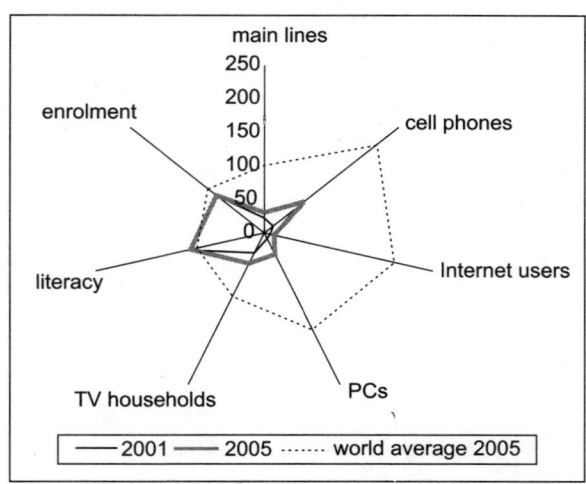
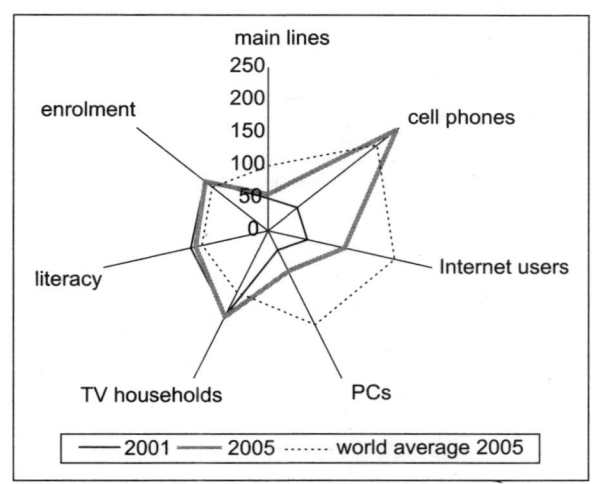
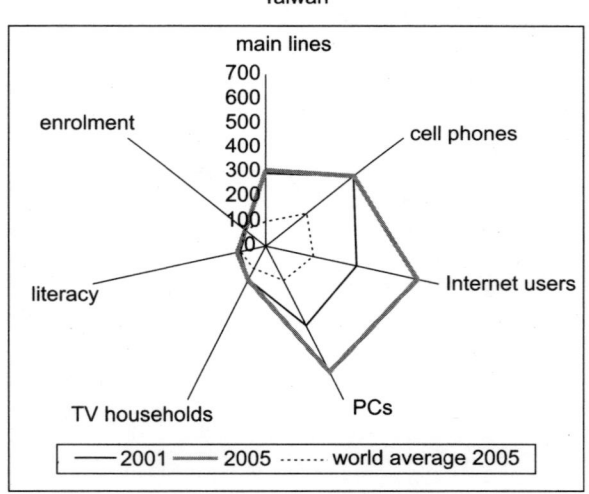
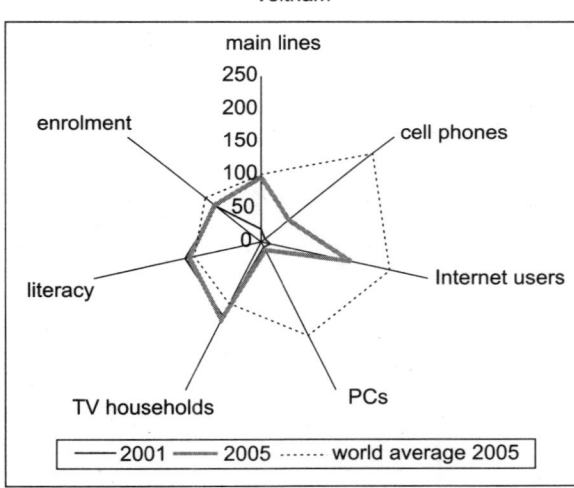

Review of individual economies

.af	Afghanistan
.au	Australia
.bd	Bangladesh
.bt	Bhutan
.bn	Brunei Darussalam
.kh	Cambodia
.cn	China
.hk	Hong Kong
.in	India
.id	Indonesia
.ir	Iran
.jp	Japan
.la	Lao PDR
.my	Malaysia
.mv	Maldives
.mn	Mongolia
.mo	Macau
.mm	Myanmar
.np	Nepal
.nz	New Zealand
.kp	North Korea
	Pacific Island Countries
.pk	Pakistan
.ph	Philippines
.sg	Singapore
.kr	South Korea
.lk	Sri Lanka
.tw	Taiwan
.th	Thailand
.tl/.tp	Timor-Leste
.vn	Vietnam

Afghanistan

Muhammad Aimal **Marjan**

Afghanistan in the past four years has been trying to recover from the trauma of war and instability. A successful presidential election followed by successful parliamentary elections has put the country on the road towards democracy and prosperity. The ICT sector has improved, teledensity grew from 0.08 per cent to 8 per cent and access to information has become easier.

Technology infrastructure

Both the government and the private sector have put together the telecom infrastructure, which has made it possible to achieve a reliable infrastructure in three years based on international best practices. By the end of 2006, 31 out of 34 provincial capitals, 160 major cities, 10 highways and 180 districts had voice and data connectivity contributed by three GSM operators, 14 ISPs and one CDMA operator with roaming service in 124 countries.

As of today, all of the international communications use a satellite connection. However, in 2006, the Ministry of Communications and Information Technology (MCIT) contracted the installation of 3,600 km of fibre through ZTE Corporation China. The fibre connection will form a ring connecting most of the major cities of the country. The aim is to be able to connect the country's major cities with TAE (Trans/Asia Europe) and SeMeWe (South East Asia Middle East, Western Europe). This will enable the country to enjoy good quality communication at reasonable cost.

Internet penetration has increased from 0.08 per cent to 1 per cent. It is hoped that this will increase with the completion of the District Communications Network (DCN) project and the fibre optic connection.

The Afghanistan National Data Centre (ANDC) will be operational by the end of 2007. It will provide a central location for hosting the government electronic data and will provide collocation space for the private sector. The Centre will also host the e-Afghanistan project of the MCIT.

Key institutions dealing with ICTs

The MCIT is the government entity working for the promotion of telecoms and IT in Afghanistan, in partnership with various international and local organizations. The private sector is importing a lot of ICTs into the country.

Indeed, ICT use in Afghan society is growing. Unfortunately, the government has not yet adopted ICT as a tool for national reconstruction and economic development. This may change with the new mandate of the MCIT. Also, the National ICT Council of Afghanistan (NICTCA) established in 2006 will be fully operational in 2007. This will boost coordination of ICT efforts by the government and the private sector.

In order to contribute to the development of national ICT policies, standards and procedures, the private sector, IT professionals and IT departments in academe are establishing the National ICT Association of Afghanistan (NICTAA). This entity, which will be fully operational in 2007, will have permanent seats in the NICTCA.

The telecom sector is regulated by the Afghanistan Telecom Regulatory Authority (ATRA), an independent regulator established in 2006 as the successor to the Telecom Regulator Board (TRB), which also covers spectrum monitoring in the country.

Legal and regulatory environment for ICT

On 18 December 2005, the new Telecom Law of Afghanistan was put in place. This law governs only telecom services. However, the MCIT has drafted an ICT Law that will address issues such as IP, digital signatures, e-commerce, e-government, IPR and cyber security.

ATRA has started working on different regulations to encourage the sector's competition, growth and new telecom services (for example, WiMAX, VOIP).

Digital content initiatives

There are a number of initiatives underway for developing a platform to meet local computing needs. The MCIT in collaboration with the Afghan Computer Science Association (ACSA) has developed a Pashto language version of MS Office 2003 and MS Windows XP, to be launched by Microsoft in mid-2007. There are plans to localize MS Office 2007 and MS Windows Vista in 2007. This will enable the 64 per cent of the population who are Pashto literate to make use of computers in their daily life. ACSA, in collaboration with the PAN Localization Project, finalized the character set, keyboard layout and collation sequence in 2006. The project will continue with font development, lexicon, spell-check and machine translation in 2007.

All of the newly customized software applications developed for the government is in the official languages of the country, Dari and Pashto.

ACSA is working to put in place a task force for the introduction of open source in Afghanistan.

Online services

Online services are not yet popular in Afghanistan due to the lack of electricity and the lack of local content. But there are a number of websites that provide information about policies, regulations and development projects to the public and to the international community. Some good examples are the website of the Office of the President (www.president.gov.af), Ministry of Foreign Affairs (www.mfa.gov.af), Afghanistan Reconstruction and Development Services (www.ards.gov.af), Afghanistan National Assembly (www.nationalassembly.af) and Ministry of Communications and Information Technology (www.mcit.gov.af).

The government has put together a unified development strategy called the Afghanistan National Development Strategy (ANDS) (www.ands.gov.af). It covers security, governance, rule of law, human rights, and economic and social development. The MCIT has developed a concept called e-Afghanistan, which focuses on the utilization of ICT to achieve the goals set in the ANDS. e-Afghanistan covers the development of e-government, national portals, e-commerce and ICT governance.

ICT and ICT-related industries

The ICT industry is growing very slowly due to the lack of physical security in the country and the lack of electric power. In spite of this, the telecom industry attracted over USD 700 million in the last three years. Some statistics shows that in 2005, the government spent more than USD 70 million on procuring ICT equipment. This shows the potential of the market for investors. Also, the number of SMEs in the country is growing.

The government has established a separate entity called the Afghanistan Investment Support Agency (AISA) to attract foreign direct investment. In addition, the MCIT has plans to set up an ICT Park. This facility will improve the state of the ICT industry in the country.

Education and capacity building

Capacity building has been the primary focus of the government from the day it came into being after the Taliban regime was deposed. There are a number of national projects addressing the issue. The Civil Service Commission (CSC), an independent body, is tasked with implementing programmes such as training civil servants and re-engineering the business process of the government administration.

There are a number of projects focusing on ICT-based training, such as the Cisco academies and the ICT training centres established by the MCIT.

The new educational policy of the government is to open the field to private sector investment, which has increased the number of institutions contributing to human resource development in the country.

Challenges

The country is on the road to recovery and is making great progress in different areas. The government and the international community have identified three main challenges to development activities in Afghanistan—terrorism, narcotics and corruption. The MCIT believes that adoption of ICT and the e-Afghanistan project will help reduce corruption and bring in administrative reforms and transparency.

Other challenges confronting the ICT sector are the lack of trained human resources, lack of awareness and acceptance of ICT, lack of electricity and political instability. Despite these challenges, different sectors are committed to the development of Afghanistan. The penetration of telecom services in Afghan society is remarkable, compared to its neighbouring countries. However, the penetration of the concept of ICT4D will take some time because of the lack of local content.

The Islamic Republic of Afghanistan and the international community are determined to strengthen their partnership to improve the lives of the Afghan people and to contribute to national, regional and global peace and security. Thus, they affirm their shared commitment to continue, in the spirit of the Bonn, Tokyo, Berlin and London conferences, to work towards a stable and prosperous Afghanistan, with good governance and human rights protection for all under the rule of law.

References

Afghan Computer Science Association Website. Available online at http://www.acsa.org.af

Afghan Telecom Regulatory Authority Website. Available online at http://www.atra.gov.af

Afghan Wireless Communication Company Website. Available online at http://www.awcc.af

Afghanistan Investment Support Agency Website. Available at http://www.aisa.org.af

Afghanistan National Development Strategy. (2007). Retrieved from http://www.ands.gov.af

Ministry of Communications and Information Technology, Islamic Republic of Afghanistan Home Page. (2007). Available online at http://www.mcit.gov.af/default.asp

Pan Localization Website. Available online at http://www.panl10n.net/

Roshan Website. Available online at http://www.roshan.af

.au
Australia

Lelia **Green** and Axel **Bruns**

Total population	20,264,082 (July 2006 estimate)
GDP	USD 612.8 billion (2005 estimate)
Key economic sectors by contribution to GDP	Agriculture (3.8 per cent), Industry (26.2 per cent), Services (70 per cent) (2004 estimate)
Computers per 100 inhabitants	72 (2006 estimate)
Fixed-line telephones per 100 inhabitants	57 (2005)
Mobile phone subscribers per 100 inhabitants	91 (2005)
Internet users per 100 inhabitants	72 (2006)
Domain names registered under .au	5,351,622 (2005)
Broadband subscribers per 100 inhabitants	28 (2005)
Internet domestic bandwidth	up to 24 Mbps (depending on exchange and telco)
Internet international bandwidth	365 Gbps (actual)/1560 Gbps (potential) (2006)

Sources: CIA 2006; ABS 2005a; Computer Industry Almanac 2006; TIAC 2006; Braue 2006.

Technology infrastructure

Australians are reluctantly realizing that, unlike South Koreans, they do not have a world-class digital technology infrastructure. This has been spelled out to them in three ways. First, Kim Beazley, then leader of the opposition Labour Party, pledged that a future Labour government would deliver a 'high-speed, fibre-to-node broadband network across the country' (Beazley 2006), which implies that Australians are lagging behind. Second, the recently privatized Australian telecommunications giant Telstra, has just announced that it will not be proceeding with the proposed AUD 4 billion super-fast infrastructure to connect consumers in major cities because of the regulatory conditions associated with the on-selling of access to competitors. Third, Australian media have found that a 650 Mb DivX movie file would take 105 minutes to download in Australia, but only one minute in South Korea, using average broadband speeds (Jenkins and Colley 2006, p. 8). Most Australians still use dial-up access.

On the other hand, Australians in urban environments are generally well-connected, with 91 mobile phone subscribers per 100 inhabitants and 57 per cent of them with domestic access to fixed phone lines. Access to the Internet for most Australians is made possible by a high-speed network to a telephone exchange and then copper cable for the local loop connection from the telephone exchange to their homes. This means that the distance between the home and the exchange can become critical in terms of final Internet speed delivered. While this wire-based infrastructure has many limitations, the costs of using wireless technologies to deliver domestic broadband services in Australia are thought to outweigh the benefits (LeMay 2005). None of the telecommunications players has committed to building fibre-optic networks, and people living in remote, rural and regional areas generally have fewer choices and higher communication costs than city dwellers.

The 2005 Australian Bureau of Statistics (ABS) report (ABS 2005b) notes that 67 per cent of 15,500 households surveyed had a computer at home while 56 per cent had domestic access to the Internet (69 per cent of those with dial-up Internet access and 28 per cent with broadband connections). The figures suggest that over the period 1998–2005, the rate of take-up has slowed down slightly, which would impact on the adoption of new services such as VoIP. More information on Internet usage in Australia is likely to emerge from the 2006 national census which includes, for the first time, a question about Internet access. The first findings are to be released in 2007.

Key institutions dealing with ICTs

The production and consumption of ICT goods and services are vital functions of Australia's commercial and industrial sector. According to the most recent information available, at the end of June 2003 there were more than 25,500 Australian businesses classified within the ICT industry grouping, with a total income of almost AUD 90 million and employing over 235,000 people (ABS 2004).

Since ICTs are deemed important to Australian productivity and prosperity, all three levels of government—Federal, state and local—are deeply involved in ICT policy and development. Regulation is at a Federal level (Australia-wide) but policies promoting 'Smart States' and aiming to attract ICT investment are enthusiastically enacted at the level of states and territories (of which there are eight). These often compete among themselves to entice digital companies to their capital city or Technology Park.[1] Local government—at the level of the Shire, Town or City—is responsible for services such as libraries and community centres, which often provide ICT access to users who may not have domestic-based facilities. The Federal government also funds ICTs for the armed forces and for homeland security initiatives. These expenses are rarely open to scrutiny but constitute an important component of the ICT research and development programme.

While there is no specific government organization tasked with the development and regulation of ICT industries, the Federal Department of Communications, Information Technology and the Arts (see section on ICT industries) deals with enabling policies. The Australian Competition and Consumer Council (ACCC 2006) and the Australian Communications and Media Authority (ACMA 2006) are given responsibility for regulating ICT industries and for ensuring that competitors are given access to the once-publicly-owned Telstra network fairly and at competitive on-selling rates.

Digital content initiatives

Australia's majority language is English, which means that much of the world's digital content is highly accessible to Australian citizens. However, Australian policymakers and digital content providers worry about the huge impact of US-generated digital content on Australian citizens, particularly on Australian children. The Federal government believes it is important that Australians have access to Australian-produced content that deals with matters that are relevant to Australian audiences. The Australian Film Commission's Policy and Research website carries a range of government and other reports and speeches dealing with the importance of having Australian-developed content for new media (AFC 2006). However, as noted earlier, Australia has not invested sufficiently in its telecommunications infrastructure to enable the majority of its citizens to participate in an online interactive streaming media environment.

Australia's premier arts funding body, the Australia Council for the Arts (Ozco 2006), used to have a New Media Arts Board to encourage innovative Australian content production in new media. This board was disbanded in 2005 following ongoing controversy and friction with the Howard government and its ministers over a AUD 25,000 (2003) grant to artists seeking to develop a video game entitled *Escape from Woomera*, apparently referring to the immigration detention centre deep in the South Australian desert that has now been closed and relocated to the nearby Baxter detention centre (Google 2006). New media arts applications these days are predominantly assessed by either the Visual Arts or the Music Boards, and there are very few opportunities for funding Australian new media products if developers cannot attract commercial backing.

In addition to being mainly English-speaking, Australia is a nation built on immigration and it prides itself on its cultural and linguistic diversity. But some sections of society have been accused of discrimination (such as surfaced in the 2005 Cronulla riots). Some Australian Muslims feel marginalizsed by the post-9/11 environment and the increasingly intrusive legal and surveillance attempts to prevent the possibility of 'home-grown terrorism'. Australia also attracts regular criticism from overseas about its failure to keep its obligations under the 1951 UN Refugee Convention. Nevertheless, Australian media are comparatively inclusive and there is a digital content regime that enables the nationwide, government-funded Special Broadcasting Service (SBS) to develop expertise in translating over 60 languages, with 68 languages broadcast on SBS radio and available for podcast downloads. Non-English speakers outside Australia might find some SBS (2006) online material interesting and accessible. The broadcaster keeps alive positive views of Australian multiculturalism, such as those contained in the 2006 report *Connecting diversity: Paradoxes of multicultural Australia* (Ang et al. 2006).

The cultures of Australia's indigenous people are supported through a variety of online initiatives. In addition to the Australian government's Indigenous Portal (http://www.indigenous.gov.au/), there is the online newspaper *National Indigenous Times* (http://www.nit.com.au), specialist indigenous Web design and cyber services (such as http://www.Cyberdreaming.com.au), and online initiatives such as Digital Songlines (http://songlines.interactiondesign.com.au/), which aims to promote 'the collection, education and sharing of Indigenous cultural heritage knowledge'.

Online services

As with all countries that have a significant land mass and a scattered population outside the major cities, the relative degree of accessibility of websites and other ICT services for remote and rural residents is a highly political issue in Australia. As the government and commercial service providers invest more heavily in online services, access and infrastructure, they also cut back on accessible face-to-face services. Thus, the proportion of the population that finds it difficult to access online information also have increasing difficulty accessing services that were once delivered in a variety of modes. For example, many consumers who are unable to use online banking are charged heavy fees for face-to-face transactions in an environment where many bank branches have been closed and there are fewer cashiers.

Online services are provided by all levels of Australian government. Government websites contain links to other websites with authoritative materials that complement or add to core government services. Communications, education, health and security are specialist Commonwealth government areas with online resources that may be of interest to people in the wider Asia Pacific region. The website of the Australian Broadcasting Corporation (http://www.abc.net.au/), Australia's primary public service broadcaster, is a gateway to a large amount of information on a wide variety of subjects.

The Federal government's education website (http://www.education.gov.au/) is a gateway to over 5,000 websites concerning education and training in Australia, as well as a range of policy documents and a site for educational policymakers. The educational spectrum is covered—from early childhood education through to higher education—and diverse subjects are included. The portal can also be used to access state-specific education websites.

Educational institutions increasingly expect students to engage fully in multimode flexible delivery models of education. Some universities, for example, require students to submit their essays digitally so that they can be checked for online plagiarism (that is, the use of online resources without crediting the original source of the comment or idea) via software such as Turnitin.com. Blackboard, which bought out rival WebCT at the start of 2006, provides the online infrastructure and tools for many of Australia's higher education institutions. Whether the subject is Shakespeare or Internet studies, students are expected to use high-level digital tools. Moreover, schools, universities and education departments increasingly deliver information via the Web to their respective communities of interest. Students enrol in courses online, find their exam marks online, and pay fees and fines using secure online websites.

There are many interesting heath-related online services. One is the Australian National University's MoodGYM (http://moodgym.anu.edu.au/), which provides, free of charge, self-paced, self-instructional, online materials teaching cognitive behaviour techniques for the prevention and reduction of anxiety and depression. The Department of Health and Aging website (http://www.health.gov.au/) offers access to a variety of interesting topics and its search function links to a large range of reliable independent resources about matters of international relevance, such as bird flu and HIV/AIDS.

Australia has a dedicated portal for national security issues—the Australian National Security website (http://www.nationalsecurity.gov.au/). This keeps interested Australians (and civil liberties and human rights lawyers) up to speed on the government's views about potential threats to national security. Taking a bigger picture, the Australian travel advisory and consular assistance website (http://www.smartraveller.gov.au/) offers guidance on the perceived safety of Australians travelling to other countries and regions. Australians are encouraged to register their travel plans on this site so that they can be contacted quickly in case of a natural disaster, civil disturbance or family emergency (DFAT 2006). The online service has been credited with helping the government organize the evacuation of Australian citizens from Lebanon in mid-2006.

ICT and ICT-related industries

Australia operates a trade deficit in terms of ICT goods and services (ABS 2004). Effectively, Australia does not have a consumer-based ICT hardware manufacturing industry with a global presence. ICT goods and services, while extensive, can sometimes be compromised by a regulatory environment which (arguably) gives too much power to industry players who work to maintain the status quo. For example, the existing commercial television services stated, at the start of digital broadcasting, that they would only invest in the necessary technological infrastructure if they were guaranteed that no new commercial free-to-air (FTA) television licenses would be issued. Although the moratorium on new licenses ended in 2006, it has held back the development of the industry, and discussions around the further development of digital television broadcasting continue (see below).

The Department of Communications, Information Technology and the Arts (DCITA) has a website dedicated to broadcasting and online regulation (http://www.dcita.gov.au/broad) which details Australia's policy responses to the evolving ICT environment. A recent DCITA report (April 2006) characterizes the Australian software industry as 'globally competitive, domestically undervalued'. The implication is that Australia has decided not to participate in ICT hardware industries but is keen to be recognized for a growing influence in the global software arena. In particular, the Australian computer games industry has been identified as a key focus for future development.[2]

Enabling policies and programmes

The Connect Australia initiative is driven by DCITA as part of the government's undertakings to provide telecommunications services to remote, regional and rural areas that are of the same quality as services enjoyed in the cities. However, although significant amounts of money (partly from the 1999 sale of 16 per cent of the government-owned telecommunications company, Telstra) have been invested in this project, rural services are generally not as good and are more expensive than those available to city dwellers.

A 2001 government project, Backing Australia's Ability (BAA), started with an 'Innovation action plan'. In 2004 this was extended to include 'Building our future through science and innovation', and funding and timelines were increased so that a total AUD 8.3 billion commitment is extended over the 2001–11 period. The BAA project focuses on three main areas: the generation of new ideas (research and development), the commercial application of ideas, and developing and retaining skills. Although key performance indicators (and progress against these) are not readily accessible, four reviews of government-funded research have been undertaken. The government has provided its response to the key recommendations arising out of three of these reviews (DEST 2006).

The Federal commitment to fostering innovation goes hand in hand with a state-based initiative to promote the creative industries, particularly centred upon 'Queensland—The Smart State'. Queensland's strategy, timeline 2005–15, aims to communicate and develop a vision of Queensland as a place where 'knowledge, creativity and innovation drive economic growth to improve prosperity and quality of life'. Implicitly, the policy uses the income from the current resources boom to fund investment in environmentally and culturally sustainable creative industries that have the potential to drive future wealth creation via the development of a knowledge economy (Queensland 2005). Performance highlights for Queensland's Smart State Strategy focus on investment, economic growth and rising skills levels (Queensland 2006).

Legal and regulatory environment for ICTs

The legal and regulatory environment for ICT and media industries in Australia is undergoing significant change and uncertainty. The Federal government has forecast and implemented a number of policy changes that impact on these industries, including changes to the 'Cross-media ownership laws' which used to restrict the range of media that could be owned by any one media company. According to then Prime Minister Paul Keating, an organization had to decide whether it wanted to be 'a prince of print' or 'a queen of the screen': in those days it could not be both. In July 2006, Senator Helen Coonan, the Communications Minister, announced further ICT and media reforms. In the main, these lift restrictions on cross-media and foreign media ownership, making it easier for foreign companies to enter the Australian market. They also allow an even smaller number of companies to own more of Australia's media.

These changes follow a raft of other reforms. In July 2005, ACMA assumed the combined responsibilities of the Australian Broadcasting Authority (ABA) and the Australian Communications Authority (ACA) as the chief regulatory body for media and communications industries and providers. The ACMA (2006) website covers subjects currently on the communications and media agenda in Australia, and their archives offer access to a range of deliberations and outcomes. Australia tends to operate a 'light touch' regulatory regime in these areas. Industry players set up codes of conduct (self-regulation) which may be discussed and amended prior to lodgement and registration. After that, the regulator is mainly interested in responding to complaints only if the complainant has evidence that an industry player has breached their code of conduct. Critics argue that the imbalance of power between an industry player and a consumer means that legitimate concerns may not get a fair hearing.

Questions remain over the power and effectiveness of ACMA, and its huge area of responsibility. On the infrastructure side, the Australian telecommunications industry is dominated by Telstra, which some have suggested is 'too big to be regulated' (ABC 2004) and which is also subject to regulation via competition legislation (ACCC 2006). On the content side, the long-held policy preference for self-regulation of media industries continues to cast ACMA in a relatively ineffective role.

These changes have not heralded a vibrant marketplace for local digital content. While existing digital television broadcasters are now able to offer more programming variety by the lifting of a requirement to simulcast content on standard definition (SD) and high definition (HD) channels, critics have noted that because all digital TV channels are currently operated by existing analogue TV providers, there is little incentive for the development of innovative and attractive new content offerings. Channels would undermine their own market and fragment their audiences by developing such competition to existing services. This regulatory environment is likely to further delay the switch to digital television broadcasting by Australian audiences (Coonan has already revised the analogue TV switch-off date

> **The privatization of Telstra: Deregulation, reregulation or a total mess?**
>
> Australia's once-monopoly telecommunications provider, Telstra, had a responsibility to provide a minimum standard of telecommunications services to all citizens. Being the only communications carrier made it possible for Telstra to charge more for highly profitable services, such as the provision of communications services to inner city locations, so that they could cross-subsidize very unprofitable services, such as those for remote and outback Australia.
>
> With the 1996 election of a conservative federal government (the Liberal-National Coalition), the scene was set for a radical redrawing of telecommunications policy. There was one problem—the conservative parties in Australia are traditionally well-supported in rural areas and rural voters suspected that a privatized Telstra would give up rural services. In a competitive environment it would be difficult for city profits to balance losses in the rural areas. The government promised that Telstra would be governed by legal obligations to keep providing services to the bush. It also set up a regulation regime to make sure that Telstra was not unfairly charging small competitors to use its networks, which had originally been built with public funding.
>
> This bargaining between the Australian government and rural inhabitants allowed for the sale of one-third of Telstra in 1997—the T1 float—and for the introduction of competition between a range of telecommunications and Internet service providers. Some of the money earned was earmarked for improving regional and remote services. By 1999, the value of Telstra shares had tripled and the government wanted to sell its remaining stake. However, it did not control the Senate which insisted that Telstra remain majority government-owned. A further 16.6 per cent of Telstra—T2—was sold off. In the face of criticism from non-metro consumers, much of the capital from T2 was used to 'future proof' rural telecommunications services.
>
> The Howard government had to wait until it controlled the Senate to get permission to sell all of Telstra. By mid-2005, with Senate control finally achieved, Telstra CEO Ziggy Switkowski had been replaced by Solomon Trujillo. The relationship between the corporation and the government quickly became fraught, with Telstra saying it was over-regulated and would not invest in new networks if the regulator was going to insist on 'unrealistic' rates of return. Meanwhile, the government criticized the company for 'talking down' the share price, as the T2 issue had more than halved in value. Shareholders, on the other hand, accused the government of withholding market-sensitive information that would have informed buy/sell decisions had it been widely available. The Australian government clearly had a conflict of interest in its various roles of legislator, shareholder and regulator. Prime Minister John Howard said Telstra's 'half-privatized' status made no more sense than the concept of being 'half-pregnant'.
>
> While Telstra board members continue to complain that the regulatory regime prevents them from maximizing shareholder equity, the government suggests that these problems were solved in late 2006 when the Telstra T3 float saw over 80 per cent of the company finally devolved into private ownership. About 17 per cent of Telstra's shares were not taken up by the market and these have been lodged in a 'future fund' set up to finance the unfunded pension costs of Australia's federal civil servants. These shares will be sold down in future years. Some commentators claim that this cache allows the government to continue its influence at arm's length. Accusations of interference were strengthened just before the T3 sale with the controversial appointment of Prime Minister John Howard's policy advisor Mr Geoffrey Cousins to the Telstra Board, achieved only through the government's use of its then-majority share-holding. The conflict seems set to continue.

from 2008 to as late as 2012). In the meantime, Australian TV audiences may continue to explore digital content alternatives on the Internet.

Copyright law

The Australia-US Free Trade Agreement (AUSFTA), which came into effect on 1 January 2005, impacts on the ICT sector especially through the requirement to harmonize copyright laws between Australia and the United States. While both countries are said to be equal partners in this agreement, there are significant concerns that Australia will now be required to 'import' US legal frameworks. These may impact on copyright legislation, and on legislation governing related rights such as moral rights and access rights. In the realm of copyright, Australia will be required to extend its term of copyright to the US standard of 70 years after the author's death. Commentators object that this may stifle the ability of authors and inventors to build on the work of their forebears. They specifically argue that the provision mainly benefits the US content industries (such as Hollywood), limiting Australian opportunities to compete on an equal footing. There are also concerns over whether the Australian rights to the use of copyright material, for example for the purpose of fair dealing, will be maintained when these are configured quite differently under US law.

Furthermore, the US uses copyright law to defend technological protection measures (TPMs, such as region coding in DVDs, or anti-copying systems for other media forms). An importation of such legal approaches into Australian law could put Australian copyright law at odds with other applicable laws, such as those governing access rights. In the context of DVD region coding and elsewhere, US content industries have been accused of utilizing TPMs to control markets and stifle competition rather than simply to prevent unauthorized copying. This may directly impact on Australian consumers' access to products as the Australian market is likely to be seen by US companies as a region of lesser importance.

The outcome of the legislative processes surrounding the AUSFTA remains unclear, and is a matter of some concern to Australian media producers. A cross-party parliamentary committee published a highly critical report from its *Inquiry into Technological Protection Measures (TPM) Exceptions*, explicitly stressing the importance of maintaining 'the balance between copyright owners and copyright users achieved by the *Copyright Act of 1968*' (House Committee 2006, p. 17). The committee noted a number of areas where exceptions from copyright enforcement should be ensured. (See Australian Copyright Council, http://www.copyright.org.au, for ongoing coverage of these developments.)

The future legality of the ways users get around TPMs for the purposes of time- and place-shifting of their media consumption (for example, by copying content to mobile devices) are of key importance in this context. It is also uncertain whether the provision of devices that circumvent TPMs is legal in Australia. An opportunity for a court decision on such matters was lost when the case against Sherman Networks, provider of the peer-to-peer (P2P) filesharing software Kazaa, was settled out of court in July 2006. Observers had hoped that a precedent in this case would have helped spell out the circumstances and conditions under which tools such as Kazaa could be legally used in Australia.

Open source and open content

As in many other countries, alternatives to copyright models continue to spread in Australia. Creative commons licences have been translated into the Australian legal context by the iCommons.au group (http://www.creativecommons.org.au/), a member of the international iCommons project. Creative Commons Australia (CC-AU), based at Queensland University of Technology, now organizes further research and advocacy around the creative commons project.

Creative commons and other open content licences are widely used. For example, AEShareNet is a company established by the Australian State and Territory ministers for education and training, which provides shareable learning and teaching materials under its own licence scheme. This includes licences such as 'Free for Education', 'Unlocked Content', 'Share and Return' and 'Preserve Integrity'. Based on a collaborative framework, the licenses involve a large number of universities and other educational providers. Australian Creative Resources Online (ACRO) is a repository for audio, video and still images content that is made available under both AEShareNet and creative commons licences. ACRO's mission is to provide source materials especially for amateur and grassroots content creators, as well as to study the creative work that draws upon this resource.

However, not all institutions in the Australian creative industries are predisposed towards the creative commons approach. In a widely publicized case in 2005, the Media and Entertainment Arts Alliance (MEAA), the union of workers in the media industries, argued strongly against its members' participation in a 'remixable' short film project by MOD Films (2005) that was to be released under a creative commons licence (APC 2005). With support from the Australian Film Commission (AFC), a government-sponsored body, the film was eventually shot in March 2005. Even so, MEAA, which exercises considerable influence in the film and television sector, has not revised its position on creative commons-licensed projects.

Education and research & development

The Australian higher education system is in a state of flux, which poses certain challenges to ICT-related research. Key to the government's new policy of measuring the quality and impact of publicly-funded Australian research is the Research Quality Framework (RQF) to be introduced in 2009, following data collection in 2008. Critics of the RQF model (such as the National Tertiary Education Union) argue that the assessment panel system may result in a small group of experts dictating

> **An Australian creative commons**
>
> Creative commons licensing schemes have become a widely accepted alternative to traditional copyright licences. They offer copyright holders the opportunity to open up particular forms of use that otherwise would be denied without explicit permission. A range of licences is available (variously combining restrictions such as the attribution of the original author; limiting use to non-commercial purposes; denying the right to create derivative works; or requiring the sharing of derivative works under an identical licence scheme). Such licences exist in three forms: a human-readable version, a machine-readable version, and a 'lawyer-readable' version that spells out licence requirements and restrictions in legally binding language.
>
> This last version must be translated into applicable legal frameworks for each country in which the licence is to be used. The project of translating this legal licence code into local frameworks is coordinated by the global iCommons group. (In 2006, some 32 national translations were in existence, with another 10 in progress.) Strictly speaking, it is possible that creative commons licences may not be binding in national jurisdictions not covered by one of these translations.
>
> The Australian versions of the creative commons licences were developed by the Creative Commons Australia group based at Queensland University of Technology, and launched in 2005 at the Open Content Licensing Conference in Brisbane. The development team included staff from the university's Law School as well as lawyers from the Blake Dawson Waldron legal firm. Creative Commons Australia now continues to maintain the Australian licence legal terms. This is an ongoing responsibility as Australian copyright law continues to evolve and as the overall creative commons licences themselves mature further. The group also advocates in favour of a broad adoption of the licences by private and public institutions.
>
> Creative commons licences were already widely used for Australian content even before the development of an Australian-law version of the legal licensing code. However, the availability of this translated licence ensures the legally binding nature of these licences in Australian jurisdictions, and provides further peace of mind for individuals and organizations wishing to use such licences. As a result, government and educational institutions in particular have adopted these licensing options, thus contributing significantly to the development of an intellectual and creative commons in the country. Further, such developments are likely to gain momentum as the Creative Commons group's Science Commons project (and potentially a Business Commons project) gathers speed.

national research priorities, which may undermine experimental and esoteric research. This may be particularly problematic in the field of ICT-related research where investigators experiment with new and emerging technologies.

Under then Education Minister Brendan Nelson (replaced in 2006), there was considerable concern about the independence of the RQF, as well as of the Australian Research Council (ARC), the key national body administering competitive non-medical research grants. In 2005, Nelson had vetoed a number of 'controversial' grant projects even after they had passed the ARC's rigorous peer review process, which is designed to protect researchers from government interference. Such concerns have eased under the new Education Minister, Julie Bishop, who has promised not to repeat the ministerial interventions. Even so, the impact of Brendan Nelson's actions continues to raise questions about the independence of Australian research funding agencies.

Less controversially, the Carrick Institute for Teaching and Learning in Higher Education was launched in August 2004 and has been widely welcomed for its quality-improvement agenda. It has outlined 'innovation in learning and teaching, especially in relation to new technologies' as a priority area for its competitive grants scheme, and has provided a significant amount of funding to innovative teaching projects using ICTs (Carrick 2006).

Australian government research funding schemes encourage universities to cooperate with industry players, with industry partners providing a significant proportion of the resources. Among the key education-related research bodies in ICT fields are the Smart Internet CRC (Cooperative Research Centre), the Australasian CRC for Interaction Design (ACID) and the ARC Centre of Excellence for Creative Industries and Innovation (CCi). Launched in 2001, the Smart Internet CRC is a joint venture between government organizations, key universities and industry partners. It published its major report, *Smart Internet 2010*, in August 2005 and has since released a number of updates dealing specifically with open source and social networks (see http://smartinternet.com.au/).

ACID is a similarly constituted body in the field of interaction design, working with major partners in industry. Its research programmes cover areas such as Smart Living, Digital Media, Multi-User Environments and Virtual Heritage. Of particular importance are its projects working with Indigenous Australian communities (see discussion earlier), and with town planners and housing developers to establish smart suburban communities (see http://acid.net.au).

In 2005, the Federal government part-funded an AUD 10 million research initiative, the ARC Centre of Excellence for Creative Industries and Innovation (CCi). It is based in Queensland with a range of associated organizations and researchers from around Australia (CCi 2005). It is the first such ARC Centre outside the science, technology, engineering and medicine (STEM) sector—which highlights an increasing government focus on the creative industries as a major contributor to the national economy. There are six research programmes investigated by CCi, including the citizen-consumer, crisis in innovation, creative workforce, and legal and regulatory impasses and innovations.

The establishment of this centre also points to a wider trend in Australia's engagement with ICTs: the continuing embedding of ICTs as tools of everyday life rather than as a separate technological category. This development is also indicated in current undergraduate enrolment trends in Australian universities, which have seen a marked decline in traditional, strongly discipline-based ICT courses in favour of combined, interdisciplinary double degrees pairing business, law, creative industries, humanities or arts with ICT degrees (Rood 2004). IT faculties at a number of Australian universities have been drastically downsized in the process, while IT education finds its way into a variety of other degree options (Dreyfus 2006).

Australia and the region

Australia's complicated relationship with its neighbours impacts on its ability to engage in ICT-related projects in the region, including collaboration for ICT-related purposes with regional practitioners, scholars and students. On a political level, Australia's relationships with its neighbouring countries have become increasingly difficult following Australia's decision to join the 'Coalition of the Willing' and go to war in Iraq. The widespread perception in Asia Pacific of the Australian government as the neighbourhood 'deputy sheriff' of the US has soured relations with a number of countries. While police and military interventions were welcomed by locals in troubled nations such as Timor-Leste and the Solomon Islands, for example, they also generated significant regional opposition which may limit Australia's ability to involve itself in ICT4D projects in the region.

On the other hand, Australia continues to welcome China and India as current and growing trading and political partners and Australian industry, researchers and governments are strongly involved with both countries on all levels. This is true especially for ICT4D projects, where Australian researchers are working closely with UNESCO and other world bodies in developing locally-based solutions to ICT challenges. Such projects include work on promoting digital storytelling (Tacchi 2006) as a means of generating local content and developing ICT skills in local (and especially rural and underprivileged) populations, and the development of community-based local and regional multimedia centres (UNESCO 2006) to boost ICT literacy and information access (Tacchi et al. 2003a). In particular, ICT4D projects are also seen as a crucial component in the fight against poverty (Slater and Tacchi 2004). A core research tool in this context is ethnographic action research (EAR), a research methodology developed by Australian and British researchers in collaboration with south and central Asian participants and UNESCO (Tacchi et al. 2003b). Beyond this, Australia has also become an important exporter of creative industries theory and policy, especially to regional economic leaders such as China and India, where a creative industries approach is seen as an important longer-term strategy beyond the current boom in manufacturing industries. Australian researchers have been instrumental in raising awareness about this approach in the region.[3]

Notes

1. Insofar as they are regulated, Technology Parks are the responsibility of state and local governments. An example is Western Australia's Technology Park at Bentley near Perth (see http://www.techparkwa.org.au/index.shtml).
2. See for example http://www.cultureandrecreation.gov.au/articles/digitalgames/; http://www.gdaa.com.au/; http://www.mmv.vic.gov.au/Games; http://www.aph.gov.au/house/committee/cita/film/subs/sub078.pdf and http://www.queenslandgames.com.au/
3. See especially Special Issue No. 9.3 (2006) on 'Creative Industries and Innovation in China' of the *International Journal of Cultural Studies*. Moreover, it is important to note that on 5–7 July 2007, Queensland University of Technology hosted the 2007 China Media Centre Conference. See http://cea.cci.edu.au/

References

ABC. (2004). *7.30 Report*. Retrieved from http://www.abc.net.au/7.30/content/2004/s1217795.htm

Australian Bureau of Statistics (ABS). (2004). *8126.0—Information and communication technology, Australia, 2002–2003*. Retrieved from http://www.abs.gov.au/

Ausstats/abs@.nsf/0e5fa1cc95cd093c4a2568110007852b/197d7c1f38fc3f81ca2569a4007c29a0!OpenDocument

Australian Bureau of Statistics (ABS). (2005a). *Household use of information technology*. Retrieved from http://www.ausstats.abs.gov.au/ausstats/subscriber.nsf/0/CA78A4186873588CCA2570D8001B8C56/$File/81460_2004-05.pdf

Australian Bureau of Statistics (ABS). (2005b). *Household use of information technology, Australia, 2004–5*. Retrieved from http://www.abs.gov.au/Ausstats/abs@.nsf/0/acc2d18cc958bc7bca2568a9001393ae?OpenDocument

ACCC. (2006). *Australian Competition and Consumer Commission: Communications*. Retrieved from http://www.accc.gov.au/content/index.phtml/itemId/3881

ACMA. (2006). *The Australian Communications and Media Authority Homepage*. Retrieved from http://www.acma.gov.au/ACMAINTER:HOMEPAGE::pc=HOME

AFC. (2006). *Australian Film Commission Website: Policy and Research*. Retrieved from http://www.afc.gov.au/policyandresearch/policy/broadband.aspx

Ang, L., Brand, J., Noble, G., and Sternberg, J. (2006). *Connecting diversity: Paradoxes of multicultural Australia*. SBS retrieved from http://www20.sbs.com.au/sbscorporate/media/documents/8243eweb2.pdf

APC. (2005). *Association for progressive communications*. Retrieved from http://rights.apc.org.au/culture/2005/03/meaa_halts_worldfirst_film_project_in_australia.php

Beazley, K. (2006). *Real family values*. Speech to the Tasmanian ALP State Conference. Retrieved from http://www.alp.org.au/media/0806/speloo060.php

Braue, D. (2006). *ADSL2+: Turbo-charged broadband, at last*. Retrieved from http://www.cnet.com.au/broadband/adsl/0,239035934,339272359-1,00.htm

Carrick Institute. (2006). *Competitive Grants Program, Carrick Institute for Learning and Teaching in Higher Education*. Retrieved from http://www.carrickinstitute.edu.au/carrick/go/home/grants/pid/49

CCi. (2005). *ARC Centre of Excellence for Creative Industries and Innovation, Queensland University of Technology*. Available online at http://www.cci.edu.au/programs.php

Central Intelligence Agency (CIA). (2006). Australia. *CIA world factbook*. Retrieved from https://www.cia.gov/library/publications/the-world-factbook/geos/as.html

Computer Industry Almanac. (2006). *PCs in-use surpassed 900M in 2005: USA accounts for over PCs in-use* [sic.]. 22 May 2006. Retrieved from http://www.c-i-a.com/pr0506.htm

Department of Communications, Information Technology and the Arts (DCITA). (2006). *The Australian software industry and vertical applications markets: Globally competitive, domestically undervalued*. Retrieved from http://www.dcita.gov.au/communications_and_technology/publications_and_reports/2006/february/the_australian_software_industry__and__vertical_applications_markets_globally_competitive,_domestically_undervalued

DEST. (2006). *Government's response to recommendations of the three research related reviews*. Retrieved from http://www.dest.gov.au/sectors/research_sector/policies_issues_reviews/reviews/previous_reviews/government_response_to_research_reviews.htm

DFAT. (2006). *Department of Foreign Affairs and Trade Online Registration*. Available online at https://www.orao.dfat.gov.au/orao/weborao.nsf/homepage?Openpage

Dreyfus, S. (2006). The must-have job ticket. In *The Age*, 16 May. Retrieved from http://www.theage.com.au/news/technology/the-musthave-job-ticket/2006/05/15/1147545259526.html

Google. (2006). *Google satellite map: Baxter Detention Centre*. Retrieved from http://perljam.net/google-satellite-maps/id/9480/Australia/South_Australia/Port_Augusta/Baxter_Detention_Centre

House of Representatives Standing Committee on Legal and Constitutional Affairs. (2006). *Review of technological protection measures exceptions*. Retrieved from http://www.aph.gov.au/house/committee/laca/protection/report/front.pdf

Jenkins, C. and Colley, A. (2006). Stragglers in speedy world of broadband. In *The Weekend Australian*, 8, 12–13 August.

LeMay, R. (2005). ADSL2 won't cut it, says baby wireless telco. In *CNet*, 4 July. Retrieved from http://www.cnet.com.au/broadband/wireless/0,239035991,240055680,00.htm

MOD Films Network. (2005). Available online at http://modfilms.com/forums/viewtopic.php?t=123 and http://creativecommons.org/video/mod-films.

Ozco. (2006). *Australia Council for the Arts Homepage*. Available online at http://www.ozco.gov.au/

Queensland Government. (2005). *Smart Queensland, Smart State Strategy 2005–15*. Retrieved from http://www.smartstate.qld.gov.au/strategy/strategy05_15/index.shtm

Queensland Government. (2006). *Smart Queensland, Smart State Progress Report Highlights 2006/07*. Retrieved from http://www.smartstate.qld.gov.au/strategy/ss_highlights/performance.shtm

Rood, D. (2004). Uni job cuts as students flee IT. In *The Age*, 30 November. Retrieved from http://www.theage.com.au/articles/2004/11/29/1101577420439.html

SBS. (2006). *SBS Corporation overview*. Retrieved from http://www20.sbs.com.au/sbscorporate/index.php?id=1202

Slater, D. and Tacchi, J. (2004). *Research: ICT innovations for poverty reduction*. New Delhi: UNESCO. Retrieved from http://cirac.qut.edu.au/ictpr/downloads/research.pdf

Tacchi, J. (2006). *New forms of community access*. Final Speech at Thematic Debate: Giving Voice to Local Communities—From Community Radio to Blogs. UNESCO. Retrieved from http://portal.unesco.org/ci/en/files/21712/11453460101jo_tacchi_speech_final.pdf/jo_tacchi_speech_final.pdf

Tacchi, J., Slater, D., and Lewis, P. (2003a). *Evaluating community based media initiatives: An ethnographic action research approach.* UNESCO: Paper for IT4D, 18 July. Retrieved from http://portal.unesco.org/ci/en/files/15731/10843655801Evaluating_Community_based_media_initiatives.pdf/Evaluating%2BCommunity%2Bbased%2Bmedia%2Binitiatives.pdf

Tacchi, J., Slater, D., and Hearn, G. (2003b). *Ethnographic action research.* UNESCO, Retrieved from http://cirac.qut.edu.au/ictpr/downloads/handbook.pdf

Technology and Industry Advisory Council. (2006). *Big pipes: Connecting Western Australia to the global knowledge economy.* Western Australian Information and Communication Technology Industry Development Forum. Retrieved from http://www.tiac.wa.gov.au/ictforum/bigpipes/bigpipesreport-06.htm

UNESCO. (2006). Communication and information: Community multimedia centres. Retrieved from http://portal.unesco.org/ci/en/ev.php-URL_ID=1263&URL_DO=DO_TOPIC&URL_SECTION=201.html

Bangladesh

Ananya **Raihan** and Shah M. Ahsan **Habib**

GDP per capita	USD 482 (USD 1 = BDT 65)
Key economic sectors	Agriculture, textile and clothing, leather and leather goods, frozen fish, construction, transport, overseas services by temporary migrant workers
Computers per 100 inhabitants	1.2
Fixed-line telephones per 100 inhabitants	2.63 per cent
Mobile phone subscribers per 100 inhabitants	2.03 (2004)
Internet users per 100 inhabitants	0.22
Domain names registered under .bd	1,500
Broadband subscribers per 100 inhabitants	not available
Internet domestic bandwidth	>122 Mbps
Internet international bandwidth	Current capacity: 10 Gbps Capacity up to 80 Gbps Current utilization: 300 Mbps

Sources: ITU 2006; UNDP 2005; Bangladesh Bureau of Statistics; Mumit et al. 2007.

Overview

Bangladesh has become more visible on the global ICT map during the last decade. Both the government and the private sector have become more active in harnessing ICT to facilitate economic growth and development. The government of Bangladesh has declared ICT as a priority sector and the country's ICT industry is characterized by increasing availability of computers and Internet use, expansion of the telecommunications network, and a growing number of software development houses, joint-venture companies and institutions working for ICT4D. Particularly noteworthy is the increasing use of ICT in developmental activities for the underprivileged.

This chapter is an analysis of ICT regulation and policies, developments and applications, and issues related to a technology-enabled environment for socio-economic development in Bangladesh.

Technology infrastructure

Computers were first introduced in Bangladesh in the mid-1960s mainly for research and data processing purposes. In the 1980s, the printing and publishing industries promoted the use of computers but high prices restricted their general and commercial use. Personal computers began to gain popularity in the 1990s and became widespread in 1998 after the government imposed a tax exemption for computers and ICT accessories, which coincided with substantial price reductions in the global market.

Competition policy in telecommunications, particularly in mobile telephony, contributed to an exponential growth in the telecommunications infrastructure. Five mobile companies are now in operation: Grameen Phone, Banglalink, Aktel, CityCell and Teletalk. Grameen Phone has more than 60 per cent market share across the country. Mobile teledensity increased to 8.04 per cent in April 2006 from only 0.58 per cent in 2001. It is believed that 10 per cent mobile teledensity will be achieved much earlier than 2010. Mobile coverage improved to 85 per cent by the end of 2005 from only 36 per cent in 2003 (ITU 2006). This growth is remarkable for a least developed country where one-fifth of the population lives on an income of less than one dollar a day.

Bangladesh could easily have a 95 per cent telecom coverage if the government will allow telecommunication companies to expand their network in three hill-tract districts that are now restricted due to the insurgency problem. It is expected that the new government of 2007 will allow telecom operators to expand their network, which would pave the way for Bangladesh to achieve 100 per cent mobile telecommunication coverage.

There has been high growth of Public Switched Telephone Network (PSTN) telecommunication in Bangladesh as well. In 2000, there were 491,303 PSTN subscribers. By June 2006, this figure had grown to 1.01 million, the average growth rate being 15.07 per cent per annum (BTRC 2006).

As of April 2006, both fixed and mobile telephone subscribers had reached 12.6 million from only about 1.2 million in 2001.

Bangladesh's international telecommunication used to be satellite-dependent, relatively slow in speed and costly. In May 2006, a submarine cable was installed. Though late, it is a significant step in ICT infrastructure development in the country. The infrastructure being laid down will enable ISPs to provide high bandwidth connectivity as the demand grows. Some ISPs have also been assigned the spread spectrum radio frequency band in the range of 2 GHz for point-to-point wireless communication between computers.

However, despite the remarkable growth in certain elements of information infrastructure, Bangladesh is still struggling to move forward with respect to other important components of a national information infrastructure, such as security and privacy of network and information, legal and financial infrastructure, and technology convergence.

Key institutions dealing with ICTs

In 1997, the government formed an ICT Task Force under the Prime Minister's Office (PMO) to foster ICT mainstreaming. A Support to the ICT Task Force (SICT) was formed in 2001 to identify and implement e-government projects. Five years later, in May 2006, an e-Governance Cell was initiated with the approval of the Prime Minister. In each ministry, a mid-level government official (at the level of Joint Secretary or Additional Secretary) has been appointed to act as the ICT Focal Point to coordinate e-governance activities and priorities within the ministry. Secretarial support to the National ICT Task Force is provided by the Planning Division of the Ministry of Planning. The latter hosts the SICT Programme to implement the objectives of the ICT Task Force, particularly in e-government, and is the hub for inter-connectivity among the PMO, Planning Commission and Secretariat.

It is the Ministry of Science and ICT, established in 2001, that is primarily responsible for mainstreaming ICTs in economic growth and development. It formulates ICT policies and pushes for ICT-related laws, such as the Intellectual Property Rights Act which was enacted by the Parliament in 2003 and amended in 2006, and the ICT Act which has been awaiting Parliamentary approval since 2002. The Ministry also facilitates the computerization of government institutions and schools.

The Bangladesh Computer Council (BCC) under the Ministry of Science and ICT is the main institution for promoting ICTs. It provides ICT training to government officials and citizens, incubates software companies, provides advisory support to government institutions regarding ICT, provides connectivity to ISPs and works for standardization through such projects as the development of a local language keyboard.

The Ministry of Post and Telecommunication is responsible for building and maintaining telecommunication infrastructure.

The Ministry of Education develops the curriculum for ICT education and spearheads the computerisation of schools.

The Bangladesh Telecommunication Regulatory Commission (BTRC) is the licensing authority and regulates telecom service providers, while the Ministry of Law, Justice and Parliamentary Affairs reviews ICT-related laws.

In the private sector, the Bangladesh Association for Software and Information Services (BASIS) plays a key role in promoting the ICT industry. It organizes an annual exposition of software and applications titled SOFTEXPO; many foreign software companies attend this exposition, helping to boost the reputation of the Bangladesh software industry. BASIS likewise acts as a lobbyist with the government for fiscal incentives and promotional support, such as income tax waiver, VAT waiver, and higher foreign exchange retention facilities. BASIS also organizes capacity building programmes for its members, particularly in product marketing. Meanwhile, the ISP association, Bangladesh Computer Samity, played a key role in the elimination of import duties on computers in the early 1990s, which facilitated ICT penetration.

Private universities and institutions lead human resource development by offering advanced courses on ICTs.

Several institutions work in the area of ICT for poverty alleviation. Grameen Phone, BRACNET.NET and Development Research Network (D.Net) are three prominent institutions championing ICT access in the rural areas. Grameen Phone built a national GSM network which is now available for Internet connection anywhere in Bangladesh through EDGE and GPRS technology. D.Net is aiming to bring ICTs to the doorstep of poor people in the rural areas. It has developed a comprehensive volume of local language content on livelihood, making the Internet relevant to the common people.

Digital content initiatives

Digital content has become a major issue as PC penetration and Internet access have increased across the country. Without locally relevant content, ICTs are of no use to people. Content development is now a priority not only of the private sector and civil society organizations but also of the government. The content issue has been highlighted in the draft Broadband Policy.

The Government of Bangladesh in collaboration with UNDP, Bangladesh has published many government forms in digital format, both on the Web (http://www.forms.gov.bd) and in CD-ROM format. Several forms can now be downloaded free of charge. However, out of 40 ministries listed in the government website (http://www.forms.gov.bd/eng/ByMinistry.aspx), only eight ministries have partially released their forms. The downloadable forms include passport application, visa application, citizenship form, pension form, Internet connection (Bangladesh Telegraph and Telephone Board [BTTB]), birth registration, income tax return and driving license. The availability of these forms online helps citizens access government services in less time and minimizes opportunities to bribe government officials. The website is bilingual and can thus be used by any literate person. Those who cannot read can get the forms from telecentres, which are now becoming popular in rural Bangladesh.

In 2003, D.Net started research on content development targeting the rural poor. Since then, a huge content base in Bangla has been developed. D.Net initially focused on the CD-ROM version of the content since Internet connectivity was not available in the rural areas at that time. But with the availability of access to the Internet through EDGE or GPRS from almost anywhere in Bangladesh, the Web version (www.jeeon.com) is scheduled for release in 2007.

The largest Bangla website at present is www.abolombon.org. The website is dedicated to human rights issues and provides legal practitioners with access to the full text of laws, the explanation of these laws, addresses of institutions for legal redress, and the like. Another local language website is www.gunijan.org, which features eminent citizens of Bangladesh for the young to get to know them.

Online services

Both government and non-government institutions offer online services, which range from information services to e-commerce.

The government's SICT programme has initiated and in some cases completed over 40 e-governance projects of varying sizes across many government agencies. The SICT programme is scheduled to end by 2006. However, there is talk of extending the project to 2011 with a new budget of about BDT 950 million (USD 14.3 million).

Among the more successful e-government projects is the innovative Ministry of Religious Affairs website (www.bdhajjinfo.org), which provides information-based services to pilgrims, their relatives and friends, travel agents and government officials. The interactive website, which was launched in 2002, can be used for searching information about individual pilgrims (including their current location and status), for sending and receiving messages from individual pilgrims and for accessing various information regarding rules and regulations.

Another successful e-government project is the Rajshahi City Corporation's (RCC's) Electronic Birth Registration System

---

D.Net's demand-driven digital content: Unleashing the potential for poverty alleviation of access to information

Content development at D.Net is unique in many ways. First, the approach is research-based, focusing on the information needs identified by rural communities and the cognition level of end-users. The research identifies two types of users: those who browse content for themselves and infomediaries who browse content for illiterate end-users. Then, raw content collected from various domain institutions is converted into easy-to-understand form and supplemented with pictures. D.Net also develops animated and audio-visual content because in many cases text and pictures are not enough to explain something to end-users.

D.Net's research, which is ongoing, focuses on the whole value chain of livelihood issues to be captured in the content. The content areas are agriculture, health, education, income-generating activities, disaster management, awareness, employment, and directory information. The rural folk visit rural information centres and browse content that addresses their livelihood problems. Through livelihood-focused digital content, hundreds of users might reduce livelihood costs, enhance income-generation opportunities, or protect themselves from potential loss or damage (see the case studies at http://www.pallitathya.org/en/case_studies/index.html).

This demand-driven approach towards content development opens up a whole new area of social entrepreneurship. Rural development organizations now buy the content for dissemination in their intervention areas.

Furthermore, the content-based approach gives a new direction to the global telecentre movement. A telecentre is traditionally a technology learning and communication centre. With D.Net's approach, telecentres in Bangladesh are now able to provide the core service—information and knowledge service. The content plays an important role in improving access to information, which is an economic resource. As poverty is an outcome of lack of access to resources, access to information is the new dimension in poverty alleviation discourse. Digital local language content and its dissemination through ICTs are thus directly linked with poverty alleviation.

(EBRS), which provides citizens with a unique identity card that they can use for various services, such as education and health care. Since the card helps them get certain social services and benefits, citizens are now encouraged to register births, which was previously considered by many to be a worthless hassle. The electronic ID is used for immunization purposes and also for getting admission to government primary schools in Rajshahi. The EBRS helps to keep track of each child registered through the system, starting from immunization requirements to school enrolment status. Since 2001, a total of 45,222 citizens have received birth registration services using EBRS, or 15.38 per cent of the total metropolitan population. In the Bangladesh context, this is a success: it means that almost all newborns are now registered with the electronic system. But there is a need for a campaign that would bring all citizens under the system.

Another laudable e-government initiative is the publication of the salary status of school teachers (http://www.dshe.gov.bd/search.php). School teachers can now check online whether their salary has been sent to the bank by the Directorate of Secondary and Higher Secondary Education.

While some e-government projects have succeeded, others have failed. An example is the Voter ID Card Project (see boxed article below).

Moreover, while providing easy access to up-to-date information is a crucial service the government can provide to its citizens, the information on government websites in Bangladesh is unfortunately not current in most cases.

Private sector online services perform better. An example is www.bdjobs.com, which was established in 2001 and which now has a monthly page view volume of 800,000 and 14,000 daily unique visitors. More than 140,000 résumés are posted on the portal, which has over 2,500 corporate clients. More than 2,500 employers in Bangladesh have recruited more than 35,000 professionals at different levels through the bdjobs.com service.

The most popular online information service provider is www.bangladeshinfo.com, a Web portal for researchers,

The Voter ID Card Project: A disappointing conclusion

The Election Commission Secretariat of Bangladesh maintains a list of eligible voters which is updated every five years. But under this system, citizens are not given a unique identifying number. To correct this inadequacy, a project was undertaken by the Election Commission Secretariat in 1995 to generate laminated ID cards for every eligible voter in the country.

However, five years later the Voter ID Card Project reached a disappointing conclusion. The project was technically completed and some voter ID cards were distributed. But many of the cards had crucial mistakes that rendered them useless. It was a case of management and strategic failure, not technological failure. The media pointed out that the project would have opened up dimensions of government accountability around elections and service delivery not preferred by policymakers and law makers.

The important lessons to be learned from the 'failure' of the Voter ID Card Project are the following:

- Lack of adequate political will and support: From its inception, there seemed to be little political and bureaucratic commitment to the project. The issue of providing ID cards to uniquely identify voters is a politically sensitive one, and without the unequivocal support of all stakeholders, the venture cannot succeed.
- Poor planning: The Voter ID Card Project suffered from poor planning and execution. Many of the mistakes that compromised the project could have been avoided if appropriate strategies had been carefully crafted prior to implementation.
- Over-ambitious target: The project was overly ambitious, with impractical deadlines set and unrealistic expectations.
- No pilot projects: A nationwide programme worth BDT 1,870 million (USD 46 million) should not have been undertaken on a massive scale right from the start. The project should have been carried out in phases, through medium-scale projects that incrementally expanded the Election Commission's earlier successful pilot initiatives, in order to identify risk factors and other lessons before extending the project to a national level.
- Inadequate cultural consideration: The project did not take into account cultural issues, such as the resistance of some women to having their photographs taken. An awareness campaign and other special measures should have been undertaken to persuade women to sit for the photographs.

academics and policymakers. It currently hosts more than 2,000 papers, articles and book chapters on Bangladesh and South Asia published by prominent research and publication houses. The website has incorporated an innovative mechanism of selling research online through prepaid cards.

ICT and ICT-related industries

The ICT market in Bangladesh is estimated to be worth approximately BDT 11,000 million (USD 165 million) per year (excluding the telecom sector). Of the total domestic ICT market in Bangladesh, computer and network hardware has the lion's share with BDT 6,000 million (USD 90 million), while the software segment amounts to BDT 1,700 million (USD 26 million). The rest is to the account of Internet and network services and other ITES (ICT-enabled services) segments.

In 2006, the number of hardware and software companies increased to 2,500 and 350 respectively, from 1,200 and 100 respectively, in 2000. The number of Internet service providers increased fivefold (from 30 to 150) and the number of training and similar service providers reached 150 in 2006 from 100 in 2000 (BCS 2006).

ICT outsourcing is gradually gaining ground in Bangladesh. Recent export statistics show that over the last two years, export volume grew by about 70 per cent per year. In financial year 2006, the total volume of export of software and ICT-enabled services was USD 26 million, which was more than double that of the previous year. In financial year 2002, the volume of such exports was only USD 2.8 million (EPB 2006). Although the total volume of export is still very low compared to the export figure of the software giant India, this sector is becoming mature and has begun to compete with Indian outsourcing companies. Also, many Indian off-shore companies are setting up offices in Bangladesh. Bangladeshi software and ITES providers have also been able to attract foreign direct investments through joint-venture projects with European partners. In 2005–06, delegations from as many as 40 European and North American companies visited Bangladesh, resulting in 12 joint venture projects.

Enabling policies and the regulatory environment for ICTs

Public policies and private programmes make up an environment for ICT-led economic growth and development. In many cases, the government's attitude of non-interference has facilitated growth.

In keeping with the designation of ICT as a 'Thrust Sector', the Ministry of Science and Technology was renamed as the Ministry of Science and Information & Communication Technology and the National Policy on Information and Communication Technology was adopted in 2002.

The government abolished import tax and VAT on computer hardware, software and accessories in 1998, bringing down significantly the cost of computers at the retail outlets. Now even low-income households can afford to have PCs. PC importation grew more than 35 per cent during 2000–2005 and the current number of PCs stands at 1.5 million (BCS 2006).

The National Telecommunication Policy 2000 paved the way for competition, which facilitated the achievement of a teledensity of 8.4 per cent within five years.

The establishment of the Bangladesh Telecom Regulatory Commission (BTRC) in 2002 was also a step in the right direction. BTRC has full authority to grant licenses to all providers of telephony, data, networks and content services. However, some of its regulatory functions overlap with those of the BTTB, and market distortion has not yet been fully corrected.

A draft Information and Communication Technology Act was formulated and sent to the Parliament for approval in 2002. Now known as the IT Act 2006, the law contains provisions for an Electronic Certification Authority which will be established to issue licenses to entities for electronic signature authentication. Under the law, electronic documents will be accepted for various types of transactions.

A weak point in the draft law is that it is not mandatory for the government and its agencies to accept, or prepare, electronic records or deeds (Chapter 1, Article 11). This will create problems for institutions intending to deal in electronic deeds, particularly the financial institutions. It also creates an opportunity for government agencies to avoid adopting e-governance.

The Act also stipulates a maximum of 10 years of imprisonment for violation of security and privacy and damage to information systems and networks, as well as for sending spam e-mail. The penalty for electronic publishing of information is five years. However, this runs counter to the freedom of the press and of expression, which are fundamental rights under the Constitution.

The draft Act was first presented to the Council of Ministers in 2002. It was returned to the Law Commission for review. It was then presented to the Council of Ministers as IT Act 2006. However, the Council has failed to enact the Law. Thus, the potential for e-commerce remains unexplored due to the lack of a legal framework for e-transactions.

To protect the ICT sector and the country's software products, the government has amended Copyright Act 2000. However, due to lack of law enforcement, software piracy continues to be a problem.

Education and capacity building programmes

The National Education Policy of Bangladesh recommends compulsory computer courses at the secondary school level. The policy mandates ICT integration by 2010 in Bangladesh's 5,694 and 15,748 junior and secondary level institutions, as well as 922 colleges, 347 professional institutes and 1,462 mid-level technical and vocational institutes. It also recommends the introduction of ICT education in the 12 new science and technology universities being established by the government, as well as the introduction of a nationwide central examination system to maintain quality standards in both formal and non-formal ICT education provision (BCS 2006).

Between 2000 and 2006, the number of ICT professionals in Bangladesh increased from 11,440 to 25,200 (BCS 2006). Public universities play an important role in supplying the market with ICT professionals. The Bangladesh University of Engineering and Technology (BUET) first introduced formal education in Information Technology in 1984. At present, 10 public universities are offering postgraduate diploma programmes in information technology. Private universities have been offering various ICT degrees since 1995. However, in terms of quality, only a few universities out of more than 50 can supply professionals who meet industry requirements.

There are more than 350 training institutions in the private sector producing different categories of ICT professionals. Many of them are franchised institutes of the National Institute of Information Technology (NIIT), APTECH, CMC, TULEC, NCC and many other foreign institutes. However, it has been reported that many of these institutes charge high fees from students while failing to maintain minimum standards of course delivery.

Government employees are increasingly becoming ICT-literate. About 28 per cent of officials and 29 per cent of Ministry/Division staff have received ICT training. At the Department/Corporation level, about 23 per cent of officers have received ICT training (BBS 2005).

There are also several private and civil society initiatives for developing the ICT skills of rural children. The Computer Literacy Programme run by the Volunteers' Association for Bangladesh, New Jersey (http://www.vabonline.org/vabnj/) and D.Net, as well as the School Online programme (www.ri.org/countries.php?cid=4) operated by Relief International, are particularly noteworthy.

Open source and open content initiatives

The Bangladesh Open Source Network is leading the open source movement. A number of institutions, including the Bangla Innovation through Open Source (BIOS) and Ankur (www.bengalinux.org), have played a pioneering role in popularizing the concept of open source.

Open content has also become popular. D.Net was the first organization in Bangladesh to launch Bangla websites built on the open content concept. Other noteworthy websites with open content are http://www.pallitathya.org/, www.abolombon.org, www.meghbarta.com and www.gunijan.org. Bangla wikipedia (http://bn.wikipedia.org/wiki) already has more than 15,000 entries.

Research and development

There are sporadic R&D initiatives related to ICT. Most have to do with localization and Bangla computing. Ankur (http://www.bengalinux.org/new/content/view/53/35) is one of the most prominent initiatives. The Bangladesh Open Source Network is the apex body, leading localization and open source initiatives. In addition, BRAC University is implementing a localization project under the PAN Localization Project supported by the Pan Asia Network of IDRC. BRAC University is developing the Bangla OCR, a spell checker and search engine.

There are ongoing robotics projects at the BUET. Some of these have received international recognition, such as the 2005 Robocon Panasonic Award in Beijing, China.

D.Net is conducting research on reduction of telecentre operating costs and enhancement of income opportunities in order for telecentres to be financially independent while serving the rural community.

Challenges

The digital opportunity index (DOI) for Bangladesh is only 0.20, and the country's overall ranking is 139th, which is better only than Nepal among its South Asian neighbours (ITU 2006). This proves that Bangladesh still has a long way to go in putting ICTs in the mainstream of efforts to achieve economic development and social well-being.

Investment is the biggest challenge both for the private sector-led ICT industry and ICT for development. Government allocation for ICTs remains small and scattered. Although the national ICT policy announced an allocation for ICTs of 2 per cent of the national budget, in 2006 the allocation was only 0.8 per cent. It is also unclear what the nature of the spending on ICT might be, due to ambiguous budget line items. Investment in telecommunications was relatively better. However, investment in human resource development is still missing.

Expenditure for rural ICT infrastructure is still not on the radar of development partners. The private sector is not likely to invest in ICT infrastructure in the rural areas until it becomes a licensing requirement. It is not clear to these groups that rural Bangladesh can be changed through ICTs, if proper infrastructure is made available with an appropriate pricing policy. However, grassroots initiatives show great potential.

Addressing the gender dimension of access to information through ICTs is also a challenge. The successful functioning of D.Net's Help Line in rural areas could contribute to women empowerment through ICTs. Through the project, the educated women in the villages, called 'mobile ladies', became the symbol of empowerment in the villages by providing access to livelihood information and advice of experts using mobile phones. The concept, which received the 2005 Gender and ICT Award at the World Summit on Information Society (D.Net 2005), must be replicated across the country. But once again, financing is a problem.

The institutions and government mechanisms are also a major challenge. The Ministry of Science and ICT is not an 'important' ministry for the government. A residual attitude to science and ICT still prevails among policymakers, and the Ministry is often a government dumping ground: people within the government who do not fit elsewhere are assigned to the Ministry. As a result, the Ministry has difficulty taking crucial decisions. It also does not receive adequate budgetary allocation to make its policy commitments a reality. Multiple centres of authorities add to the problem. As mentioned, the Prime Minister's Office took over a number of responsibilities regarding ICT promotion, which literally made the Ministry redundant. Institutional restructuring is urgent, but policymakers seem reluctant to pursue this. If telecommunication is not merged with the Ministry of Science and ICT, then it would be difficult to address many important issues, such as convergence.

The private sector, NGOs and civil society organizations are vibrant in Bangladesh. They are the driving force of economic progress, despite all odds. With effective and efficient government, Bangladesh could become a middle income country within a decade.

References

BBS. (2005). *Data collection and dissemination of ICT statistics: The Bangladesh experience*. Dhaka: Bangladesh Bureau of Statistics.

BCS. (2006). *Industry profile and statistics: Bangladesh*. Retrieved 25 October 2006 from http://www.asocio.org/member/BCS/Bangladesh-Profile.pdf

BTRC. (2006). *Press release of BTRC*, September 2006.

Chowdhury, M.R. (2001). *ICT education in Bangladesh: A review*. Dhaka: Bangladesh University of Engineering and Technology.

D. Net. (2005). *Pallitathya Help Line: A precursor to people's call centre*. Dhaka: D.Net, November 2005.

D. Net. (2006). *Twenty most important services and information for citizens assessment of Bangladesh*. A report prepared for the World Bank, Dhaka, June.

Export Promotion Bureau (EPB) Bangladesh. (2006). *Bangladesh Export Statistics 2005–2006*.

ITU. (2006). *World Information Society Report 2006*. Geneva: International Telecommunications Union, August.

Mumit, K., Munir, A.A., and Raihan, A. (2007). *Bangladesh country report on local language computing policy initiatives*.

The Law Commission, Government of Bangladesh. (2002). *Law of Information Technology*. Working paper.

UNDP. (2005). *UN human development report 2005*. Retrieved from http://bangladeshictpolicy.bytesforall.net/?q=node/210

Bhutan

Sangay **Wangchuk** and Gopi **Pradhan**

Total population	634,982 (2005)
GDP per capita (USD)	USD 1,320.90[a]
Key economic sectors	Hydropower, Tourism, Minerals, Forest products
Computer per 100 inhabitants	1.5[b]
Fixed-line telephones per 100 inhabitants	4.7[c]
Mobile telephones per 100 inhabitants	12.19[d]
Domain names registered under .bt	247[e]
Internet domestic bandwidth	2 Mbps[f]
Internet International bandwidth	30 Mbps[g]

Sources: [a]Cabinet Secretariat; [b]DIT, MOIC; [c–g]Bhutan Telecom and Druknet.

Overview

Bhutan is located on the eastern foothills of the Himalayan mountain range, nestled between two Asian superpowers—India and China. Covering an area roughly the size of Switzerland, the country is divided into 20 districts called *dzongkhag* and further subdivided into 201 communities or *geog*. Bhutan is a mountainous country, with rugged terrain that is often difficult to traverse. As of the 2005 census, it had a population of 634,982.

Despite the difficult terrain and sparsely populated human settlements, people in Bhutan have access to terrestrial, satellite and mobile telephones, radio, print media, television and the Internet. Having recently introduced modern ICTs, Bhutan is an example of a nation that is 'leap-frogging' over older and often obsolete technologies in the ICT sector. However, there is still a marked disparity between ICT access and usage across Bhutanese society. While urban citizens are making use of the latest ICT gadgets, rural folk have only basic telephone services.

Technology infrastructure

Analog radio was introduced in Bhutan on 11 November 1973. Today, radio remains the main source of information and entertainment in Bhutan's remote villages. Bhutan Broadcasting Service (BBS) is the only broadcasting agency in the country, providing both short wave (SW) and frequency modulation (FM) services. Radio reaches all parts of the country and is the most prolific form of media in all 20 districts. News, entertainment and development-related programmes in health, agriculture and education are aired daily for 12 hours in four languages—Dzongkha, Nepali, Sharchop and English.

In the last two decades, Bhutan's telecommunication network has evolved from a physical wire network to a digital network. The first television network and Internet Service Provider (ISP) were launched on 2 June 1999. Within six years of its introduction, television, particularly cable television, has reached all districts. There are 35 cable operators and about 38,000 cable television subscribers, with the majority concentrated in the capital city of Thimphu and the urban town of Phuentsholing, near the Indian border. BBS broadcasts local news and national programmes. With INSAT4A satellite broadcasting, BBS can be received in 42 countries throughout Asia Pacific. This satellite broadcasting was launched on 20 February 2006 with financial support from the ITU.

Bhutan's first telephone was installed in 1963 while mobile telephone services were introduced in 2003. Overall teledensity (including both fixed and mobile subscribers) as of December 2006 was 17.9 per cent—32,123 fixed telephone lines (5 per cent of the population) and over 82,000 mobile subscribers (12.9 per cent of the population). Within a year, the number of fixed telephone subscribers increased by 4,685 lines, while the number of mobile phone subscribers increased fourfold from 19,000 in 2004. Mobile services are currently provided by B-Mobile, which is fully owned and operated by Bhutan Telecom (BT). B-Mobile uses GSM technology and provides voice and SMS services. The number of mobile users continues to increase at a steep rate. According to B-Mobile, its service covers 16 districts and 12 satellite towns and it plans to provide service to the remaining seven districts by 2007.

At present, there are three ISPs in Bhutan: Druknet, Samden Tech and Drukcom. Druknet is fully owned by BT and is the oldest ISP in Bhutan. As of September 2006, it had 6,000 dial-up and 55 leased line account holders, mostly from government, semi-government and international agencies. Druknet's main Internet node is located in Thimphu with nine other points of presence (POPs) in different parts of the country (Trashigang, Mongar, Bumtha, Chukha, Samtshe, Samdrup Jongkhar, Paro, Wangdi and Trongsa). In May 2006, Druknet doubled its international bandwidth from 10 Mbps to 20 Mbps to accelerate Internet connectivity and overcome peak hour congestion. Druknet has three satellite upstream links with KKDI Japan, British Telecom and Lrel Skynet. SamdenTech Private Limited was established in May 2005 and provides a bandwidth of 2 Mbps (512 Kbps upstream and 1.5 Mbps downstream) through a satellite upstream link with Europe Star. Among its 50 customers, there are 30 leased lines and 20 small volume users. With the substantial increase in the number of mobile subscribers, SamdenTech aims to provide Internet services through mobile phones.

According to Druknet, there are approximately 20,000–25,000 Internet users, 13,000 e-mail accounts, 139 websites and 247 domain names registered in Bhutan as of 1 September 2006. Of the 10,000 computers in the country, it is estimated that 3,000–3,500 computers are connected to the Internet using leased lines. Two private VSAT operators were licensed in 2004 to provide Internet and wireless broadband services with a view to introducing competition in the value-added services market.

Key ICT institutions

The Ministry of Information and Communications (MoIC), which was established in July 2003, is the lead agency for Bhutan's information and communication sector, including telecommunications and traditional and new media. Under MoIC are the Department of Information Technology (DIT), Department of Information and Media, Road Surface Transport Authority, and Department of Civil Aviation. DIT is the national focal agency for the development, promotion and coordination of all ICT-related activities in the country. Specifically, it is responsible for policy formation, coordination and implementation of donor-assisted projects, and governmental communication planning, including standardization of ICT products. The Department of Information and Media is responsible for planning and research in support of mass media. It is also responsible for the drafting of policies and regulations related to information contents and media.

With the enactment of the Bhutan Information, Communications and Media Bill (also known as the ICM Bill or Act) in July 2006, the erstwhile Bhutan Telecom Authority has become the Bhutan Infocomm and Media Authority (BICMA). The new ICM Act gives BICMA full independence in exercising its functions and responsibilities, which covers regulation of ICT facilities, ICT services, spectrum management and radio communications, content and media.

Apart from these major ICT institutions, all ministries and important government agencies have ICT divisions that play a strategic role in promoting the development of ICT in their respective ministries and agencies. The ICT divisions also work with the MoIC to coordinate ICT activities across the country.

ICT industries

Although computers were introduced in Bhutan in the early 1980s, it was only since 1999, after the introduction of the Internet, that a handful of private sector firms, most dealing with computer supplies, came into the market. With the government policy to outsource most of its ICT developmental activities to the private sector, the lifting of import duties for ICT products, and tax holidays for new ICT businesses, the growth of ICT businesses looks promising. While there is no noticeable increase in the number of firms, the size and dimension of existing businesses have improved. The 15 firms previously dealing only with supplies have now expanded their services, providing both software development and network solutions (*Bhutan E-Readiness Study* 2003). The same is true with the 13 training institutes. There are also a few repair and maintenance shops located mostly in the urban centres.

Furthermore, as of October 2006, the government has licensed two contact centres (or call centres) to private companies—one to serve as a call centre and another to provide medical transcription services. The selection and recruitment of employees and their training for these two businesses are being supported by the government through the DIT. The centres will commence operations in March 2007. Call centres are considered a potential growth industry, with the capacity to generate employment.

ICT policies and regulatory frameworks

Bhutan's ICT policy repertoire has expanded considerably in the past five years. The Bhutan Information and Policy Strategy (BIPS) issued in October 2004 and the ICM Act of 2006 are strong policy and regulatory instruments for the promotion of ICT in Bhutan today.

The BIPS addresses five key areas—Policy, Infrastructure, Human Capacity, Content and Applications, and Enterprise. Underpinned by the philosophy that 'with people at the centre of development, Bhutan will harness the benefits of ICT, both as an enabler and as an industry, to realize the Millennium

Development Goals towards enhancing Gross National happiness', BIPS serves as a road map for ICT development in Bhutan. It covers three overall policy objectives:

1. to use ICT for good governance;
2. to create a Bhutanese info-culture; and
3. to create a 'High-Tech Habitat'.

The Information, Communication and Media (ICM) Act enacted in July 2006 at the 85th National Assembly provides a modern technology-neutral and service sector-neutral regulatory mechanism to implement convergence of information, computing, media, and communication technologies, and to facilitate privatization and competition in the establishment of ICT and media facilities, interconnections, universal services, e-services, activities related to cyberspace, and media operations. The Act also provides the new regulatory regime for BICMA to independently exercise its power to regulate all aspects of ICT activities and resolve disputes between operators. The Act facilitates the formalization of interconnection and infrastructure-sharing frameworks to enable converged data, voice and video services, and thereby meet universal service and access obligations. Other service providers can now secure converged service licenses to provide voice, data and video traffic over Bhutan's modern networks. This translates to reduced costs for ICT services.

The MoIC is also preparing other guidelines and regulations for electronic signatures, e-business and e-commerce, content regulation, security policies, and information management.

Enabling ICT projects

Thimphu WAN

A wide area network for Thimphu is currently under construction. This fibre network will connect all of the government agencies in the capital city. Completion of the project will facilitate the use of existing and future government information systems. It will also provide a fast, reliable and secure platform for e-governance and intra-government communication and information sharing.

District LAN

The DIT has also constructed, in each district, a local area network with 64 Kbps leased lines to connect the districts to the Internet. These leased line connections are leveraged to connect all of the surrounding offices with wireless technology. District Local Area Networks (LANs) have greatly enhanced the efficiency and effectiveness of the district administration offices and improved communication between the districts and the central government. However, much more needs to be done in developing and managing the content of the district websites.

Community Information Centres (CIC)

Several community information centres are being set up to provide integrated ICT access to rural communities and thus mitigate the negative effects of the digital divide. The 10th Five-Year National Development Plan states that the DIT shall establish at least one CIC in each of the 201 *geog*. Each CIC will be equipped with computers, Internet connectivity, a telephone, a fax machine, and photocopying facilities. The intent is to use these tools to improve access to relevant information that would enhance the health, education and livelihood of the villagers. The Bhutan Portal (http://www.bhutan.gov.bt) would be a main source of information, and its contents—which will include e-health, e-education, e-agriculture, and e-commerce/e-business services—are being developed based on the results of the information needs assessment at each CIC.

To date, the DIT in collaboration with Bhutan Post and the Ministry of Agriculture, has established 30 CIC with support from the United Nations Development Programme (UNDP), Government of India, International Development Research Centre (IDRC) of Canada and Microsoft Unlimited Potential. An additional 50 CIC are to be established before June 2007 mainly through the financial support of the Asian Development Bank, UNDP, Microsoft and IDRC.

Online services

The Royal Government of Bhutan has set forth a policy to provide 75 per cent of its services online by 2010. In order to meet this goal, all government and semi-government agencies are exerting considerable effort to publish information about the services they offer on their respective websites. Almost all of the government ministries and agencies have a Web presence, including the National Elections Commission, which is providing online voter registration forms in preparation for the first democratic election in 2008. The majority of government acts, publications, reports, statistics, forms, guidelines and application procedures are available online.

However, most of the current online applications in Bhutan were developed to address the internal needs of organizations and do not support inter-operability or communication with the systems of other agencies. Even if systems support internal efficiencies, they do not fully exploit the opportunities for interconnection and integration. Also, the majority of the applications

> ### The Tangmachu community information centre
>
> The CIC established in Tangmachu (located in the Lhuentse district of Eastern Bhutan) with the support of IDRC is among the most notable of CICs in Bhutan to date. The centre is located in the remotest district in the east. It has five computers connected to the Internet via IP Broadband VSAT with a bandwidth of 64 Kbps and an asymmetrical connection (downlink 64 Kbps and uplink 32 Kbps) and low-cost wireless phones (802.11 b/g specification). Approximately 100 households, representing about 90 per cent of the community in this locality, are connected with IP phones using Wi-Fi technology.
>
> The inhabitants of nine remote villages in Tangmachu use the CIC for telephone services (VoIP and wireless loop-based telephony), Internet access, printing of documents, photocopying, and use of business planning and management tools. The centre is used not only for information sharing among the communities in the district, but also for attending to emergencies such as in health care. An illustrative case is that of Ap Tandin, who was found seriously ill by field engineers when they went to his house to install an IP phone. The nearest health facility (a Basic Health Unit or BHU) was two hours walking distance from his house. The project team made a call to the BHU for emergency help using the IP phone. Following instructions by the health assistant to bring in the patient for treatment, the engineers carried Ap Tandin on their back and brought him to the BHU. Ap Tandin was to be kept under observation for one night. In her anxiety, his wife forgot to bring food and blankets. She was about to leave her ailing husband to walk the two hours home to get the supplies, when the engineers advised her to use the newly installed Wi-Fi phone in her home to call her daughter and have her bring the necessary items to the BHU.
>
> The CIC also connects the BHU, the village headman's office, the middle secondary school, the primary school, and the Renewable Natural Resources Centre. Content and applications for use at the CIC are being developed in accordance with the information needs of the villagers and include agriculture and education, in addition to health. Applications like e-learning and online marketing are also incorporated in the system. The interface is in Dzongkha, the local language.

are government-centric and do little to ease the problems of the general public in accessing these services.

Some good examples of e-governance initiatives are as follows:

- The Automated Border Management System (ABMS) aims to achieve an integrated cross-sectoral approach to border management with efficient and secure flow of data/information between the various stakeholders.
- The Bhutan Civil Registration System (BCRS) was developed by the Ministry of Home and Cultural Affairs to collect, store and update data about every Bhutanese citizen. The system facilitates the issuance of new citizenship ID cards which are not only handy and convenient but also valuable for use in other information systems, including the banking, health care and education systems.
- A machine-readable passport system was developed by the Ministry of Foreign Affairs to facilitate immigration checks, minimize errors and reduce the chances of forgery.
- The Central Bank of Bhutan, Bank of Bhutan and Bhutan National Bank joined SWIFT in 2005. The latest version of SWIFT, running at the Royal Monetary Authority (RMA), is linked to the SWIFT centre in Chennai, India via leased line. The implementation of SWIFT at the RMA was carried out with assistance from Scandent Solution, a SWIFT service bureau in Mumbai, India.

Recently, Bhutan's print media welcomed two new entrants. *Kuensel* (www.kuenselonline.com), the only national newspaper until June 2006, was joined by two government-licensed newspapers, *Bhutan Times* (http://www.bhutantimes.com/modules/headlines/) and *Bhutan Observer* (www.bhutanobserver.com). *Kuensel* is published twice a week while the other two newspapers are weeklies.

The most frequently visited websites at present are the Bhutan Portal, Kuzoo, Kuensel, *Bhutan Times*, The Job Portal and Druknet. Most of the tourist and travel companies also have an online presence to further their businesses.

The Bhutan Portal (http://www.bhutan.gov.bt) was developed through a UNDP-sponsored project titled 'Pilot Public Access to Information and Service'. The website contains a host of useful information and materials, including forms, publications and government legislation. It also contains links to the websites of all government, semi-government, non-government and private organizations. The Bhutan Portal is a one-stop source for information about the country and is a window for providing e-services online.

The Job Portal (http://www.molhr.gov.bt/DHR/) was launched by the Ministry of Labour and Human Resources on 11 March 2005, to provide a platform for job seekers and job providers to obtain information and interact for mutual benefit. It allows registered employers to post job vacancies, which are categorized according to occupations and posts. Job seekers are able to receive notifications of the vacancies that match their preferred occupations or job category.

Kuzoo (www.kuzoo.net) was put up by the present King of Bhutan to serve as a platform for Bhutanese youth to meet and keep in touch. Young people can manage their own online profiles, connect with their friends, post their own blogs and announce and track upcoming events and social gatherings, among others. The website is becoming quite popular and is featured on Kuzoo FM 90, a live radio station airing youth-oriented programmes, including music.

Applications for development

The rural telecommunication project

Providing access to basic information to all Bhutanese is a key concern in the roll-out of ICT infrastructure and services. Despite the increasing rate of rural–urban migration, the majority of Bhutanese still live in the rural areas. The Rural Telecommunication project being implemented by Bhutan Telecom aims to provide at least 10 telephone lines in each of the 201 *geog* by end of 2007. As of December 2006, 21 per cent of the communities still do not have any form of electronic connectivity (Pradhan 2005).

There is also a plan to upgrade the current telecommunications network by 2010 to 155 Mbps from the current 34 Mbps digital microwave transmission system.

Agriculture

The Ministry of Agriculture, with support from the Food and Agriculture Organization (FAO), is developing the Virtual

ICT and rural development

Two-thirds (69.1 per cent) of Bhutanese live in rural and remote areas where they face many hardships. Only about 8 per cent of the country's land is cultivable. This small percentage of cultivable land is also due to stringent environment regulations that aim to keep 62 per cent of the land under forest cover at all times. The reverse effects of the strict conservation policies on farming are evident in the growing public outrage at wildlife destruction of farmlands and livestock. When 85 per cent of the rural population depend on agriculture and livestock, the smallest impact on productivity can have significant repercussions.

Rural–urban migration is another issue confronting Bhutan. Educated youth are leaving villages in increasing numbers for better opportunities in towns and cities. The net rural–urban migration is 14.45 per cent. A number of policies have been initiated to promote rural development. One promising solution is innovative exploitation of ICT at all levels of society. Bhutan has already implemented a number of community-focused ICT projects. The e-Post project (a combination of traditional postal services and electronic communications) has benefited some of the most remote and isolated communities. Microsoft's CIC project, implemented in partnership with the Ministry of Agriculture and Bhutan Post, has been enhancing the IT skills and knowledge of communities. This will go a long way in building a critical mass of users with sufficient knowledge to be able to benefit from the information revolution in the future. IDRC has piloted a very successful rural connectivity project in Tangmachu, giving much hope to other communities. The UNDP for its part, supported some of the first rural telecentres in Bhutan.

However, despite its commitment and consistent efforts, the Government of Bhutan will continue to face challenges in rural development. Illiteracy is an important aspect of the challenge. No matter how sophisticated or affordable the solutions provided, people should have a sufficient level of literacy. The affordability of facilities and services in rural areas is also a key concern. While the average Bhutan farmer earns only BTN 10 an hour (about USD 0.238), the cost of a national call is BTN 100 (about USD 2.38) per hour. Third, content and applications need to be useful and understandable. At present, most if not all digital content is in English, a language that has very little use in the rural areas. Fourth, coordination among public sector offices in the rural areas is essential for the provision of a one-stop shop for poor people who cannot afford multiple access points to information.

Extension Research and Communication Network (VERCON), which aims to use the Internet to strengthen communication linkages among the research and extension service components of the national agriculture knowledge and information system. The overall goal of VERCON is to improve agriculture advisory services to Bhutanese farmers in order to increase agricultural production through improved research-extension linkages. VERCON will have two inter-dependent components: (a) the human component, consisting of a network of people committed to communicating, sharing information and supporting agricultural producers; and (b) the technology component, consisting of an Internet-based tool for information development, sharing, storage, retrieval, dissemination and communication. VERCON is being piloted in three regional research centres and in the headquarters of the Ministry of Agriculture. There is a plan to expand VERCON to RNR Extension Centres at the community levels.

Health

The Ministry of Health is integrating the different components of its health services under one system called Bhutan Health Management Information System (BHMIS). The aim is to facilitate monitoring of changes in disease incidence/prevalence and thus be able to prioritize interventions at all levels for both modern and traditional medicine. The BHMIS will have the following components:

1. Referrals outside Bhutan: A database with administrative, clinical and financial information relating to patients referred outside the country for medical care.
2. Telemedicine: To rationalize the use of scarce specialized resources within the country or specialized services from outside the country using IT. While telemedicine has been introduced in six hospitals, it has not been fully operationalized due to lack of bandwidth and proper equipment. E-mail is being used to send ultrasound and X-ray images.
3. Individualized Patient Records (IPR): To store and easily retrieve administrative and clinical information relating to a patient within a health care facility. This system has been developed and is being piloted at the Thimphu General Hospital.
4. Laboratory Information System: To monitor diseases like HIV/AIDS and to keep track of clinical tests. This component is under development.
5. Pharmacy Information System: To facilitate the management of the pharmacies within health facilities, including tracking the movement of drugs. This component has been developed and is being used by the pharmacy department of the Jigme Dorji Wangchuck National Referral Hospital in Thimphu.

Education

The Ministry of Education is currently implementing the Information Technology in Education Master Plan. Through an annual government fund and donor contributions, all 23 higher secondary schools and 16 middle secondary schools have been provided with computers. In March 2006, the DIT, with financial support from the Government of India, provided computers, printers, and Internet access and training to 100 primary schools across the country. This project complements the efforts of the Ministry of Education towards 'ICTization' of schools and implementing ICT curricula at all levels of basic education by 2010. However, more work needs to be done. Although the schools can be considered to be 'connected', the rather high student-to-computer ratio translates to limited exposure and use.

In terms of teacher training, Sherubtse College and the IT Education section of the Ministry of Education are offering a Post Graduate Certificate in Teaching Information Systems (PGCTIS) for in-service teachers in middle and higher secondary schools. The National Institute of Education (NIE) has likewise been offering IT as an elective subject in the Bachelor of Education programme since 2004 to ensure a steady flow of teachers who can teach computer courses in schools. Moreover, all teachers, regardless of specialization, are required to undergo an IT literacy programme called Functional IT, which was introduced in 2002.

The two principal ICT training providers in Bhutan are Sherubtse College and the Royal Institute of Management (RIM). Sherubtse College offers a Bachelor of Science Honours programme in Computer Science while RIM provides a two-year Diploma in Information Management Systems (DIMS) for those who have completed class XII. In addition, both institutes offer short-term and ad-hoc courses. Approximately 100 IT personnel from civil service and private organizations received training between 2005 and 2006 in the CISCO Certified Network Academy jointly managed by the DIT and RIM.

In addition to the state-run training institutes, there are 13 private IT training institutes offering standardized curricula in basic computer courses such as network administration, Web development, and desktop publishing. Although there are currently no institutes for more advanced vocational ICT skills such as telecommunications, mass communications and media studies, there are a wide range of courses offered, from basic to diploma-level courses in communication and information management system.

At the moment, there are approximately 500 ICT personnel in Bhutan, including around 300 certificate holders, which indicates a severe shortage of qualified ICT professionals. The Ministry of Education's IT Master Plan and the academic programmes mentioned earlier aim to address this problem.

Open source initiatives

Although a written policy on open source does not exist, in practice, open source software like the Linux operating system is the *de facto* choice for servers in Bhutan. Most of the government-funded application projects seriously consider free and open source software (FOSS) and standards as a viable option. FOSS such as Apache for Web hosting, Squid proxy for Internet connection, and BIND for DNS are becoming almost a standard. PHP for Web scripting and MySQL and Postgresql for database are also commonly used.

A significant illustration of the open source option coming to fruition is the successful development and launch of *Dzongkha Linux*. Funded by IDRC, this project forms part of a regional network examining tools and technologies to 'localize'—that is, to develop fonts for commonly spoken languages.

Use of FOSS is also a feature of the digital library project of the DIT in collaboration with the University of Virginia. Preservation and promotion of tradition and cultural values is mandated by the Bhutan Information and Policy Strategy thus: 'By 2008, Bhutan will use ICT to preserve and promote its cultural heritage and boost the creation of local content.' Bhutan's digital library is based on FEDORA (Flexible Extensible Digital Object Repository Architecture), an open source digital library developed by the University of Virginia. According to the DIT, this project aims to develop an invaluable record of Bhutanese culture for Bhutanese students, researchers, tourists and anyone interested in Bhutanese culture. It aims to bring all stakeholders together to develop one digital library encompassing bibliographies, folktales, digital reprints of religious and ritual texts, journals, and audio and video collections. It will create online cataloguing records, transcripts, and analysis of audio-video collections of all Bhutanese cultural contents and other artefacts. There will also be an online *Dzongkha–Dzongkha* dictionary with 23,000 entries and an online *Dzongkha*–English dictionary with 13,000 entries (Dzongkha Localization Project 2006).

In another open source project, the DIT has completed the localization of the open source *Drupal* content management system, which allows individuals and communities to publish, manage and organize a variety of content on a website. The department is planning to organize training sessions for users in March 2007.

Research and development

There is a keen appreciation for the importance of research and development in the ICT sector in Bhutan.

Since 2005, a great deal of research on localization has been carried out under the PAN Localization Project (see www.panl10n.net) of IDRC. *Dzongkha Linux*, an operating system with *Dzongkha* desktop, is a product of two years of R&D during which many new technical terminologies were coined and about 90,000 text messages were translated. Sorting rules, which were non-existent, had to be researched and embedded into the system. A whole range of open source engines, such as rendering engines for proper display of *Dzongkha* text, was developed. *Dzongkha Linux* supports all open source office applications, including Web browsers, chat programmes and an image manipulation programme equivalent to Adobe Photoshop, and all these have a *Dzongkha* interface. Other R&D projects to be pursued over the next three years are a text-to-speech system, a spell checker for *Dzongkha*, a morphology analyser, a word segment algorithm, letter-to-sound rules for *Dzongkha*, and an online lexicon. All will be developed using an open source platform.

To integrate ICT into the education system, a distance education service is under pilot implementation at Samtse district. Beginning in 2003, the project supported a teacher education programme using Web-based technologies. To date, the project has helped upgrade the NIE's internal networking system, created a website to allow students to access academic information online, and installed new software for supplementary online learning materials, tutorials and academic counselling. Moreover, findings from the project have informed the development of a broader Distance Learning Technology (DLT) project (www.pandora-asia.org) consisting of nine sub-projects researching DLTs in 11 countries across Asia.

On the technology side, there is a limited amount of research on new products being carried out at the DIT to determine the right choice of technology for adoption as a standard. Other research worth mentioning are ICT impact studies and information needs assessments that are conducted for policy formulation.

The challenges and opportunities ahead

Much has been achieved in the ICT and telecommunications sector of Bhutan. However, the sector needs to address several challenges to achieve more.

The first challenge is the cost of infrastructure. Basic telecommunications infrastructure needs to be deployed throughout the country. However, due to difficult mountainous terrain and sparse human settlements, the cost of providing basic ICT facilities and infrastructure to every village and community is huge. Bhutan Telecom estimates that it invests as much as BTN 200,000 (about USD 4,761.90) to install a telephone in a remote village. Despite its plan to provide at least 10 telephones in each community by end of 2007, Bhutan Telecom will still not be able to connect six remote communities that need more than USD 30,000 each to have a telephone system installed.

A second challenge is addressing 'capacity voids'. Specifically, the country requires significant investments in building capacity among its ICT and telecom personnel. Today Bhutan has only around 500 IT personnel in the public and private sector with a sufficient level of skills and knowledge of ICT. As the networks grow, businesses flourish, and opportunities expand, the capacity void is felt more and more. It is not only that skilled people should join the market to fill the vacuum; they also need to come with new and better skills and knowledge to keep up with the demand.

The third challenge is lack of ICT standards. Though ICT has proliferated in the last two decades, they are bifurcated and isolated from each other. Many of the systems currently being developed and deployed by different agencies adhere to different standards, resulting in lack of interoperability and eventually, limited efficiency and scalability. The cost of interconnecting existing applications and networks will be huge.

The fourth challenge is slow private sector growth. Most of Bhutan's few qualified ICT personnel are employed in the public sector, which is a contributing factor to the fledgling state of the country's private sector. Unless the government takes firm action to uplift and promote the private ICT sector, ICT development in the country will be confined to the public sector.

Despite these challenges, the future of ICT and telecommunications development in Bhutan is very bright. Bhutan is a peaceful and politically stable country that has been able to avoid many of the disadvantages that accompany legacy systems and obsolete infrastructure and applications. In addition, the government is aiming for 100 per cent coverage of electricity and telecommunication infrastructure by 2020. Offshore businesses, such as business process outsourcing (BPO) and contact centres, offer an opportunity for growth. In setting up these businesses, many of the multinational companies consider security and political stability—which Bhutan has—as key factors in choice of location.

The plan to establish approximately 300 CICs primarily in the rural areas over the next five years highlights an important opportunity for bridging the rural–urban divide and for bringing rural communities closer to the information society. With the licensing of the second mobile telephone company in 2006, the use of ICT services across the country, overall connectivity and access will increase. Revenue sharing and infrastructure sharing schemes are being explored to reduce costs and maximize efficiency on the telecommunication front. On the governance side, ICT is being used to support transparency and good governance.

References

Department of Information Technology, MOIC. (2004). *National ICT centre of excellence project plan*. Thimphu, Bhutan.

Dzongkha Localization Project. (2006). Homepage available at http://www.hongnu.org/dzongkha_gnome/

GMCT Consulting Services, Asian Development Bank and UNESCAP. (2007). Empowering rural areas through community e-centres. *Bhutan baseline and pre-feasibility study*. Thimphu, Bhutan.

Kuenselonline: Kuensel Corporation. (2006). *Open ground wire*. Retrieved 11 December 2006 from http://www.kuenselonline.com/modules.php?name=News&file=article&sid=3775

Ministry of Information and Communications. (2003). *Bhutan e-readiness study*. GMCT Consulting Services, Thimphu, Bhutan.

Office of the Census Commissioner. (2005). *Population and housing census of Bhutan*. Thimphu, Bhutan.

Policy and Planning Division, Ministry of Information and Communications. (2006). Project proposal for establishing community information centers in geogs. *Final report of the national stakeholder meeting*. Thimphu, Bhutan.

Pradhan, G. (2005). Bhutan country chapter. *Digital review of Asia Pacific 2005/2006*. Penang, Malaysia: Southbound Publications.

Rai, B. (2006). *Dzongkhalinux—made in Bhutan*. Retrieved 11 December 2006 from http://www.idrc.ca/es/ev-98489-201-1-DO_TOPIC.html.

United Nations Economic and Social Commission for Asia Pacific. (2006). *Mid-term review of the Brussels Programme of Action* (Bhutan country chapter). Bangkok, Thailand: UN Publications.

.bn
Brunei Darussalam

Yong Chee Tuan and Hj Abd Rahim **Derus**

Introduction

The 8th National Development Plan (NDP) of Brunei Darussalam came to an end in the first quarter of 2006. However, the implementation of unfinished development projects will continue well beyond 2007. While the details of the next NDP have not been officially released, the primary objectives are likely to remain the same. These include diversifying the economy to become less dependent on oil and gas resources, enhancing the quality of life of the people, and strengthening the capacity for greater foreign direct investment (FDI).

The per capita GDP in 2006 for the population of 383,000 was BND 48,000 (USD 1 = BND 1.54). As the country enjoys the surge in oil prices and numerous associated opportunities, it also faces new challenges in moving towards diversification, improved living standards and increased FDI. The urgency and importance of meeting these challenges may be somehow sidelined when the dangers of depleting revenues from natural resources, in particular oil, are temporarily being turned around. But the young and new generations will have to become more prepared in order to continue to enjoy peace and prosperity. People between the ages of 20 and 54 comprise 54.3 per cent of the total population while the in-school population (between ages nine and 19) comprises another 26.7 per cent of the total population. The new NDP will unravel a roadmap for harnessing this young and dynamic workforce.

Meanwhile, the 12 ministries and the Prime Minister's Office (PMO) have been quick to formulate new strategic plans and roadmaps to lead the country into the new era. For one, the investment environment will be strategically enhanced by leveraging the opportunities provided by the improved communication infrastructure, innovative technologies and a highly educated workforce. More innovative approaches and strategic development programmes will be introduced to steer the country toward more peace and prosperity, with a strong national identity and a greater role in regional development.

These were some of the expected outcomes of the e-government initiative defined in the 8th NDP. More than BND 850 million (about USD 555.9 million) was set aside for the e-government initiative. Despite the sluggish pace of imple-mentation, the majority of the requested projects have already been awarded or tendered out. However, at the beginning of 2007 many of these projects are just at the initial phase of implementation. Thus, the overall impact of e-government is not yet visible or significant.

Technology infrastructure

For a small population density of only 66 persons (less than 20 households on average) per square kilometre, the challenges of providing high quality technology infrastructure can be slightly daunting. The economic returns on investment from household subscriptions alone do not look attractive enough for more than two service providers to compete. Some governmental interventions may be required to ensure sustainable quality services. One way to look at the situation is the implementation of government-wide EG-Bandwidth which is a Virtual Private Network (VPN) Line connecting all government agencies via several data centres.

TelBru Sdn Bhd (Brunei Telecom privatized as of 1 April 2006) took up the Internet service provider role. In the last quarter of 2006, a Next Generation Network (NGN) at the back office was successfully installed and commissioned. Domestic broadband services of 512 kb/s now cost BND 98 (about USD 63) per month. Currently, the telecommunication industry is driven by two major providers, DST (the first GSM provider) and B-Mobile (offering 3G services).

The number of Internet subscribers in 2006 was 131,141 (34 per cent of the population) while the number of mobile subscribers rose to 281,704 (74 per cent of the population). The landline subscriptions remained steady at 23.2 per 100 inhabitants.

Key institutions

The Ministry of Communications (MoC) is the central agency leading and shaping ICT development in the country. The Minister of Communications, Pehin Abu Bakar, is also the Chairman of the Brunei IT Council. During his three-year ministry, he has initiated several collaborative initiatives with international players such as Microsoft, Oracle, SAP and the various IT centres in Canada and India. The Ministry is also pressing for the rapid acceleration of the local ICT industry, and the establishment of an Innovation Centre managed by the Brunei Economic Development Board (BEDB).

Reporting to the MoC is the Authority for Infocommunications Technology Industry of Brunei Darussalam (AiTi), another important agency regulating the ICT industry. It grants industry players one of two principal licenses: the Infrastructure Provider for the Telecommunication Industry License (InTi) and the Service Provider for the Telecommunication Industry License (SeTi). An InTi License is required of any operator who owns and provides infrastructure, systems, networks, facilities and other related equipment for telecommunication services. The SeTi License is meant for operators who sell services to consumers or corporate customers. The operator does not own any infrastructure outside its premises, but uses the infrastructure provided by the InTi Licensee.

Besides its regulatory role, AiTi is also the agency that looks after the Brunei Computer Emergency Response Team (BruCERT), which is the nation's first trusted one-stop referral agency dealing with computer-related and Internet-related security incidents. BruCERT also coordinates with local and international Computer Security Incident Response Teams (CSIRTs), network service providers, security vendors, government agencies and other related organizations to facilitate detection, analysis and prevention of security incidents on the Internet. The roles of AiTi were expanded recently to include management of e-government development and leading the growth of the ICT industry.

The Prime Minister's Office (PMO) plays a crucial role in the implementation and monitoring of e-government initiatives. The Deputy Minister of the PMO chairs the E-Government Leadership Forum (EGLF) which meets regularly to provide policy directions on e-government development. The members of EGLF are the Permanent Secretaries of each Ministry; they lead the campaign for various e-government flagship applications and policy changes.

Although its role in coordinating and monitoring e-government has been shifted to AiTi, the Department of IT and State Store (ITSSD) of the Ministry of Finance still functions as the primary coordinator for procuring or outsourcing major government-wide IT hardware, software and services. ITSSD recently signed an agreement with Microsoft to procure Microsoft-related products for government use.

Digital content initiatives and online services

The official language of Brunei Darussalam is Malay, but English is widely used in the country. The education system is bilingual. From Kindergarten to Primary 4 (Year 9) classes, all of the subjects except English are taught in Malay. However, from Primary 5 (Year 10) onward, all of the main subjects are taught in English. Therefore, most of the government websites are in English or Malay, or both.

The major initiatives for developing digital content are driven by the Ministry of Education under two main projects, namely, the e-Curriculum and Knowledge Management Systems. The first project involves the digitization of English, Science, Mathematics, Malay, ICT and Islamic Religious Studies for primary classes. The digital works are to be uploaded to a new Learning Management System (LMS) that will be accessed by students and teachers. The first phase of the e-Curriculum project will be completed in October 2007, while the LMS project will be tendered in 2008. The Knowledge Management System project aims to collect information on crisis management, conferences and asset maintenance, among others. Various communities of practice and staff members of the Ministry of Education are involved in the development of the content to be made available later at the ministry portal.

Since most of the e-government projects implemented to date are related to back office applications and automation, there are many notable online services offered by government agencies. The much talked about next wave of e-government is about the integration of these back-office systems and the implementation of online services.

ICT industries

As the number of mobile telecommunication subscribers increases steadily, Brunei society gradually becomes more ready to adopt new technologies and move away from landlines. A natural progression from letter to e-mail, voice and digital telephony is beginning to take place in some major corporations in the country. Sales of mobile handsets are promising, and an unofficial study reveals that more than 75 per cent of young executives (aged between 21 and 30) have changed their handsets over the last 12 months.

Another promising growth industry is the sale of computer notebooks and desktops. The ratio of computer notebooks to desktops sold is nearly 5:1. There are many PC exhibitions where discounted computers and accessories can be bought. These exhibitions are popular because although consumers can buy computers tax-free, all electronic components or accessories are subjected to a 5 per cent tax.

Locally assembled PCs are being heavily squeezed by the influx of major international brands such as Acer, Dell, HP, Asus, Apple and Lenovo. There is no sign that any of the hardware manufacturing or assembly plants will be established or located in the country despite some competitive advantages, such as having the lowest tariffs on electricity, fuels and factory rentals in the region. The limited local market consumption and the lack of highly skilled human resources, coupled with the absence of major supporting services, appear to be the external inhibiting factors. However, although these factors were critical during the industrial age, they have been gradually suppressed by opportunities offered in an increasingly globalized society. The flow of people and capital across borders is addressing the problem of lack of skilled human resources. Also, the expected 70:30 ratio of local consumption versus the export market is becoming obsolete as the cost of supply chains and logistics are now insignificant compared to production costs.

The internal factors are becoming the key considerations of major ICT investors. One of the most crucial questions raised is how to increase investor confidence in sustainable business growth. The lack of non-oil and gas plants and the limited success stories from the current small pool of investments do not communicate the right message to investors. All locally incorporated companies are required to have at least 50 per cent local directors. Since foreigners are not allowed to own land, investors ask their local directors or local friends to purchase property. Government does not interfere in the business arrangements of private sector organizations. When conflicts occur among the directors, foreign partners tend to feel threatened by the insecurity of land ownership. This becomes one of the weak links in the effort to boost investor confidence in sustainable growth. It is even more difficult to attract assembly plants that require heavy capital investment with rates of return lasting longer than a decade.

While assembly plants for ICT hardware require heavy capital investments, the software development industry requires a constant influx of innovative human resources. An internal factor that may dampen the growth of the software development industry in Brunei Darussalam is the uncertainty of being able to employ highly skilled workers and support staff. For example, the issuance of work permits/quotas for engaging accountant clerks is being suspended as the Labour Department puts more pressure on companies to hire unemployed accountants. The employment of foreigners is subjected to scrutiny and a lengthy processing time. Nevertheless, many software development companies have expressed interest in setting up branch offices in the country in order to enjoy the amenities, the light traffic and the clean environment.

Enabling policies and programmes

As noted by Naseem (2003), Kaname Akamatsu's 'Flying Geese Theory' is still relevant in explaining the behaviour of ICT vendors (manufacturers, developers, suppliers) who tend to follow the relocation/investment moves of key industry players. For this concept to become a reality, three aspects need to be addressed:

1. The intra-industry aspect that looks at product development in terms of Import, Local Production and Export: In a highly successful IT-driven economy, these three elements—import, production and export of IT-related products/services—are very active and proportionally higher than for other products. Consider the case of Korea and Finland (the two early adopters of communications technology) where the overall figures of local consumption of ICT components, local production and value-added services, together with export of technology to various overseas manufacturing plants, are relatively higher than for agricultural products. In the case of Brunei Darussalam, the rollout of e-government projects in the 8th NDP has created a surge in the local demand for ICT products and services together with a rise in the importation of these products. Note that prior to the e-government projects, the ICT industry could not have taken off simply because of the limited ICT adoption by the private sector.

2. The inter-industry aspect that looks at the sequential appearance and development of industries in the country moving towards diversification and upgrading of industries: The late adopters generally go into developing relatively newer products instead of starting with less sophisticated products. Brunei Darussalam is going in the right direction in terms of diversification of the oil and gas-based economy into downstream and service industries. In this effort, IT can play the role of an enabler as well as a competitive differentiator. The

government is on the lookout for potential collaboration with players who can help develop niche products (for example, educational content, health care system, etc). Various ICT-related promotions and competitions such as the Brunei ICT Award (BICTA) and Inforama are heavily funded by the MoC and the Ministry of Education, respectively.
3. The international aspect that looks at the relocation (migration) process of industries from advanced to developing countries: Currently, there are several clusters of IT industries geographically distributed across Asia Pacific. Many of the manufacturing plants have been relocated to less expensive locations such as China and Indo-China, while the technology and intellectual property rights are retained by the companies in the advanced countries. The emergence of globalization and of superpowers such as China and India (and possibly Indonesia) has drastically affected the industry landscape. Most countries take advantage of largely populated countries by outsourcing low-cost and low-value services and supplying them with valued-added goods and services in volume (low margins). In this regard, the Brunei Darussalam Ministry of Trade and Foreign Affairs has been lauded for maintaining a very cordial relationship with all international partners, including China.

Open source and R&D initiatives

Although the government through various ministries and initiatives has allocated funds for R&D in technology advancement, there are no significant takers. Activating the funds requires several bureaucratic processes and information about the funds is not well published. The Innovation/Incubation Centre managed by BEDP is opening the floodgate for local 'technopreneurs' to focus on R&D work instead of merely playing the role of system integrators or customizing applications. This landscape may therefore be turned around within a few years.

Any research work on open source is encouraged, but there is no specific policy on the use or promotion of open source technology in government organizations. In fact, many of the e-government project proposals that are based on open source applications are not usually lower in cost than those based on proprietary systems. The lower initial cost of open source systems is sometimes offset by the implementation/customization and the long-term maintenance costs.

Relatedly, the use of open source software/applications is generally taken up in operations that are not mission-critical or in experimental activities. Because these operations are not properly budgeted in 'project implementation', they suffer from lack of financial support and they are at the mercy of the few technology-savvy staff for systems maintenance. Thus, the success stories that can be shared or showcased to others are few and far between.

Making a change in total productivity and research orientation requires a shift in management styles, entrepreneurs' mindsets and business models. Also needed is enhancement of support services, such as outsourcing capabilities, logistics and marketing. While many local companies are not familiar with R&D and innovation, some forms of assistance/facilitation from academic institutions and large corporations are required as the final deal depends on their available technological products. While creating strong Bruneian brands in ICT products may seem far-fetched to many, this process is a prerequisite to taking local products abroad. It also means that these solutions must first be piloted nationally and preferably taken up by government agencies. Success in international business depends on the ability of local companies to form a consortium or partnership with others for sizable projects.

Conclusion

Brunei Darussalam has been blessed with abundant oil and gas resources and the capacity to lead the industry to world-class standards. There are several remarkable breakthroughs in championing new processes or coming out with innovative products in the Brunei Shell Petroleum Company Sdn Bhd. This economic driver is so efficient that it often dwarfs the development of other industries in the country.

Nevertheless, diversifying the economy away from heavy dependence on oil and gas is still a serious priority. There are several paths that the country can take. Some potential clusters of industries in which the country has demonstrated strategic advantages include tourism, health care, education, logistics and services industries. Regardless of any combination of strategic implementation, developing a strong information technology industry is a prerequisite to enable the other industries to take-off.

Developing the ICT industry and using it to strengthen diversification holds much promise. However, although the external factors discussed earlier are very positive, many of the internal factors need to be resolved quickly in order to develop investor confidence. The issues of land ownership, work permits, dependent pass, company registration and sustainable funding for national ICT-related projects must be comprehensively addressed in order to attract and retain foreign investment. Rolling out the e-government initiative in the 8th NDP was a good start. But more needs to be done so that the long-term benefits of ICT can be realized.

The e-government projects are the best hope for developing the ICT industry, with local SMEs taking up the opportunity to participate in these projects. Although in the initial five-year period there will be a lot of 'sorting' among the local SMEs through open bidding for projects, the industry will quickly settle into various categories of players with some forms of specialization in the later years. The categorization of ICT players will become an important process in the pre-qualification of tenders. This will also ensure that the SMEs will not be spread too thinly in many areas of ICT development, but will instead focus on their and their partners' core competence. The companies can then develop and become more specialized in their expert domain.

Government agencies continue to provide opportunities for local specialized companies to meet their basic growth targets and to foster research, collaboration and innovation. The current e-government projects were designed with so-called Tier-1 standards and requirements, meaning that the solutions sought can hardly be conceived by regional SMEs (let alone local ones). Some forms of project grouping must be introduced, to assist and recognize local innovation. Each ministry should plan and allocate several small-scale projects that are not so mission-critical but which are strategically important for developing a local ICT industry. For example, an exploratory RFID project and courseware development projects may be good choices for locally incubated solutions.

Fostering rapport between local ICT players and government can create win-win solutions. Some forms of private–public partnership programmes should be introduced. And while government continues to promote local SMEs, the Tier-1 enterprises (such as Microsoft, Oracle, IBM, SAP) should not be made to feel 'unwanted'. They are, in today's terms, the leading geese that can influence the other flying geese. Some of the e-government projects should be allocated to these Tier-1 companies to attract large corporations and develop their confidence in building bases in Brunei Darussalam.

References

Department of Statistics, Department of Economic and Development, Prime Minister's Office, Brunei Darussalam. (2006). *Brunei Darussalam key indicators*.

Naseem, S.M. (2003). Rethinking the East Asian miracle. *Journal of Economic Studies*, 30(6), 636–44.

Cambodia

Brian **Unger** and Naomi **Robinson**

Overview

From Phnom Penh, the development of Cambodia's ICT sector appears blindingly fast. One year ago there were no ATM machines in the country. Now one bank has 26 in Phnom Penh. Moto (the local name for motorcycles for hire) drivers in the big cities often have cellphones. Cellphones are always answered, even when in the middle of a kiss, a sandwich, a shower or a speech to 50 people. Yet little has changed in rural areas where electricity is still rarely available.

Cambodia's population at the end of 2005 was 14 million, with a density of 78 people per square kilometre. The GDP per capita was USD 440 and 8 per cent of the population had a landline or cellular phone.[1] Cambodia was one of the first countries in the world in which mobile phone subscribers surpassed fixed-line subscribers. In 2005, 97 per cent of all telephone numbers provided by eight telecom service providers were for mobile subscribers. Yet even with this penetration, only 1.1 million people had cellphones at the end of 2005 (see Table 1). In 2007 the *Cambodian Daily* reported that there are 1.5 million mobile subscribers or 11 per cent (World Bank 2007). This suggests that the number of cellular subscribers grew by nearly 25 per cent in 2006.

Although mobile phone penetration is increasing rapidly, the number of landlines per capita has actually decreased. Also remarkable is the fact that though phone ownership is low, many Cambodians can make calls using portable cellphone booths that are usually managed by one woman.

There are roughly 40,000 Internet users (0.28 per cent of the total population) in Cambodia.[2] Most of these users access the Web through one of the more than 340 Internet cafes (*Cambodia Yellow Pages* 2007) since less than three of every one thousand people have access to a home computer (ITU 2007). There has been almost no growth in computers and Internet access per capita during the past year.

Digital Subscriber Line (DSL) access during working hours is expensive—for example, USD 99/month for 256 Kb/s and up to 1 Gb/month total upload and download volume. Charges for data transfer (upload and download) per month range from USD 0.07 to USD 0.10 per Mb (Savage 2007). There are approximately 200 computer hardware companies, 100 software companies, and 10 Internet Service Providers (ISPs) listed in the May 2007 online *Cambodian Yellow Pages*.

Internet access costs as well as electricity costs in Cambodia are among the highest in the world. Given this and the country's recent history of conflict, it is not surprising that Cambodia also has one of the lowest electrification rates in Asia, with only 12 per cent of its population connected to an electric power supply. Outside the provincial towns, electrical power is rare. About 6 per cent of Cambodia's rural households have access to electricity and another 3 per cent own some type of individual power generating unit. Of the remaining 91 per cent of the rural population, some 55 per cent use automobile batteries (costing USD 2–3.5 per kWh) for occasional and limited use, and the rest (36 per cent) do without electricity completely (World Bank 2007).

Television and radio are very popular in the rural areas, as they serve as one of the few forms of entertainment. TV and

Table 1
Basic ICT indicators

ICT indicators	Value	Notes
Main phone lines	0.23/100 in 2005[1]	0.24 in 2000[1] (decreasing)
Cellular subscribers	7.55/100[1]	1.12 million[1] (1.5 million in 2007)[2]
Local network usage	72.8 per cent[1]	of total network capacity
Main phone line cost	USD 3.0/month[1]	residential, USD 0.03/local call[1]
Main phone line cost	USD 6.5/month[1]	business
International phone traffic	0.9 minutes	per capita[1]
Internet users	0.28/100[1]	.26/100 in 2004[4]
Computers	0.26/100[1]	~ same as 2004[4]
DSL subscribers	0.04/1,000[1]	1.1 per cent of phone lines[1]
Internet access cost	USD 0.50	Dial-up[4]
DSL Internet access cost	USD 350/month	128 Kb/s unlimited data volume[3]
	USD 39/month	64 Kb/s unlimited, off peak hours[3]
.kh domains	320	NiDA 2006[4]
Domain registration	USD 40	$30 annual renewal[4]

Sources:
[1] ITU (2007) (statistics refer to 2005 year end)
[2] Welsh and Thul (2007)
[3] Savage (2007)
[4] Miyata (2006)

radio have the broadest reach among the different forms of mass media. While only eight of every one thousand Khmer own a television set, this form of mass media reaches more people than this statistic indicates, as whole villages will crowd around one set to watch a sporting event or favourite television series on one of the nine Khmer channels. A good example of the efficacy of this medium is that the UK's BBC weekly television drama *Taste of Life*, which dealt with issues such as HIV/AIDS, traffic safety and women's rights, was viewed by over 80 per cent of Cambodians during the two seasons in which it aired.

Radio is even more far reaching and accessible. Eleven per cent of the Cambodian population own radios, many of which are useable in areas off the electrical grid. Cambodia has two national radio stations and 14 local radio stations. There have been some attempts to set up community-run radio stations, but these efforts have been disallowed by the Ministry of Information.

Cambodia's Opportunity Index, which uses 10 indicators to measure ICT capital (network infrastructure), ICT skills, ICT uptake or usage, and the intensity of ICT usage,[3] grew from 20 in 2001 to 29 in 2005—a growth rate of 42 per cent. In 2005, Cambodia ranked 164th out of 183 countries. At first glance the growth rate might seem good. However, the speed at which ICT is developing throughout the world indicates that Cambodia is developing more slowly than most. To compare Cambodia with her neighbours: from 2001 to 2005 Myanmar's OI grew from 9 to 19 (ranking 177th), Lao PDR's OI grew from 24 to 39 (ranking 151st), Vietnam's OI grew from 44 to 77 (ranking 111th), and Thailand's OI grew from 75 to 99 (ranking 87th). The growth rates during these five years for Myanmar, Lao PDR, Vietnam and Thailand were 112 per cent, 49 per cent, 76 per cent and 32 per cent, respectively. Cambodia's ICT growth rate is lower than that of all of its neighbours except Thailand whose OI is three times that of Cambodia.

Key organizations

The government ministries, organizations, companies, educational institutions and NGOs that are involved in ICT can be categorized as illustrated in Figure 1. The categories used here are 'policy and regulation', 'infrastructure', 'content', 'enterprise' and 'human resources'.

In the past, the Ministry of Post and Telecommunications (MPTC) was responsible for all government policy, regulation and infrastructure. Government efforts to reform the telecommunication and ICT sector have resulted in a plan to divide these three functions into separate, independent organizations. In 2005–06 the responsibility for telecommunications infrastructure was moved from MPTC to Telecommunications Cambodia (TC). MPTC currently carries responsibility for both policy and the regulation of Cambodian telephony and ICT. In the next year or two, it is expected that MPTC will divest itself of responsibility for regulation of the telephony and ICT sector.

Figure 1
Key organizations involved in ICT

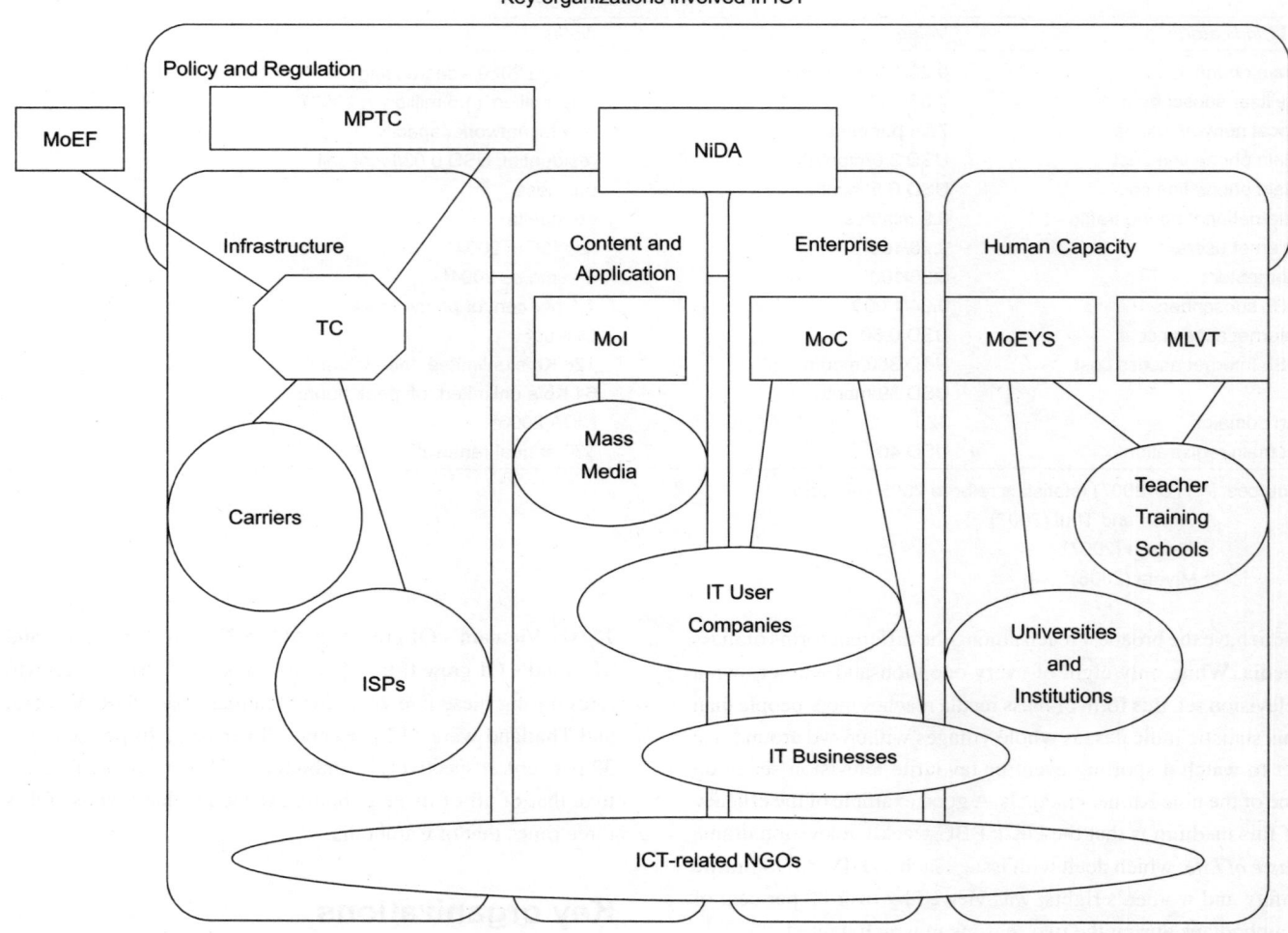

TC now has full responsibility for Cambodia's telecom and ICT infrastructure. This includes the major backbone landlines, international connections, connections to mobile phone service providers, and connections to the new IPstar satellite national network. TC's current backbone and planned extensions are illustrated in Figure 2. TC revenue last year was over USD 25 million, employing roughly 600 people operating in eight of Cambodia's 24 provinces. TC is jointly owned by MPTC and MoEF (Ministry of Economics and Finance).

In 2000, the National ICT Development Agency (NiDA) was created to promote ICT. It reports directly to the Office of the Council of Ministers and is chaired by the Prime Minister. Although NiDA was initially intended to formulate ICT policy for short, medium and long-term development, this responsibility clearly overlaps with MPTC, which has resulted in conflict and confusion. NiDA currently has five divisions: infrastructure, policy, human resource development, enterprise, and content and applications. Partly in response to the confusion, NiDA has begun focusing on the content and enterprise areas, particularly toward the development of e-government (Miyata 2006).

The role of the Ministry of Information (MoI) is the development and regulation of media and publications. This also overlaps to some extent with the roles of NiDA and MPTC. Thus the MoI has been, and continues to be, actively involved in ICT policy discussions. The MoI's role also includes censorship of the media and control of information sharing. For example, the MoI appears to have stopped the licensing of community radio in 2006–07. The MoI does support the dissemination of important information about issues such as health and government activities.

The Ministry of Commerce (MoC) is responsible for enterprise development, import and export regulation and development, foreign and domestic trade, intellectual property issues,

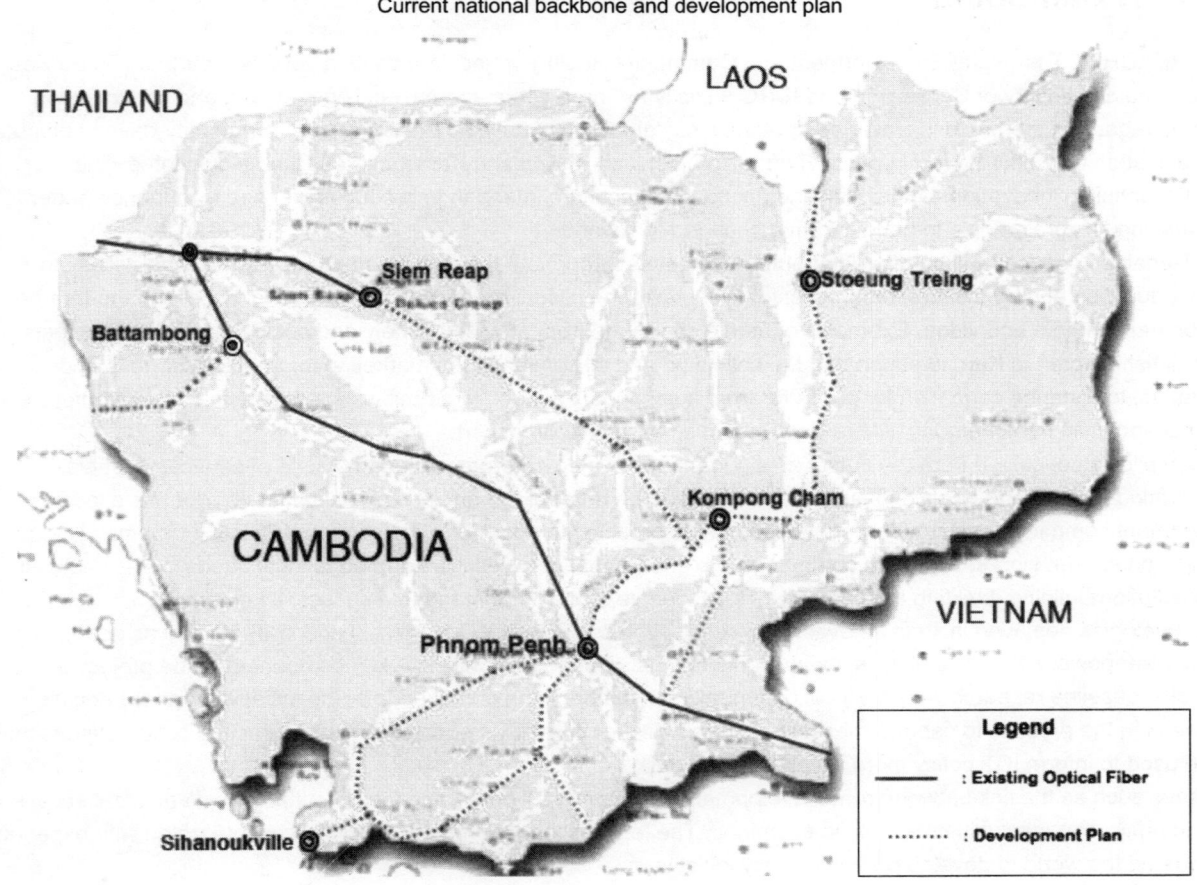

Figure 2
Current national backbone and development plan

and supporting the development of rural Cambodian small and medium-sized enterprises (SMEs). MoC is also leading ICT for Rural Empowerment and Community Health (iREACH), which aims to develop local enterprises to sustain rural e-communities (see the box 'iREACH Cambodia').

The Ministry of Education, Youth and Sport (MoEYS) plans to integrate ICT within its various initiatives towards its long-term vision of 'education for all'. The ICT in education policy focuses on teacher training: the Ministry gives priority access to all teachers and to secondary schools, and emphasizes the role of ICT as a teaching and learning tool. MoEYS also endorses the use of various types of media technology in outreach education and the use of ICT to improve education administration.

The Ministry of Labour and Vocational Training (MLVT) is also involved in education and training, as well as in labour conflict resolution.

Two NGOs focusing on ICT are the Centre for Information Systems Training (CIST) and Digital Divide Data (DDD). According to Jessamy (2001),

DDD was established in Cambodia through a partnership between an international team of advisors and local people with a wide range of proven experience in the business, non-profit and social entrepreneurial sectors. DDD hires talented workers from some of Cambodia's disadvantaged groups, such as land mine victims and women, and trains them to provide data entry and digitization services outsourced from the developed world. DDD runs as a self-sustaining cooperative, with all profits going back into the business to provide fair salaries, ongoing training and health services for its employees.

The Centre for Information Systems Training (CIST) is an NGO that funds education for disadvantaged children and helps to employ young adults in Southeast Asia. CIST aids in the education of Khmer youth so they may contribute to the development of their country.

> ### iREACH Cambodia
>
> The iREACH (ICT for Rural Empowerment and Community Health) project launched in 2005 is a partnership between the Cambodian Ministry of Commerce and IDRC. Initially for three years, the project is generating and documenting evidence regarding the process, value and potential sustainability of building multi-purpose, community-driven computer-communication networks for rural people. Two 'e-communities' in Kep and Kamchai Mear are being piloted. The project's technical innovations and social and economic effects are intended to produce evidence to influence better ICT policymaking for rural people to benefit from a changed ICT world.
>
> ICT-enabled services will be offered at numerous points throughout these two communities, including access to health, education and agricultural resources; low-cost telephony; local-area video conferencing; access to the Internet; and community radio and video. Examples of usage can range from providing timely and accurate weather forecasts for small fishing craft in Kep, to supporting the collection and dissemination of marketing information for fruits and vegetables, to distance education for pupils far from a secondary school, to a multi-village interactive video conference on maternal and child health among village women and a remote health expert.
>
> Each pilot is conceived from the outset as a community-owned enterprise offering a suite of services capable of generating income and becoming sustainable. Building on an assessment of local needs, each pilot has a local management committee selected through a democratic process. In addition to the core staff, iREACH is employing and training 'content developers' from the communities to ensure that services are adapted to local needs. A key challenge is local institution-building, building the capacity of local people to plan and manage a successful enterprise.
>
> Technologies deployed include innovative low-cost wireless such as WiMax, Wi-Fi and VoIP, as well as eco-friendly solar power since electricity is not available in most of the communities. A key component of the project is research—baseline research with the local community to help track the contribution of the initiative to poor people's lives; documenting the process to capture the lessons from mistakes as well as successes; and extracting policy lessons that can be used to inform ICT policy more widely in Cambodia.
>
> Issues such as the link between rural development policy and ICT policy and the potential role of universal access funds in replicating such initiatives will be examined. The lessons learned from iREACH will be integrated with experience from around the world to develop potential policy initiatives.

Policy and regulation

The National Strategic Development Plan (NSDP) is Cambodia's comprehensive five-year plan for 2006 through 2010. It lists ICT as one of the government's 16 highest priority areas. The NSDP states the following in the section on key strategies and actions:

> ICT: The long-term development vision is to develop a cost-efficient and world-class post and telecommunications system that has a nation-wide coverage. The realization of this vision would require high levels of investment to build the backbone infrastructure of the telecommunications systems, especially high-speed optical fibre cables for the development of rural telecommunications systems. The immediate challenge is to bring down the cost of telecommunications to help businesses and people at large. Telecommunications and Information Technology (IT) should be made to work for the betterment of the poor. Priorities during NSDP, 2006–10 are:
>
> - Expand the telecom network in urban areas and extend them to smaller cities and towns.
> - Strengthen postal services and the capacity of concerned institutions to improve the coverage and quality of the services.
> - Strengthen and improve efficiency and quality of Radio and TV broadcast networks.
> - Continue to follow an open policy in promoting a high level of private sector participation.
>
> Emphasis will continue on promoting extensive use of IT in all aspects of governance and government to improve efficiency and effectiveness in maintenance of records, data bases and websites which will provide easy access to the public at large on all matters of their concern. Each ministry or agency will host its own website and keep it

fully updated every six months or more often as needed. Such websites will contain all data and information pertaining to the ministry or agency.

The intent of the Royal Government of Cambodia (RGC) with respect to ICT policy and regulation is positive. However, the practice and implementation of this policy is very limited. For example, the MPTC has entered into joint ventures with foreign private investment companies to construct the mobile network. While this has helped to build infrastructure with little government funding, the implementation has been marked by a lack of consistency and transparency. The licensing of private mobile carriers has been dealt with case by case, without a clear legal framework. Currently, the sector is undergoing a major reform, as it is a condition of a Japanese soft loan for a fibre optic network between Sihanoukville and Kampong Cham.

The Telecommunication Sector Policy Statement is 'a national policy for the development of the sector' for 'the delivery of efficient, cost-effective and affordable telecom services to the people of Cambodia'. The statement primarily aims at reforming the sector by dividing MPTC into three independent entities responsible for policy, regulation and infrastructure. As noted earlier, MPTC would retain its 'policy' functions, TC has been created to deal with 'infrastructure', and a Telecommunications Regulator of Cambodia (TRC) will be created (Miyata 2006).

Consistent with the above policy statement, a new law on telecommunications has been developed over the past three years and in 2006 was sent to the National Assembly. This law is expected to establish the rules by which the TRC will regulate the ICT sector, including all telecom service providers (operators) (Miyata 2006).

In 2003, recommendations for a national ICT policy were developed within a study of the sector performed by NiDA with the assistance of UNDP-APDIP. The scope of this study included policy, infrastructure and access, content and applications development, human capacity development, and ICT enterprise development. It has been under review by the Council of Ministers for several years. According to Miyata (2006),

> [the] Prolonged process of formulating ICT policy revealed a serious strain of coordination and cooperation among key players, in particular, between MPTC and NiDA. Coordination and cooperation are crucial as ICT cuts across all sectors .... It seems that it is the Deputy Prime Minister ... who acts as a mediator and a coordinator among key agencies, namely MPTC, NiDA and MoI.

The intent in creating TC and moving responsibility for infrastructure out of MPTC was to separate this function from policy and regulation. However, although TC is now a separate corporation, it is state-owned, 100 per cent by MPTC and the MoEF. Thus MPTC still has a significant influence on the ICT infrastructure and operational functions.

It is clear that the transparency and consistency of the overall telecom and ICT policy and regulation environment have a major impact on sector development. The government is struggling with conflicting responsibilities for policy, for example, MPTC and NiDA's overlapping roles with respect to policy, and MPTC ownership of TC on infrastructure. This lack of clarity significantly retards development.

Achieving separate, independent, arms-length responsibilities for policy, regulation and infrastructure, which is the current government's plan, would have a significant positive impact on ICT sector development, including e-government, e-health, e-commerce and e-learning.

Education and capacity building

During the recent 30 years of internal conflict, primary, secondary and higher levels of education were all seriously weakened. The education system has been operating fully only since 1998. As a result, many Cambodians within the current working generation have not had adequate education, and many in the 20–45 age group lack sufficient primary education (Miyata 2006).

Recent enrolment increases in university level computing and IT courses are now producing many degree holders in Computer Science and Information Technology. However, many employers seeking ICT professionals have complained that these graduates have low levels of knowledge and inadequate skills for employment. Companies report that a great deal of time has to be invested in on-the-job training. There is often a lack of basic skills such as understanding how to plan and schedule work, complete tasks and meet deadlines. Currently, very few education programmes target these basic skills, and they are not adapted to meet ICT market needs.

On the positive side, in 2006, CIST performed a market study in Cambodia interviewing over 200 organizations and companies in both, the ICT sector and sectors that utilize ICT services. Specific skills were identified that ICT graduates need for employment in construction, consumer goods and services, education, energy and utilities, finance, health, NGOs, manufacturing, transportation and logistics, as well as ICT suppliers. CIST has plans to create an education programme that would be tailored to develop the skills that are most needed in these areas.

Many universities have developed ICT-related courses. With assistance from UNESCO, teacher training centres have been outfitted with second-hand computers and an introductory computer class has been added to all teacher training courses. MoEYS has also slowly worked towards improving the quality of higher education. For example, MoEYS introduced a first-year foundation course in every university and has set up the Accreditation Committee of Cambodia.

The Institute Technologie du Camboge (ITC) in Phnom Penh which has a strong focus on IT has received substantial support from the French government. ITC also has an important role in addressing human capacity in applications development within the Khmer PAN Localization Project (see the box 'PAN Localization Cambodia').

Donor support of ICT

There has been, and continues to be, substantial donor support for the development of Cambodia's ICT sector. Donor-supported ICT4D projects are listed by 'Activity' and 'Theme' in Table 2. 'Activity' refers to the main project output category

Table 2
Current and recent donor ICT4D projects

Activity	Theme	Donors	Partners
Infrastructure Development	Telecom	KfW or Germany (east-west optic fibre network)	MPTC
		Japan (Local exchange, optic fibre network)	
		IDRC (first Internet connection)	
	E-Government	Korea (optic fibre in Phnom Penh)	NiDA
	Community centre	Korea (Internet Plaza), India (10 Internet Kiosks)	NiDA
		USAID (22 Community Information Centres)	Asian Foundation
Policy Development	Telecom	World Bank, ADB ITU-UNDP	MPTC
	ICT General	UNDP-APDIP (ICT Policy)	NiDA
		JICA (IT Action Plan for Cambodia)	
	E-Commerce/SME	ADB (e-commerce law)	MoC
		UNDP (e-trade strategy/e-assessment)	
		ASEAN (e-ASEAN initiative)	
		WIPO (IPR Laws)	MCFA, MIME, MoC
		IDRC (survey on the use of ICT in SMEs)	NiDA (planned)
	ICT in Schools	UNESCO (ICT Policy in Education)	MoEYS
	Rural Access	IDRC (2 Pilots) (see iREACH boxed article)	MoC
Human Capacity Development	ICT General	UNDP-UNESCO-IBM (IT awareness)	NiDA
		InWent, Intel, Microsoft (basic IT skill trainings)	
	E-Government	Korea (IT Forum on e-government GAIS centre)	NiDA
	E-Commerce/SME	UNCTAD (training for government officials)	MoC
	ICT Skill Training and Education	Cisco Systems-UNDP-APDIP (Cisco Academy)	NiDA
		Korea (National Polytechnic Institute of Cambodia)	MLVT
		India (Cambodia-India Entrepreneurship Development Centre)	
		Singapore (Cambodia Singapore training centre)	
		France (Institute of Technology of Cambodia)	MoEYS
		France, Private IT Companies (establish Center for Information System Training: CIST)	Enfant du Mekong
	ICT in Schools	UNESCO (use of ICT in EFA)	MoEYS
		Private and Individual contribution (Village Leap)	NGO
	Khmer Scripts and Application	InWent (OpenOffice training)	NiDA/Openforum
Enterprise Development	E-Commerce/SME	World Bank, IFC (promoting SMEs using ICT)	MPDF
		UNESCAP (e-biz development service for SMEs)	ICT Association
		GTZ, USAID (private sector promotion with ICT)	
		GTZ-UNDP (village phone f/s)	n/a
		UNDP (support of local enterprise for job creation)	ODD
		Private and Individual contributions (Village Leap)	NGO
Content and Application Development	E-Government	Korea (Government Admin. Info. System: GAIS)	NiDA
	ICT in Schools	UNESCO (use of ICT in EFA: creation of content)	MoEYS
	Khmer Scripts and Application	UNDP-APDIP (KhmerOS)	NiDA/Openforum
		CICC, Japan (Workshop and Seminars on FOSS)	
		IDRC (Pan Localization) (see boxed article)	MPTC
	Access to Info.	UNDP (support CIC pilot outreach)	Local NGO/CIC
		USAID (election information outreach via CIC)	Asian Foundation

Note: Projects which only support attending international conferences, workshops and trainings are not included.

> ## PAN Localization Cambodia
>
> Access to most computing tools, and to most Internet and Web content and services, requires some level of English literacy. Unfortunately, very few Cambodians can read English, which severely limits the number of people who can take advantage of ICT tools, content and services. Khmer is the national language and is the official language of communication, official documents and government documentation in Cambodia.
>
> The PAN Localization Cambodia (PLC) project aims to develop Khmer language access to computing. It was established in May 2004 with support from the International Development Research Centre (IDRC) of Canada through its Pan Asia Networking (PAN) programme and the National University of Computers and Emerging Science (NUCES) based in Lahore, Pakistan, and in partnership with the National Committee for Standardization of Khmer Script in Computers (NCSKSC).
>
> The PLC project is addressing a number of aspects of Khmer language localization, including linguistic standardization of Khmer script for use in computing, applications, development platforms, content publishing and access, effective marketing and dissemination, and intellectual property right strategies. More specifically, the PLC project has explored both Linux and Microsoft platforms for a number of applications. During the past three years, the project has completed the following localization software and utilities:
>
> - Khmer Encoding Conversion Utility
> - Khmer Collation and Sorting Utility
> - Khmer Basic Lexicon
> - Khmer Word Segmentation Utility
> - Khmer Spell Checker Utility
> - Khmer Text Corpus (around 500,000 words)
> - Khmer Mobile Interface Terminology (in progress)
>
> Since its creation in October 2006, the PLC website has been accessed over 30,000 times. Khmer Unicode fonts and some of these tools can be freely downloaded from the website.

while 'theme' refers to the main subject area. This list was created from projects mentioned in interviews and existing reports with an ICT4D focus. It does not include projects with non-ICT goals but which may have an ICT component. Some projects appear more than once in the table, since they have many activities (outputs) under a single topic. These groupings enable a comparison of the focus of donors. Most donors, for example, have focused on human capacity development. Infrastructure development has been through bilateral cooperation (Miyata 2006).

Challenges

The Kingdom of Cambodia and the RGC have been in existence since 1994, only 18 years. The previous 25 years of conflict destroyed nearly all of the country's infrastructure, including educational systems, health care, land titles, telecommunication systems, social institutions, civil society, financial systems and government structures. In the context of this tragic recent history, elements of Cambodia's ICT sector have developed rapidly, usually due to exceptional efforts by a few Khmer individuals, and at times, by the government.

However, Cambodia's start from near zero in 1994 and the lack of clear, consistent, forward-looking policy and regulation place the country's ICT sector far behind those of its neighbours. Cambodia's ICT sector is developing at a slower rate compared to ICT sector development in its neighbouring countries and most other LDCs. There is a lack of decisive and progressive government action towards implementing modern telecommunication and ICT policy and regulation, which in turn stems from a lack of progressive knowledge and expertise in the government teams in this rapidly growing field.

The RGC, in concert with many committed donors, has developed good plans and it has drafted laws that if implemented, would truly boost Cambodia's rate of development. The plan to divide MPTC's policy and regulation functions and to privatize infrastructure development is positive. However, implementation of this plan is proceeding very slowly and, in some cases, also appears to subvert the intent (for example, MPTC's ownership of TC).

On the policy and regulation front, little has changed since the 2005–06 DirAP report (Klein 2005). Restrictive VoIP and community radio licensing continue to be examples of backward-looking policy and regulation that is hampering progress. The implementation of progressive ICT policy and regulation would unleash many Khmer private companies and entrepreneurs, with a corresponding substantially improved ICT sector rate of development. The impact of this should not be underestimated. It is not only the ICT sector that would improve much more rapidly; the much larger business, education, health and government sectors that depend on access to effective, efficient and economical ICT services would also hugely benefit.

Acknowledgements

This report borrows many ideas and specific data from the April 2006 UNDP report written by Mayumi Miyata who worked in 2005–06 as a short-term UN Volunteer for the ICT Sector, UNDP Cambodia. Quoting from that report:

> The UNDP report was prepared under the direct guidance of Ms Yoko Konishi, ICTD Focal Point/Governance Specialist, Governance Cluster, UNDP Cambodia, and with valuable advices from Dr Brian W. Unger, Executive Director of GRID Research Center/Professor of Computer Science, University of Calgary, and Mr Dara Bunhim, IT Analyst, UNDP Cambodia. (Miyata 2006)

Mayumi acknowledges many other people in her report who have contributed to this chapter as well. Valuable suggestions have also been made by Maria Ng Lee Hoon and our editor, Patricia Arinto.

Notes

1. Compare this with Laos at 12 per cent, Vietnam at 30 per cent and Thailand at 34 per cent.
2. Laos, Vietnam and Thailand have 0.42 per cent, 13 per cent and 11 per cent, respectively.
3. The 2007 *International Telecommunications Review* titled Measuring the Information Society defines an ICT Opportunity Index (OI) according to 10 indicators: main telephone lines per 100 persons, cellular phone subscribers per 100 persons, international Internet bandwidth, adult literacy rates, enrolment rates, Internet users per 100 persons, proportion of households with TV, computers per 100 persons, broadband Internet subscribers per 100 persons, and international outgoing phone traffic (min) per person.

References

Cambodia Yellow Pages. (2007). Retrieved 10 May 2007 from http://www.yellowpages-cambodia.com/

ITU. (2007). *Measuring the information society: ICT opportunity index and world telecommunication/ICT indicators* (statistics refer to 2005 year end).

Jessamy, D. (2001). *Digital divide data.* Retrieved 10 May 2007 from http://www.digitaldividedata.com/about/about_mngTeam.htm

Klein, N. (2005). Cambodia. *Digital review of Asia Pacific 2005/2006.*

Miyata, M. (2006). *Situational analysis of ICTD in Cambodia.* UNDP-Cambodia Internal Document. April.

National strategic development plan 2006–2010. (2005). English draft of Cambodia's comprehensive five-year plan (2006–2010); Khmer version launched 2006.

NiDA. (2006). *Information communication technology in Cambodia.* Royal Government of Cambodia.

Savage, R. (2007). Waiting for speed. *South eastern globe.* March 2007, pp. 22–27.

Welsh, J. and Thul, P.C. (2007). Mobile phone usage now at 1.5 million. *Cambodian Daily*, 10 May.

World Bank. (2004). *Seizing the global opportunity—Investment climate assessment and reform strategy for Cambodia.* June. Retrieved 10 May 2007 from http://siteresources.worldbank.org/INTPSD/Resources/336195-1092412588749/cambodia.pdf

World Bank. (2007). *Cambodia and energy.* Retrieved 10 May 2007 from http://web.worldbank.org/WBSITE/EXTERNAL/COUNTRIES/EASTASIAPACIFICEXT/EXTEAPREGTOPENERGY/0,,contentMDK:20490346~pagePK:34004173~piPK:34003707~theSitePK:574015,00.html

.cn

China

Zhang Guoliang, Zhang Xinhua and Deng Jianguo

Total population	1.30756 billion (2005)
Key economic sectors	Energy, Mining, Manufacturing, Construction, Transport, Storage, Post and Telecommunications, Wholesale and Retail Trade, Catering Services, Tourism, Finance and Insurance, Real Estate
Computer per 100 urban households	approximately 31 (2005)
Fixed-line telephones per 100 inhabitants	approximately 27.99 (2006)
Total computer hosts	54.50 million
Internet hosts per 10,000 inhabitants	4.2
Cell phone subscribers per 100 inhabitants	approximately 33 (2005)
Total bandwidth of international connections	214,175 Mbps (2006)

Sources: China Bureau of Statistics, China Ministry of Information Industry.

Overview

The Chinese government has been working to transform China into an Information Society since the early 1990s, primarily through the nationwide 'Three Golds Projects' in the customs, banking and taxation sectors. The first national informatization conference was held in 1997. The importance of developing China's information sectors has been increasingly emphasized in various sessions of the National Congress of the Communist Party of China (CPC). During the 10th Five-Year Period (2001–05), significant progress in this regard has been achieved, as evidenced by the following:

- The information network has expanded rapidly, and is the essential infrastructure for the country's economic and social growth. The Chinese telephone network and subscriber-base are now the world's largest; China's population of Internet and broadband users rank second in the world and radio and TV networks now cover most Chinese villages.
- The information industry has experienced rapid growth, contributing significantly to the nation's economic growth. In 2005, the ICT industry's contribution to GDP reached 7.2 per cent, electronic and information product exports accounted for over 30 per cent of China's total exports and Chinese enterprises owned more proprietary key technologies.
- Information technology has been applied in a widening set of areas, with notable results. An agricultural information service system has been completed; traditional IT industries like energy generation, traffic and transportation, machinery and chemistry are being reformed; the information service sector is thriving; and information development in finance and banking sector has boosted innovation and a modern finance service system has taken form. E-commerce is booming and information development in science, education, culture, medicine and health, social security and environment has been accelerating.
- E-government development has been expanding, enabling governments at various levels to transform their functions, increasing administrative efficiency and promoting transparency. Using information technology, governments at all levels have been expanding public access to information, resource sharing and administrative coordination. Use of IT in the customs, banking and taxation departments has achieved notable results and use of IT in public security and government approval bodies is progressing steadily.
- Online information in China has been increasing and information processing ability has been enhanced.
- A national information security strategy has been formulated and an information security administration and work mechanism has been set up. Internet information safety management has been consolidated.
- National defence and military information development is being pursued.
- Information development legislation and standardization work is moving forward.
- ICT awareness and training have improved significantly and the number of ICT human resources is increasing.

Technology infrastructure

In 2005, there were 38.68 million newly registered fixed-line telephone subscribers and 58.6 million newly registered cellphone subscribers in China, bringing the total fixed-line telephone subscribers to 350.43 million and the total cellphone subscribers to 393.43 million.

The *18th Statistical Survey Report on Internet Development in China* released in July 2006 shows that as of 30 June 2006, China had approximately 2.9 million domain names (including names registered in .CN ccTLD and gTLDs); .CN domain names reached more than 1.1 million. The top five provinces and cities in terms of number of websites are Beijing (144,800 websites, or 18.4 per cent of the total), Guangdong (141,105 websites, or 17.9 per cent), Zhejiang (73,304 websites, or 9.3 per cent), Shanghai (64,704 websites, or 8.2 per cent) and Jiangsu (63,933 websites, or 8.1 per cent). Total bandwidth connecting to the United States, Russia, France, the United Kingdom, Germany, Japan, Korea and Singapore reached 214,175 Mbps.

By 31 December 2005, there were 2.4 billion Chinese Web pages, representing a 269 per cent year-on-year increase; 64 per cent of these are dynamic Web pages. Web page bytes increased from 46,763 gigabytes in 2004 to 67,300 gigabytes in 2005, which indicates that Internet content and resources in China have increased substantially. However, since 2004 the number of online databases decreased slightly from 295,400.

As of 30 June 2006, there were 123 million Internet users (netizens), representing an 18.1 per cent (17 million) increase from 2004 figures. In terms of accessing methods, 26.8 per cent use leased lines, 47.5 per cent use dial-up and 77 per cent use broadband.[1] About 13 million access the Internet using mobile phones, and about 6.10 million Internet users have other types of accessing facilities, such as mobile terminals and information appliances.

China's Internet penetration rate increased from 8.5 per cent in December 2005 to 9.4 per cent in June 2006, and is increasing both in urban (from 16.9 per cent to 18 per cent) and rural areas (from 2.6 per cent to 3 per cent). The June 2006 data show that the urban penetration rate is six times the rural area penetration rate. It is expected that this digital gap will remain for some time due to the sharp difference in economic development between urban and rural areas and across different regions in the country. The coastal areas, special zones and central cities in the hinterlands have prosperous business communities and have access to advanced science and technology. The western and middle parts of the country are underdeveloped but possess tremendous potential for development.

There is also a big gap between east and middle-west in the volume of domain names and websites per 10,000 people. In terms of the volume of domain names per capita in late 2005, the east grew more quickly than the middle-west. However, the middle-west developed better than the east in terms of website volume per capita.

During the 10th Five-Year Period (2001–05), major innovations were achieved in integrated circuits, computing, network and telecommunications, software, and digital audio and video generation and dissemination. The National High Technology Research and Development Programme (also called the National 863 Programme) has given birth to a range of innovative key technologies and a line of strategic products and systems for technology integration and applications, improving the competitiveness of China's high technology industry. Some of the major innovations during the 2001–05 period are given below.

China-made central processing unit chips

China is one of the top 10 high-performance computer manufacturers in the world. The development of a proprietary CPU series has laid the foundation for the development of core technologies for China's information industry. China-made Longxin and Zhongzhi chips have been massively produced. China has also achieved a breakthrough in making its own computer operating system (such as Kylin) and office software (such as WPS Office, Yongzhong Office and EduOffice) and has produced a line of products with some market share.

IPv6: China's next generation Internet

In September 2006, China announced the launch of the world's largest pure next-generation Internet using IPv6. The Chinese network, called CNGI-CERNET2/6IX or CERNET2 for short, and broadly referred to as the China Next-Generation Internet (CNGI) project, currently links 167 institutes in 25 universities in 20 different cities. It also has links to telecom operators China Telecom, China Unicom, China Mobile and China Tietong, as well as partner equipment providers ZTE, Tsinghua Unisplendor and Tsinghua Tongfang. CERNET2 uses Chinese IPv6 routers rather than the foreign routers that support the current IPv4 network around the world. Chinese experts say that it would take about 10 years to make the full transition from IPv4 to IPv6.

China Next-Generation Internet (CNGI) will:

- move data at about 100 times current Internet speeds;
- support online streaming video at unprecedented levels;
- allow over 160 departments and institutions on CERNET2 to set up experimental labs and conduct research into new applications that people may not have seen before;
- drive new technology deals and innovations;
- allow China to develop new standards for the Internet Engineering Task Force (IETF), which develops and promotes Internet standards; and

- support an infinite number of IP addresses, providing the platform for what many call the Internet of Things, a world in which objects have their own IP addresses and can share data.

3G mobile phone networks

China has made remarkable achievements in the 3G system with its TD-SCDMA (time division synchronous code division multiple access) standard. Under the direction of the Ministry of Information Industries, trials of 3G/UMTS (Universal Mobile Telecommunications System) radio access networks and terminals were completed in 2005. W-CDMA technology has been found to be quite satisfactory on the network side and more dual mode handsets are expected for further testing. With the final results of these trials expected soon, the UMTS Forum anticipates that China will signal its intention to award 3G licenses in due course. The Chinese central government has yet to decide when to license operators to build 3G networks on the mainland. Many industry observers now expect the 3G licensing to occur in the first half of 2007 at the earliest.

Digital content

The *Survey Report on Quantities of China Internet Information Resources* released by the Informatization Office of the State Council on 15 May 2006 shows that 60.4 per cent of Web content in the Chinese mainland are company websites and 21.9 per cent are personal websites. These originate mainly from Beijing, Shanghai, Guangdong, Fujian and Zhejiang, which indicates a digital gap between east and west China.

Portal websites

Portal websites serve as a one-stop place for information and services. The service information available includes information on education, training, jobs, daily-life consumables, leisure, entertainment and tourism, health, sports, hospitals, industry and business.

The *2005 Survey Report on Quantities of China Internet Information Resources* shows that enterprise websites accounted for most (60.4 per cent) of the 694,000 websites in China. But for over 50 per cent of the enterprise sites, daily page views are under 50 due to lack of links to other sites. Most enterprise sites provide only product introduction and display, and are not fully integrated into the business.

E-commerce

In 2005, there were six major B2B e-commerce companies in mainland China, including, in order of strength, alibaba.com, hc360.com, 123trading.com, 8848.com, sparkice.com and meetchina.com; the first three have most of the market share. Chinese B2B e-commerce websites still lack experience in integration into the global procurement system and in marketing clients' products to the international market.

The major B2C e-commerce companies are joyo.com, ebay.com.cn and dangdang.com. Due to a deep pocket and good management, e-commerce companies with foreign investment seem to be more competitive.

IT Information

IT information websites offer information related to highly technical programming knowledge or popular IT software and consumption electronics. But websites with content that is too technical have begun to turn off visitors while those with less esoteric content attract more users. Currently, Chinese language Web page content is being copied heavily across sites, with a 'repetition rate' of as high as 25–40 per cent. China's first IT information website, PChome.net, was acquired by American CNETNetworks for USD 11 million in March 2005.

Government affairs

As of the end of 2004, most of China's central government departments had gone online, with over 90 websites established. Local government websites totalled more than 20,000. According to a survey conducted by Singaporean newspaper *Lianhe Zaobao* in 2003, China ranks 74th in the world in e-government readiness. The United States tops the world list, while Singapore ranks 12th, taking the Asian lead. The weaknesses of China's government websites include non-standardization of URLs and site names, incompleteness of contact information, lack of professionalism in page design, lack of timeliness in content updating and lack of interactivity.

Agricultural information

Agricultural websites provide information on agricultural technology, policy and administration, market analysis and prediction, meteorological and environmental information, horticulture, fishery, planting and agricultural machinery, and the like. A China Ministry of Agriculture survey shows that as of 2001, the Chinese mainland has 2,175 agricultural websites, 2,000 more than in 1998.

In 2005, the China Electronic Commerce Association and China Agriculture Web jointly compiled China's Top 100 agriculture websites in the hope of promoting the growth of these sites.

The weaknesses of Chinese agricultural sites at present are: nearly half of them (49.27 per cent) are based in Beijing and in coastal provinces rather than in more rural western areas, most of these sites are run by governments and some lack two-way information communications with users, information is repeated across sites, and some information is impractical and outdated.

Personal websites

There are more than 30 million personal websites, accounting for 21.9 per cent of all websites in China. Many have attracted venture capital.

The top 10 portal websites in China

Baidu.com, Inc. (http://www.baidu.com/) is a leading Chinese-language Internet search engine. According to Wikipedia, Baidu has an index of over 740 million Web pages, 80 million images and 10 million multimedia files. It also offers blogs and other services. Baidu.com had its initial public offering (IPO) on 5 August 2005.

Sina (http://www.sina.com.cn) is the largest Chinese-language Web portal, providing news, entertainment, e-mail, search and related services. Alexa ranks Sina as the 7th biggest Web property in the world, just behind MySpace. It is said to have 94.8 million registered users and more than 10 million active users of fee-based services, with an estimated three billion page views every day.

Founded in August 1996 and developed from the first Chinese search engine, Sohu (http://www.sohu.com) claims to be 'China's premier online brand and is indispensable to the daily life of millions of Chinese'. It is in Chinese and is ranked 16th in the world in terms of traffic volume by Alexa.

Also called NetEase (http://www.163.com), this famous Chinese-language website is a two-time winner of the Best Ten Chinese Websites Award from CNNIC. NetEase's daily page views in September 2005 exceeded 614 million.

Alibaba.com Corporation (http://www.alibaba.com) is China's leading e-commerce company, operating the world's largest online market for international and domestic trade and China's most popular online payment system, AliPay. Alibaba.com also owns and operates Yahoo! China, which it acquired in October 2005.

CE.Net (http://www.ce.net.cn), in Chinese and English, is the largest Chinese B2B e-commerce website.

The bilingual (Chinese and English) ChinaChamber.com (http://www.chinachamber.com.cn) was established by the All-China Industrial and Commercial Union and partnered with CE.Net in May 2000.

Rongshu.com (http://www.rongshu.com) is the biggest Chinese-language cultural and artistic website with an average of 5,000 new articles submitted daily and about 3 million articles in its archive. In October 2005, Rongshu had 4.5 million registered users, with over 700 page views daily, ranking the world's 400th in daily page view numbers.

Ctrip.com (http://www.ctrip.com), in simplified and traditional Chinese and English, introduces scenic spots all over China and provides online reservation services.

The electronic version of the Chinese newspaper *People's Daily* (http://www.people.com.cn) reports on China in seven languages: Chinese, English, Japanese, French, Russian, Spanish and Arabian.

Run by the official Xinhua News Agency, http://www.xinhuanet.com provides worldwide news 24 hours a day in seven languages: Chinese (simplified and traditional), English, French, Spanish, Russian, Arabian and Japanese. About 5,000 news items are published daily.

A national academic computer network (http://www.edu.cn) has been established principally to serve educational and research institutions. It is in Chinese and English.

CCIDNET (http://ccidnet.com) is the largest Internet and IT information service and commercial service provider in China. By 30 June 2006, the site had 1.52 million registered users and its daily page views reached 5.88 million; the number of daily visits reached 1.5 million.

Established in 2000, Joyo.com (http://www.joyo.com) offers online sales of various books, audio-visual products, software, toys and gifts, and general merchandise. In August 2004, Amazon.com acquired a 100 per cent stake in Joyo.com.

Web2.0 contents

These fast growing websites, copied mostly from their US equivalents, provide very good user experience due to good ideas and their use of the latest Web technologies. Some offer rich content consisting of text, music, pictures and video. Others give users a platform to collaborate online or build social networks.

Online services

Search services

As Internet content increases exponentially day by day, the need to retrieve the right information when it is needed has intensified. Hence the popularity of the major Internet search service providers in China, which include Google China, Baidu, Yahoo, Zhongsou, Sohu, Sina and Skynet. Blog search engines for personal online content are also popular. Chinese blog search engines include feedsearch.net, 8fang.net, grassland (http://www.cnblog.org/cnblog.html) and Blogcn (http://www.blogcn.com/search/index.shtml) Among the challenges faced by Chinese search engines are that users are currently limited to searching for key words and category browsing, which limits accuracy of search results, and Web content in Chinese is limited.

E-government services

The Chinese government portal site (http://www.gov.cn/) received 40.5 million hits and 5.19 million page views on the first day of its official operation on 1 January 2006. The Chinese Ministry of Foreign Affairs website has 3.20 million daily visits. By November 2005, 138,400 new enterprises in Shanghai completed licensing, quality control and tax registration via e-government sites. In Hangzhou, over 9 million data exchanges between government bodies had been recorded by 20 December 2005, suggesting increased government administration efficiency and service quality. Chengdu, a western city, completed in November 2005 a comparison of historical data on 90,900 enterprises, and data sharing between government departments on the registration, licensing and information modification of 19,000 new enterprises.

According to the *China Informatization Report 2006*, the top 10 provinces and cities in the e-government performance evaluation in 2005 were Shanghai, Beijing, Jilin, Zhejiang, Hebei, Anhui, Jiangsu, Yunnan, Shanxi and Heilongjiang. The list shows an even distribution between developed and less-developed areas.

E-commerce

The *18th Statistical Survey Report on Internet Development in China* released in July 2006 shows that as of 30 June 2006, China had approximately 1.8 million websites registered under .com or com.cn. These account for 81.5 per cent of all the websites registered in the Chinese mainland. In 2005, China's e-commerce transactions, most of which were B2B transactions, reached 600 billion yuan, a 40 per cent increase over 2004 figures. China's e-commerce transaction volumes are expected to exceed the 800 billion yuan threshold in 2006. E-commerce models in the Chinese mainland include B2B, such as Alibaba (http://www.alibaba.com.cn), Huicong (http://www.hc360.com) and Jinyindao (http://www.315.com.cn); C2C, such as dangdang (www.dangdang.com); and B2C, such as Ebay China (www.ebay.com.cn).

E-community

According to the *China Online Community Research Report* of the China iResearch Inc., 68 per cent of 3,094 respondents surveyed visited online community websites in 2004. Major online community services in the Chinese mainland include comprehensive community websites, such as Xici (http://www.xici.net/) and Xilu (http://www.xilu.com/); BBSs affiliated to major portal websites, such as NetEase BBS (http://bbs.163.com/) and Sina BBS (http://people.sina.com.cn/); hobby websites, such as Rongshuxia (http://www.rongshuxia.com/); and online game sites, such as 17173 (http://www.17173.com/).

ICT industries and services

According to the *China Informatization Development Report 2006*, total sales in the electronic information sector was CNY 3,841.1 billion (about USD 502.35 billion) in 2005, representing a 24.8 per cent increase year-on-year. Profits reached CNY 130.7 billion (about USD 17 billion), a 5.2 per cent increase; the sector paid CNY 43.5 billion (about USD 5.7 billion) in taxes, a 10.5 per cent increase; and exports totalled CNY 268.2 billion (about USD 35 billion), a 29.2 per cent increase.

After nationwide sectoral restructuring, China now ranks No. 1 in the world in the manufacture of computers, mobile phones and TV sets. New audio-video products, telecommunication networks, equipment and electronic displays are the new growth points of the country's economy. Computers have become a top earner, with a sales volume of CNY 1,064.4 billion (about USD 139.2 billion), a 21.7 per cent increase since 2004.

In the past five years, China's policy of mindfully fostering ICT giant companies has had notable effects. In 2005, the number of ICT companies with sales of over CNY 10 billion (about USD 1.3 billion) increased to 22. The eastern coastal area, especially Jiangsu, Beijing and Shanghai, have seen a growth rate of over 30 per cent. In 2005, the whole industry's production–sale ratio was as high as 98 per cent.

Pushed by increasing demand for high-end products, the production growth rates for LCD and plasma TV sets were 418 per cent and 215 per cent respectively. In the computer sector, laptop computers accounted for nearly 60 per cent of all newly made computers.

In 2005, China's electronic and information product exports totalled USD 268.17 billion, topping the exported product categories.

The software sector has also experienced rapid growth. By November 2005, there were over 11,000 software enterprises, 200 of them with over CNY 100 million (USD 13 million) in sales. Over 200 enterprises have passed CMM2 or above standards.

In 2005, total sales in the Chinese software industry was CNY 390 billion (USD 51 billion), a 40.3 per cent increase year-on-year. Software exports reached USD 3.59 billion, a 28.2 per cent increase year-on-year. By September 2005, there were 28,401 registered software products in China. There are 11 national software industrial bases and six national software export bases.

While getting stronger at home, China's information industry companies have been expanding overseas, with notable feats by computer giants, such as Lenovo's acquisition of IBM's PC branch in May 2005. After the acquisition, Lenovo's yearly sales are expected to reach USD13 billion at a yearly sale of 14 million PCs.

Enabling policies

The CPC Central Committee's Recommendations about the Making of the 11th Five-Year Plan on National Economic and Social Development, passed at the Fifth Plenary Meeting of the 16th CPC Central Committee held on October 2005, defines the main tasks and directions of Chinese informatization thus:

- advancing the national economy and social informatization;
- developing distance education and projects to 'extend radio and TV coverage to every village';
- improving rural communications and communications networks;
- vigorously developing core industries such as integrated circuit and software, focusing on fostering an information industry based on digital audio and video, a new generation of mobile communications, computers and high quality network equipment;
- enhancing the development and sharing of information resources and promoting the dissemination and application of information technologies;
- strengthening the construction of information infrastructure such as broadband Internet, digital television network and next generation Internet, integrating the three as a whole; and
- improving information security.

Development Strategies of National Informatization 2006–2020, published by the National Leading Group of Informatization in November 2005, emphasizes promoting e-government to enhance the government's ability to manage state affairs, informatization of national defence and military affairs to maintain national security, social informatization to create a harmonious society, development of core technologies through independent innovative capacities to improve the quality and competitive power of the information industry, accelerated development of technological standards, establishment of a legal framework for ICT and improved training for employees.

To match these macro policies, government has formulated a series of special measures in rural informatization, e-government, e-commerce and information security. A conference on 'Promoting Rural Informatization to Create a Harmonious Society' was held in Beijing on 22 September 2005. The State Department's *Guidelines to Construct China E-Government* published in August 2002 defines the basic principles, development goals, main tasks, guarantee measures and infrastructure for e-government for the next five years. In early 2005, the Office of the State Department published *Ideas about Quickening the Development of E-Commerce*, which defines a policy orientation for e-commerce and focuses on operative measures to respond to the *Electronic Signature Act*. The Central Bank later issued *Regulations of Internet Banking*. The China Banking Regulatory Commission also passed *Regulations of Electronic Banking, Guidelines to Assessment of Electronic Bank Security* and *Rules of Qualification Certification of Security Assessment Institutions of Electronic Banking*, which require supervision agencies at all levels to encourage financial entities to launch electronic banking by simplifying examination procedures.

With respect to information security, the National Information Leading Group passed in 2003 *Suggestions for Enhancing*

Information Security, and the Information Office of the State Department, together with the Ministry of Public Security and others, began to draft an *Information Security Statute* one year later. The State Administration of Radio, Film and Television succeeded in drafting the *Act Guaranteeing Transmission of Radio, Film and Television*. The Minister of Public Security issued a circular on *Operative Measures for the Protection of Information Security* on 15 September 2004 while the National Certification and Recognition Supervision Commission issued the *Notice of the Establishment of the National Certification and Recognition System for Information Products* in October 2004.

Regulatory environment

China's informatization campaign is supported by the Special Programme of Informatization in the 10th Five-Year Plan released in January 2004. A series of related laws and regulations has been released recently, including amendments to the Criminal Law, Contract Law, Custom Law and others, to cover information network servicing, Internet security, information rights and electronic transactions, thereby legally ensuring the healthy development of informatization. The *PRC Electronic Signature Act* came into effect on 1 April 2005, and legislation of the *Telecommunications Act* is in progress. The *Statute of Governmental Information Openness* has been listed in the first-class legislation schedule of 2006 by the State Department. In addition, research for legislation regarding the protection of minors online has been completed and submitted to the legislature.

The *PRC Electronic Signature Act* confirms the legal power of e-signatures, regulates the use of e-signatures and protects the legal rights of all parties concerned, ensuring the security of e-commerce and laying a firm foundation for the creation of a safe certification system and national Internet trust system. Under this legal authorization, the *Management of E-Certification Service* and *Management of E-Certification Code* have been implemented. By the end of January 2006, 17 e-certification services had acquired a License of e-Certification Service from the Ministry of Information Industry.

Administrative regulations relevant to Internet input servicing, Internet information servicing and Internet logging servicing are likewise in place. The *Temporary Rules of International Networking Administration* issued on 20 May 1997 is the most important legal document for regulating international networking of the Internet in China. *Management Measures of Internet Information Service* came into effect on 25 September 2000, while the *Notice for Further Enhancing Regulation of Sites for Internet Access Service* was issued by the Office of the State Department on 3 April 2001. In August 2002, *Management Measures of China Internet Domain Names* first came out; this was modified and reissued by the Ministry of Information Industry in November 2004, and implemented since 20 December 2004.

The *National People's Congress Standing Committee's Decision on Maintaining Internet Security*, passed at the 19th Conference of the 9th National People's Congress on 28 December 2000, is China's first legal decision to protect information security. In 2005, the Informatization Office of the State Department began to write the *Draft Information Security Statute*.

Regarding protection of information-related rights, the State Department modified in 2001 the *Statute of Protection of Computer Software* first issued in 1991 and published it on 20 December of the same year. On 27 October 2001, the 24th conference of the 19th National People's Congress passed the decision to modify the *PRC Copyright Law* to take into account the new Internet economy. On 30 April 2005, the Ministry of Information Industry and National Copyright Bureau issued *Approaches of Administrative Protection of Internet Copyright*, improving the system of copyright protection in the Internet environment. The *Statute of Collective Management of Copyright* was also put into practice on 1 March 2005. The *Regulations of Management of Internet News Information Service*, which regulates activities involving Internet news services, was issued jointly by the News Office of the State Department and the Ministry of Information Industry on 25 September 2005.

Moreover, guided by the *CPC General Offices and State Department's Suggestions about Further Promotion of Administration Openness,* by the end of 2005, Central governmental agencies had developed 30 legal documents involving information openness, and 75 local party and administrative agencies had issued related regulations. Handan city in Hebei province and Jiaxing city in Zhejiang province developed rules on openness of governmental information. To regulate the collection, publication and use of business and personal credit information, Shanghai, Hunan and other places issued credit information management measures. Beijing, Jilin, Anhui, Yunnan and others formulated approaches to manage intellectual property rights protection in an Internet environment.

Education and capacity building programmes

Great importance is being attached to the education of citizens, especially the younger generation, regarding ICT applications.

Colleges and universities have produced many information science professionals trained in courses such as information and communications engineering, cybernetics and engineering, computer science and technologies, and electronics. ICT-related education and training is required of all students in universities. By the end of 2004, 389 of 1,731 colleges and universities had computer software majors and 550 had majors related to IT and software. In the same year, there were 67,454 graduate students and 1.794 million undergraduates in these specialized subjects nationwide. There were 31,683 students in 35 demonstration software colleges and 62,550 in 35 demonstration professional software colleges. Currently, there are 127,468 students in the departments of information science of colleges and universities. The number of undergraduate computer and information science programmes has increased from 505 in 2003 to 771 in 2005. The percentage of such programmes relative to all undergraduate programmes has increased from 4 per cent to 18 per cent.

Meanwhile, IT training in the secondary vocational schools has been strengthened and an IT subject is now a required course in 98 per cent of such schools. There are 1,400 specialized IT vocational schools in the country, and another 5,000 schools have IT majors with a total of 1.5 million students enrolled in IT-related subjects.

On-the-job training is an important way for China to cultivate IT talents. The CPC Central Committee and government ministries and commissions have organized various types of training classes on informatization and e-government, developing a team of professionals to lead, organize and promote informatization and e-government. In 2005, 683,635 government employees were trained through the National IT Training Programme; about a third (218,144) of them participated in qualification tests and over 20 per cent acquired qualification certificates of different grades. Another third (274,112 staff) took part in IT self-study tests, and a little less than a third (191,379) in other tests for authentication. In the National IT Training Programme, the training areas include software, hardware, Web design, Internet and professional English.

The authentication of professional skills in the electronic industry is in full swing. The Ministry of Information Industry has established 50 authentication stations in the country, while 15 provinces with developed electronic industries have set up provincial guiding centres for authentication. Bases for training in advanced IT skills, numbering 109 in all, have been established in the whole country, and 39 national professional standards have been developed and released. To date, 320,568 technical workers, students of technical schools and professionals of various types have been trained and tested; 63,881 of them have acquired qualification certificates as high-grade workers and 4,679 were granted the qualification of Technician or High Technician.

Chinese provincial and municipal governments are also promoting informatization. In Shanghai for example, the Municipal Association of Women, Municipal Commission of Information, Municipal Office of Civilization, and Municipal Association of Science and Technology jointly organized a project to help one million families to surf the Internet within the next 3–5 years. The project was ranked one of 12 practical undertakings by the municipal government in 2003. Projects like these, which involve the cooperation of foreign educational institutions and domestic IT businesses, are indeed noteworthy.

In the years to come, China will continue to enhance IT knowledge and training, developing the information capabilities of leaders, public servants and professionals. It will continue to carry out the '653 Programme' to develop a large base of human resources skilled in IT, software and integrated circuits.

Open source movement

Open source software (OSS) is increasingly being used in China. Recently, Linux of China and OSS companies forged a mechanism for cooperation in product orientation, R&D, marketing and training, creating a primary ecosystem of open source software, from applications at the business level to tabletop products, from embedded Linux software to hardware and development tools supporting Linux.

Since the 10th Five-Year Plan, the Chinese government has strengthened support for the development of OSS, including the Linux operating system and its applications. The Ministry of Information Industry has vigorously supported R&D work on Linux operation systems and office software for servers, desktop computers and other equipment that embeds such software, contributing to the emergence in the market of products such as Zhangke Red Flag Linux, Zhongbiao Puhua Linux, Gongchuang Linux and Turbo Linux. All these make the Chinese Linux market a rich one.

In May 2005, the Ministry of Information Industry set up a ministerial centre for promoting software and integrated circuits and a public service platform for national software and integrated circuits, and established Linux laboratories with HP and Freescale. With guidance from the Ministry, the Chinese Open Source Software Promotion League was created on 22 July 2004, and now has 80 members. On 10 May 2005, the China Linux Industry Strategic Alliance was officially founded; it has 60 members.

As open source technology matures, more and more Chinese businesses are beginning to think highly of its commercial value. Studies show that 23 per cent of the Top 100 Chinese Commercial Technologic Businesses in 2005 have deployed Linux or other OSS. However, open source technology still has low penetration in small and medium-scale businesses, open source communities are relatively small and have little influence, open source businesses are small-scale and lack core competencies, and technical support is also lacking.

International cooperation is a promising development. In April 2004, the ministries for telecommunications of China, Japan and South Korea signed the *China, Japan and South Korea: Cooperative Memorandum of Open Source Software*, which defines 10 cooperative projects and sets three governmental IT meetings and three Southeast Asian forums for promoting OSS. China and France also signed the *China-France Cooperative Memorandum of Open Source Software*.

Research into ICTs

As mentioned earlier, various IT innovations have been realized, especially in integrated circuits, computers, networking and communications, software and digital audio and video. The integrated circuit technology of central processors such as Zhongzhi and Longxin are highly effective, the Shu Guang 4,000A super-computer ranks among the Top 500 in the world, new high-speed routers are being used in the construction of Chinese next-generation Internet, industrial standards of TD-SCDMA have been formed, and 3G mobile technology is advancing rapidly.

More investment is being put into the R&D of typical IT businesses, resulting in many innovations. In 2004, the Top 100 in China's electronics industry spent CNY 311,000 million (about USD 40,673.27 million) on R&D, representing a 17 per cent increase on R&D spending in 2003 and 3.8 per cent of sales revenues. There were more than 3,600 patent applications in 2005. The Huawei Company, which has the most number of patent applications, has applied for over 3,000 3G patents and owns 5 per cent of the basic patent of WCDMA. In June 2005, the Haixing Digital Video Processing Chip, the first video chip in China developed by the Haixing Group, passed qualification by the concerned agencies, freeing the Chinese colour TV industry from its dependence on the import of core chips. The Yongzhong Science and Technology Co. Ltd. in Wuxi, Jiangsu has succeeded in solving technological puzzles worrying the global software sector by relying on an independent innovation: the Yongzhong Integrated Office which integrates word processing, worksheet and briefing using digital object storage base technology development.

Current efforts aim to establish and improve an innovation system combining production, study and research. The guideline to develop high-performance 64-bit CPUs has been defined and the Longxing Industry Alliance has been set up. The construction of a public information platform for home-made hardware and software and a platform for national auto computation is advancing. China is also speeding up its implementation of standards and intellectual property rights and encouraging large domestic corporations to participate in international standard R&D.

Future trends

China's 11th Five-Year Plan commenced in 2006. With *Development Strategies of National Informatization 2006–2020* and *Special Program of Informatization for the 11th Five-Year Plan* in place, informatization in China in the next 5–15 years will enter an entirely new development phase marked by the trends discussed below.

According to the new requirements for rural development, information servicing through radio, television and communications in Chinese rural areas will gradually improve. This will contribute greatly to reducing the digital divide between urban and rural communities, and between east and west. IT will serve as an important means of saving energy, water and materials, and protection from pollution caused by such industries as metallurgy, oil chemistry, construction materials production and paper making. IT will play an important role in reducing land pollution and resource waste arising from urban development and exploitation of mineral resources. IT will also improve the productivity of new industries through such services as networking, e-finance, modern logistics, chain operations, special information services and consultancies.

Social informatization plays an important role in promoting the construction of a harmonious socialist society. There have been signification developments in terms of social security, employment, digital radio, digital television, digital publication, video games and public cultural information service. Informatization will be systematically deepened in education, health, epidemic prevention and monitoring, emergency response and environmental protection.

With the development and implementation of the *Overall Framework of National E-Government*, the construction and integration of national e-government networks will accelerate. E-government applications, interdepartmental information sharing and business cooperation with a focus on basic information sharing, e-port creation, comprehensive governance of taxation and regulation of the market economy will progress further. The system of governmental websites will be improved. The building of applied and integrated environments, such as an administrative information resources directory, exchange system, simulation examination of e-government, promotion of home-made products and the like will take place.

With the implementation of *Suggestions for Strengthening Development and Exploitation of Information Resources*, more policy work in this area will be conducted. Research on policies regarding the development of the digital content industries, supervision of information resource markets, wider and better uses of governmental information and management of information assets will be enhanced. To optimize the structure of information resources, efforts are being made to improve the exploitation of information resources in fields such as production, distribution, science and technology, population and environmental resources, and speeding up the digitization of information resources in education, culture and literature.

To improve the IT industry's ability to innovate core competencies independently, the government is focusing on the following efforts: enhancing close cooperation between sectors to study and promote IT innovations, increasing financial investment in IT businesses, strengthening R&D on information security technologies and on the basic and common technologies involved, speeding up the implementation of standards and intellectual property rights, promoting the IT innovation system with business as the main body, with a market orientation and a combination of production, learning and research, and formulating new rules and regulations on IT innovation. To accelerate the structural adjustment of the information industry and its optimization and upgrade, the state agencies concerned will develop related policies and measures that will further encourage the development of the software and integrated circuit industries.

Guaranteeing information security is a continuing priority. Based on the tasks proposed in *Suggestions for Enhancing Information Security* and *Suggestions for Further Enhancing Management of the Internet*, the basic infrastructure for information security continues to advance and the Internet environment will be developed further.

Note

1. The percentages do not add up to 100 per cent as Internet users who adopt multiple accessing methods are recounted.

References

Alibaba.com. *Alibaba company profile*. Retrieved 9 October 2006 from http://www.alibaba.com/aboutalibaba/index.html

Baidu.com. *Baidu company report*. Retrieved 7 August 2006 from http://moneycentral.msn.com/investor/research/profile.asp?Symbol=BIDU

Business model comparison and analysis of China's B2B websites. Retrieved 13 October 2006 from http://www.qjy168.com/forum/article.php?articleid=7751

CCIDNET.com. *About us*. Retrieved 9 October 2006 from http://image.ccidnet.com/help/aboutus.html

China Association of Children's Science Instructors. (2006). China Informatization Work Office. (2006). *Informatization development report 2006*. Retrieved 9 October 2006 from http://www.china.com.cn/chinese/PI-c/1254023.htm

China.com. *Unicom gets nod to develop 3G network in Macao*. Retrieved 11 October 2006 from http://english.china.com/zh_cn/business/telecom/11024502/20060818/13552023.html

Chinese e-Government Centre. (2006). *China ranks 74 in world e-government list*. Retrieved 19 October 2006 from http://www.ccw.com.cn/cio/research/zf/htm2005/20050104_1572L.asp

China Informatization Work Office. (2006). *Informatization development report 2006*. Retrieved 9 October 2006 from http://www.china.com.cn/chinese/PI-c/1254023.htm

China Internet Network Information Centre. *The 18th statistical survey report on Internet development in China*. Retrieved 9 October 2006 from http://www.cnnic.net.cn/download/2006/18threport-en.pdf

China Internet Network Information Centre. (2006). *Statistical report on Internet development in China 2006/7*. Retrieved 16 September 2006 from http://www.cnnic.cn/uploadfiles/pdf/2006/7/19/103651.pdf

China iResearch Company. *China online community research report*. Retrieved 9 October 2006 from http://down.iresearch.cn/Reports/Charge/882.html

Fen F. (2004). The key problem and the thought of policy of ICT development. *China economic time* (20040220), 18–20.

Fortuneage Technology Company. *China e-government research report*. Retrieved 11 October 2006 from http://www.fortuneage.com/mycore/researchcontent.aspx?cid=632979859838906250&id=633050805761562500

He J. (1999). A comparison of China-US backbone nets. *Cyber World Magazine*, 29.

He Y. (2003). A study of the current situation and development of China's agriculture information websites. *Chinese Agricultural University Journal*, 3.

Li X. (2006). An examination of China's e-government. *China Management Informatization* (August).

Liu W. (2006). *Specialized IT websites turn off users*. Retrieved 16 October 2006 from http://net.chinabyte.com/332/2114832.shtml

Liu Z. and Feng X. (2006). *China Webpage number skyrockets and CN domain name now world's sixth largest*. Retrieved 16 October 2006 from http://www.cctv.com/news/science/20060516/103133.shtml

Ministry of Science and Technology, People's Republic of China. (2006). *China science and technology newsletter*. Retrieved 10 November 2006 from http://www.most.org.cn/eng/newsletters/2005/200512/t20051229_27301.htm

Rongshuxia.com. *Ronshuxia company profile*. Retrieved 9 October 2006 from http://www.rongshuxia.com/channels/gy/new/index.html

Sohu. *Sohu company profile*. Retrieved 8 October 2006 from http://corp.sohu.com/companyprofile-en.shtml

Virtual China. *IPv6: China's next generation Internet*. Retrieved 5 October 2006 from http://www.virtual-china.org/2006/10/ipv6_chinas_nex.html

Wikipedia. *Sina*. Retrieved on 8 October 2006 from https://secure.wikimedia.org/wikipedia/en/wiki/Sina.com

Wikipedia. *Sohu*. Retrieved 9 October 2006 from https://secure.wikimedia.org/wikipedia/en/wiki/Sohu

WWW.3G.CO.UK. *China's mobile industry readies for 3G/UMTS*. Retrieved 11 October 2006 from http://www.3g.co.uk/PR/October2004/8481.htm

You Y. *CNET to buy Pchome for US$11million*. Retrieved 13 October 2006 from http://tech.tom.com/1121/1794/2306/20050421-184457.html.

Zhao D. and Zhu Q. *The current situation and prospect of Chinese search engines*. Retrieved 4 October 2006 from fttp://www.agri.ac.cn/agri_net/02/2-04/bd160zw.htm

Zhao S. *The present situation and trend of construction of Chinese informatization*. Retrieved 11 October 2006 from http://tech.ccidnet.com/art/7/20060119/419079_1.html

Zhu Y. (2005). Review of Chinese information policy research from 1994 to 2003, *Library Development*, 3.163.com. *Company profile*. Retrieved 9 October 2006 from http://corp.163.com/eng/about/overview.html

Hong Kong

Geoff **Long**, John **Fung** and Charles **Mok**

Population	6.9 million (2006)
GDP per capita	HK$ 214,710 (2006) (about USD 27,500)
Key economic sectors	Banking and finance, tourism, import/export trade, logistics
Fixed-line telephones per 100 inhabitants	95.6 (2007)
Mobile phone subscribers per 100 inhabitants	135 (2007)
Domain names registered under '.hK'	138,952 (March 2007)
Broadband subscribers per 100 inhabitants	72.8 (2007)
International bandwidth	698,106 Mbps (2006)
International Internet bandwidth (not including to Mainland China)	59,522 Mbps (2007)

Sources: GovHK portal (www.gov.hk) and Office of the Telecommunications Authority.

Overview

Hong Kong's information and communication technology (ICT) sector is one of the most developed not just in the region but also globally, as evidenced by a number of recent reports. For example, in the World Economic Forum's latest *Global Information Technology Report*, published in March 2007, Hong Kong was ranked 12th out of 122 economies in terms of Network Readiness. The report's Network Readiness Index examines the preparedness of countries to use ICT effectively on three dimensions: the general business, regulatory and infrastructure environment for ICT, the readiness of the three key stakeholders—individuals, businesses and governments—to use and benefit from ICT, and their actual usage of the latest information and communication technology available.

Hong Kong fared even better in the Economist Intelligence Unit (EIU) e-readiness rankings, where it jumped from 10th place in 2006 to fourth in 2007, behind only Denmark, Sweden and the US. The EIU places more importance on connectivity and the quality of infrastructure, with more weight given to broadband access than dial-up, for example, while it also rates the affordability of such communications. According to the report, Hong Kong owes much of its e-readiness success to strong government policy.

This is not to suggest, however, that Hong Kong is rigidly regulated. One of the main success stories of the regulatory environment has been to create a liberalized and competitive market for services, something that is easier to do in this highly urbanized environment than it might be in less densely populated countries. Situated on the eastern side of the Pearl River Delta, bordering the South China Sea and Mainland China, Hong Kong has 6.9 million people living in an area of just 1,104 square kilometres, making it one of the most densely populated places on earth. As a result, it provides a concentrated and economically viable market that has been seized upon by ICT providers to quickly and efficiently build a critical mass of customers while lowering prices.

First-grade infrastructure is also required by the country's business sector, particularly as Hong Kong's GDP is now predominantly derived from the provision of services. Hong Kong is the 11th largest trading economy in the world and operates the busiest container port in the world in terms of throughput. It is the 10th largest banking centre in terms of external banking transactions and its stock market is Asia's second largest in terms of market capitalization and the eighth largest in the world. These international logistic, trade and finance services rely on Hong Kong's excellent telecommunications infrastructure.

ICT indicators

Since 2000, the Census and Statistics Department has published annual ICT statistics, with the most recent, titled *Hong Kong as an Information Society*, coming in November 2006. According to the report, 71.7 per cent of Hong Kong's 2.3 million households had PCs at home, compared with 70.1 per cent in 2005. The PC penetration rate has hovered around the 70 per cent mark since 2004. The number of households with PCs connected to the Internet increased from 64.6 per cent in 2005 to 67.1 per cent in 2006, with the majority connecting through a broadband connection.

The highest rate of computer use was in the 10–14 age group, with 98.2 per cent using a PC at least once in the 12 months before the survey. Usage among those aged 65 and over was lowest, with just 5.3 per cent of those surveyed having used a PC in the 12 months prior to the survey, although this was up from 2.2 per cent in 2003. The increase in the number of senior persons using computers was a result of intervention at the community level. Now, almost all centres for the elderly within the territory offer computer equipment or training classes. Gender disparity among the senior citizens was high in terms of computer usage—almost two men to one woman. This could be because of the comparatively low literacy rate of women in that age group.

The report suggested that computer users are likely to be more educated and economically more active. Utilization of electronic business such as the Octopus card, ATMs and e-cash was also quite high. Around 97.4 per cent of those aged 15 and above claimed that they had used such services before the survey. However, only 8.8 per cent of all persons in that age group had made an online purchase. E-government services had also been used by 34.6 per cent of those aged 10 and over.

The report did not indicate the utilization of PCs and Internet by disadvantaged members of the population. However, such a survey was conducted by the University of Hong Kong and the Hong Kong Council of Social Service between April 2005 and February 2006. Six disadvantaged groups were identified and investigated: Single Parents with at least one child aged below 18, adults aged 60 or above, children in households with income lower than half of the median household income, new arrivals, female homemakers whose highest level of education was primary school, and Persons with Disabilities and/or Chronic Illness (PWD/CI). The data was collated to create a Comprehensive Digital Inclusion Index (CDII), where a rating of 1.0 is the level of access for the mainstream or non-disadvantaged population in society. The figures are shown in Table 1.

Due to the compulsory nine-year education system in Hong Kong and the fact that all schools now provide access to the Internet, the situation of children in low income families was better compared to other disadvantaged groups. Older people in society were the least 'included' compared to other disadvantaged groups, followed by people with disabilities.

The findings of the Digital Inclusion Index (DII) study have been quoted by the Hong Kong Government in its Digital 21 Strategy public consultation paper, which suggested that further investigation would be required to understand the specific needs of and barriers experienced by different groups of people with disabilities. The DII study was also the first of its kind in Hong Kong and if continued could be used as a tool to monitor and narrow the digital gap.

ICT infrastructure

One of the hallmarks of Hong Kong's ICT infrastructure is the breadth of competitive service providers. As of April 2007, there were 11 fixed network operators (one of which offered fixed wireless services), five mobile operators, six providers of satellite service, 22 cable operators, 255 external telecommunications operators and 179 Internet Service Providers (ISPs).

Despite widespread competition, incumbent operator PCCW-HKTC has a universal service obligation to provide a good, efficient and continuous basic service to consumers anywhere in Hong Kong within a reasonable period of time. The other local fixed-line providers are New World Telecommunications, Wharf T&T, Hutchison Global Communications, Hong Kong Broadband Network, Towngas Telecommunications, CM TEL (HK), TraxComm, HKC Network and Hong Kong Cable Television (HKCTV).

Also remarkable is the quality of the network, with the majority (72.8 per cent) of Hong Kong households accessing broadband services rather than dial-up, while in the mobile sector four of the five operators have widespread 3G (W-CDMA) coverage and are starting to offer higher-bandwidth mobile access through a 3G upgrade known as High Speed Packet Access (HSPA). The regulator has also announced its decision to go ahead with the assignment of spectrum for a CDMA 2000 network in 2007.

The ADSL network of incumbent operator PCCW covers 98 per cent of households, while cable modem service covers

Table 1

Sub-indexes	Older people	New arrivals	Single parents	Women	Children (low income)	Disabilities &/or chronic illnesses	All disadvantaged persons
Accessibility Sub-Index	0.50	0.70	0.71	0.75	0.72	0.53	0.62
Usage Sub-Index	0.04	0.50	0.19	0.06	0.88	0.17	0.26
Knowledge Sub-Index	0.04	0.52	0.41	0.05	0.92	0.14	0.26
Affordability Sub-Index	0.50	0.71	0.48	0.62	0.00	0.54	0.52
CDII	0.27	0.61	0.45	0.37	0.63	0.35	0.41

Source: CDII

more than 90 per cent. The customer access networks of other operators cover 71 per cent of households. Due to the widespread broadband coverage, services such as Voice over IP (VoIP) and IP-TV are popular. Hong Kong boasts the highest penetration of IP-TV in the world with a household penetration of 25 per cent.

Throughout Hong Kong Wi-Fi hotspots are common in coffee shops, office buildings, public transport facilities and so on. The Office of the Telecommunications Authority (OFTA) has a search service to locate registered access points on its website, which can be searched by name, area or district. It lists 29 licensed organizations that offer 1,299 Wi-Fi access points across Hong Kong.

In terms of satellite and submarine cable facilities, Hong Kong is one of the major hubs in Asia Pacific. Satellite-based telecommunications and television broadcasting services are provided via a multitude of satellites in the region with more than 50 satellite earth station antennas operated by Reach Networks Hong Kong, Reach Cable Network and Reach Global Services, Asia Satellite Telecommunications, APT Satellite and a number of external fixed operators and broadcasters.

The Hong Kong Government is also taking steps to prepare its infrastructure for the next generation of Internet Protocol, IPv6. The Internet Engineering Task Force (IETF) has proposed IPv6 because of its far greater address space, which will be needed to connect a multitude of mobile devices and consumer electronics equipment in the future. The Government believes it will generate new business opportunities for the ICT and other sectors.

As a result, the Government will take the lead in migrating to IPv6. The Internet backbone of the local universities has already been upgraded to a high-speed network of 10 Gbps in support of IPv6. The new protocol will be adopted in the Government's internal network by 2008. The Government will also encourage Internet service providers to prepare for the migration.

In 2007, Hong Kong also started the transition to digital terrestrial television (DTT). While digital television stations are already available through cable and satellite networks, the two terrestrial TV broadcasters, Asia Television Limited (ATV) and Television Broadcasts Limited (TVB), will migrate their analogue services to digital. Digital services offer better reception, the ability to offer high-definition TV (HDTV) and new interactive services.

At the end of 2006, ATV and TVB submitted a proposal to the Broadcasting Authority to adopt the national standard announced by the Standardization Administration of China (SAC) called 'GB 20600–2006: Framing Structure, Channel Coding and Modulation for Digital Television Terrestrial Broadcasting System'. Under the transition framework, the two stations will broadcast both analogue and digital signals in 2007 and aim to extend the coverage of their digital networks to at least 75 per cent of Hong Kong by 2008. Subject to further market and technical studies, the Government will direct ATV and TVB to switch off analogue broadcasting within five years after the commencement of dual broadcasts.

The Broadcasting Authority has approved ATV and TVB's investment plans for digital TV programme service and network rollout. ATV has committed an investment totalling more than HKD 400 million (about USD 51 million) up to 2009 to provide a hybrid digital service of HDTV and multichannel broadcasting, while TVB has committed an investment totalling more than HKD 400 million up to 2009 to provide an HDTV channel starting from end 2007. This HDTV channel will include at least 14 hours of HDTV programmes per day.

Key national initiatives

Digital 21

In July 2004, the Office of the Government Chief Information Officer (OGCIO) was set up to provide leadership for the development of ICT within and outside of the Hong Kong Government, as a single focal point with responsibilities for ICT policies, strategies, programmes and measures under the Digital 21 Strategy.

First drawn up in 1998, the Digital 21 Strategy for information technology is the highest-level policy initiative by the Hong Kong Government in the area of ICT development. The strategy is currently under review and a new version will be published in 2007. Previous reviews of Digital 21 were undertaken in 2001 and 2004. The most important areas of focus in the new strategy will include expanding the use of advanced IT in citizen services like a 'one-stop portal' for e-government services, health care and transportation, as well as further investment to achieve better digital inclusion.

In a consultation paper on the Digital 21 Strategy, the government notes that innovation and technology will continue to play a key role in helping Hong Kong to compete by enabling businesses to transform and provide goods and services of increasing value and harnessing the role of Hong Kong as a business hub for Mainland enterprises to attract foreign investments and participate in the global economy. There is also a clear vision of moving towards an inclusive, knowledge-based society in which the benefits of ICT adoption are widely available to different segments of the community. In addition, the consultation paper points out that issues relating to data standards, information management and intellectual property rights protection are becoming areas of increasing focus.

Aimed at making Hong Kong a world digital city, the Digital 21 Strategy has initiatives around four areas:

1. Promoting advanced technology and innovation, including initiatives such as the strengthening of the Cyberport, Science Park and newly established R&D centres.
2. Developing Hong Kong as a hub for technological co-operation and trade, including initiatives to boost the ICT workforce by developing competency standards and strengthening training leading to professional recognition.
3. Enabling the next generation of public services, including initiatives such as the establishment of a new government portal, GovHK, as the single entry point to online government information and services.
4. Building an inclusive, knowledge-based society through provision of broadband connectivity for every citizen and other initiatives.

Digital solidarity fund (DSF)

The Digital Solidarity Fund (DSF) was established as a platform to engage government, the private sector and NGOs in the work of narrowing the digital divide. The government and private sectors contribute financially and are also heavily involved in direction setting and project selection. NGOs contribute their expertise in project implementation and are also engaged in fundraising and fund management.

Of around 120 applications for funding, DSF has granted more than USD 256,000 to 13 community programmes. Projects supported are mainly for providing access and capacity building services such as training and technical support for specific groups. Mobilization of volunteers was a major part of almost all programmes.

The response from stakeholders has been positive. DSF sponsors have noted that their direct involvement in the management of the fund and occasional participation at the community level on a voluntary basis was a rewarding experience. Community organizations have found it encouraging to learn that their expertise is being acknowledged and that disadvantaged people in the community would be able to participate in the information society effectively through their help. The Government has already proposed to include its support for DSF in the Digital 21 Strategy.

However, despite its achievements, financial support for DSF remains a big constraint. The USD 130,000 raised annually is slightly below 50 per cent of the expected target. DSF is expected to do better at attracting public and private funds in the future.

Public Wi-Fi services

In May 2007, the Legislative Council approved the release of USD 217.6 million to fund the provision of Wi-Fi facilities at about 350 government premises for free use by the public. Priority sites will be set up at premises frequently visited by members of the public by mid-2008. These premises include libraries, public enquiry service centres, community halls/centres, parks and government buildings. The OGCIO will centrally oversee, coordinate and manage the programme, but installation, provisioning and operations will be outsourced to the private sector.

The Government will specify the security requirements in the tender document to ensure that the contracted service providers will provide the necessary hardware, software and technology with appropriate security features. The Government will also require service providers to install various security measures to safeguard user data, such as encryption, intrusion prevention and detection systems, and filtering software. Security consultants will be engaged to perform security risk assessment on the Wi-Fi network designs and conduct security audits after the networks have been put into full operation.

The Government expects the initiative to stimulate the development of wireless and mobile applications that will be conducive to the development of the ICT industry and the wider economy.

Research and development

Hong Kong has two key centres for ICT R&D: the Cyberport situated in Telegraph Bay in the Southern District of Hong Kong island, and the Science Park located in the Tai Po area of the New Territories.

The Cyberport, a USD 2 billion project managed by Hong Kong Cyberport Management Company Limited and wholly owned by the Hong Kong government, is seen as the territory's main ICT hub. It is a strategic cluster of more than 100 IT companies and more than 10,000 IT professionals. The clustering of local and overseas companies and professional talent is envisioned as a catalyst and hub for the growth of local and regional IT digital entertainment industries, with particular emphasis on IT applications, information services and multimedia content creation. Cyberport also provides IT education for the broader community.

The Science Park is dedicated to applied research and development. It is also being developed along a clustering concept, with four clusters of electronics, IT and telecommunications, biotechnology and precision engineering. The first phase of the Science Park was opened in 2002, while phase two is to be completed in stages in 2007 and 2008.

In the latest Digital 21 consultation paper, the Government stated that it would continue to strengthen both the Cyberport and the Science Park. It has also set aside USD 2 billion under the Innovation and Technology Fund to set up a further five R&D centres which are described as dynamic hubs forging partnerships among multiple players, including the ICT industry, different industrial sectors, academia and overseas/ Mainland enterprises in the development, application and commercialization of new technology. In particular, the new R&D centres will undertake industry-oriented research in technologies demanded increasingly in Mainland China. The five focus areas are R&D for ICT, automotive parts and accessory systems, textiles and clothing, logistics and supply chain management enabling technologies, and nanotechnology and advanced materials research.

Regulatory environment

One of the strengths of Hong Kong's ICT sector, as noted in the EIU e-readiness survey, is its enlightened and forward thinking regulatory policies. Key to this is a fully liberalized market where service providers are free to adopt the technology that best meets market demands. All sectors of Hong Kong's telecommunications market have been liberalized and there are no foreign ownership restrictions.

In the fixed network sector, there is no preset limit on the number of licences issued, nor a deadline for applications. There is also no specific requirement on network rollout or investment. The level of investment is determined by the market. Similarly, in the satellite sector Hong Kong adopts an 'open sky' policy in regulating the provision of satellite services. The only limits on operators are in wireless services where there is a physical constraint such as spectrum availability.

Number portability has been introduced in both the fixed and mobile sectors to promote competition by allowing consumers to retain their existing telephone numbers if they change providers.

Charged with overseeing the sector and enforcing such competition measures is the Office of the Telecommunications Authority (OFTA), which was established as an independent government department on 1 July 1993 as the executive arm of the Telecommunications Authority, the statutory body responsible for regulating the telecommunications industry in Hong Kong. The Government also plans to set up a Communications Authority by merging the existing Telecommunications Authority and Broadcasting Authority. This is to recognize the convergence of the telecommunications and broadcasting industries.

OFTA is vested with all of the necessary powers under the Telecommunications Ordinance to administer licence conditions, make determinations and impose sanctions for breaches. Its role can be broken down into six main areas: regulating public telecommunications services, enforcing fair competition in the telecommunications sector, tracking down illegal telecommunications activities, managing the radio frequency spectrum and coordinating satellite orbital positions, advising the Government on telecommunications matters, and representing Hong Kong in international telecommunications organizations and forums.

OFTA is also looking at new measures to ensure a more competitive environment. In April 2007, it published a new spectrum release plan that will make available further radio spectrum over the next three years through an open bidding or tendering process. OFTA said the publication of the plan, which will be updated annually, is also for increasing the transparency of the supply of radio spectrum.

The new spectrum is expected to be used for the deployment of broadband wireless access (BWA). In May 2007, OFTA announced proposals for the allocation of the 2.3 GHz band for BWA and also requested feedback from interested parties to assess the potential demand for spectrum in the 2.5 GHz band. An auction is planned for 2008.

Further measures are intended to remove restrictions on incumbent operator PCCW with reference to unbundling the local loop and pricing tariffs. This is because the Government thinks that there is now effective competition in all sectors of the market, both through a variety of network operators and network technologies (ADSL, cable, satellite, etc).

In the past, the regulator required approval on tariffs set by the incumbent so as not to reduce competition through pricing pressure. As of 2005, however, this was removed and all operators are free to set their own tariffs. OFTA is also in the process of removing requirements on unbundling the local loop, whereby the incumbent is required to provide new players access to its telephone exchanges. Due to adequate network coverage by others and the future likelihood of fibre being deployed, OFTA no longer regards the local loops of the incumbent as essential facilities. To promote facilities-based competition, OFTA has set the end of June 2008 as the termination date for phasing out mandatory unbundling of local loops at the telephone exchanges of the incumbent. After this date, unbundling can still be available through commercial agreement or as mandated by regulation in the small number of locations where the local loops remain as 'essential facilities'.

OFTA has also reviewed its regulations regarding VoIP. It allows such services, both from facilities-based and services-based operators. Telephone numbers will be made available to both types of operators. Additionally, broadband access providers are not allowed to block their customers' access to VoIP services. Moreover, with the introduction of VoIP services and new radio technologies that can economically reach customers in remote areas, OFTA is currently reviewing its universal service obligations to make sure that its system is efficient and more fairly shares costs between operators.

Content and services

DigitalCopyright.hk

With the support of the Innovation and Technology Fund, a Digital Rights Management (DRM) infrastructure employing state-of-the-art technologies was set up at the Cyberport in November 2005 to provide a channel for digital content creators to distribute their products to consumers efficiently and at low cost. The two-year programme starting from June 2006 promotes the use of DRM among ICT system developers, digital content developers and consumers, particularly young people, to cultivate a legal software download culture in the community.

DigitalCopyright.hk aims to provide the infrastructure to facilitate protection and distribution of digital content. In other words, it provides the backend infrastructure for content protection, license issuing and tracking for audio-visual and Flash-based multimedia content. For content owners/distributors, a portal at www.iresource.hk has been built for distribution of protected digital content and is aimed at content partners with no existing e-commerce network.

The project aims to facilitate the distribution of legitimate digital content across all platforms and devices and accelerate the adoption of intellectual property protection technology by lowering the technological, investment and management barriers. For digital artists and independent content producers, the platform acts as a safe place for permanent archival of their works and as a creative space to deliver innovative products. The project also strives to stimulate more research and development in digital asset management and related technologies.

IP service centre

An Intellectual Property Servicing Centre was established in June 2006 as part of the Integrated Circuits Design Centre at the Science Park. The Intellectual Property Servicing Centre provides a platform to support and facilitate the wider use of semi-conductor intellectual property and to protect the technological investment of integrated circuit design companies.

Software

A key component of the IT services sector is the software industry. Although most of the local software developers are small firms with less than 20 employees, they are able to produce competitive customized software to support local and foreign clients in various sectors such as finance and banking, transportation and logistics, supply chain management, transportation, trading and telecommunications.

Some recent examples of Hong Kong's successful and pioneering software development include the Smart Identity Card, which is given out to all residents in Hong Kong—the first of its kind in the world—and the Octopus card, the largest cash payment card in circulation in the world. Octopus was introduced in 1997 to provide a simple and unified way to pay fares on public transport in Hong Kong. Soon after, Octopus extended its reach into simple micropayments for purchases in retail outlets and became a simple way for cardholders to gain access to buildings and schools and to identify themselves. Today, over 500 service providers accept Octopus, and there are more Octopus cards in circulation than there are residents in Hong Kong.

HK ICT awards

There are 38 professional bodies related to the ICT industry in Hong Kong. Most have sought funding from the Government to organize events such as awards nights. In 2006, the OGCIO consolidated Government support for these events and sponsored a single HK ICT Awards programme. The programme was considered a success and will be continued in 2007.

The six categories for the first ICT Awards were a Digital Entertainment Award won by a project called Moving Music whereby images of human movement are used to control sounds produced in games; an e-Business Award won by a Next Generation Terminal Management System; an e-Government Award won by the Immigration Department's e-Channel, an automated passenger clearance and vehicle clearance system that uses biometrics to enhance public service; the e-Learning Award won by GoChinese, an online Chinese learning platform; an e-Youth Award won by 'Flaber', a Flash-based Web building tool; and the Digital Inclusion Award won by the IT Inclusion Community Project where a secondary school student mobilized community resources to help new migrants and other disadvantaged families learn to use ICT.

Electronic Health Records (e-HR)

The Hospital Authority (HA) is a government statutory body established in 1990 to manage all public hospitals in Hong Kong. Among the successful IT implementations at the HA is its Clinical Management System (CMS), which links the entire HA network, allowing seamless access to a centralized IT system and database for health professionals. More recently it has been piloting a scheme for electronic patient records (e-PR). Patient participation in this pilot scheme is voluntary and a number of encryption and other security measures have been set up to ensure information privacy.

Public and private hospitals are now discussing whether to use the e-PR service as the basis for a territory-wide electronic health record (e-HR) system. Subject to confidentiality and security safeguards and the patient's consent, e-HR could be accessed by a health care professional in public and private hospitals, clinics and residential care homes for the elderly. The availability of comprehensive records will enable timely and informed decisions to be made at the point of care. However, before it is introduced a number of questions need to be addressed. These include a body to oversee or regulate the e-HR operation, whether legislative backing is needed, financing of the capital investment and recurrent costs, ownership of the records and limitations on access to these records, security and privacy protection of individual data and the entire system, and whether any penalty should apply to proven cases of unauthorized use of the data. In the meantime, the HA has embarked on a pilot project to share its e-HR system with a number of private hospitals and private medical practitioners.

Open source community

In the area of open source software (OSS), the Hong Kong Government and the industry in general take the practical approach of adopting open and interoperable standards, as opposed to having any officially imposed preference for a particular technology platform. Since 2002, the Government has set up an Interoperability Framework (IF) to reflect major changes in the industry and set out government standards and practices to ensure data and technical interoperability among its IT systems and e-government services.

The Government's policy on procurement of software products is based on objective criteria, such as value for money, functionality, security, system compatibility and the availability of reliable technical support. However, the Government has stated in its policy that it will promote the use of OSS technologies and solutions within the Government in order to widen product choice and maximize the potential for cost savings.

Future trends

While it already boasts one of the best ICT infrastructures in Asia Pacific, Hong Kong looks set to improve its digital outlook. It continues to introduce measures designed to improve the competitiveness of the ICT sector, such as freeing up radio spectrum and looking at new areas of technology, such as fixed-mobile convergence. For example, the regulator is examining the existing regulations, including interconnection charges and carrier licensing arrangements, to facilitate the convergence of the two sectors.

The Government is also progressive in providing electronic access to its services. A new Web portal, GovHK, was introduced in September 2006 to replace the Government Information Centre (www.info.gov.hk) as the entry point to online government information and services. The portal, which provides access to some 1,200 existing government electronic services, is undergoing constant development. One enhancement planned is the provision of geospatial information to underpin information and services. Efforts are also under way to develop a youth portal that will provide access to a range of public services for youth aged 15–24.

Another potential strength is Hong Kong's links with Mainland China. The Government has established channels for cooperation with the relevant Mainland authorities and Guangdong province in areas such as innovation and technological development. Identified areas for cooperation include software development, wireless and mobile technology, automotive parts and accessory systems, integrated circuit design, digital entertainment, digital certificates cross-recognition and the development of standards and applications in emerging technologies such as RFID and next-generation Internet. The Government is also looking to participate in the Mainland's technology development plans and the formulation of national standards through the Mainland/Hong Kong Science and Technology Cooperation Committee. The ICT industry, professional bodies and academics from both sides are expected to be involved in these initiatives.

If it can also continue to act on 'digital inclusion' measures through the Digital Inclusion Index and other measures outlined in its latest Digital 21 Strategy proposals, Hong Kong will remain one of the most significant ICT players in Asia Pacific.

References

Census and Statistics Department. (2006). *Findings of the 'Household Survey on IT Usage and Penetration' and the 'Annual Survey on IT Usage and Penetration in the Business Sector' for 2006*. Retrieved from http://www.censtatd.gov.hk/press_release/press_releases_on_statistics/index.jsp?sID=1810&sSUBID=7766&displayMode=D

Commerce, Industry and Technology Bureau. (October 2006). *Public consultation on digital 21 strategy*. Retrieved from http://www.info.gov.hk/digital21/eng/strategy_consultation/D21ConsultationPaper(E).pdf

Digital TV, Commerce, Industry & Technology Bureau, HKSAR Homepage. Available at http://www.digitaltv.gov.hk

Economist Intelligence Unit. (2007). *The 2007 e-readiness rankings*. Retrieved from http://www.eiu.com/site_info.asp?info_name=eiu_2007_e_readiness_rankings

HKSAR Homepage. Available at www.gov.hk.

Hong Kong Domain Name Registration Company Homepage. Available at www.hkdnr.hk

Office of the Telecommunications Authority, HKSAR Homepage. Available at www.ofta.gov.hk

Wong, Y.C., Law, C.K., Fung, Y.C., and Lam, J. (2007). *Study on the digital inclusion index of Hong Kong*. University of Hong Kong. Unpublished report.

World Economic Forum. (2007). *The global information technology report (GITR) 2006–2007*. Retrieved from http://www.weforum.org/pdf/gitr/rankings2007.pdf

.in India

Frederick **Noronha** and Sajan **Venniyoor**

Total population	1,112 million (2006)
GDP per capita	USD 3,547 (PPP)
Key economic sectors	Agriculture (18.6 per cent), Industry (27.6 per cent), Services (53.8 per cent) (2005 est.)
Fixed-line telephones per 100 inhabitants	4.1 (2004; up from 0.6 in 1990)
Mobile phone subscribers per 100 inhabitants	4.4 (2004)
Internet users per 100 inhabitants	3.2 (2004)
Sources: IMF 2006, UNDP HDR 2006.	

Overview

India has an estimated 45 million Internet users. But only about 10 million are 'power users', that is, those who regularly use the Web for research and e-commerce. Government efforts to loosen restrictions on access, and the entry of private companies into the Internet market, are said to be driving the growth (Kaufman Bros. 2006).

Figures vary. According to the Internet and Mobile Association of India (IAMAI) and IMRB International, Internet users in India reached 37 million in September 2006, up from 33 million in March 2006. During the same period, the number of 'active users' (those who have used the Internet at least once in the last 30 days) has risen from 21.1 million in March 2006 to 25 million in September 2006. The survey was conducted in early 2006 among 16,500 households covering 65,000 individuals across 26 major metros and small towns in India, with an additional coverage of 10,000 business and 250 cybercafé owners. The survey did not include the rural areas.

India's potential in shaping the Internet globally—because of its large population—should not be overlooked. Of the seven Asia Pacific nations in the top 20 countries in terms of number of Internet users, three—China, Japan and India—are in the top five (Make-IT-Safe April 2005).

However, the quality of usage can be questioned. Also, the lack of sufficiently widespread local language solutions could hamper future growth. Wikipedia points out that '[n]otably absent from … [the languages used on the Internet] is Hindi, one of the most commonly spoken languages of the world, as well as the national language of India, the second most populated country in the world.' The reasons cited are lack of access to the Internet by the large majority of the Indian population, and a preference for English among those with Internet access.

According to India's National Readership Survey 2006, there are currently 9.4 million Internet users who log in every week, up from 7.2 million users in 2005. But this figure, says the study, constitutes only 1.2 per cent of India's 12 years plus population. Urban India has shown faster growth in Internet reach—from 2.3 per cent to 3.4 per cent, says the survey. Mobile phones are reaching a critical mass. The reach of this medium, as measured by the proportion of the population accessing value-added-service (VAS) at least once a week, grew from 1.1 per cent in 2005 to 2.7 per cent in 2006—translating to nearly 22 million individuals (Hindu Business Line 2006).

As for mobile subscribers in India, the estimate as of July 2006 was 111 million mobile subscribers, representing some 10 per cent of the total population. Short Messaging Service (SMS) is believed to be the leading method of communication (Kaufman Bros. 2006) in the country. But while the figure might seem large, in the context of India's total population (1.1 billion) the coverage is obviously far from sufficient. ITU (2007) argues that countries that are leaping ahead in Internet use, such as India, may have a sluggish mobile sector.

In September 2006, the Manufacturers' Association for Information Technology (MAIT), the lobby group of the hardware, training and R&D services sector in India, said that sales of desktops and laptops together exceeded 1.2 million units in the second quarter of fiscal year (FY) 2006–07. They expect desktop sales in FY 2006–07 to exceed 5.6 million units. According to MAIT, IT consumption in the country continues to be dominated by industry verticals and corporate sectors such as telecom, banking and financial services, manufacturing and IT-enabled services. Apart from these traditional sectors, high consumption was also witnessed in SMEs, education, retail and other computer-centric small enterprises. Aggressive pricing by PC vendors has also helped improve the PC penetration, especially in the households and the SME segments (MAIT 2006).

Internet growth

The youth are the main drivers of Internet usage in India. College students and those below the age of 35 are the biggest segments on the Internet. Both these segments have the highest proportion of conversion of 'ever' users to 'active' users of the Internet. According to an IAMAI-IMRB study, the 'ever' user base is estimated to grow from 37 million in September 2006 to 42 million in March 2007 and 54 million in March 2008, while the active user base will hit 28 million in March 2007 and 43 million in March 2008. Smaller cities and towns recorded a whopping 142 per cent year-on-year growth, with 25 per cent of users coming from smaller towns.

Kaufman Bros. (2006) suggests that 'due to the limited online population and the smaller subset of researchers and purchases on the Web, the concept of assisted Internet has taken root in India'. This refers to Internet-based companies opening 'retail storefronts' throughout the country. Kaufman Bros. (2006) says that India still awaits a 'killer application that drives consumers not only to the Web, but [also] to obtain a broadband connection'. Such an application could be in education, commerce, entertainment or gaming.

According to a NASSCOM analysis (2007), Indian IT vendors are increasingly turning their attention to the domestic market. NASSCOM is the trade body for India's software and service companies. The Indian user industries are outsourcing parts or entire IT infrastructure to specialized vendors. Recognizing the growing importance of the domestic market, NASSCOM put the domestic software and services segment at USD 3.9 billion in 2003–04, up from USD 3.0 billion in 2002–03.

NASSCOM also sees potential in the rise of smaller cities as new investment destinations for the IT-ITES sector, with the industry taking a keen look at these locales that are now gradually taking the pressure off the already infrastructure-strapped, saturated metros.

Some like Mohan Krishnan, Vice President and Country Manager of the eTechnology Group@IMRB, have argued that, 'The next round of growth will be driven by new and innovative applications such as blogs, P2P, video on demand and online gaming while the old favourites such as email, Chat and IM will drive first time users to the medium' (IAMAI September 2006).

Indeed, 51 per cent of Indian Internet users come from the low-middle class. Checking blogs is the second most preferred online activity, according to a study which notes that Indian languages, small towns and broadband are gaining prominence. The Internet is making deep inroads in the lives of Indian people, spreading horizontally and vertically, perhaps leading to the probable reduction of class, language and regional barriers (Srivastava 2006).

Digital content

As India grows more accustomed to the digital world, local content initiatives are increasing. But the lack of local language solutions and widespread acceptability is limiting potential growth.

Some initiatives help keep track of what's happening in diverse areas. For example, India's Manthan Awards (http://www.manthanaward.com/) encourages 'the development of e-content at every level and enhancing the e-content production capabilities and inducing poverty alleviation exercises'. Among the 2006 winners were the Agriwatch portal (www.agriwatch.com), meant to be a knowledge hub for the agricultural sector in India; www.sumul.coop, covering milk procurement, cattle feed management and more; the bhojpuria.com portal for people speaking the regional Bhojpuri language; indianheritage.com launched in 1997; anandautsav.com which focuses on the regional Durga Puja festival of Bengal; and namiindia.com which builds awareness to reduce the stigma among families and persons affected by mental illness. Other awardees came from the fields of telemedicine projects for rural areas, e-governance (deployed for 'panchayati raj' or rural governance, grievance handling, ICT initiatives for local governance, utility-driven websites for municipalities), and a prison management system. In e-learning, there were vernacular-language Braille solutions, IT training lessons, multimedia content for high schools and a knowledge-sharing website.

Other categories of awards were 'e-inclusion', e-news, e-localization, e-environment, e-youth and e-content. Community broadcasting became a new category for Manthan Award 2007. The award panel noted that community broadcasting 'is the most effective way of empowering the masses at the grassroots level in India where oral communication is a medium of information and knowledge sharing'. There have been plans to open up non-commercial, non-state community broadcasting for a decade now. But while the authorities have been cautiously opening on-campus radio projects, the arrival of 'community radio' is still happening only very slowly.[1]

Meanwhile, the Traditional Knowledge Digital Library (TKDL) is a Government of India initiative based in New Delhi that aims to build a database of traditional knowledge that 'enables the protection of such information from getting misappropriated'. TKDL says it has completed the transcription of 36,000 Ayurvedic formulations into English, German, French, Spanish and Japanese since an inter-disciplinary team started working in October 2001. It has also made a presentation at the IPC Union in Geneva. One of the issues that TKDL addresses is the erroneous application of patents to traditional knowledge. According to TKDL director V.K. Gupta, 'A majority of the patents (taken on Indian knowledge) is by expatriate Indians or multinational corporations. There are about 2,000 patents which have been wrongly issued, in our view. Each takes 11 years to fight.'

In mid-August 2006, a digital knowledge centre set up at a cost of INR 7 million (about USD 173,913) was inaugurated at the central library of Anna University. The digital centre aims to help students, scholars and faculty access the Internet, the online journals and periodicals subscribed to by the university, and digital libraries. There are also plans to digitize selected books and journals in the main library. Alumni from the batch of 1979 contributed INR 2.5 million (about USD 62,111.80) towards the infrastructural needs of the project. They also undertook the installation of the systems. Anna University funded the purchase of computer systems at a cost of INR 4.5 million (about USD 111,801.24) (*The Hindu* 2006b).

In another initiative, the Natural Disaster Information System (NDIS) was launched as 'a first of its kind pilot project aimed at alerting people about any impending natural disaster' (*The Hindu* 2006a).

At the Indian Telecentre Forum held in New Delhi on 23–25 August 2006, representatives and leaders of telecentre networks from countries around the world had an opportunity to meet and discuss common concerns, issues and opportunities.

Networks such as BytesForAll (www.bytesforall.net and http://groups.yahoo.com/group/bytesforall_readers) and India-GII (https://ssl.cpsr.org/pipermail/india-gii/) deal with ICT or ICT4D issues in South Asia and India.

Elsewhere, a new initiative hoped to become the craigslist (www.craigslist.com) for India, especially for places other than the largest Indian cities. Allahabad-based bitsTek launched vargikrit.com (http://www.vargikrit.com) to 'provide a classified platform to community'. In this website all postings are free and there are no banner or pop-up ads.

The spirit of sharing knowledge and information also appears to be catching on. For example, there is some discussion about a 'dollar one encyclopedia', or a wikipedia-on-a-CD for easy and inexpensive distribution (Thejesh 2006).

Other online ventures like YouTube.com also encourage the sharing of local content. Tools like the Wikipedia (http://en.wikipedia.org), which is among the 20 most visited globally, help give Indian content and issues global visibility. For example, Wikipedia has been highlighting on its home page a number of Indian issues, including the Indian Institutes of Technology, a group of seven autonomous engineering and technology-oriented institutes built to create a 'skilled workforce to underpin India's economic and social development after independence in 1947...' (Wikipedia).

During his visit to India in late August 2006, Wikipedia founder Jimmy Wales noted that the volume of volunteer contributions to the Kannada Wikipedia has been growing 22 per cent per month while the volume of volunteer contributions to the Bengali Wikipedia has been growing 35 per cent per month. This is significant because although other Asian languages are spoken by millions, there tends to be little sign of them in the English-dominated cyberspace. 'We still have an enormous amount of work left to do. India has 23 official languages. English [is the only language used in India with] more than 10,000 articles. We aim to have 200,000 articles for every language spoken by a million people. That covers 94 per cent of the people on the planet,' Wales said. By the end of September 2006, the Bengali Wikipedia, built both in Bangladesh and India, crossed the landmark of 10,000 articles. It became the 50th language to do so, and only the second from South Asia. Bengali is spoken by almost 220–250 million people; it is the seventh largest language in terms of total speakers. The Telugu Wikipedia, representing the south Indian language, now has over 15,000 articles.

Meanwhile, an 18-month-old 'Wikipedia for India' network on the social networking site Orkut has some 215 members. The network aims to become 'the place where u can find info on anything from Hindu mythology to Besant Nagar beach...or Connaught Place'.

Knowledge commons issues have been debated. There is a deepening debate on how knowledge is shared or controlled in this new information-dominated century, and it is a debate of vital relevance to a country that is making an increasingly visible global impact through its brain power and at the same time has one of the most impressive collections of traditional medicines and knowledge. There are diverse views on how these kinds of issues should be tackled, as was obvious at the 'knowledge symposium' held in New Delhi on 24–25 August 2006. The issue of the need for 'IP-unencumbered software' has also come up (Noronha 2006a).

Open Access is a new trend in India. This refers to free on-line availability of research-oriented scientific and scholarly journal articles. It picked up globally since around 2002. According to Chennai (South India)-based information scientist Subbiah Arunachalam, 'Nearly a hundred journals (in India) have already taken the Open Access route.' In early 2006, the Bangalore-based information company Informatics (India) Ltd, launched Open J-Gate (www.openj-gate.com), a portal covering more than 3,500 English-language journals. Some 2,000 of these are peer-reviewed. Open J-Gate claims to be the 'world's biggest Open Access English language journals portal'.

But more action in this area is needed. As Arunachalam argues: 'Research performed in India, funded by Indian taxpayers, is reported in a few thousand journals, both Indian and foreign. Since some of these journals are very expensive, many Indian libraries—including sometimes the author's own institutional library—are not able to subscribe to them.' Arunachalam estimates that Indian researchers publish approximately 20,000 papers a year in 2,500–3,000 journals in 130 countries, 'including in (small countries like) Pakistan, Bangladesh, Sri Lanka or Croatia.' Because of the current situation, researchers cannot even read their peers. Besides, most Indian journals have poor circulation: only six of the crucial Council of Scientific & Industrial Research's 20 odd journals have over 1,000 subscribers. Few Indian researchers reach high-impact journals abroad, while roughly half of all Indian research is published abroad. Thus, Indian research work does not reach a wide audience, which affects both its visibility and its impact.

India's Open Access journals include 11 journals published by the Indian Academy of Sciences, four journals published by INSA, one journal published by the Indian Institute of Science, one journal published by the Indian Council of Medical Research, and three journals published by the Calicut Medical College. In addition, the National Informatics Centre of the Government of India operates the Indian Medlars Centre, which makes available electronic versions of 38 Indian biomedical journals, mainly published by professional societies. Indian Medlars Centre also has an ePrints-based archive called OpenMED where biomedical researchers from anywhere in the world can deposit their papers. IndianJournals.com, a Delhi-based company, publishes eight Open Access journals dealing with subjects like forensic medicine, fire engineering, neonatology, agricultural sciences and veterinary sciences. medind.nic.in offers free access to 38 biomedical journals. The Institute of Mathematical Sciences (IMSc) was the pioneer in Open Access archiving in India. Also gaining increasing attention is the GNU EPrints archive at the Indian Institute of Science in Bangalore.

LexLibre aims to contribute projects or articles as working papers to the public domain. Promoters say 'this will enable the creation of a large database of legal resources, something which is sorely needed in India'. There is also an ongoing consultation on making theses available online. The University Grants Commission (UGC) of India is inviting public comments on its draft consultation paper 'Electronic Thesis Online (India) UGC (Submission of Metadata and Full-text of Doctoral Theses in Electronic Format) Regulations, 2005' (Digital Opportunity Channel 2005a).

Proponents point out the many advantages of Open Access institutional archives. They not only make Indian research work more visible, but also help Indian research papers win more citations. Such archives are easy to set up as the required software is free.

Some major global commercial publishers had promised to offer access to countries with less than USD 1,000 per capita incomes. But they went back on their word in India, arguing that they enjoyed sizeable subscriptions in the country. On the positive side, in November 2006 44 scientists and policymakers from Brazil, China, Ethiopia, India and South Africa met in Bangalore to set guidelines for developing countries to freely access publicly funded research. The success of their draft national policy will depend on whether the relevant governments, funding groups and research institutes will adopt their recommendations (Noronha 2006b).

Enabling ICT policies and programmes

One of the big plans currently being unveiled is Mission 2007 (http://www.mission2007.in/), which aims to connect a targeted 25,000 villages in the first year by pooling resources from various states, government agencies and corporations, and using affordable and accessible technology. According to the project planners, 'Mission 2007 will initiate formation of consortia of content developers to provide content and ensure that local livelihood needs are met and available content resources are pooled for achieving the common goal... The endeavour will...make it possible for local communities to collect, access and use data on their livelihoods and assets using these applications for local, regional and national planning.'

In connection with the plan, there are efforts to 'influence policy issues such as low tariff and de-licensing of last mile ICT applications, especially wireless spectrum and community media'. The project seeks to include training and capacity building of 'village entrepreneurs'. Other issues raised include peer-to-peer learning, sharing of knowledge at the village level, working towards a suitable legal and regulatory environment, Open Source and content, security issues, educational programmes, and building the technology infrastructure. Because it is an ambitious project, there are doubts about the degree of effectiveness and timeliness of implementation, end-use deployment of practical results and its actual impact on rural India.

In May 2004, Union Minister for Communication and Information Technology Dayanidhi Maran announced a 10-point agenda, including achieving convergence of ICTs; bringing about transparency in administration and making government functioning more 'citizen-centric'; providing broadband connectivity to all 'at the most reasonable prices'; leapfrogging to next generation (4G) mobile wireless technologies; connecting all ISPs in India to a national Internet exchange; significantly improving the Indian Internet Domain Name service; migration to the IPv6 protocol; cyber infrastructure protection and security and digital signatures; promoting Media Lab Asia work in rural connectivity, healthcare, literacy through distance education and development of low-cost PCs; local language computing; outsourcing skilled human resources and an R&D thrust.

Referring to the 10-point plan, Gartner Research suggests that India's ICT Ministry cannot achieve most of these goals on its own 'because the power to make the necessary decisions is distributed among many players, including the Telecom Regulatory Authority of India, the federal economic and other ministries and the country's state governments'. Gartner Research also notes that the Indian ICT Ministry has a limited budget. Moreover, the two government-owned telecom providers, BSNL (Bharat Sanchar Nigam Ltd) and MTNL (Mahanagar Telephone Nigam Ltd), fund most of their capital outlays through internal accruals, and this is unlikely to change. The IT section of the Ministry of Communication and Information Technology likewise has limited resources for IT projects.

Gartner Research notes that the Ministry can make a difference only through policies that promote investment and competition and remove market distortions. It also suggests that the Indian government 'promote the use of IT by using it aggressively, both for internal processes and for citizen services. Because of the scale of government operations, this can "kick-start" the domestic IT market, a segment that has lagged.' Also suggested is a reduction in the cost of IT for all user segments through fiscal and other incentives, to spur domestic demand. 'This will help with the legitimate social goal of bridging the Digital Divide', Gartner argues (Kumar et al. 2004).

There are other ICT initiatives, such as the Natural Disaster Information System (NDIS) launched in mid-February 2006 by the Union Minister for Science, Technology and Ocean Development Kapil Sibal. The project seeks to alert people about any impending natural disaster using the local language, mobile phones and a specially set up wireless public address system. It was developed by a private–public partnership between the Technology Development Council (TDC) and Bangalore-based Geneva Software Technologies. Data security is assured with a 128-bit exception using dedicated leased lines. Alerts will then be sent out to mobile phones in the language of the local community concerned. The voice message will be streamed as an outbound call and sent to the wireless public address system for direct audio alert[2] (*The Hindu* 2006b).

In October 2006, the AirJaldi Summit was held in India, with the intention of addressing 'some of the ways that wireless solutions can be used to provide affordable Internet access in rural communities'. The conference focused on 'showing the advantages that wireless networks can provide, by enhancing the quality of education, governance and health care, increasing economic development, and promoting cultural exchange'. The spotlight was on the Dharamsala Community Wireless Mesh Network developed and managed by the Tibetan Technology Center (TibTec) to provide Internet connectivity to rural communities, schools and institutions in the Dharamsala region. This relatively large-scale experiment saw a combination of low-cost yet robust technology and community-based implementation. The Summit organizers said the Dharamsala project is an appropriate model for many rural areas around the world.

Some projects at the state (regional)-level include: Akshaya (Kerala), Bhoomi (Karnataka), CARD (Andhra Pradesh), e-district (Tiruvarur), FAST (Andhra Pradesh), FRIENDS (Kerala), Gyandoot (Dhar, MP), Kalyan-Dombivli Municipal Corporation (Maharashtra), Koshwahini (Maharashtra), Sarita (Maharashtra), Community Information Centres (Lakhimpur, Assam), Drishtee kiosks (Assam), Bhulekh (Orissa), commercial taxes (Bihar), e-computerized operations for police services (eCOPS), electronic data interchange of the NIC, electricity power billing (Bihar), e-procurement (Andhra Pradesh), Oswan (Gujarat), Integrated Financial Information Systems (Andhra Pradesh), Kaveri (Karnataka), Khajane (online treasure computerization project) in Karnataka, Lokmitra (Himachal Pradesh), Saukaryam (Andhra Pradesh), SETU (Maharashtra), SmartGOV (renamed as CaringGOV) in Andhra Pradesh, STAR (Coimbatore, Tamil Nadu), Sukhmani (Punjab), Tarahaat (Punjab) and the Automatic Vehicle Tracking System (Delhi).

Also among the regional initiatives is the launching of the Goa Knowledge Commission website (http://www.knowledgeforgoa.com/). In addition, the National Association for the Blind in New Delhi has been conducting a weekly programming workshop online moderated by Arun Mehta. Different people are learning different languages at different speeds.

> ## Media Lab Asia: Building on ruined plans?
>
> Media Lab Asia appears to have come unstuck. This much-hyped project between the MIT Media Lab and the Government of India, among others, failed to work as planned and, amidst a lot of dissatisfaction, both partners went their own way. However, here is a list of projects that have been supported under this initiative. It would be interesting to track the attainments of each.
>
> **aAQUA:** An agro information system being tested in Maharashtra
> **Ashwini:** A Broadband Wireless Network for delivery of virtual services being tested in rural areas of Andhra Pradesh
> **Ca:sh:** Hand-held device-based health data collection and management being tested in Haryana
> **CHIC:** Craft revival: A CAD tool for helping the Chikan embroidery artisans, being deployed in Lucknow, Uttar Pradesh
> **Digital Gangetic Plains:** Long range Wi-Fi tested over a distance of 75 kms in the Lucknow–Kanpur belt
> **Digital Mandi:** An electronic trading platform for agro commodities, being tested in Uttar Pradesh
> **e-Sagu:** An IT-based tool for agricultural extension, being deployed in Andhra Pradesh
> **GramPatra:** Store-and-forward messaging system being tested in Karnataka for land records delivery over the BHOOMI project
> **Polysensors:** Portable and affordable water quality test kit being tested in Maharashtra
> **Sahayika:** A system for supplementing the knowledge requirements of school students, field-tested in West Bengal
> **Sanyog:** A communication system for the speech impaired and for people affected with cerebral palsy, being tested in West Bengal and Delhi
> **Sehat Saathi:** Portable/Mobile health-care delivery being tested in the Lucknow–Kanpur belt and in Andhra Pradesh
> **Shruti:** An embedded Indian language text-to-speech system being tested in West Bengal
>
> Source: Media Lab Asia is dead; Long live Media Lab Asia. (2003).

Industry initiatives

One of the more impressive networks—because of its breadth of vision, the persistence of its initiatives, and the number of initiatives taken up—with an ICT4D agenda is the Telecommunications and Computer Networking Group (TeNeT), linked to the IIT-Madras (http://tenet.res.in/). Its mission is to transform communication in India by enabling 200 million telephones and 50 million broadband connections; to connect every village in order to double rural per capita GDP; to set next-generation wireless standards; and to develop high-quality distance education for rural areas.

TeNeT and its partners have developed a number of technology solutions for the rural areas such as:

- The Cable Wireless Internet Triple-play Unified System (indigenous broadband access solutions such as CitiusTM) of Midas Communication Technologies.
- Broadband corDECT, a wireless local loop solution with per-user always-on data speeds of up to 336 Kbps and a spectrum efficiency approaching 1.66 bits/second/Hertz, also by Midas Communication Technologies.
- CygNet, a product of NMSWorks, another TeNeT partner, designed to provide network management solutions for large telecom operators.
- An 802.11 b/g based mesh network for rural communities.
- Gramateller, an ATM designed to enable a low-cost model of delivering banking services in rural areas.
- An affordable telemedicine solution which includes a Remote Diagnostic Kit developed by Neurosynaptic. The kit, which can be installed at villages and other remote locations with Internet connectivity, connects rural patients to a doctor in the city via a video conference link.
- Online Tutorials to enable rural students to pass examinations.
- Indic Computing which aims to extend the reach of computing to India's non-English speaking masses.

TeNeT says it is exploring the possibility of using the ICT infrastructure established by n-Logue through its Chiraag network of villages, to enhance income generation. Several crafts initiatives have been initiated and outsourcing to rural areas is planned. In collaboration with the Indian Society of Agribusiness Professionals (ISAP), an NGO based in New Delhi, they are offering online consultancy to farmers in Tamil Nadu, Gujarat and Karnataka through videomail using the software MV4.

Ashok Jhunjhunwala of TeNeT says they have been setting up Internet kiosks in India for INR 55,000 (about USD 1,356.69) per kiosk. The only requirement is that the kiosks be set up in a cluster of nearby villages within a 25 km radius.

The TeNeT Group incubates R&D companies and collaborates with 'like-minded organizations' such as Midas Communications in telecom infrastructure (www.midascomm.com); Banyan Networks Limited (www.banyannetworks.com), which merged with Midas Communication Technologies in 2004; NMSWorks (www.nmsworks.co.in), which works on integrated network management systems for emerging convergent networks; NexGe Technologies Pvt Ltd, which develops IP communications products targeted towards carriers, broadband operators and ISPs; iSoftTech, a product development company in the networking space; iSoftTech (www.isofttech.com) for embedded solutions and enterprise solutions in Data Networking, VoIP and WLan; Nilgiri Networks (www.nilgirinetworks.in); and Amdale (www.amdale.com), which is working to create what is probably the only open standards-based packaged CTI suite which can be used to build a vast array of computer-based telephony solutions.

Others in this network of companies are Novatium (www.novatium.com); N-Logue (www.n-logue.co.in) for evolving technically superior and cost-effective solutions for countries like India; CK Technologies Pvt Ltd which is currently involved in promoting Indian language computing as a subject (www.shaktioffice.in); Vortex Engineering Pvt Ltd for low-cost electro-mechanical systems for rural areas; Neuro Synaptics (www.neurosynaptic.com) for medical devices, software and solutions for telemedicine; OOPS or Object Oriented Programming Services for network telephone applications; Lattice Bridge (www.lbinfotech.com) for speech technologies; and Tekriti (www.tekritisoftware.com), which is engaged in outsourced product engineering services in the domain of social networking, rich Internet application development and media publishing, the aggregate of which is referred to as Web 2.0.

There have been some difficulties in implementing TeNeT projects on the ground, for example in villages that get power for only two hours a day. Yet, this network continues to inspire many to carry on the dream.

Regulation and security

The Ministry of IT's e-Security Division 'deals with technical matters related to [the] Internet, e-regulation and e-security and implements a programme that promotes research and development in the area of e-security'. An official note says that the division promotes R&D through grant-in-aid support to recognized autonomous R&D organizations and academic institutions proposing to undertake time-bound projects in the thrust areas identified by the Working Group on e-Commerce and Information Security.

The Working Group provides full advisory support to the information security sector through analysis of technology trends, identification of thrust areas, preparation of technology development plans, as well as evaluation of submitted project proposals for execution with financial support from DIT. Project proposals are accepted via http://www.mit.gov.in/R&D/projects/index.asp. Projects already carried out by various organizations have led to the development of network security tools, cyber forensic tools, a virtual private network security solution, a biometric identification and authentication system, a public key infrastructure solution, a payment gateway for e-cheque or credit card, an intrusion detection system, an information security management system, network monitoring tools, a brain mapping technique to examine suspects, a secure print document tool, tools for enterprise system security management and tools for steg-analysis.

Current 'thrust areas' identified by the Working Group are cryptography and cryptanalysis, biometrics for identification and authentication, network security, systems security and security architectures, risk assessment and assurance, monitoring, surveillance and forensics (Department of IT 2007).

Free/Libre and Open Source Software

India's interest in Free/Libre and Open Source Software (FLOSS) is being closely watched around the world. Across India, there are over 130 GNU/Linux, Linux and Free Software User Groups, an indication of the interest in this field (visit http://wikiwikiweb.de/LugsList for a listing of GNU/Linux and FLOSS user groups in India). A number of FLOSS techies maintain blogs and some of these can be accessed at http://planet.foss.in.

Ironically, while FLOSS has grown rapidly on the ground and among technical and academic networks, it has received little support from the government.

A handful of regional governments, more the exception than the rule, have been pushing for wider FLOSS adoption, as has happened in the south-western pocket of India. An example is the plan of the south Indian state of Kerala to adopt FLOSS for the computers used in some 12,500 high schools in the state.

The use of computers across the country is being encouraged by the distribution of free CDs that contain localized versions of popular open source applications. For example, the government has started distributing CDs containing Tamil-language versions of various open source applications (Marson 2005). At Vigyan Bhavan New Delhi, a Hindi Software Tools CD was launched by political leader Sonia Gandhi. The CD contains open source software that runs both on Windows and Linux. It includes Open Office, Firefox, Gaim, Columba E-mail and Limewire (CDAC 2006).

There are also low-cost 'CD outlet options' where FLOSS software is downloaded, replicated and sold at a reasonable rate of INR 50–250 or about USD 1–6 (depending on whether CD or DVD). See one example of such operations at http://linuxdvdsale.tripod.com.

The Technology Development for Indian Languages (TDIL) programme of the Department of Information Technology (DIT) aims to develop information processing tools and techniques that remove language barriers from human–machine interaction. Officials say efforts are being made to provide these language tools to the masses through the Indian Language Data Centre (http://www.ildc.in/index.aspx).

The Asia OSPA (Open Solutions is Public Administration) Forum is intended to analyze and support the use of Open Data Standards (ODS) and Open Solutions (OS) for e-Government and Public Administration (PA) in Asia Pacific. It was scheduled to be held along with the South Asia e-Government Summit in New Delhi in October 2006.

Meanwhile, Indian techies and campaigners have been involved in the drafting of the GPLv-3, the General Public License of the Free Software Foundation (FSF). GPLv-2, released in 1991, is the most widely used free software license. Jaldhar H. Vyas, who is unusual in being both a Hindu priest and Debian geek, has pointed out that one of the problems facing GNU/Linux users in countries like India is low bandwidth. Therefore he argued, a distribution on CD or DVD is preferred over large network installations or updates. He has found a sponsor and is looking for a 1-DVD version of the Debian system.

Sarai.net's Project Resource Centre is a network for students interested in discussing GNU/Linux and FLOSS projects. Some networks are offering support to students wanting to undertake FLOSS projects (see http://ekalavya.it.iitb.ac.in/brochure.do). The main objective of the e-GURU programme is to assist students in their final year in a Bachelor's or Master's degree course in computer science, information technology or electronics who find it difficult to carry out a project, a major component of their curriculum, due to lack of resources and mentors. NRCFOSS (National Resource Centre for Free and Open Source Software) offers a common page for student project proposals, fellowship proposals and other FLOSS projects (NRCFOSS). Other NRCFOSS plans include:

- Getting the technical universities and colleges to offer two FOSS elective courses to undergraduate students.
- Arranging for or conducting Teacher Training Programmes (TTPs) for faculty of colleges of engineering.
- Supplying engineering faculty with teaching materials, books, lab ware and the like.
- Assisting colleges to migrate to FOSS equivalents wherever possible (personal e-mail from Professor C.N. Krishnan 2006).

Finally, an open source simple computer for agriculture in rural areas (called OSCAR) was launched in January 2004. It aims to develop an open source weed identification software for the major weed species of the rice-wheat crop systems which can be deployed on computers or simputers. The application is intended for agricultural extension workers, farmers/farmer groups, and students in the Indo-Gangetic plains (Digital Opportunity Channel 2005b).

Local language solutions

Clearly, this is an issue where a breakthrough is still awaited. In recent years, technology solutions that would make computing in Indian languages easier on both proprietary and FLOSS platforms have been announced. But large-scale and effective Indian-language computing is still a challenge, for a range of reasons. This is evident from the lack of mailing lists and blogs in Indian langauges, and the difficulty of building websites or wikis in the Indian languages.

There are a few initiatives underway. HP Labs in Bangalore have been engaged in developing a Devanagiri input device which works on the basis of partial handwriting recognition. During preliminary studies it was revealed that in spite of a large demand for a Hindi keyboard, the ones available in the market (QWERTY keyboards with Hindi labels) have not been accepted. See other HP proposals at http://www.hpl.hp.com/india/.

Bangla/Bengali OpenOffice.org 2.0: Ankur group is the official team for Bangla OpenOffice.org. Translation work is reported to have started in 2006. Their goal is to have Bangla as a supported language. OpenOffice.org 2.0 is already out with most of the menu entries translated in Bangla. Future versions of OpenOffice.org will have more translated modules (Angkur Supporting Bangla [Bengali] on GNU/Linux).

C-DAC Mumbai is working on *Project Janabhaaratii for Localisation of Free/Open Source Software: Development, Deployment and Community Building*. Initiatives include contributing to community efforts in developing a software suite based on GNU/Linux and made available in Indian languages. The project is funded by the TDIL group at the DIT, Ministry of Communication and Information Technology.

The DIT says its mission is to 'proliferate the use of Indian languages on computers, to overcome language barriers that restrict the nation's path to knowledge and development'. It has invited individuals, public, private agencies and academic institutions to participate in a national public–private partnership to launch and distribute applications, tools, utilities and products developed for Indian language computing.

Dr G. Nagarjuna, Chairman of the Free Software Foundation (FSF) of India and scientist at the Homi Bhaba Centre for Science Education, has been focusing on the issue of 'self-reliant e-governance in Indian languages'. He argues:

> Keeping in mind that only about 5 per cent of the Indian population can manage with English-speaking computers, the need for e-governance, by default, in a multilingual system is clearer than daylight and no additional argument is required. Since public documents must live for eternity, the need of open standards for e-governance will be underlined. If these values of self-reliance in information technology and local language computing with open standards are to be followed in letter and spirit, then it is mandatory that Governments must legislate the use of Swatantra Software and open standards.

On another level, an Israeli firm is teaming up with CK Technologies to produce Shakti Office Suite (http://www.shaktioffice.in), an 'indigenous, affordable bilingual alternate to popular office suites'. It will be the first to integrate FastKeys, software developed by the Israel-based FTK Technologies. Its interface 'intuitively' toggles between English and one of the 18 Indian languages and vice versa. The integration of FastKeys in the Office Suite is meant to make the product more user-friendly. When typing in Hindi or Tamil, most users would have to constantly glance between the keyboard and the monitor which makes the process difficult. Other existing solutions are prohibitively expensive. FastKeys addresses the problem by projecting onto the monitor a real-time simulation of the keyboard and user's hands as they type. The keys on the virtual keyboard would show the characters in the vernacular, simplifying the process for a new user who might otherwise have to plough through scripts of character positions on the keyboard.

Meanwhile, the Center for Research in Urdu Language Processing (CRULP) at the National University of Computer and Emerging Sciences, Lahore, Pakistan has announced the release of an updated version of the open source character-based Nafees Web Naskh Open Type Font for writing Urdu in Naskh script based on Unicode standards. Such initiatives from across the border could benefit India, which has a significant Urdu writing population.

Jitendra Shah announced via the Indic computing users mailing list that http://203.199.16.202 contains a few GIS applications and a few Web-based database applications using FLOSS. This includes tenders of the Maharashtra government

ICT tools for diverse groups

- The Chief Minister of Andhra Pradesh, Y.S. Rajasekhara Reddy, announced the release of a CD comprising an e-Governance solution for the Drug Control Department. The CD details 336 court judgements (*Business Line* 2005).
- In a pilot installation in a village near Mumbai, India, students use PCs donated by Via Technologies to perform geometry homework, while local women use computers to track their savings in a micro-payment programme (Kanellos 2005).
- Encore Software Ltd has announced plans to launch a range of cheap desktops costing around USD 230–280, three years after it launched the USD 200 Simputer. These desktops are targeted mainly at basic users like students, small shop owners and educational institutions (Roy 2007).
- Parivartan.net is growing into a full-fledged portal with 25 different services, under the Maharashtra Knowledge Corporation. Its goal is ICTs for agriculture. It has a network of infomediaries, offers information to farmers' queries, and promotes courses like a certificate course in good agricultural practices. They also have some half dozen CDs on mango cultivation, bio-fertilizers, mushroom cultivation, medicinal plants and dairy management.
- A special type of foot-operated PC-based communicating tool for children with cerebral palsy has been developed by the Industrial Design Centre of the Indian Institute of Technology, Mumbai in cooperation with the Happy Hours Centre from suburban Khar. This special tool can be used as an effective communication device by children with physical disabilities, including those affected by cerebral palsy.

for rate-contracts where OpenOffice.org with Marathi (Gargi) fonts are compulsory on either platforms (proprietory or open source). The Maharashtra government buys an estimated 5,000 PCs annually.

Education and R&D initiatives

India-based Digital Learning Asia (http://www.digitallearning.in/index.asp) aims to take stock of the progress of Asian countries in utilizing ICTs to enhance the quality and reach of education to develop human capital that will respond to the needs of a globalized world.

Another attempt to make IT relevant in education is Laxman Mohanty and Neharika Vohra's *A Guide For School Administrators*, a book that recognizes the potential of ICTs to make the school curriculum more relevant and purposeful.

The Centre for Science, Development and Media Studies (http://www.csdms.in/), an NGO located at Noida near India's capital city of New Delhi, has been publishing what is probably the world's first ICT4D monthly. Founded in 1997, the Centre is 'committed to advocacy and developing solutions for under-privileged societies through the use of innovative and effective ICTs and Geographic Information Systems (GIS)'. The Centre used to be known as the Centre for Spatial Database Management and Solutions.

Changes have also been reported regarding the delayed plans for building the Simputer, a low-cost alternative to the personal computer that allows for sharable computing. Geodesic Information Systems (GIS), a Mumbai-based Internet product company, has acquired Bangalore-based PicoPeta Simputers, a company founded by the co-inventors of the Simputer. However, although the Simputer has been in the news in recent years in India, its delivery has been delayed because of a number of financial, technology obsolescence and production challenges.

Conclusions

India continues to be a land of contrasts. Positive trends like the lowering of bandwidth costs and mobile services are expanding access. Yet many hundreds of millions are totally untouched by the benefits of IT and cyberspace. Because of its size, even a small percentage of Indians getting access to the Internet could change its complexion and orientation. Moreover, India has become a 'poster boy' of sorts for ICT4D. However, ensuring wider coverage for the bulk of its 1.1 billion population is obviously still an unmet task.

Notes

1. On 16 November 2006, a long-awaited development materialized when the federal Indian Cabinet decided to grant non-profit organizations and educational institutions permission to set up community radio stations under certain terms (Press Information Bureau 2006). In an unrelated development, engineer Vickram Crishna wrote that a 'suitcase' radio station for very local broadcasting (within a radius of 400 m roughly) could be available for under USD 120.
2. For an overview of the official take on this project, see the annual report for 2005–06 of the Ministry of IT at http://mit.gov.in/download/annualreport2005-06.pdf

References

AirJaldi Homepage. Available at http://drupal.airjaldi.com/About

eGURU. (2006). Kanwal Rekhi School of IT. eKalavya Homepage. Retrieved from http://ekalavya.it.iitb.ac.in/eguruHome.do

Angkur Supporting Bangla (Bengali) on GNU/Linux Homepage. (2003–07). Available at http://www.ankurbangla.org/projects/ooo/

Business Line. (2005). Drug control dept e-governance (CD Released). Retrieved from http://news.com.com/Rural%20Indias%20rough%20road%20to%20computer%20literacy/210\0-1047_3-5700701.html?tag=nefd.lede

BytesForAll mailing list. Available online at http://tech.groups.yahoo.com/group/bytesforall_readers/

CDAC. (2006). *BharateeyaOO.o: OpenOffice.org in Indian Languages*. Retrieved from http://www.ncb.ernet.in/bharateeyaoo

CDAC Homepage. (2007). Available at http://www.cdac.in

Centre for Research in Urdu Language Processing Homepage. Available at http://www.crulp.org/

Content Sutra Homepage. Available at http://www.contentsutra.com/

Department of Information Technology, Ministry of Communications & Information Technology, Government of India Website. Available at http://mit.gov.in/

Department of Information Technology, Ministry of Communications & Information Technology, Government of India. (2007). *R&D in e-commerce and cyber laws*. Retrieved from http://www.mit.gov.in/default.aspx?id=711

Digital Opportunity Channel. (2005a). *India starts consultation on making thesis online*. Retrieved from http://www.digitalopportunity.org/article/view/112326/1/

Digital Opportunity Channel. (2005b). *Open source simple computer for agriculture in rural areas*. Retrieved from http://www.digitalopportunity.org/article/view/117977/1/

eGovWorld 2006 Website. (2006). Available at http://www.egovworld.org/asiaospa/index.html

eKalavya Homepage. (2006). Available at http://ekalavya.it.iitb.ac.in/brochure.do

Goa Knowledge Commission Homepage. Available at http://www.knowledgeforgoa.com/

HP Labs Bangalore, India Homepage. Available at http://www.hpl.hp.com/india/

india-gii. *India's bumpy progress on the global infohighway*. Retrieved from http://lists.cpsr.org/lists/info/india-gii.

India PR Wire. (2005). Availabe at http://www.indiaprwire.com/

Indian Institutes of Technology. *Wikipedia*. Retrieved from http://en.wikipedia.org/wiki/Indian_Institutes_of_Technology

Industrial Design Centre, Indian Institute of Technology (IIT Bombay) Homepage. Available at http://www.idc.iitb.ac.in

International Monetary Fund (IMF). (2006). World Economic Outlook. Available online at http://www.imf.org/external/pubs/ft/weo/2006/02/

International Telecommunication Union (ITU) Homepage. (2007). Available at http://www.itu.int/

Internet & Mobile Association of India (IAMAI). (2006). *IAMAI-IMRB study pegs urban Internet users at 37 million for September 2006*. Press release. Retrieved from http://www.iamai.in/section.php3?secid=15&press_id=1202&mon=9

Internet World Stats, India. Retrieved from http://www.internetworldstats.com/asia/in.htm

Kanellos, M. (2005). Rural India's rough road to computer literacy. *CNet News.com*, 9 May 2005. Retrieved from http://news.com.com/Rural+Indias+rough+road+to+computer+literacy/2100-1047_3-5700701.html?tag=item

Kaufman Bros. (2006). *Equity Research Report on the State of the Indian Internet Market*.

Kumar, P., Desai, K., Data, R., and Jethanandani, J.H. (2004) New Indian Government's ICT Plan Will Need Broad Support. *Gartner*, 2 June 2004. Retrieved from http://www.gartner.com/DisplayDocument?doc_cd=121207

LexLibre. Retrieved from http://www.nalsartech.org/tikiwiki/tiki-view_faq.php?faqId=1&highlight=IT%20law%20India

Make-IT-Safe. (April 2005) *Fact Sheet #2: IT Use—World and Asia*. Retrieved from www.make-it-safe.net/eng/pdf/World_and_Asia.pdf.

Manthan Award Homepage. Digital Empowerment Foundation. Available at http://www.manthanaward.com/

Manufacturers' Association for Information Technology Homepage. Available at http://www.mait.com/

Marson, I. (2005). Free CDs spread open source in India. *CNet News.com*. Retrieved from http://news.com.com/Free+CDs+spread+open+source+in+India/2100-7344_3-5720008.html?tag=cd.top

Media Lab Asia is dead; Long live Media Lab Asia. (2003). *Express Computer*. Retrieved from http://www.medialabasia.org/index.php?option=content&task=view&id=28&Itemid=37

Mission2007.org Homepage. Available at http://www.mission2007.in/

NASSCOM. (2007). *Indian Domestic Market: NASSCOM Analysis*. Retrieved from http://www.nasscom.in/upload/5216/Indian_Domestic_IT_Market_Factsheet.doc

National Association of Software and Service Companies (NASSCOM) Homepage. (2007). Available at http://www.nasscom.in/

Noronha, F. (2006a). India at the forefront of knowledge commons debate. *Intellectual property watch*. Retrieved from http://www.ip-watch.org/weblog/index.php?p=389&res=1280_ff&print=0

Noronha, F. (2006b). Scientists push open access for developing nations. *Science and development network*. Retrieved from http://www.scidev.net/gateways/index.cfm?fuseaction=readitem&rgwid=4&item=News&itemid=3251&language=1

NRCFOSS. *Project ideas*. Retrieved from http://nrcfosshelpline.in/code/wiki/ProjectIdeas

Parivartan.Net Homepage. Available at www.parivartan.net

Press Information Bureau, Government of India. (2006). *Grant of permission for community radio broadcasting*. Retrieved from http://pib.nic.in/release/release.asp?relid=22064

Roy, A. (2007). Indian firm plans cheap desktop worth $230. *Yahoo India news*. Retrieved from http://in.news.yahoo.com//050510/137/5yhow.html

Srivastava, R. (2006). Internet taking root in India. *The hoot*. Retrieved from http://www.thehoot.org/story.asp?storyid=Web5917639139Hoot74409%20PM2183&pn=1

Telecom Regulatory Authority of India Homepage. Available at http://www.trai.gov.in/

The Hindu. (2006a). Bangalore: Kapil Sibal launches Natural Disaster Information System. Retrieved from http://www.nalsartech.org/tikiwiki/tiki-read_article.php?articleId=10938&highlight=NDIS

The Hindu. (2006b). Digital Knowledge Centre Inaugurated. Retrieved from http://www.hindu.com/2006/08/17/stories/2006081720580400.htm

Thejesh, GN. (2006). Dollar one Encyclopedia. *TechMag*. Retrieved from http://www.techmag.biz/dollar_one_encyclopedia

The Telecommunications and Computer Networking Group Homepage. Available at http://tenet.res.in/

United Nations Development Programme. (2006). *Human Development Report 2006*. Retrieved from http://hdr.undp.org/hdr2006/pdfs/report/HDR06-complete.pdf

Wikipedia. (2007). Global Internet usage. Retrieved from http://en.wikipedia.org/wiki/Internet_users

WiMAX Report. Retrieved from http://www.wimax.com/commerce/catalog/wimax_india_report (paid access)

.id
Indonesia

Donny **B.U.** and Rapin **Mudiardjo**

Total population	222 million (2006)
GDP per capita	USD 1,663/year (IDR 15 million/year)
Key economic sectors	Manufacturing, Trade and Agriculture
Computers per 100 inhabitants	2
Fixed-line telephones per 100 inhabitants	6
Mobile phone subscribers per 100 inhabitants	30
Internet users per 100 inhabitants	11
Domain names registered under .id	32,849
Broadband subscribers per 100 inhabitants	0.2
Internet domestic bandwidth	Peak 1.46 Gbps (end 2006)
Internet international bandwidth	Peak 4 Gbps (end 2006)

Overview

Indonesia's population of 222 million, the fourth largest in the world, are scattered in more than 17,504 islands (7,387 named; 10,117 unnamed). Thus, the Indonesian government, through Presidential Decree No. 6 issued in 2001 and titled 'Information and Communication Technology (ICT) Development and Usability in Indonesia', has asserted the need to make optimal use of ICTs to reach the Indonesian public and to unify the nation.

However, although it is expanding, teledensity in Indonesia is still not enough to serve the entire population. Fixed telephone line teledensity is 14 million lines or about six lines for every 100 citizens. Mobile phone teledensity is higher at 66.5 million active numbers or 30 numbers for every 100 citizens in mid-2006. The average annual growth rate of mobile phone teledensity from 1999 to 2005 is 63.7 per cent.

According to the 2005 National Social Economic Survey conducted by the National Statistics Bureau, 7.7 million of 58.8 million households (13.11 per cent) have a telephone. Of this total, 6.6 million are located in urban areas. About 11.7 million households (19.9 per cent), 9 million of which are in the urban areas, have mobile phones. About 2.2 million households, or 3.68 per cent of the total 58.8 million households, have computers; 2 million of these households are located in the city. For every 100 households with a computer, only 27 have an Internet connection.

According to the Directorate General of Post and Telecommunication (Postel), there were about 25 million Internet users (about 11 persons for every 100 citizens) in Indonesia in 2006. However, the number of Internet subscribers is only 6 million, suggesting that most people access the Internet at work or at Internet kiosks. Between 2000 and 2004, only 6 million people (or 2.7 persons for every 100 citizens) used the Internet. Internet penetration increased when the government officially liberalized 2.4 GHz of wireless technology in January 2005. Private non-commercial use of the 2.4 GHz frequency without a license is allowed. However, operators must register with Postel, which will check whether the equipment used is standard. Internet service providers (ISPs) use the 2.4 GHz frequency not only as a last-mile infrastructure but also as a backbone network to haul Internet traffic over large distances.

ICT penetration increased further with the opening of cyber-cafés or Internet kiosks (locally known as *warung Internet* or *warnet*) in many Indonesian cities. There are about 4,000 *warnet* around the country, each with an average of 10 computers that can be used by turns for 12 hours (at an average of two hours per user).

The Indonesian Telecommunications Regulatory Body (*Badan Regulasi Telekomunikasi Indonesia*—BRTI) was formed in December 2003, which demonstrates the government's willingness to put in place the appropriate regulations to improve telecommunications infrastructure and services. Efforts to lower the cost of telecommunication and Internet bandwidth in Indonesia are being pursued. There is also a plan for fibre optic backbone development and the promotion of healthy competition among telecommunication businesses. These are expected to make quality ICT services available to all Indonesian citizens at an affordable cost.

Technology infrastructure

PT Telekomunikasi Indonesia (Telkom) launched the Telkom-2 satellite in November 2005. However, three to six more satellites with a 24-transponder capacity are still needed to connect every Indonesian region, consisting of about 43,000 villages, in one telecommunication infrastructure network, assuming a 64 Kbps connection (two voice channels and 32 Kbps Internet) for every village. PT Telkom is now preparing for the operation of the Telkom-3 satellite in 2009.

In February 2006, the government chose three telecommunication operators to provide 3G telecommunication services via a tender process. In October 2003, two other operators were granted a license to provide 3G services: PT Natrindo Telepon Selular (Natrindo) and Hutchison CP Telecommunications (formerly Cyber Access Communications). Two of the new operators, Telkomsel and Excelcomindo (XL), have built 3G infrastructure in several big cities. As of October 2006, XL estimated about 18,000 customers actively using 3G services. Telkomsel says there are 240,000 customers registered as 3G users. Telkomsel, which has the biggest cellular phone market share in Indonesia, has allocated about USD 300 million to build its 3G network. XL has allocated about USD 50–100 million while Indosat is ready to invest USD 200–300 million.

Currently under discussion is unified access licensing for telecommunication services. This means that the 'frequency used right fee' (*Biaya Hak Penggunaan*—BHP) between fixed wireless access (FWA), cellular and data communication operator, and fibre optic will be the same, which in turn means lower prices as competition among operators becomes more intense.

According to Onno Purbo, an Indonesian ICT expert, the cost of Internet bandwidth in Indonesia nowadays is about USD 3,000 per Mbps per month. Postel puts the bandwidth cost at around USD 5.07 per 100 Kbps. The Indonesia ISP Association states that the cost for 1 Mbps of bandwidth taken from satellites in other countries is around USD 3,000–4,000, while it costs up to USD 6,000 from local operators. The biggest component of Internet bandwidth pricing in Indonesia is the fee for the international backbone connection and local backbone access to the local Network Access Provider (NAP) network. This accounts for 60–80 per cent of the monthly income of ISPs. Although theoretically ISPs can use other NAPs, including those in other countries, which are cheaper, government regulation limits ISPs to purchasing bandwidth only from the local NAP.

Recently, there has been some discussion of Dirjen Postel allowing a minimum international bandwidth service application in implementing NAP service. The international bandwidth price can be reduced if the NAP operator can bargain with the foreign hub owner to get a cheaper price. To get that bargaining position, the NAP operator must purchase at least 45 Mbps of international bandwidth.

The fibre optic network for the national Internet backbone is only 12,000 kilometres, according to the Department of Communication and ICT. At least 35,000 kilometres of fibre optic network is needed to serve all islands in Indonesia. The government has announced a fibre optic network development project called Palapa Ring, which is estimated to cost USD 500 million to USD 1 billion. Palapa Ring is a 25,000-kilometre undersea cable network in an integrated ring shape spread out from Sumatra to West Papua. Every ring will transmit broadband access of about 300–10,000 Gbps. It is hoped that the Palapa Ring backbone network will significantly lower telecommunication and Internet costs. Sofyan Djalil, the Communication and ICT Minister, is aiming to make the Internet fee in Indonesia as cheap as in Singapore by mid-2007.

There are two national interconnections—Indonesia Internet exchange (IIX) and National Interconnection Exchange (NICE). Thus, inter-ISP traffic in Indonesia is no longer dependent on the international Internet interconnection.

According to the Association of Wireless-LAN Internet Indonesia (IndoWLI), the use of the 2.4 GHz frequency in 2006 was higher by 70 per cent compared to 2005 when the license for this frequency had just been released. However, according to IndoWLI, more than 50 per cent of 2.4 GHz users, which include government institutions and state-owned corporations, have violated regulations, using power beyond the maximum limit, using uncertified radio and refusing frequency arrangement by community.

The government and the Indonesia Telecommunication Regulatory Body BRTI are still deciding the frequency allocation for Broadband Wireless Access (BWA), also known as WiMax. Fearing interference, PT Telkom, one of Indonesia's satellite operators, is not in favour of a 3.5 GHz frequency for BWA. They say that 3.4 GHz is suitable only for satellite in S-Band and extended C-Band because Indonesia has a tropical climate with high rainfall. Indonesia is an archipelago, which makes satellite necessary for telecommunication.

On the other hand, the Indonesia Broadband Wireless Association (Abwindo) and all WiMax vendors are asking for a 3.5 GHz frequency because based on economies of scale, it will be cheaper to develop WiMax at 3.5 GHz than on the alternative frequency of 5.8 GHz (as in the US). The WiMax Forum is already issuing certification for equipment designed for a frequency of 3.5 GHz. WiMax with a 50 km reach is expected to be in position to substitute the Digital Subscriber Line (DSL) once a 3.5 GHz frequency is assigned by Postel.

The government is studying the feasibility of making the 3.5 GHz frequency available for both WiMax and satellite. If an agreement is not reached, the government will uphold satellite as the primary occupant of the 3.5 GHz frequency and relocate BWA to another frequency. If this happens, BWA operators will expect the government to provide compensation, such as investment cost compensation for the BWA infrastructure acquired for the 3.5 GHz frequency.

ICT and ICT-related industries

The Indonesian ICT market sector grew by 22.1 per cent in 2005 to about USD 1.7 billion. Of this about USD 0.5–0.75 billion came from the banking sector. The following sectors comprised the Indonesian ICT industry in 2006: 50 middle and large-scale computer companies, 5,000 computer assemblers, 154 middle and large software companies, 214 small software companies, 12 telecommunication tools companies and 150 animation companies.

The software companies are concentrated in Jakarta, Bandung, Surabaya and Medan. Some of the software products are finance applications, geographical information systems (GIS), inventory systems, executive information systems, office automation, animation, multimedia presentation, intranet/Internet, integrated LAN-WAN, and consultancy services. According to the General Directorate of Intellectual Property Rights, 133 applications for software copyright were made between 2002 and 2005. But as of August 2006, 104 software copyright were registered, suggesting that appreciation for IPR is a recent development.

ISPs comprise another ICT industry. As of 2005, 232 ISPs were registered with the Ditjen Postel (up from 50 in 1999). Several are members of the Indonesia Internet Service Providers Association (*Asosiasi Penyelenggara Jaringan Internet Indonesia*—APJII) which was founded in May 1996 and which has successfully campaigned for a national interconnection among ISPs in Indonesia to enable users to communicate easily regardless of the ISP, and at a lower price.

Internet kiosks, of which there are more than 2,500 not counting Internet kiosks managed by schools and universities, comprise a segment of the ICT industry in Indonesia. According to a study conducted by ICT Watch in September 2003, in general, Internet kiosks are not really profitable. To acquire an Internet backbone lane, an Internet kiosk needs at least 35 separate connections (hops). To have a proper bandwidth, an Internet kiosk needs to pay about IDR 4 million (about USD 453) a month. So only Internet kiosks with a minimum of 10–20 PCs charging about IDR 3,000–5,000 (USD 0.3–0.5) per hour and located around a campus, tend to be profitable. Many Internet kiosks are either just a hobby or are subsidized by another business of the owners. Internet kiosks face other problems, such as the high cost of Internet bandwidth, extortion, and pornography and other cyber crimes.

According to the Department of Communication and ICT, the level of PC ownership in Indonesia is very low at five million units for about 220 million citizens. On the other hand, the Indonesian Computer Business Association (*Asosiasi Pengusaha Komputer Indonesia*—APKOMINDO) states that PC sales have grown 20 per cent annually. In 2005, PC sales reached 1.2 million units; the projection for 2006 was 1.44 million units. Sixty-five per cent of the total PCs bought consists of local PCs; the rest are imported. The corporate sector is the biggest consumer, accounting for 30–35 per cent of the total market segment. Government accounts for 25 per cent of consumers; the retail sector and households comprise 10–15 per cent, while the educational institutions account for the rest. IDC research, on the other hand, states that the small and medium-sized enterprise (SME) sector is the biggest PC consumer with 34 per cent of the market share, followed by big companies, government, households and educational institutions.

Key ICT institutions

Several institutions and organizations are responsible for ICT development in Indonesia. They can be classified into three: regulator, industry and civil society. Coordination among the three groups is crucial to the development of ICT in the country.

National ICT Council

Indonesian President Susilo Bambang Yudhoyono officially declared the formation of the National Information Communication and Technology Council on 11 November 2006. Its task is to formulate general policies and set the direction of strategic development using ICT. It is also responsible for coordinating the ICT development efforts of local and central government agencies, state-owned and private companies, ICT communities, entrepreneurs, professional institutes and the general public. The Council has authority over the implementation of ICT programmes across departments.

The Council has three years to accomplish its mission. President Susilo Bambang Yudhoyono himself heads the Steering Committee, with the Coordinating Economic Minister as head deputy. The Chairman on Duty is the Minister of ICT. Ten ministers or ministerial-level officers comprise the Council membership. There is also an Implementation Team which includes the Minister of ICT as *ex-officio* chairman and ICT experts as members. The Advisory Team consists of the Rectors of the Bandung Institute of Technology, University of Indonesia, Sepuluh November Institute of Technology and Gadjah Mada University. The Partner Team consists of all active participants in the ICT industry.

Regulator/Government agencies

Dirjen Postel (www.postel.go.id), along with the Indonesia Telecommunication Regulatory Body, is responsible for formulating and implementing ICT regulations and for standardizing the technical aspect of the telecommunication industry. Dirjen Postel includes the Directorate of Telecommunication, Directorate of Radio Frequency Spectrum Satellite Orbit, Directorate of Post and Telecommunication Standardize, and Directorate of International Post and Telecommunication Institutional.

The Indonesia Telecommunication Regulatory Body BRTI (www.brti.or.id) is an independent regulatory body that aims to protect the interest of telecommunication users and to foster competition in the telecommunication industry. It coordinates with Postel and reports to the Minister of Communication and ICT.

The Ministry of Research and Technology (*Kementerian Riset dan Teknologi*) (www.ristek.go.id) has as one of its programmes the expansion of ICT infrastructure to foster economic activity. The priority areas include telecommunication development, Internet services, energy-saving and low-priced computers, digital technology and open source applications.

Finally, the Indonesia Security Incident Responses Team on Internet Infrastructure (ID-SIRTII) is tasked with ensuring network safety and security in accordance with Indonesian law. The team includes representatives of the Central Bank, academe, ICT experts and law enforcement. The team is also expected to assist the Minister of Communication and Information Technology in discharging the planning, coordinating, supervising and controlling functions of ID-SIRTII.

Industry/Business

The Indonesia Infocomm Society (*Masyarakat Telematika—MASTEL*) (www.mastel.or.id) is a non-profit institution that functions as a bridge between government and ICT industries. Focusing on telecommunication, multimedia and information technology, MASTEL has the support of about 12 associations in the ICT sector. It has 63 company members, about 215 individual members, 27 not-for-profit organization members and 14 special members.

The Indonesia Information Technology Federation (*Federasi Teknologi Informasi Indonesia*—FTII) (www.ftii.or.id) consists of several associations in ICT-related and other fields striving for industrial development and the expansion and integration of information technology applications.

The Indonesia ISP Association APJII (www.apjii.or.id) is the umbrella organization of several ISPs. It aims to develop the Internet network in Indonesia through affordable Internet service fees, management of the Indonesia-Network Information Center (ID-NIC) and Indonesia Internet eXchange (IIX), and negotiation of the telecommunication service infrastructure fees. APJII also provides Network Information Resources (NIR) to its members, gives advice to government, and organizes seminars and training programmes.

Civil society/Consumer groups

The Air Putih Foundation (Air Putih) (www.airputih.or.id) started out as a group of humanitarian volunteers with knowledge of ICTs who wanted to open communication lines to Nangroe Aceh Darrusalam which became isolated after the 2004 tsunami. It now serves as an ICT Emergency Response Team mediating and accelerating the distribution of information for disaster management.

The Indonesia Telecommunication Users Group (IDTUG) (www.idtug.net) is an independent organization serving as a bridge between consumers, operators and government and advocating for better telecommunication services in Indonesia. IDTUG is affiliated with the International Telecommunications Users Group (INTUG) based in Belgium.

The Center for ICT Studies Foundation (www.ictwatch.com) is involved in humanitarian programmes in the ICT field, such as the Society Self-supporting Computer Laboratory for middle-to-low-income groups and Internet-based health campaigns. Also known as ICT Watch, the foundation undertakes communication programmes and research about the state of ICT4D in Indonesia.

The following organizations are also involved in the development of ICT in Indonesia:

- ICT Business Software Association of Indonesia (ASPILUKI—*Asosiasi Piranti Lunak Telematika Indonesia*).
- Business Computer Association of Indonesia (APKOMINDO—*Asosiasi Pengusaha Komputer Indonesia*).

- Higher Education on Information Technology and Computer Association (APTIKOM—*Asosiasi Perguruan Tinggi Informatika dan Komputer*).
- Internet Kiosk Association of Indonesia (AWARI—*Asosiasi Warung Internet Indonesia*).
- Wireless LAN Internet Association of Indonesia (IndoWLI—Indonesia Wireless LAN Internet).
- Cellular Provider Association of Indonesia (ATSI—*Asosiasi Telepon Selular Indonesia*).
- Broadband Wireless Association of Indonesia (ABWINDO—*Asosiasi Broadband Wireless Indonesia*).
- Satellite Association of Indonesia (ASSI—*Asosiasi Satelit Indonesia*).
- Telephone Kiosk Association of Indonesia (APWI—*Asosiasi Pengusaha Warung Telepon Indonesia*).
- Association of Telecommunication National Company Association (APNATEL—*Asosiasi Perusahaan Nasional Telekomunikasi*).
- Association of Community of Internet Center (APW Komitel—*Asosiasi Pengusaha Warnet Komunitas Telematika*).

Enabling policies and programmes

National information system

The development of the National Information System (*Sistem Informasi Nasional*—Sisfonas) was started in 2002 and is expected to reach completion in 2010. Under the system, the Department of Communication and ICT, through its Universal Service Obligation (USO) programme, aims to provide between 2006 and 2010 telecommunication services to 43,000 villages that are not yet reached by the telecommunication network. A comprehensive Sisfonas blueprint has been drawn up and it is expected to become the main standard in the business world. Thus, one of the main concerns among ICT experts is interoperability of government business processes, work schemes, content management, document management, information standards, back office applications, language, search engines, payment gateway, knowledge management and information scheme analysis.

e-Indonesia initiatives 2006

The May 2006 e-Indonesia Initiative Conference organized by the Indonesia ICT Institute, National Planning Body, and Department of Communication and ICT brought together in Bandung 500 academics, practitioners, government officials, IT observers and entrepreneurs. Among the conference recommendations to the government are:

- Encourage initiative and leadership in the development of inexpensive ICT infrastructure.
- Create the position of Chief Information Officer (CIO) in government institutions.
- Establish the single identity policy as soon as possible.
- Coordinate regulation functions of the Department of Communication and ICT and the Ministry of Research and Technology.
- Develop local human resources, especially for the Department of Communication and ICT, Department of National Education, Department of Labour and Ministry of Research and Technology.

Bandung High Tech Valley (BHTV)

The Bandung High Tech Valley (BHTV) is an initiative to foster technology-based businesses and industries in the Bandung region, which extends beyond Bandung city (the capital of West-Java, Indonesia) to the greater Bandung area. The BHTV initiative originated from the Ministry of Industry and Trade as a means to increase electronics exports. It was started in 1986 but was forgotten when Indonesia faced an economic crisis in 1997. It was subsequently restarted by some people in the Bandung Technology Institute (ITB).

BHTV has most of the ingredients for a successful technology area:

- During the Dutch occupation, Bandung was designed to host many government supporting institutions, including research centres. This is why many government agencies (especially those related to technology) have their headquarters in Bandung.
- Bandung is considered as a 'student city': many students from various parts of Indonesia are studying in Bandung.
- Universities operate technology research centres in Bandung.
- Bandung is known as a centre for the traditional and modern arts. Many musicians, bands, painters and other artists come from Bandung.

According to Budi Rahardjo, one of the BHTV founders, the only ingredient that is missing in Bandung is a technology-based multinational company that will attract and keep talented people (science and engineering graduates) in the region. Current BHTV initiatives are focused on helping small technology companies to start in Bandung and attracting multinational companies to invest in research and development in the region. A BHTV expo in which 70 companies participated was held in 2004. On 10 February 2006, the BHTV Foundation was established by four ITB faculty members to further oversee the development of BHTV.

Indonesian Higher Education Network (INHERENT)

The Indonesian Higher Education Network (INHERENT) aims to connect 32 institutions of higher education in Indonesia as part of a distance education programme. To establish a reliable connection among the universities, PT Telekomunikasi Indonesia (Telkom) built and maintained the backbone infrastructure for the 'Smart Campus' programme. Universities in the network receive a grant of more than IDR 500 million (USD 50,000) for network support from Ditjen Dikti. Universities serving as network hubs receive assistance in the form of equipment worth IDR 2 billion (USD 200,000). The university that is selected to develop e-learning content and the digital library could get a IDR 2.5 billion (USD 250,000) fund.

The long-term goal is to make all of the 200 universities in Indonesia part of INHERENT. However, different universities enjoy different connection speeds. In Java the connection uses a STM-1 infrastructure that provides 155 Mbps. In outer Java the connection speed is 8 Mbps. For Papua and Moluccas the connection speed is only 2 Mbps due to infrastructure limitations.

Digital content initiatives

There are about 20,000 third-level domains under the country code Top Level Domain (cc-TLD).id. About 1,000 new domain names are registered every month.

The Department of Communication and ICT has published guidelines for central and local government institutions regarding digital content distribution through the Internet. These include the Government Infrastructure Development Portal Guidance, Making Local Government Website Guidance, Management of Electronic Document System Guidance and Compose e-Government Development Master Plan Guidance. According to the Department of Communication and ICT, about 48 per cent of the 472 provinces and regencies/cities have a website.

A national magazine, *Warta Ekonomi*, gives an e-Government Award annually to government institutions that are evaluated on the following criteria: change process, leadership, e-government investment strategy, coordination with other parties and focus on social services. Websites are also evaluated in terms of design, content, accessibility and responsiveness. The 2005 winners, the fourth set since the contest was established, were the following: Yogyakarta—www.jogja.go.id (Regency/City Administration), East Java—www.jatim.go.id (Province Administration), Department of Public Works—www.pu.go.id (Department), and Institution of National Survey and Mapping Coordination—www.bakosurtanal.go.id (Non-Department Government Institution).

The Indonesia ICT Institute also publishes infometric rankings of government websites using size, links and Web impact as criteria. The top 10 government websites are: DKI Jakarta Province website (www.jakarta.go.id), Surabaya City Administration—East Java (www.surabaya.go.id), East Java Province (www.jatim.go.id), Bali Province (www.bali.go.id), Indonesian Bank (www.bi.go.id), Department of Industry and Commerce (www.dprin.go.id), Bantul Regency Administration—Yogyakarta (www.bantul.go.id) and West Kalimantan Province (www.kalbar.go.id).

The national portal, www.indonesia.go.id, is managed by the Office of the Secretary of State and the Department of Communication and ICT. President Susilo Bambang Yudhoyono also has an official website, www.presidensby.info, which is managed by the Office of the Presidential Spokesperson and an independent editorial team. There is some debate among ICT experts about 'PresidenSBY' as the website's name (some think it is not formal enough for a presidential website) and the domain '.info' (the official state domain name is '.gov.id'). The website contains information on the agenda and activities of the President. It has an English-language section and provides information services through SMS, audio streaming, podcast feed and syndicated news through RSS. Launched in February 2006, the website cost IDR 84 million (USD 8,400) to put up; its monthly maintenance costs amount to IDR 42 million (USD 4,200).

Online services

Many online services are available to the Indonesian public. One of these is e-banking services such as those provided for BCA clients (www.klikbca.com) and Bank Mandiri (www.bankmandiri.co.id). Some universities, such as Bina Nusantara University (www.binus.ac.id), use integrated virtual management systems and online study materials. There are also many e-commerce websites, including computer store websites like Bhinneka (www.bhinneka.com) and tourist service websites like Indo.com (www.indo.com). Airline tickets can be bought online from Air Asia (www.airasia.com) and Adam Air (www.flyadamair.com).

Also available is a Yellow Pages service on the Internet (www.yellowpages.co.id), as well as various online media services such as Detikcom (www.detik.com), Kompas (www.kompas.com) and Bisnis Indonesia (www.bisnis.co.id).

> ## Tax directorate serves tax payers better through IT
>
> The Tax Information System (*Sistim Informasi Pajak*—SIP), which has been operational since 1990, is the 'oldest' IT application of the Directorate General of Tax. The plan is to gradually reduce and then replace this SIP application with the Integrated Tax Application System (*Sistim Aplikasi Pajak Terpadu*—SAPT) under the Directorate General of Tax Information System (*Sistim Informasi Direktorat Jenderal Pajak*—SIDJP). SAPT will be used by the Tax Area Office to handle large tax obligators (or large tax payers locally known as *Wajib Pajak Besar*). SIDJP will be used by the Tax Area Bureau to handle tax obligators such as state-owned corporations, foreign institutions or expatriates, foreign investment and company exchange. Both the SAPT and SIDJP are already being used in Jakarta.
>
> According to KlikPajak.com, the Directorate General of Tax's information portal is very useful, providing an early warning service to tax obligators who have not paid their taxes or posted a Tax Payment Announcement Mail after specific dates. Other applications are an e-filling and e-registration application. Both of those applications also connect to SAPT, SIDJP and SIP through software bridges.
>
> The use of IT by the Directorate General of Tax has already increased tax income collection. According to Taxation Director General Hadi Purnomo, the total income tax collected between 2001 (when the Directorate started using the Internet) and 2004 reached about IDR 660 quintillion, which is more than the total income tax of IDR 600 quintillion collected over 30 years before the Directorate started using the Internet.
>
> IT applications at the Directorate include Yearly Announcement Mail (*Surat Pemberitahuan Tahunan*—SPT) with Tax Obligator ID Number (*Nomor Pokok Wajib Pajak*—NPWP) and online tax obligator registration (e-Registration or e-Reg). Taxes can be paid through any payment channel, including e-banking. Thus, tax payers now enjoy the benefits of e-government, such as better service and information available 24 hours a day seven days a week. Hadi said that with IT, tax evasion is easier to detect and the taxation process is simpler, quicker and cheaper.

Open source and open content initiatives

Open source movement

The Indonesian Go Open Source! (IGOS) movement was launched on 30 June 2004 by the Ministry of Research and Technology, Ministry of Communication and Information Technology, Ministry of Justice and Human Rights and Ministry of National Education. It aims to:

- Promote the use of Open Source Software (OSS) in Indonesia;
- Prepare guidelines for the development and use of OSS in Indonesia;
- Establish a training centre, competency centre and open source-based business incubator centre in Indonesia; and
- Improve the government and society's ability, creativity and participation in the use of OSS.

Some of the open source applications developed by the Indonesian ICT community are Windows in Indonesian language (*Windows Bahasa Indonesia*—WinBi), BlankOn Linux Distro, IGOS Desktop and Application System and *Waroeng* IGOS for Internet Kiosk Client-Server Application.

Computer knowledge for free

IlmuKomputer.Com (IKC), which means computer knowledge, is a website that contains free material and lectures on computer technology in Bahasa. Free materials with Open License Content are available and ready to download in PDF and CD-ROM format. Established on 17 April 2003, IKC received a World Summit on the Information Society (WSIS) award as one of 'The 21 Continental Best Practice Examples in the e-Learning Category.'

IKC materials were written by hundreds of volunteer contributors from cities in Indonesia and abroad. The materials include tutorial/lecture materials, translations, reviews and various practical advice. Online consultations are also available through Yahoo Messenger and virtual seminars.

Blogs

Enda Nasution, who is known as the Father of Indonesian bloggers, estimates that there are currently 70,000–90,000 Indonesian blogs. Indonesian blogger communities may be built around a geographic area; an example is the Loenpia.net community for bloggers from Semarang (Central Java). Other blogger communities have members from a wider area, such as the BlogFam.com community.

In Indonesia, it is not only the young people who keep blogs. Prominent senior people who are known bloggers are former World Marketing Association President Hermawan Kartajaya, Minister of Defence Juwono Sudarsono and IT journalist Budi Putra. Books about blog-making techniques have been published and mainstream online media group detikINET (www.detikinet.com) frequently gather the blogger community to hold discussions or training sessions on how to make blogs for students, teachers, housewives and even activists.

A thousand books for the blind

The Thousand Books Project, spearheaded by the Mitra Netra Foundation (www.mitranetra.or.id), is an opportunity for volunteers to transform popular books in Indonesia into e-books that can be read by blind people through the Mitra Netra intranet. To date, more than 280 people have joined the project. They encode the books on their own and then send the files by e-mail or post service to Mitra Netra which then edits the files and posts them on the Mitra Netra intranet. Blind people are able to 'read' the e-books using screen reader software. The Mitra Netra website contains a list of the available e-books, as well as books that are still being re-encoded. Volunteers include people from outside Indonesia.

National computer camp for the blind

On 23 November 2006, a competition titled 'The First National Computer Camp for the Blind' was organized by Yayasan Mitra Netra to encourage blind people to make use of computers. As many as 100 people with sight disability competed in poetry writing using Microsoft Word, calculating profit and loss using Microsoft Excel and browsing the Internet. Contestants used notebooks in which a screen reader software called Job Access with Speech (JAWS) was installed.

Education and capacity building

One of the obstacles to the success of the e-government initiatives in Indonesia is the lack of human resources adequately trained in the use of IT. Data from the Re-Registration of Government Employees in 2003 conducted by the National Government Employee Body show that only 27 per cent of the 3.6 million government employees completed college and 55.8 per cent of the total are 40–56 years old. This profile suggests a lack of skills necessary for IT-based government work as well as a lack of willingness to adopt IT-based changes.

According to the Human Resources Blue Book published in October 2003 by the Bandung Technology Institute and the Department of Industry and Commerce, by 2010 Indonesia will need about 327,813 personnel for the ICT industry. Around 32.5 million personnel are needed to support public services and other industries by 2008, according to the Department of Communication and ICT. However, based on calculations by the ICT and Computer Higher Education Association (APTIKOM), there are only 20,000 ICT graduates every year.

To address the need for trained ICT personnel, some ICT-related industry associations have formed the Indonesia National Work Competence Standard (*Standar Kompetensi Kerja Nasional Indonesia*—SKKNI) which lists 27 competencies for computer operators and 91 competencies for programmers.

Several educational institutions provide capacity building for ICT human resources. According to the latest research conducted by Tempo Data Research and Analysis, the top 10 universities with ICT programmes in 2006 are:

1. Bandung Technology Institute—Bandung (www.itb.ac.id)
2. Bina Nusantara University—Jakarta (www.binus.ac.id)
3. Gunadarma University—Jakarta (www.gunadarma.ac.id)
4. University of Indonesia—Jakarta (www.ui.ac.id)
5. 10 November Technology Institute—Surabaya (www.its.ac.id)
6. University of Gadjah Mada—Yogyakarta (www.ugm.ac.id)
7. Tarumanegara University—Jakarta (www.untar.ac.id)
8. Trisakti University—Jakarta (www.trisakti.ac.id)
9. Telkom Technology College—Bandung (www.stttelkom.ac.id)
10. Pelita Harapan University—Jakarta (www.uph.edu)

At its November 2006 National Congress, the Association of Higher Education on Information Technology and Computers APTIKOM in Bandung declared its support for the implementation of the e-Learning Joint Content Programme in 2007. The academic institutions that are members of APTIKOM are requested to offer online a minimum of one subject worth three credits as part of the e-learning programme. This means that students of one university can enrol in the online courses offered by other universities.

Research and development

VoIP is a main focus of ICT research and development in Indonesia. ICT Center, Jakarta (www.ictcenter.net) successfully piloted the use of Maverick VoIP with a group of vocational high schoolteachers. The technology will soon be implemented

> **Multichannel learning by Bina Nusantara University**
>
> Bina Nusantara University (UBiNus) (www.binus.ac.id) in Jakarta combines on-campus and off-campus course delivery methods in a programme called Multi Channel Learning (MCL). First implemented in 2002, the MCL programme offers 13 sessions in one semester, of which four are delivered via e-learning mode. Around 78 per cent of UBiNus courses can be accessed through the Internet (*e-Indonesia Magazine*, January 2006).
>
> The MCL programme uses a learning management system (LMS) developed by UBiNus. Free Internet services are given to students living within a 5 km radius of the campus. There are hotspots at the Syahdan and Anggrek campuses providing a wireless connection to students with the appropriate PDAs or notebook computers. Classrooms are equipped with an LCD projector and a multimedia computer that is connected to the Internet.
>
> On 25 July 2006, UBiNus signed a memorandum of understanding with the Department of Communication and ICT for the development and implementation of strategic communication and information technology for national development.

in more than 4,000 schools. Maverick VoIP technology can be used free of charge and interconnects easily with government and private telecommunication networks, whether fixed phone or mobile phone.

The liberalization of the 2.4 GHz band for Wi-Fi in 2005 stimulated the spread of VoIP activity. The liberalization of the 2.4 GHz band was the result of concerted efforts by the Indonesian ICT community, especially those who joined the mailing-list indowli@yahoogroups.com. An innovation by the mailing-list moderators is the '*wajanbolic*' and '*pancibolic*' technology, a 2.4 GHz antenna made out of a wok or cooking pan and which can provide connection speeds of 11–54 Mbps within a 3–4 kilometre radius at a cost of only IDR 350,000 (USD 35). Using RT/RW-net technology, or metro LAN, households can connect at a speed of 100 Mbps Fast Ethernet for only IDR 200,000 (USD 20). With Wi-Fi, 24 hours of Internet access costs only about IDR 350,000 (USD 35) every month.

Several other R&D efforts in Indonesia are directed at making ICTs more accessible to rural communities. An example is the award-winning e-Pabelan telecentre.

Challenges

Given the size of the Indonesian population and the relatively low ICT penetration rate at present, it is clear that Indonesia can become a huge market for ICTs in the future. For the moment, however, it ranks 60th among 65 countries in the 2005 e-Readiness Rankings published by the Economist Intelligence

> **Observing the price of chilli at the e-Pabelan telecentre**
>
> The National Development Planning Board (*Badan Perencanaan Pembangunan Nasional*—Bappenas), in coordination with UNDP, has put up six telecentres under the Partnership for e-Prosperity for the Poor (PePP) programme. One of the six, the telecentre in Pabelan, received the APEC Digital Opportunity Center (ADOC) Award for ICT Best Practice for Bridging the Digital Divide category in Taipei in July 2006. The e-Pabelan telecentre project, as it is called, bested similar programmes in Chile, the Philippines, Peru and Vietnam.
>
> The e-Pabelan telecentre, which opened on 23 April 2004, is a 7 x 12 metre house providing Internet facilities for farmers. At the start of the project, 10 farmer groups of 15–25 farmers each used the Internet for two hours every day for 10 days. The farmers gathered information on the plant diseases that were afflicting their crops. The chilli farmers in Pabelan II orchard, Pabelan village, Mungkid sub-district, Magelang regency, Central Java used the Internet to find information on the market price of chilli. Armed with this information, they no longer get cheated by middlemen. The farmers also get online information on seeds, planting techniques, pests and pesticides and fertilizers.
>
> The Islamic boarding students, called *santri*, of Pondok Pesantren Pabelan also make use of the telecentre. The facility charges the general public a fee of IDR 3,000/hour (USD 3) while farmers pay only IDR 1,000/hour (USD 1). Using the telecentre as a 'workshop', one *santri* won second place in a computer application competition by Microsoft and earned a training slot in South Korea.

Unit. Indonesia was ranked lowest in terms of the legal and policy environment and social and cultural environment.

The lack of appropriate legal and policy controls is underlined further by Indonesia's having been identified as one of three countries with the highest piracy rates in 2005 in a piracy study published by BSA and IDC Global Software. The 2005 piracy rate was 87 per cent, lower by only 1 per cent than the 2003 piracy rating. According to BSA, piracy is most rampant in the business sector, especially among small and medium-sized enterprises (SMEs). The Indonesian government estimates a financial loss of around USD 280 million (about USD 31,758) every year because of piracy. Because of the weakness of law enforcement against piracy, Indonesia has been included in the United States Trade Representative (USTR) priority watch list, hampering trade with the US and making Indonesia vulnerable to sanctions from the World Trade Organization (WTO).

While it may not be fair to ask ICT producers and vendors to sell their products at prices lower than the prevailing price in developed country markets, the fact is that the purchasing power parity in Indonesia is much lower compared with that in developed countries. Thus, one finds that even in Internet kiosks, violations of intellectual property rights, such as use of pirated software are rampant. One reason is that the price of 10 sets of software, for use in an Internet kiosk equipped with 10 computers, could be as high as the price of the computers themselves.

The Indonesian government recognizes that Internet kiosks play an important role in promoting the benefits of using the Internet to the public. Presidential Decree No. 6 Year 2001, on ICT Development and Usability in Indonesia, points out that Internet kiosks could broaden the range and content of public information services, including medical and education services, serve as a public service centre in cities and villages, and provide e-commerce services for small and middle enterprises. One solution to software piracy in Internet kiosks is use of open source software. However, it is not easy to change the behaviour of businessmen and their customers.

The government continues to work towards increasing teledensity in Indonesia. At the Indonesia Infrastructure Conference and Exhibition (IICE) held in November 2006, the Department of Communication and ICT optimistically projected growth of telecommunication teledensity by as much as 20.1 per cent for fixed-lines, 26.7 per cent for mobile phones, 30.8 per cent for Internet use and 71 per cent for bandwidth by 2010. Clearly, much work needs to be done by government, the business community, academe and the ICT community in Indonesia for these projections to be realized.

References

ACNielsen. (2005). *Detikcom online audience profile*.

Asia Foundation. (2002). *SMEs and e-commerce*.

Bangun, R. (2006). *Pengembangan Industri Telematika Nasional (National ICT Industry Development)*. Discussion Paper, Roadmap Pengembangan Industri Telematika Indonesia, Jakarta, September 2006.

Basuki, D.R. (ed.). (2006). *Panduan Memilih Perguruan Tinggi 2006 (Choosing higher education guidance)*. Jakarta: Pusat Data dan Analisa TEMPO.

Berita Resmi Statistik. (2006). *Profil Pemanfaatan Teknologi Informasi Oleh Masyarakat (Profile of information technology use by society)*.

Bisnis Indonesia Newspaper. Available at www.bisnis.co.id

detikINET ICT Online Media. Available at www.detikinet.com

Djiwatampu, A. Ph. (2005). Braving the challenge of satellite technologies: National breakthroughs and indonesia's role in international forums. *Online journal of space communication*. Retrieved 1 October 2006 from http://satjournal.tcom.ohiou.edu/issue8/

Driyo, A.D. (2006). Prospek bisnis telekomunikasi di indonesia (Business prospects of Indonesian telecommunication). *Economic Review*, 204, June.

Economist Intelligence Unit. (2005). *2005 e-readiness rankings*.

Editorial Team. (2006a). Geliat industri software tanah air (Awakening of the national software industry). *PC plus tabloid*, 269/VI, 20–21.

Editorial Team. (2006b). Kebijakan TI SBY perlu komitmen (SBY IT policy requires commitment). *e-Indonesia magazine*, 1/2006, 10–45.

Editorial Team. (2006c). Mengetik buku untuk tunanetra (Typing book for blind people). *Warta kota newspaper*, 13 September 2006, 11.

Editorial Team. (2006d). Nikmatnya layanan 3G (The delicious taste of 3G services). *Tren digital tabloid*, 86/IV/11, 7–9.

Editorial Team. (2006e). Peluang kerja TI (IT job field opportunity). *Info komputer magazine*, 08/2006, 80–92.

Editorial Team. (2006f). Penetrasi TI terhambat dana (IT penetration obstructed by funds). *Biskom magazine*, 06/06, 56.

Editorial Team. (2006g). Petani dusun di warnet plus (Village farmer in the telecenter plus). *Gatra magazine*, 38/XII, 70–75.

ICT Watch. (2002). *Penelitian profil industri warnet Indonesia (Indonesia internet kiosk industry profile research)*.

IDC and BSA. (2005a). Indonesia chapter. *Expanding the frontiers of our digital future*.

IDC and BSA. (2005b). *Third annual BSA and IDC global software piracy study*.

Investor Daily Newspaper. Available at www.investorindonesia.com

LSP Telematika. Available at www.lsp-telematika.or.id

Minges, M. (2002). *Kretek Internet: Indonesia case study*. International Telecommunication Union.

Pattiradjawane, R.L. (2006). Teknologi VoIP mendorong produktifitas bangsa (VoIP technology, pushing national productivity). *Kompas*, 6 September, 33.

Purbo, Onno W. (2006). *Sejarah Internet Indonesia* (*Indonesian Internet history*). Retrieved 1 October 2006 from http://wikihost.org/wikis/indonesiainternet/wiki/start

Rahardjo, B. (2002). The story of the Bandung High-Technology Valley. Paper presented at the Seminar Nasional Industri Berbasis Teknologi Informasi dan Telekomunikasi. Bandung.

Riset dan Teknologi, Kementerian. (2006). *Indikator telekomunikasi dan informatika di Indonesia tahun 2005 (Indicator of telecommunication and information technology in Indonesia)*.

Rusli, A. (ed.). (2003). *Teknologi informasi, pilar bangsa Indonesia bangkit (Information technology, national pillar for Indonesian awakening)*. Jakarta: Indonesia Ministry of Communication and Information.

Soendjojo, H. (2005). *Proceedings from Konferensi Nasional Teknologi Informasi dan Komunikasi Indonesia: Implementasi e-government sejumlah pemerintah daerah (e-Government implementation in some local governments)*. Bandung.

Sunggiardi, M. (2006). Mengembangkan VoIP di Jaringan RT-RW-Net (VoIP development in the RT-RW Net). *Kompas*, 6 September, 40.

Telkom Indonesia. (2005). *Laporan tahun 2005 (Yearly Report 2005)*.

Telkom Indonesia. (2006). 2006 First Quarter Results. *Info memo*, May.

World Bank. (2006). *Indonesia: Economic and social update*.

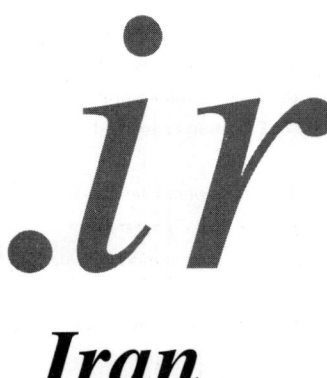

Iran

Masoud **Davarinejad** and Massood **Saffari**

Population (October 2006 census)	70,490,262 (urban 69.9 per cent; rural 30.1 per cent)
GDP	USD 187.9 billion (current prices), USD 2,730 per capita
Key economic sectors (per cent of GDP)	Petroleum & Gas (27 per cent), Agriculture (10 per cent), Industry (16 per cent), Services (47 per cent)
Computers per 100 inhabitants	7.5
Fixed-line telephones per 100 inhabitants	29.7 (20.34 million lines)
Mobile phones per 100 inhabitants	12.5 (8.51 million subscribers)
Internet users per 100 inhabitants	12 per 100 inhabitants (8.25 million subscribers)
Domain names ('.ir') registered	46,442 as of December 2006
Broadband subscribers per 100 inhabitants	0.38 (260,000 subscribers)
Internet domestic bandwidth	10,075 Mbps
Internet international bandwidth	3,720 Mbps as of December 2006

Note: All figures are as of 22 March 2006 unless otherwise stated.

Technology infrastructure

From 2000 until the end of 2005, Iran witnessed large increases in infrastructure capacity and development of information technology. Privatization and competition in the telecommunications market were the main themes of the Third Five-Year Plan 2000–2004. Although the infrastructure was off limits to private investment and competition due to restrictions imposed by Article 44 of the Constitution, major investment attractions such as the second nationwide mobile license were auctioned, and service provision in data communications, the Internet and satellite communications was opened to the private and public sectors.

Official statistics at the end of 2005 show a remarkable increase in telecommunications growth indicators. One of the main contributing factors was the implementation of guidelines in the Third National Development Plan to downsize and thus improve the efficiency of government organizations and companies. This hazardous and bumpy yet life-saving path resulted in record achievements. There are 1.48 employees for maintenance of fixed telephony per 1,000 lines compared to 4.99 per 1,000 lines at the end of 1999. The number of fixed-line subscribers jumped to 20.34 million with a penetration rate of 29.7, and the number of mobile telephone subscribers went up to 8.51 million in 990 cites or almost 100 per cent nationwide. By July 2006, there were 21 million fixed-line subscribers and 10 million mobile phone subscribers.

Parallel efforts to increase digital switches and fixed-lines in urban areas provided one million transmission circuits at the end of 2005. This was due to the increase in the number of microwave stations and the upgrade of legacy transmission facilities utilizing digital radio links.

The use of fibre optics in the telecommunications backbone is fairly recent. Official figures provided by the Telecommunications Company of Iran (TCI) show that there was no fibre optic network in 1995, nearly 7,200 km of fibre optic network at the end of 1999 and 56,000 km in 2005. TCI reports that to minimize downtimes and boost the availability of the network, the whole network is designed as three interconnected DWDM loops with a capacity of 4 STM 64 and routes ending in Tehran with 8 STM 64.

The deployment of fibre optics in other infrastructure sectors has also gained momentum. The Iran Power Generation, Transmission & Distribution Management Company, known as TAVANIR, has reported the implementation of Optical Ground Wire (OPGW) and Power Line Communications (PLC) to 2,500 km, with plans to extend to 6,500 km. This network is intended mainly to connect the five dispatch centres scattered throughout the country and to trade the excess capacity with TCI for voice channels or other services. Iran Railways also benefits from a network consisting of coaxial cable and fibre optics for its signalling and messaging needs. The network is 2,500 km long.

Telecommunications facilities available till the latter half of the Third National Development Plan included High bit-rate Digital Subscriber Line (HDSL) links using leased lines and base band modems and spread spectrum wireless links which had gained popularity due to the fact that no license or payment of fees was required. At the same time, the Data Communication Company of Iran (DCI), which was established in 1991, finally gained momentum by providing high-speed HDSL links and E1 and fractional E1 connectivity to cope with the growing demand for data communications facilities. At the end of 2005, DCI, which was renamed the IT Company after the reorganization of the Ministry of PTT (MPTT) and TCI in 2004, reported a total of 106,104 high-speed ports.

Since 1969 satellite communication has been dedicated to providing international voice connection. The first trial to enable 300 rural areas to access the telephony system via satellite failed due to technical difficulties and design flaws. In 1988, attempts to provide data communications using VSAT terminals succeeded and TCI established its first hub in Bumehen, a city 20 km east of Tehran. In 2002, TCI had more than 900 VSATs installed. In the same period, the Central Bank of Iran was issued a license to provide satellite access solely to the banks and for banking applications only. By the end of 2004, they had more than 2,500 terminals installed and a plan to deploy another 1,500 within two years. In 2005, the Communication Regulatory Organization opened VSAT and satellite services to competition and issued four five-year licenses.

Between 2003 and 2005, in accordance with regulations set by the Supreme Council of Information Dissemination (HCID), MPTT issued to the private sector, hundreds of permits to provide Internet services and a few permits to provide gateway services to the Internet (known as ICP). Those providing the connectivity to the international gateways installed their own satellite facilities and acted in accordance with mutual contracts with their partners, mainly from Europe. Official reports indicate that there were 747 ISPs and 35 ICPs at the end of 2005 in Iran.

Among several regulated services opened for licensing to the private sector were data communications service provider licenses (known as PAP in the local market) which permit licensees to provide wired/wireless data communications services throughout the country. By the end of 2004, out of 26 applicants only 11 remained in the market, providing ADSL, G.HSDHL and high-speed wireless services to around 200,000 subscribers in more than 40 cities. Competition has reduced the price of Internet access to a monthly payment of USD 16 for a minimum package of 64/128 Kbps, including an ADSL modem or CPE (Customer Premises Equipment).

Increased use of the Internet bandwidth is partly due to international calls. Although there was no specific regulation for Voice over Internet Protocol (VoIP) in 1999, there was tacit approval among the middle management of the MPTT and DCI of VoIP for calls originating from Iran while terminating calls would not be permitted. By 2005, the increase in the number of minutes for terminating calls had reached about 600 million minutes or 57 per cent of the total incoming traffic to Iran. This forced MPTT to regulate the situation and finally issue temporary (renewable annually) permits for international calls originating from the country and to declare terminating calls as illegal. This in turn has created a grey market for what has been nicknamed 'smugglers of minutes'. These unwanted small entities use unlicensed satellite equipment to connect to their minute providers abroad using Internet protocols and dedicated routers. They are known to have made fortunes. In May 2005, the Communication Regulatory Commission (CRO) produced a comprehensive model named 'international voice services' to regulate the volatile market covering both origination and termination as well as transit calls, and a draft license was prepared for the regulatory commission to approve, which is still pending.

There is lack of agreement about the number of Internet users in Iran. Sanaray, the Software Export Research & Development Co., notes the number of Internet users as of December 2005 to be 7.35 million: 4.04 million (55 per cent) home use, 1.91 million (26 per cent) government use and 1.4 million (19 per cent) business use. Internet World Stats reported on 12 August 2006 the number of Internet users in Iran in mid-2005 to be around 7.5 million. Other sources, such as IT newsletters and blogs, estimate the number of Internet users in Iran to be between 5 and 15 million as of end 2005 to mid-2006. A report circulated by the Office of the Minister of CIT in August 2005 probably presents the most accurate estimate: 8.25 million with a penetration rate of 12 users per 100 inhabitants. One reason for the differences in reported numbers is the difference in definitions of Internet users: some do not count the mere use of e-mail, while others think including those who access the Internet at Internet cafés is double-counting. However, most studies agree that there are 2.5 million dial-up Internet subscribers in 316 cities (or 31.6 per cent of the 990 cities in the country).

Key institutions dealing with ICTs

Policymaking in information technology is multifaceted and requires coordination among several sectors that have different agendas and priorities. In Iran, the issue is compromised by concurrent councils and bodies each addressing a division within IT. These include the Ministry of Communications and Information Technology (MCIT), the High Council of Informatics (HCI) which is affiliated with the Management and Planning Organization (MPO), the Supreme Council of IT within MCIT which is headed by the President, and the High Council of Information Dissemination (HCID) which is affiliated with the High Council of the Cultural Revolution. In addition to these official bodies, the High Council of Cyberspace Information Exchange Security (AFTA), which is under the Office of the President, has published general guidelines for ICT security.

HCI, composed of representatives of several key ministries and organizations, was the first body empowered by the Supreme Council of the Revolution in 1979 to make decisions on computer-related issues. During its first years of operation, HCI managed and settled claims and disputes between large international computer companies. It was not until 1990 that HCI played an active role in policymaking and guidance of the IT sector.

The Secretariat of HCI, the executive body that carries out the decisions of the HCI within the Planning and Budget Organization which was later renamed the Management and Planning Organization (MPO), was reconstituted and took the lead in developing the IT sector in Iran and in responding to the needs of the commercial market and the researchers by providing them with information and analysis of applied technologies, IT indicators and statistics, and insights into the need to implement professional standards. This laid the foundation for an era of interaction among key players.

In the years 2000–2002, when the High Council of Information Dissemination (HCID) and the Development of Information Technology Applications programme (TAKFA) began to take over policymaking in the IT sector, HCI's influence and policymaking role gradually faded away although its legal foundations and official status and duties remained unchanged.

HCID was established in 1998 as the main authority and policymaker in IT and Internet-related issues in Iran, in response to concerns expressed by the cultural authorities and influential religious figures regarding the presentation of explicit content over the new media, in particular the Internet. Its first move was to announce in 2001 the regulation of operations of the ISPs and the ICPs. An executive body was established to ensure that materials on the Internet did not undermine or violate accepted social norms of decency and modesty and national security. This body has been the object of much criticism from both its foes and friends.

Although the approach and rules set by HICD had a limiting impact on the growth of the Internet as intended at the outset, the efforts of educational institutes, research centres and universities, and the pioneer vanguard Internet service providers spurred the growth and spread of the Internet, which in turn resulted in an increase in the number of ISPs and ICPs. This made the Internet accessible around the country.

In 2003, a law stipulating the duties and powers of the MCIT was passed and a new council to govern IT in Iran was born. However, this did not eliminate the outdated, ineffective and redundant bodies and councils. The new Supreme Council of Information Technology (SCIT), chaired by the president and managed by the Minister of CIT, took over all aspects of IT policymaking but oversight of digital contents remained within the responsibility of the Supreme Council of Dissemination.

There is one regulatory body governing ICT development in Iran: the Communications Regulatory Commission (CRC) headed by the Minister of CIT and its executive body, the Communications Regulatory Organization (CRO). The CRC consists of government-appointed experts and senior officials of MCIT and other government bodies. It approves the rules and regulations for the development of the ICT sector. The CRO implements the rules and regulations approved by the CRC and is responsible for radio frequency management and planning. It has offices all over the country.

The institutions responsible for supporting and promoting ICT-related initiatives are the Hi-Tech Industries Center under the Ministry of Mines & Industries, the Electronics Fund for Research & Development (ESFRD) also under the Ministry of Mines & Industries, the Development of Information Technology Applications programme (TAKFA) and the Production and Management of Electronic Content initiative (TASMA) started by HCID in early 2006 to complement TAKFA. TAKFA, an initiative in 2001–04, was part of the National ICT Agenda of the President's special envoy in close collaboration with the secretariat of HCID.

ICT research is centred mainly in the Iran Telecommunications Research Center (ITRC), which is affiliated with MCIT, the Institute for Studies in Theoretical Physics & Mathematics (IPM), the Sharif Advanced Information and Communications Research Center and the Research Center of Informatics Industries.

Several NGOs have been active in ICT development in Iran for more than two decades. However, due to tight government control of the sector, they do not have a serious role in ICT development in the country. These NGOs include the Computer Guild Organization, Iran Informatics Companies Association, Sanaray Software Export Research & Development Co., Informatics Society of Iran, Computer Society of Iran and Union of Iranian Software Exporters.

Several legacy government organizations mandated to produce information and statistics and established before 1977, such as the Statistical Center of Iran, the National Cartographic Center and the National Remote Sensing Center, have been major users of large mainframe computers and are believed to be among the major contributors of digital data for the growth of other sectors and national planning. Also among such organizations are the Information Technology Company (ITC) affiliated with MCIT and the Iranian Information & Documentation Center (IRANDOC).

Digital content

Websites in the Persian language form part of non-Latin based digital content on the Internet. The number of Persian blogs is estimated to be around 800,000, of which one-eighth are registered with service providers like Persian Blog and Blogfa (Ziyaee-Parvar 2006). Other sources report the number of active blogs to be around 150,000. There is also a considerable volume of Persian content on Islamic-related issues available on the Internet. A large array of CD-ROMs on Islam and Iranian culture, basic education and history is likewise available in the market.

The DPI Law system, an online Web-based Persian search engine and data base established in 1999, has about 10 Gb of text and indexes that include all of the legislation, enacted laws, bylaws, government and ministerial decrees, important court rulings and regulations with cross-links, as well as references to obsolete, ineffective and amended paragraphs. However, because legal expertise is required for the evaluation of remarks and pronouncements, and because of lack of financial support, the website's operation and services have been reduced to a minimum since early 2006.

Tebyan, which is affiliated with the Islamic Development Organization, runs a website covering religion and related sciences and issues. It is expected to roll out the largest data centre in Iran by mid-2007.

TASMA, an acronym for 'Production and Management of Electronic Content' in Persian, was announced by HCID in 2006 as a substitute for the TAKFA programme in the promotion of digital content in Iran. This programme is part of a restructuring plan that led HCID to focus exclusively on Persian content and leave other IT-related issues to the Supreme Council of IT.

Use of the Persian language in computer applications, specifically typefaces on printers, keyboards and data entry systems, goes back to 1975 when IBM introduced Persian printing chains and slugs and banking applications used NCR Persian keyboards to generate bank statements and monthly reports in Persian. Through the joint efforts of SHCI and Sharif University of Technology, the Persian language is now a part of Unicode and portability of Persian data from any platform is assured.

Online services

Basic online services, such as downloading of forms to be filled in, signed and then mailed back to processing authorities, have been available to the public in Iran since 2000. However, it was not until 2003 that these basic online services were recognized and utilized by the few with access to the Internet. The lack of payment capability in these online services gravely limited their scope. In early 2003, very few websites offered electronic marketing. Only local cash or debit cards were accepted, as credit cards were not yet available. In the beginning, only locally available products, for which shipment was not an issue, were marketed, such as scarves, handicrafts and flower bouquets for feasts and funerals. Goleshahr was one of the few that pioneered in online sales in Iran; it has left the market to hundreds of newcomers.

The root Certificate Authority (CA) is now in place in Iran. Although many B2C websites are in service, shopping online is not yet popular. B2B services have not yet been developed.

Another online service is the publication of university entrance examination results. Millions log in to websites designed for this purpose, such as the Sanjesh Organization website. The period when examination results are published is characterized by the worst Internet traffic in Iran.

Most banks offer basic online services over the Internet, such as online statements of account. These services are also offered using fax and SMS. In some banks, individuals can now transfer funds between their accounts within a bank. The interbank network called SHETAB, which is affiliated with the Central Bank, provides connectivity and switching services for all of the 16 banks and credit institutes in Iran. SHETAB has gained momentum in the last two years, despite public dissatisfaction with the pricing scheme and its monopoly status. The Central Bank reports 18.542 million electronic cards issued as of September 2006, representing a 78 per cent increase from the end of 2005. The number of automated teller machines has grown 56 per cent in the same period to 7,630 nationwide. There were 115,537 point-of-sale terminals in use as of September 2006, a 210 per cent increase from the end of 2005 (Central Bank Payment Systems Bureau September 2006). Online payment of utility bills has gained popularity and account holders now require more sophisticated services from their banks. These services are being organized and advocated by the card companies associated with the banks. One factor in the late introduction of these services is said to be the incompetence of the IT bureaus of the utility companies.

The railways authority is among the few government agencies in Iran with IT plans and successful implementation of online services. Its Web-based services include presentation of detailed train schedules, reservation and sale of tickets.

Online reservation and sales of theatre tickets, CDs and DVDs, household appliances, and second-hand and used items sold by electronic thrift shops, as well as online student registration are becoming popular. However, no statistics on these services are available as of now.

ICT and ICT-related industries

Production of analog telephone switching centres in Iran goes back to 1969 when the Iran Telecommunications Manufacturing Company (ITMC) was founded in the city of Shiraz to produce switches for local, STD and transit switching centres for the telecommunications backbone. To guarantee the flow of technology and know-how, the German Siemens company was given 40 per cent of the shares of ITMC, and the MPTT and Bank of Industry and Mines a 30 per cent stake each. ITMC's nominal production capacity for analog switches was 30,000 subscriber lines. In 1990, a new production line for manufacturing digital switches was installed and production reached 260,000 subscriber lines. Later in 2003, as reported by TCI, production reached 2.5 million subscriber lines. ITMC has 70 per cent of the local market for fixed telephony switches.

In addition, there are private companies manufacturing digital telephone switches and pertinent software for small-capacity switching centres and PABX with less than 10,000 subscriber lines. Although there is a market for small switching centres below 10,000 subscribers in small cities and rural and industrial areas, these companies are unable to envision exporting their goods and they are always threatened by the active presence of major manufacturers. Thus, they are always seeking government protection to limit the import of telephones and other telecommunications equipment into the country.

Production of fibre optic cables started in 1994 in the city of Yazd with a capacity of 30,000 km of standard cable of different capacities.

The Syndicate of Telecommunications Industries has reported a membership of 50 major companies manufacturing antenna, telephone sets, small PBX, UHF/VHF radios and wireless handhelds, multiplexers, GSM BTS and antenna, copper cables and wires. The Iran IT Manufacturers Syndicate (IITMS), another strong union, has 52 members producing CRT and LCD monitors, power supplies and UPS, keyboards and other common computer peripherals such as mouse and speakers, PCs in CKD, simple SMD boards and computer accessories. Except for less than USD 10 million worth of exports, most target the Iranian market and lack competitive export capabilities.

Enabling policies and programmes

The Third National Five-Year Development Plan 2000–2004 was a landmark move for the privatization and liberalization of the telecommunications sector in Iran. Eleven licenses were granted to data communications service providers known as PAP, through which they provide ADSL and cellular wireless technologies; five licenses were issued to satellite service operators; six licenses were given to PSTN operators; and two licenses were issued to GSM mobile operators.

In July 2002, the government launched TAKFA (Development of Information Technology Applications programme, an acronym in Persian), which had seven major axes:

1. e-government;
2. education and digital skills development;
3. higher education, health, and medical therapy and training;
4. social services;
5. commerce and trade;
6. culture, arts, and Persian language and script in computer environments; and
7. ICT industry through SME empowerment, incubation centres and technology parks.

TAKFA encouraged major government agencies to review their existing plans and incorporate these seven initiatives. Agencies with plans that were fully in line with TAKFA's thrusts were granted a generous budget or subsidized finances. TAKFA initiated 110 major projects consisting of more than 5,000 sub-projects in almost all sectors. TAKFA also initiated several by-laws and decrees issued by the MPO and the President to minimize government spending and improve bureaucratic efficiency. However, TAKFA failed to produce a cohesive roadmap for sustainable growth of IT in Iran. A second phase for TAKFA is now under study.

The Fourth National Five-Year Development Plan 2005–09 envisions a knowledge-based economy in which the ICT sector, technology parks and incubators are to play a key role. Technology parks are exempted from state and local taxes and given other incentives to attract investors. The Plan highlights the following thrusts:

1. Systematic expansion of ICT applications towards the realization of a knowledge-based economy consistent with national development goals;
2. Development of human resources as a strategic priority in the expansion of ICT applications in order to create more 'value-creating' jobs;
3. Cultural development and creation of an empowering environment for creating maximum national synergy;
4. Implementing the necessary infrastructure for ICT development, including access network, security, laws and regulations, resources and facilities; and
5. Development of facilities and opportunities towards mobilization of the private sector.

Technology parks and incubator centres have been the centre of government attention since 1998 and their growth was considerable during the period of the Third Plan. There are now nine incubators authorized by the Ministry of Science and Technology. Pardis Technology Park (PTP), a major project, is located 20 km east of the capital city of Tehran, with an area of 38 hectares. The Park enjoys proximity to one of the largest telecommunications facilities in Iran, with major local and international fibre cables passing through it. PTP, which was started in 2001 as a government initiative, aims to create an environment for companies with similar missions to develop high-tech industries and to facilitate the flow of foreign investment and the transfer of technologies to its tenants. The full rollout has been rescheduled to early 2007. Out of 70 companies committed to joining PTP, 42 are in IT and telecommunications; the rest are engaged in automation, biotechnology, mechanics and chemistry. One of the major data centres in Iran is under construction at PTP. The techno-market, first conceptualized by PTP in Iran, aims to harbour a cyber market for the trading of know-how, innovative ideas and products. Having started off by building physical capacity, PTP has recently diverted attention to strategic planning and devising a clear business plan and a legal and financial framework.

Legal and regulatory environment

As previously mentioned, the Communications Regulatory Commission (CRC), consisting of seven members and headed by the Minister of MCIT and with the president of the Communications Regulatory Organization (CRO) as secretary, is empowered to restructure the telecommunications sector and to approve telecommunications regulations and by-laws. The CRO implements the policies and decisions of the CRC, issues licenses for telecommunication services, and allocates and manages the frequency spectrum. CRO is built on the former Directorate-General for Radio Communication, a major bureau within MPTT and the ITU contact point.

Unfortunately, due to ambiguities in interpretations of the Constitution, arbitration is not covered by existing laws and is a matter of voluntary engagement. Arbitration plays an important role in forming and guiding the telecommunications market.

Some of the important issues perceived to have a major impact on the growth of the software market have been addressed by the relevant councils. Copyright law, which has been in effect in Iran since 1990, is the most important in this regard. However, Iran has not yet joined any international agreements on copyright of foreign software, except for bilateral treaties in which a mutual copyright is also respected.

Tehran Software and IT Park

The Tehran Software and IT Park (TSITP) started out as an idea of the Tehran Municipality to capitalize on its 435 m tall telecommunications tower (called Milad Tower) and the surrounding 8 hectare (expandable to 15 hectares) park and forest. The MPTT, on the other hand, envisioned TSITP as a prestigious IT park with an attractive and flexible franchise to support the development of innovative and technology-driven enterprises. The Minister of PTT and the Mayor of Tehran signed a partnership agreement in January 2002, a five-member steering committee was appointed, and the study phase commenced on October 2002, with a budget and financial resources allocation by ITRC. The study recommended skipping over the Milad premises for a larger and more appropriate location.

TSITP was intended to upgrade Iran's access to technology, facilitate Iran's integration into the global economy, create a world-class environment that would attract innovative elites, pave the way for blending with the global wave of information technology, and prepare the ground for interaction between the local and world markets. As part of the consensus-building around TSITP's mission, a lot of effort was devoted to reaching a collective understanding of the country's current status and future directions for development. A pilot project, in the form of an elaborate IT tower in Tehran, was conceived to answer the short-term goals of TSITP and to provide the experience needed for the main project. In August 2004, concluding an international bid on the Internet, the main contract for the basic design and international promotion was granted to an experienced consultant from Ireland.

However, changes in the management of ITRC and the TSITP steering committee, changes in MCIT policies in mid-2003, and the change in government in July 2005 had an adverse impact on the project. In spite of the professional planning and strong inception of the project and its well-defined and documented outputs, the TSITP project has been practically shut down since mid-2004. The latest assessment indicates a grim future for the TSITP project.

The e-Commerce Law (also known as the digital signature law) was passed in 2004. However, due to lack of pertinent by-laws and decrees, the law has not yet become effective enough to govern e-commerce transactions in the country.

The Computer Crimes Act passed the preliminary approval process in the Parliament in September 2006, and its details are now being thrashed out by a special parliamentary committee. This act is in compliance with the Cyber Crimes Convention approved by the European Council in Budapest in 2001.

Education and capacity building

IT and computer engineering programmes are widely offered in Iranian universities and higher education institutes. Majority (68 per cent) of the students are enrolled in the bachelor's programmes, 28 per cent are enrolled in the post-diploma courses, and 4 per cent are enrolled in the Master's courses. The number of graduates in fields related to ICT is estimated to be more than 50,000 a year, while the annual intake of new students is more than 12,000.

Almost all of the universities and other higher education centres are connected to the Internet. Indeed, the universities were among the first in Iran to be connected to the Internet via the IPM facilities in 1995.

Computer courses have been part of the secondary school curriculum for many years now. More than 150,000 students take computer courses in different high school grades annually. There are also plans to connect up to 1,000 high schools to the Internet by mid-2007.

Following TAKFA guidelines, obtaining an ICDL certificate is now part of the qualification requirements for new government recruits and for promotion in the civil service. Many training institutes throughout the country provide ICDL training.

At present, there are few online schools and universities and distance learning centres. However, their number is expected to grow considerably.

Research and development

There are many small and medium-sized research and development institutes dealing with ICT in Iran. Those with a substantial impact on ICT development include: the Iran Telecommunications Research Center (ITRC) affiliated with MCIT, the Electronics Support Fund for Research & Development (ESFRD) affiliated with the Ministry of Mines & Industries, the Hi-Tech Industries Center also affiliated with the Ministry of Mines & Industries, the Sharif Advanced Information and Communications Research Center and the Research Center of Informatics Industries.

ITRC was established in 1970 by the University of Tehran and the Government of Japan (NTT). In 1979, the Revolutionary Council decided that for the sake of self-sufficiency the Center would be affiliated with MPTT. ITRC is the largest ICT research institute in Iran. Its projects include design and development of digital radios, transmission networks, antennae, satellite communications, cellular networks, fibre optics and laser switching and, more recently, ICT strategic planning and telecommunications consultancy services. One of ITRC's greatest contributions to ICT capacity building are many non-government communication companies that started out as ITRC research groups. However, in recent years ITRC's work has been criticized as lacking in direction and there is an ongoing debate regarding its mission and framework.

The ESFRD is a fund dedicated to supporting R&D in ICT and other fields in electronics. It was established in 1997–98 with a total paid-in capital of USD 30 million. Its goals are to promote entrepreneurship, software development, engineering services, international collaboration and export. ESFRD helps non-government ICT projects by means of inexpensive loans, guarantees and underwriting, information services and soon through venture capital investment.

With respect to open source software development, the High Council of Informatics in conjunction with TAKFA and the Sharif Advanced Information and Communications Research Center supported the development of the Persian Linux platform launched in early 2006.

Challenges

TAKFA was probably the most highly publicized IT project in Iran in the past decade. It succeeded in making ICT an important agenda for the President and influenced major government bodies to spend on ICT. TAKFA spurred government authorities and a frail private sector to respond to the challenge of sustained growth in information technology with a comprehensive plan. However, experts have pointed out that lack of consistent strategies, cohesive plans and experienced consultants has doomed TAKFA's efforts.

Indeed, in July 2005 a new government was sworn in and IT is not in its new set of priorities. Some in the IT sector believe that the privatization and liberalization of the economy will not be sustained and a more government-centric development will emerge. Iran's telecommunications market has a very promising future, due mainly to the size of the population and the current penetration rate of mobile telephony and other telecommunications services. The growth of the national IP network backbone is also a contributing factor. However, recent upheavals and the partial reversal of privatization policies, as seen in the case of the second mobile operator, do not bode well for the continued growth of the telecommunications sector in Iran.

References

BBC Persian. Available at http://www.bbc.co.uk/persian/iran/
Blogfa. Available at http://www.blogfa.com
Communication Regulatory Authority (CRA)/ Communication Regulatory Organization (CRO) Homepage. Available at http://www.cra.ir/Main.asp
Computer Guild Organization Homepage. Available at http://www.irannsr.org/
Computer Society of Iran (CSI) Homepage. Available at http://www.csi.org.ir/index.asp
DPI Law System. Available at http://www.iranlaw.ir
Electronics Support Fund for Research & Development (ESFRD) Homepage. Available at http://www.esfrd.ir/
High Council of Cyberspace Information Exchange Security (AFTA) Homepage. Available at http://www.afta.ir/
High Council of Informatics (HCI) Homepage. Available at http://www.shci.ir/
Hi-Tech Industries Center Homepage. Available at http://www.hitech.ir/
Informatics Society of Iran (ISI). Available at http://www.isi.org.ir/index.asp
Information Technology Company of Iran (ITC) Homepage. Available at http://www.itc.ir/english/index.asp
Institute for Studies in Theoretical Physics & Mathematics (IPM) Homepage. Available at http://www.ipm.ac.ir/IPM/homepage/homepage.html
Iran Informatic Companies Association (IRICA) Homepage. Available at http://www.irica.com/portal.aspx
Iran Power Generation, Transmission & Distribution Company (Tavanir) Homepage. Available at http://www.tavanir.org.ir
Iran Railway Co. Homepage. Available at http://www.rai.ir/site.aspx
Iran Telecommunication Research Company (ITRC) Homepage. Available at http://www.itrc.ac.ir/engl11.php
Iranian Information & Documentation Center (IRANDOC) Homepage. Available at http://www.irandoc.ac.ir/english/default.htm
Islamic Development Organization Homepage. Available at http://www.ido.ir/
ISNA News Agency. (2006). (news article in Persian). Retrieved from http://isna.ir/Main/NewsView.aspx?ID=News-791595
ITIRAN Online Magazine. (news article in Persian). Retrieved from http://www.itiran.com/?type=news&id=6257
ITIRAN Online Magazine. (news article in Persian). Retrieved from http://www.itiran.net/archives/002118.php
ITNA Online Magazine. Available at http://www.itna.ir/archives/news/001038.php
Mihan Blog. Available at http://itmanagement.mihanblog.com/Post-263.ASPX
Ministry of Communication and Information Technology (MCIT) Homepage. Available at http://www.ict.gov.ir/
Ministry of Science and Technology (MSRT) Homepage. Available at http://www.msrt.gov.ir/sql/asp/park.asp
National Cartographic Center (NCC) Homepage. Available at http://www.ncc.org.ir
National Technomarket of Iran Homepage. Available at http://www.techmart.ir/en/
Online Public Relation Blog. Retrieved from http://weblog.eprsoft.com/archives/005276.html
Pardis Technology Park (PTP) Homepage. Available at http://www.techpark.ir/
Persian Blog. Available at http://www.persianblog.com/
Research Center of Informatics Industries (RCII) Homepage. Available at http://rcii-ir.org/WebGenerator/PageViewEn.aspx?src=1
Sanaray Co. Homepage. Available at http://www.sanaray.com/english/Site.aspx
Sanjesh Organization Homepage. Available at http://www.sanjesh.org/
Sharif Advanced Information and Communications Research Center Homepage. Available at http://www.aictc.com/
Sharif University of Technology Homepage. Available at http://www.sharif.ir/en/
Statistical Center of Iran (SCI) Homepage. Available at http://www.sci.org.ir/portal/faces/public/sci_en
Syndicate of Telecommunications Industries Homepage. Available at http://www.irtelesyndicate.com/
Tebyan Homepage. Available at http://www.tebyan.com/
Telecommunication Company of Iran (TCI) Homepage. Available at http://www.irantelecom.ir/eng.asp
Telecommunication Infrastructure Company of Iran (TIC) Homepage. Available at http://www.tic.ir
Union of Iranian Software Exporters Homepage. Available at http://www.uiseonline.org/
Ziyaee-Parvar, H. (July 2006). (news article in Persian). Retrieved from http://www.reporter.ir/archives/85/3/003652.php

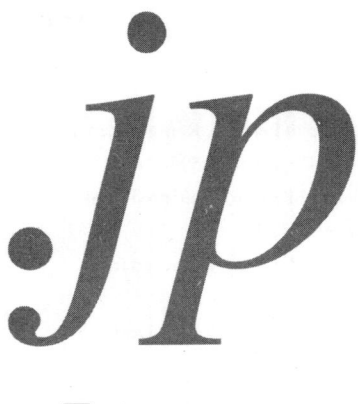

Japan

Keisuke **Kamimura** and Adam **Peake**

Population	128.08 million (ITU Database 2005)
GDP per capita	USD 35,756.532 (IMF Country statistics 2005)
Key economic sectors	Manufacturing, Services, Real Estate (MIC 2003)
Computers per 100 inhabitants	54.15 (2005, ITU Database)
Fixed-line telephones per 100 inhabitants	45.89 (ITU Database 2005)
Mobile phone subscribers per 100 inhabitants	73.97 (ITU Database 2005)
Internet users per 100 inhabitants	50.20 (ITU Database 2005)
Domain names registered under .jp	801,997 (JPRS 2006, March)
Broadband penetration rate per 100 inhabitants	19.0 (OECD, 2006 June)
Internet domestic bandwidth	139.2 Gbps (Monthly average, May 2006, MIC)
Internet international bandwidth	68.5 Gbps (Monthly average, May 2006, MIC)

Notes: MIC is the Ministry of Internal Affairs and Communications. JPRS is the Japan Registry Services Co., Ltd.

Introduction

The chapter on Japan in the 2005/2006 edition of *Digital Review of Asia Pacific* began with the slightly optimistic observation that the country might at last be emerging from more than a decade of economic stagnation. Since that chapter was written, the Japanese economy has continued on a long and slow recovery. The Cabinet Office Monthly Report for September 2006 described exports, imports, housing construction and public investment as 'flat'. However, corporate profits have been showing a year-on-year increase for 16 consecutive quarters, with the information, communication and electronics industries generally strong over this period. Real growth in 2005 was 2.7 per cent. Real GDP grew at an average of roughly 1 per cent yearly in the 1990s, compared to growth during the 1980s boom of about 4 per cent per year. The Japanese economy, the second largest in the world, is still not the driving force it once was, but we can hope that it is on the road to a sustained period of growth and prosperity (Japan Cabinet Office 2006).

All of the economic trends in the information and communications technology (ICT) industry for the first quarter of 2006 are positive. Shipments are generally up, prices are falling slightly although average household expenditure is increasing, and within the ICT industry, employment is increasing and salaries rising. The presence of the ICT industry has continued to increase and its impact on the Japanese economy as a whole is becoming greater. The last cycle of ICT growth that peaked during the period 2004–05 was based on increased ICT-related investment by corporations, as well as increasing and maturing use of the Internet and mobile phones by ordinary users. The growth cycle underway in the third quarter of 2006 is being driven particularly by demand for digital home appliances, although Internet and mobile communications remain strong in both business and home sectors, as does general growth in ICT production.

The ICT sector has been important to the recovery that has taken hold of the economy over the last few years. Information and communication devices and services are becoming pervasive in society. Mobile phones are near ubiquitous, and wired broadband is available in more than 50 per cent of households. Broadband, particularly technologies like Wi-Fi which is used in almost half of all broadband households, is changing the way people get their news and entertainment, as well as how they communicate and share information and ideas. With the pervasiveness of ICTs in Japanese society, we see a new phase in government policy shepherding the creation of a ubiquitous network society.

Technology infrastructure

Telecommunications and Internet

Broadband services in Japan are well known to be among the cheapest and fastest in the world. Open access policies have given rise to a very competitive broadband industry providing cheap, high-speed and innovative services that are available in most regions of the country. However, even after this success, the Ministry of Internal Affairs and Communications (MIC) is concerned that the market could be turning to favor the incumbent telecommunications operator, NTT.

Liberalization of the Japanese telecommunications market began in 1984, and since that time, government policy has focused on introducing competition and reducing NTT's market power. Styles of regulation have changed over the years, but NTT still remains the dominant provider in most telecommunications sector markets. After 22 years of competition, the two NTT local companies, NTT East and NTT West, still hold over 91 per cent of contracts for subscriber telephones. The NTT companies handle approximately 75 per cent of all domestic calls. In the mobile phone market NTT DoCoMo is the leading provider with over 53 per cent of handset contracts. Only the broadband market is different—it is highly competitive.

The Japanese broadband market has been built around an open access regime where competitive providers have been able to use essential elements of NTT's network, particularly low-cost access to copper lines to homes and metropolitan fibre connections running between NTT exchanges and to other locations. These elements have been the basic building blocks of the networks of competitive ADSL providers. In less than four years, the number of broadband users has grown from 0.5 million in June 2002 to over 23 million at the end of March 2006. Over 14.5 million subscribe to ADSL service, and almost 5.5 million to fibre-to-the-home (FTTH). There are more FTTH subscribers in Japan than in all other countries combined. Cable TV companies provide broadband Internet to over 3.3 million users. Half of Japan's 46 million households subscribe to broadband service (Japan Statistics Bureau 2006).

The number of new FTTH subscribers (820,000 in all) during the 1st quarter of 2006 was the highest ever. As the overall increase in new broadband subscriptions was 930,000 for the quarter, FTTH is driving continued growth. The monthly increase of FTTH subscriptions has exceeded that of ADSL subscribers in each quarter since the 1st quarter of 2005. There were less than 37,000 new ADSL subscribers during the first three months of 2006, and in Tokyo and Osaka, the largest metropolitan areas, the number of ADSL subscriptions dropped as people switched to fibre.

The trend towards more FTTH subscribers brings with it some policy concerns. In the ADSL market the NTT group of companies provides 40 per cent of lines; in the FTTH market, they control 64 per cent of lines and their share is increasing as more people take up fibre. MIC wants the migration from ADSL to FTTH services to happen, but is concerned that while the majority of users migrating service to NTT FTTH are NTT ADSL subscribers, users of other ADSL providers are also migrating to NTT at faster rates than to the fibre services provided by other operators. If this migration pattern continues, the combined NTT companies may achieve the same level of market power in FTTH as they have in the Public Switched Telephone Network (PSTN). Thus, there is a risk of losing the powerful competition that has emerged in the ADSL market.

With the convergence of fixed and mobile, and convergence of communications and broadcasting underway, concern about the NTT group's ongoing market dominance will be the subject of ongoing MIC competition reviews.

Broadband and broadcasting

All large broadband providers have begun video broadcasting services. Yahoo!BB, NTT and KDDI offer video download services for movies and television programmes, as well as live streaming television. For competitive reasons, and to ensure transmission quality, many of these services are available only to subscribers and downloads are kept within the company's network and control. Usen, the third largest FTTH provider, offers video services to its subscribers in this way, but also offers a free video broadcast service called Gyao, which is available to any Japanese Internet user.

People need to register to watch Gyao programming. The business model also dictates that the subscriber must be in Japan as Gyao finances its service through targeted advertising. Gyao programming includes movies, TV programmes, music and sports events, with live baseball proving popular. In August 2006, Gyao had 11 million registered users and total viewing hours exceeded 18 million. Advertising revenue has grown more slowly than expected, and Usen is struggling with JPY 160 billion (about USD 1.3 billion) in interest-bearing liabilities. However, Usen expects to see a profit on a monthly basis before the end of 2006.

Fixed-line phones, broadband and voice over IP (VoIP)

Fixed-line phone service covers 100 per cent of the Japanese population and both the number of customers and revenues are decreasing as people switch to newer and cheaper IP telephony services, or simply to mobile service.

There are three classes of IP-based telephony service currently in use in Japan: an IP exchange service known as 0AB/J IP telephony which has a guaranteed quality of service, 050 IP telephony with no guaranteed quality of service, and pure IP telephony such as Skype and similar services.

050 IP telephony is a service provided by an ISP in which the customer receives service via a broadband connection and VoIP adapter, and a telephone number with the prefix 050. Voice traffic in 050 IP telephony routes over a broadband connection and the ISP's network, and sometimes over the public Internet, and the quality of calls may vary depending on other traffic. 050 service might not include access to emergency calls and some other PSTN features.

0AB/J IP telephony is the potential replacement for traditional fixed-line phone. 0AB/J stands for the standard geographic prefix (for example, 03 for Tokyo), which will only be assigned to an IP telephony service that achieves the same quality of service and features as an ordinary fixed-line telephone. To get 0AB/J IP telephony, customers need broadband access and a VoIP adapter. Unlike 050 IP telephony, 0AB/J IP is provided by a network provider that underlies the user's ISP access network and the voice traffic of 0AB/J is routed through a managed network that provides the same quality of service as the standard PSTN telephone.

While the number of customers of standard fixed-line telephone is decreasing, the numbers of 050 and 0AB/J customers are increasing. In December 2003, there were 4.3 million IP telephony customers. The number more than doubled over two years, reaching 10 million in December 2005. As of March 2006, the standard PSTN telephone service had 50.6 million subscribers, 050 IP telephony had 10 million and 0AB/J IP telephony 1.4 million subscribers. Almost half of 050 subscribers are Yahoo!BB customers, although Yahoo!BB's share has declined over the past four quarters from over 55 per cent, while NTT's share has increased over the same period from 19 per cent to over 25 per cent. 0AB/J IP numbers tend to be used by FTTH customers, and during the first quarter of 2006, NTT East and West's share of 0AB/J IP numbers increased by 6.4 per cent to 69 per cent.

Mobile phones

Although the mobile phone subscription base is still growing, its growth has slowed significantly as the penetration rate has reached over 75 per cent of the population. As of September 2006, the total number of mobile phone subscribers was 93,812,400 and the number of personal handy phone system (PHS) subscribers was 4,879,500. The number of mobile phone subscribers increased only by 4.78 million or 5 per cent over the past 12 months (Telecommunications Carriers Association 2007).

The monthly average revenue per user (ARPU) for Japanese mobile subscribers is still probably the highest in the world, although it is decreasing steadily. In the 3rd Quarter of 2005, DoCoMo's 3G subscribers spent JPY 9,050/month (about USD 74/month) and 2G subscribers JPY 6,140 (about USD 50). The combined 2G/3G ARPU is JPY 7,050 (about USD 57.78). KDDI's combined 3G and 2G ARPU is JPY 7,190 (about USD 58.92), although its 3G subscribers spent an average of JPY 9,990 (about USD 81.87). These figures are going downwards due to the steady decline in prices for voice calls and increased competition as rapid growth in the number of users has slowed. Profit margins are much slimmer than in the past, although Japan still has some of the world's highest mobile call rates.

More than 80 million (that is, 81,346,000) mobile phones are able to use IP packet services, over 80 million have cameras, and approximately 57.2 million are 3G. The average turnover in phone handsets is about once per 12–18 months, and 16.2 million new phones were bought between April and July 2006. This high turnover of handsets is a good indicator that new services requiring the functionality of new phones have the opportunity to quickly achieve acceptable market penetration.

Willcom's PHS network coverage is almost nationwide, focusing on providing data access at speeds ranging from 64 Kbps to 402 Kbps rather than voice. Willcom's data services are significantly cheaper than those provided by 3G operators. Its network has better coverage and transmission speeds are more reliable. Willcom's service is popular with people who need reliable Internet access anywhere and at anytime.

Softbank is best known in Japan for revolutionizing the broadband ADSL market in 2002 through a price-slashing business strategy and introduction of new high-speed services. In 2005, Softbank received a license and spectrum to operate 3G services as part of a policy initiative to introduce new operators to the mobile market. Softbank had been widely expected to begin its new mobile service in mid-2007 with the same low-price and innovative service approach it used in the broadband sector four years earlier. The license was for W-CDMA, but Softbank had been testing handover between this 3G service to Wi-Fi and to Mobile WiMax. However, following Softbank's takeover of Vodafone for USD 15.5 billion, the new spectrum was returned to MIC and these technical innovations have been delayed. The high cost of acquiring Vodafone and the need to run an established network and provide service to existing customers has changed Softbank's business plan. Thus, the company's entry to the mobile market has not yet brought the much hoped for severe price competition.

Number portability was introduced on 24 October 2006 and the three major providers spent the earlier part of 2006 gearing up for the challenge. The total cost of switching providers is around JPY 5,000 (about USD 40.98). However, the three companies have introduced various discounts and offers to encourage users to switch. All have revamped their handset line-up: NTT DoCoMo introduced 14 new handsets in early October; KDDI announced 12 new models in August; and Softbank came out with 13 models in September.

Information provided by MIC flagged one problem facing number portability: 72 per cent of mobile phone users have never changed their phone e-mail addresses. This could be a barrier to change because e-mail addresses are not portable between providers. Downloaded content, such as ringtones, also cannot be transferred over to a new provider. On the other hand, reports in the Japanese media indicate that 10 million users have expressed interest in changing providers.

The potential for higher customer turnover should increase competition and trigger changes in market share. Softbank in particular is looking to number portability to boost its share of the market and has launched a series of new services linking its mobile customers to Yahoo! brand products and services. Through this linkage, Softbank is able to offer news, stock price information, some games and other content free of charge, unlike DoCoMo and KDDI where such services are subscription-based. However, an aggressive marketing campaign by Softbank in the lead-up to the launch of number portability may have violated laws prohibiting misleading advertising. KDDI filed a complaint about the advertisements with the Fair Trade Commission. Softbank's situation worsened when a systems error struck shortly after number portability began. System errors, customer complaints and the bad press that followed the misleading advertising campaign made number portability a setback rather than an opportunity for Softbank.

Until new services and price plans firm up, industry analysts expect KDDI's AU service, known for a range of music services not available from DoCoMo or Softbank, to gain the most customers in the early days of number portability.

Next generation networks

In 2004, the two leading telecommunications operators, NTT and KDDI, announced their intention to migrate from traditional PTSN phone networks to next generation network (NGN) technology. The migration will take place gradually. NTT began an NGN field trial in 2006 to demonstrate and verify the new network technology and its interoperability with other network operators and service providers. The company expects 30 million customers to subscribe to NGN services by 2012.

However, there are a number of potential policy concerns regarding NGN, particularly open access to NTT's infrastructure and interconnection with other operators. Operators other than NTT are not as positive about building NGN infrastructure and are concerned about whether the new infrastructure will be open to them and on what terms.

Key government institutions dealing with ICTs

As ICTs become common in all aspects of Japanese society, all branches of the government are involved to some degree in ICT policy formulation or implementation. Some of the key institutions are described below.

The IT Strategic Headquarters, established under the Cabinet and chaired by the Prime Minister aims to provide the necessary coordination among government agencies concerned with IT-related policy.

Mobile broadcasting

Phones capable of receiving high-quality digital TV broadcasts are the latest trend in mobile technologies. Japanese terrestrial digital broadcasts divide one channel into 13 segments–12 segments for high definition television (HDTV) and one segment for use by mobile devices. The mobile broadcast feature is known as 'one-seg' (one segment). The service provides high-quality digital images on small screens. Service officially began on 1 April 2006. Six hundred thousand one-seg phones were sold in the first three months of service. They receive ordinary digital TV broadcasts and are proving popular not only for niche viewing, such as horseracing, but also for general news, weather and some sports broadcasts. Marketing for Vodafone's first one-seg handset was targeted at the start of the 2006 World Cup.

Broadcasters cannot produce programmes especially for phones until after broadcasting deregulation in 2008, and the networks are using the period until then to develop new business and advertising income models that are different from those for terrestrial broadcasting. New regulations may also change the copyright structure: at present actors and producers receive no royalties for this new type of distribution. Organizers of major events may also try to sell broadcasting rights for one-seg.

To build an industry capable of demanding that new types of content be produced for it, and to compete with terrestrial broadcasters for rights to high-interest events, industry analysts say 20–30 million one-seg handsets will need to be in use by the time deregulation occurs in 2008. However, NTT DoCoCom's first one-seg phone, the P901iTV, was recalled after a few months of operation for a software upgrade to prevent people receiving TV broadcasts after they had quit DoCoMo service. The handset, which is rumoured to have cost DoCoMo as much as JPY 60,000 each but was heavily discounted to retailers for as little as JPY 20,000 to encourage initial sales, could continue to receive TV broadcasts after customers cancelled their subscription to DoCoMo. Apparently 20,000 people bought the phone and then immediately cancelled their DoCoMo service, taking advantage of the software glitch to use the phone as a low-cost mobile digital television.

One-seg is also used in car navigation systems and some PCs have one-seg tuners built in. Tuners are also available on USB devices. Japanese mobile music and video players that can display one-seg broadcasts are being sold, and the one-seg feature is something not even Apple's iPod can match.

The Ministry of Internal Affairs and Communications (MIC) has the overarching role both in policymaking and regulation of the telecommunications industry.

The Ministry of Economy, Trade and Industry (METI) is responsible for the information economy, information services industry, information security and ICT development for small and medium-sized enterprises, in keeping with its mission of industry development.

The National Information Security Center (NISC), which was established in April 2005 under the Prime Minister's Cabinet, formulates and implements information security policy within the government in a coordinated and uniform manner. NISC has responsibilities in planning and policy, international coordination, intra-government information security, critical infrastructure protection and incident response.

Enabling policies and programmes

The Japanese government's current ICT policy can be traced to the e-Japan Strategy published by the IT Strategic Headquarters in 2001. Over the last five years, the Japanese government has gradually changed focus from the development of ubiquitous broadband infrastructure to the exploitation of the infrastructure to achieve social and economic benefits.

It is widely understood that the government has a firm commitment to achieving nationwide broadband infrastructure. While this is true, the role of the government has been limited to strategic oversight, and funding and actual deployment of infrastructure is the responsibility of the private sector. In the e-Japan Priority Policy Program 2006, the role of government is summarized as pointing to the future direction that Japan should take, to push deregulation and market competition forward, to motivate the private sector, to provide a minimum level of investment as well as safeguards against the digital divide, to ensure safety and security, and to optimize the operation of the government and public sector. With a few exceptions in research and development and infrastructure development in rural and remote areas, the government has not directly funded broadband take-up.

The e-Japan Strategy adopted in January 2001 sought to transform Japan into the 'most advanced IT nation within five years'. The Strategy was access- and infrastructure-oriented, although it did mention priority areas other than infrastructure, such as electronic commerce, human resource development and electronic government. Target figures (always-on Internet access to all citizens, high-speed Internet access for at least 30 million households and ultra-high-speed Internet access for 10 million households) were set for broadband penetration. The rapid take-up of ADSL service was attributed to open access policies regarding NTT's copper infrastructure, which resulted in the budget ADSL service of Yahoo!BB.

As it became clear that the objectives of the 2001 e-Japan Strategy were being met, e-Japan Strategy II was adopted in July 2003 to create an 'energetic, worry-free, exciting and convenient' society using information technology. Through e-Japan Strategy II, the government put more emphasis on the exploitation of the broadband infrastructure, focusing on seven areas: health, food, lifestyle, SME financing, knowledge, employment and public administration. In addition, the new Strategy aimed to further develop network infrastructure that could meet the demands of the coming information society. Among the key concepts introduced were ubiquitous network, security and reliability, and research and development in information technology.

Thus, another notable development during this period was the publication by MIC in December 2004 of the u-Japan policy. Unlike e-Japan, u-Japan is a ministerial policy, which means it has less impact beyond the ministry itself and it may be interpreted as a set of action plans that should be put into a Priority Policy Program. MIC made 'ubiquitous networks' the focus of the Japanese government's contribution to the Tunis Phase of the UN World Summit on the Information Society (WSIS).

e-Japan Strategy II also revamped relevant ICT laws, regulations and policy guidelines to ensure that unnecessary or outdated legal and administrative barriers did not hinder the growth of the ICT sector and use of ICTs. The reforms undertaken at this time were extensive, and new legislation has been unnecessary in the past 2–3 years.

Every year since 2003, the IT Policy Headquarters has updated the Priority Policy Program, which outlines the action plans of each government agency to achieve the objectives set forth in the e-Japan Strategy. In January 2006, the IT Strategic Headquarters made another policy move: it published the New IT Reform Strategy, the national strategy for 2006–10. The highlight of the Strategy is structural reform of the government and of the Japanese social and economic systems, while addressing user and citizen concerns such as accessibility, the digital divide, and security and safety, and emphasizing international contributions and competitiveness.

In line with the New IT Reform Strategy, MIC published a new set of policy priorities in August 2006, called the Next Generation Broadband Strategy 2010. MIC aims to bridge the remaining digital divide between urban and rural areas. By 2010, 100 per cent of Japanese local communities (cities, towns and villages) will be covered by broadband access and 90 per cent of Japanese households will be covered by ultra-high-speed broadband access.

Unlike the previous strategies which focused more or less on the technology, the New IT Reform Strategy aims more to achieve social reform based on the benefits of ICT. The Strategy seeks solutions to the problems Japan is facing now, such as the aging of society, revitalization of the economy and improvement of national competitiveness.

Intellectual property rights

In 2003, the Prime Minister convened a Strategic Council on Intellectual Property to discuss policy to strategically protect and utilize the results of intellectual activity, such as scholarly research and creative activity, thus enhancing the international competitiveness of Japanese industries. The Council published a national plan on intellectual property, called the Strategic Program for the Creation, Protection and Exploitation of Intellectual Property, which resulted in the enactment of the Basic Law on Intellectual Property of 2003.

The Strategic Council on Intellectual Property was later replaced by the Intellectual Property Policy Headquarters, which was established under the Cabinet. Every year the Policy Headquarters publishes an Intellectual Property Strategic Program that outlines action plans for each government agency to implement in the next fiscal year. In 2006, the following actions were identified to facilitate the development and distribution of digital content:

- Promotion of broadcast via IP multicast;
- Development of content protection systems with due consideration to users;
- Promotion of content delivery via the Internet;
- Content creation based on the reuse of existing content;
- Establishment of a content business market on the Internet;
- Network for information appliances; and
- Development of a rights management mechanism suitable for the distribution of broadband and digital content.

With regard to local language and culture, the Strategic Program also identifies the protection of the typeface design of computer fonts as an area that needs policy attention. Under the current interpretation of the copyright law, typeface design was considered to be out of the scope of intellectual property protection. The Strategic Program calls for the government to consider how to protect typeface design and take appropriate measures when necessary.

Digital content and life online

According to the Digital Content Association of Japan (DCAj), the Japanese media and content industry, which includes broadcast, sell video, film, music, video games, books, journals, and newspapers and other publication, had an estimated volume of JPY 13,681 billion in 2005 (about USD 112 million), including JPY 2,527 billion (about USD 20 million) for digital content.

Online music

Music sales have been shrinking since 2001. In 2005, the total sale of music content was JPY 614 billion (about USD 5 billion), which represents only 80 per cent of the sales in 2001. CDs and DVDs are selling less. On the other hand, online distribution of music via mobile phone and the Internet keeps growing. For example, mobile operator KDDI launched a full track download service in 2004, making over 110,000 songs available and selling 30 million downloads in December 2005. However, these and other new online sales outlets are not adequate to offset losses made by traditional brick-and-mortar music retailers.

The music industry argues that illegal music sharing by peer-to-peer applications, such as Winny, contributes to the loss of music sales. However, an empirical survey (Tanaka 2004) shows that there seem to be no positive correlations between the decrease in music sales and peer-to-peer file sharing, which nullifies the music industry's argument. During the first nine months of 2005, mobile music sales in Japan reached JPY 26 billion (about USD 213 million), accounting for 96 per cent of Japan's total digital sales.

Although online music sales were not unpopular in Japan, Apple's iTunes Music Store came to the Japanese market late. iTunes Music Store was first launched in the United States in April 2003, and rapidly expanded to some 20 countries worldwide. But the service did not reach Japan until August 2005, two years and four months later. The launch of the service was delayed due to negotiations with Japanese content holders. Interestingly, one of the biggest music labels in Japan, Sony Music Entertainment (SME), has not joined the iTunes service; it has its own channel for online music distribution called 'bitmusic'.

Social networking services (SNS)

It is not yet clear what people are using broadband for, but the recent popularity of social networking services is a significant broadband phenomenon. An OECD report (2006) quotes data from Technorati, an online blogging information company, indicating that 21 per cent of the blogs worldwide that Technorati tracks are based in Japan. According to a report by MIC (2005), the net number of Japanese blog users was estimated to be 1.65 million at the end of March 2005, of which 950,000 (57.6 per cent) are active users.

Social networking services (SNS) have grown rapidly since the MIC report was published. The biggest SNS in Japan, called 'mixi', had more than 5.7 million registered users when it made an initial public offering on the Tokyo Stock Exchange

'MOTHERS' in September 2006. The service had only 300,000 registered users a year-and-a-half ago. Mixi is an invitation-only service: to join you must be invited by someone who is already a member. Members provide online diaries, photo albums, shared bookmarks and customized news, as well as basic SNS functions such as personal profiles and testimonials, online communities and bulletin boards that help users connect to other users. Mixi earns about 80 per cent of its revenue from advertising carried on the site.

There are over 5,000 smaller SNS sites typically with 100–1,000 members. They are usually constructed around open source SNS software 'Open PNE', and are managed by individuals or small groups and some specialized small companies. These smaller SNS services can grow into very tight-knit communities. Furthermore, discussions among the millions of SNS users and the ideas they express are having a significant influence on Japanese society.

Online gaming

The popularity of online games has also been enabled by broadband. In mid-2006, there were an estimated 28 million online game players in Japan. Among heavy gamers, the MMORPG (Massively-Multiplayer Online Role-Playing Game) *Final Fantasy 11* is one of the most popular games with over half a million registered players and as many as 170,000 simultaneous users. Hangame, Japan's largest Internet game portal company, provides more than 150 kinds of online multiplayer games. The portal is modelled on its parent company NHN Corporation's successful operation in Korea. In September 2006, Hangame had 17 million registered players and could handle a maximum of 125,000 concurrent users, over 200 million page views a day, and e-mail from 200,000 registered users daily. Hangame describes itself as an online gaming community.

Portable games machines remain extremely popular and these are also increasingly network-enabled, either to connect between players' machines directly or to online game portals like Hangame. Nintendo DS has sold about 16 million units since it was launched at the end of 2004. A Wi-Fi connection function was introduced at the end of 2005 and the service has reached 1.3 million users and provided over 40 million game sessions since its launch.

Education and human resource development

Although as a whole, Japan has been able to produce a sufficient number of ICT professionals, there has been some mismatch between what universities have offered students in terms of ICT education and the ICT skills that the private sector expects from young university graduates. Japanese universities tend to educate students as generalists rather than as specialists and university graduates usually need to spend a considerable amount of time in re-training to acquire practical skills and knowledge after getting a job. *Nikkei Computer*, a major computer magazine for business readers, reported in December 2005 that only 37.3 per cent of some 2,300 ICT professionals surveyed by the magazine said they acquired their professional knowledge from university education and 26.5 per cent said they learned on their own. This finding points to a significant gap between the skills universities equip graduates with and the needs of industry. In June 2005, a report published by the Japan Federation of Economic Organization (Keidanren) (2005) pointed out that the gap between universities and business was even wider in the training of highly skilled ICT professionals, and called for joint efforts among the government, industry and universities to improve the situation.

Another challenge facing the ICT sector is that computing and technology subjects are becoming less popular with university entrants. In April 2006, Nikkei Business Online reported that the Department of Electronics and Computer Engineering, which used to be the most competitive department at the University of Tokyo, was ranked one of the least popular departments by prospective students. According to Mainichi Communication (2006), 6 per cent of students in 2002 wanted to find a job in the ICT sector, but in 2006, interest in ICT dropped to 4.1 per cent.

The Japanese government, industry and universities are working to improve the education and training of ICT professionals. METI has implemented a number of policy measures, including a certification programme for IT Coordinators; the 'Exploratory Software Project', which is an awards programme for university students and young engineers; and the Skill Standards for IT Professionals (ITSS), which aims to set standards for the required level of competence of ICT professionals and which is expected to affect course design in ICT education in universities. The Ministry of Education, Culture, Sports, Science and Technology (MEXT) has been working with Keidanren since April 2006 on a programme to train 'Advanced IT Specialists'. As of September 2006, six accredited universities have joined this programme (Mainichi Communication 2006).

Human resource development in the ICT field is currently in a weak position, but the government, industry and universities are well aware of the nature of the problem and have made a number of efforts together to improve the situation. It is hoped that these improvements will pay off in the longer term.

Research and development initiatives

R&D initiatives are being undertaken by the government, private sector and academia. Among the government agencies, MIC has the most direct and comprehensive R&D programme for ICT. It provides funding for qualified projects in the following thematic areas: universal communications technology, new generation network technology, and security and safety technology. The National Institute of Information and Communication Technology (NICT), under the auspices of MIC, operates the Japan Gigabit Network (JGN), a nationwide IP network test bed.

The private sector and universities are also active in R&D activities. Building research capacity among young researchers and students is becoming a concern. One notable initiative in this regard is the Exploratory Software Project led by the Information Technology Promotion Agency (IPA). The programme aims to nurture the next generation of talented programmers and entrepreneurs.

Conclusion

With the Japanese economy at last emerging from more than a decade of stagnation, it is interesting and almost ironic that broadband, one of the great successes of recent years, may be taking a step backwards. There are indications that consumers in Japan's dynamic and competitive broadband market are beginning to favour NTT, the old incumbent operator, over the new players. This appears to be occurring naturally through consumer choice, not abuse of market power or unfair competition. It is too soon to tell if this is just an early trend in the development of the FTTH and advanced broadband market. Other providers may soon begin to gain subscribers at NTT's expense. Or it might be a more permanent trend that will require some intervention from the Ministry of Internal Affairs and Communications.

Throughout the history of telecommunications, Japan has been able to learn from and adapt successful policy strategies from overseas to suit the local market's needs. Now Japan is a world leader in broadband and she has to make her own rules to meet new demands. How Japan meets these challenges should be explored in future editions of the *Digital Review of Asia Pacific*.

References

Impress R&D Ltd. (2006). *Internet white paper 2006*. Tokyo, Japan. [The authors of this chapter recommend this work as a primary sourcebook on the state of the Internet and Internet technologies, services and applications in Japan. The *Internet White Paper* is an annual publication in Japan.]

Japan Cabinet Office. (2006). *Monthly economic report (September 2006)*. Retrieved 10 October 2006 from http://www.cao.go.jp/index-e.html

Japan Federation of Economic Organization (Keidanren). (2005). Retrieved February 2007 from http://www.keidanren.or.jp/japanese/policy/2005/039/honbun.html (in Japanese).

Japan Statistics Bureau. (2006). *Contracts for information communication services*. Retrieved 20 September 2006 from http://www.stat.go.jp/data/getujidb/

Mainichi Communication. (23 March 2006). Retrieved February 2007 from http://navi.mycom.co.jp/saponet/release/ishiki/2006ishiki/2006ishiki.pdf (in Japanese).

Ministry of Internal Affairs and Communications (MIC). (May 2005). *The present condition analysis of [burogu] SNS and in the future estimate*. Retrieved 1 October 2006 from http://www.soumu.go.jp/s-news/2005/pdf/050517_3_1.pdf (in Japanese).

Nikkei Business Online. (17 April 2006). Retrieved February 2007 from http://business.nikkeibp.co.jp/article/manage/20060417/101387/ (in Japanese).

OECD. (2006). *OECD information technology outlook 2006*. Retrieved 1 October 2006 from http://www.oecd.org/document/9/0,2340,en_2649_34223_37529673_1_1_1_1,00.html

Tanaka, T. (2004). *Does file sharing reduce music CD sales?: A case of Japan*. Hitotsubashi University Institute of Innovation Research (IIR) Working Paper, WP#05-08. Retrieved 12 October 2006 from http://www.iir.hit-u.ac.jp/file/WP05-08tanaka.pdf

Telecommunications Carriers Association (TCA) Homepage. Available at http://www.tca.or.jp/index-e.html (in Japanese).

Lao PDR

Phonpasit **Phissamay**

Total population	5.62 million[a] (73 per cent rural population)
GDP per capita	USD 490[a]
Telephone lines per 100 inhabitants	1.62[b]
Cell phone subscribers per 100 inhabitants	12.8[b]
Computer ownership per 100 inhabitants	0.88[c]
Internet users per 100 inhabitants	0.35[d]
Websites in Lao languages	30[d]
Websites in English	250[d]
National transmission backbone	2.5 Gbps via fibre, 34 Mbps via microwave and 512 Kbps via satellite[b]
International bandwidth	310 Mbps (256 Mbps via fibre, 34 Mbps via microwave, 20 Mbps via satellite)[b]

Sources: a. National Statistics Center
b. Ministry of Communication, Transport, Post and Construction
c. Science Technology and Environment Agency
d. Lao National Internet Committee

Overview

As of March 2005, Lao PDR, a landlocked country in Indochina with a land area of 236,800 sq km, had a total population of 5.62 million. Majority (73 per cent) live in rural areas, down from 83 per cent in 1995 as a result of significant rural–urban migration. The population density in 2005 was 24 persons. Of the 952,386 households in the country, almost 50 per cent have access to electricity through the national grid, another 10 per cent have electricity from generators or car batteries, and about 40 per cent have no access to any electricity supply.

More than a fifth of the population (23 per cent) has never been to school, 28 per cent are in school and 47 per cent have left school. A much higher percentage of women than men have never been to school (30 per cent of women compared to 16 per cent of men). About 16 per cent have completed primary school, 6 per cent have completed lower secondary and 5 per cent have completed upper secondary. Few are able to complete the highest level of education even with increasing basic education enrolment rates. According to the 2005 census, 73 people out of 100 are literate.

From 2000 to 2005, GDP growth averaged 6.2 per cent annually. The growth of the economy in 2005 was 1.3 times more than that of 1995, with an average annual per capita of USD 490. All economic sectors have grown both qualitatively and quantitatively. Agricultural production reached 3.4 per cent, industry reached 11.3 per cent and services reached 6.7 per cent growth in 2005.

The ICT sector in Laos is lagging behind the ICT sectors of its neighbouring countries. Even within the country the ICT sector is making slow progress compared to other sectors. This is because ICT has not yet been identified as a priority programme for national development and there is a lack of trained human resources. Consequently, Laos relies on foreign cooperation for both specialists and consultants, as well as for financing, soft loans and financial grants.

The government recognizes that ICT is essential to the country's overall social and economic growth and development, and that it is necessary to increase ICT service penetration to the levels that are available among Lao PDR's regional neighbours and major trading partners. Increased availability of modern telecommunications services and ICT services will stimulate domestic economic growth and enable greater participation in ASEAN and in the global information economy. For this reason, and in spite of its technical, financial and human resource problems, the government is working hard to narrow the digital divide between Lao PDR and other Asian countries.

Technology infrastructure

Although telecommunications has second priority status to roads, water and electrical power, Laos has an aggressive programme to establish a national information infrastructure based on optical fibre. At least 15 provinces are already connected and operable. So far, the passive fibre optic connection has been installed, connecting all 17 provinces. The national fibre backbone is already being used for domestic telecommunication, both fixed and mobile, as well as for international Internet access. The domestic part of the Internet, however, is still not developed. In order to better exploit the existing national fibre backbone, the transmission capacity of the links in the domestic part of the Internet should be increased to provide for more users, new domestic applications and new services (the so-called triple play services of data, audio and video services require a capacity that is several orders of magnitude larger than the existing). All of these will lead to a dramatic increase in domestic traffic volumes. And thanks to the existence of the local Internet exchange (called Laonix) operated by the Science Technology and Environment Agency (STEA), domestic traffic can be kept local and will not depend on international links.

The current telecommunication network infrastructure being provided by ETL (Enterprise Telecommunication of Laos), LTC (Lao Telecommunication Company Ltd), LAT (Lao Asia Telecommunication Company Ltd) and Sky Telecom enables all types of telecommunication services such as international gateway, fixed-line, mobile telephone and Internet services, including VoIP. MILLICOM provides mobile telephone service.

By 2007, under the e-government action plan, STEA is mandated to establish the national e-government infrastructure of the following networks:

- Multi-service Transport Backbone—The 2.5 Gbps greater terrestrial transport backbone to be built in Vientiane will interconnect 50 ministry offices within a 20 km radius to the National Data Center (National e-Government Service Center) in a ring and star-topology. This will be a true multi-service transport backbone capable of carrying a mix of TDM, IP, Ethernet, ATM and FR traffic. Each ministry office will be equipped with a two-fibre core (in one of four optic fibre cables) carrying the bandwidth between offices and the National Data Center. To accommodate future growth, a minimum 18-core fibre optic cable is recommended.
- Metropolitan Area Network (MAN) by Wireless Broadband Access—Ministry offices that are not located on the Multi-service Transport Backbone or have distances of more than 20 km will use various forms of wireless access technologies, such as bridge Wi-Fi, 802.16 rev d/e WiMAX, or Point-to-Point Microwave. This wireless broadband access sub-project will provide 50 ministry offices with a high-speed link to the nearest Multi-service Transport Backbone to enable these remote offices to gain access to the National Data Center.
- Communication Infrastructure for the Rural Communities— The existing network infrastructure of LAT and, if necessary, ETL and LTC (such as LAT optic fibre transmission network) will be used to deliver public services to provide last mile connectivity to 17 provincial offices, 141 district offices and 1,200 village administration offices. Depending on the current state of LAT or ETL or LTC infrastructure and the geographic areas and terrain condition, the plan is to bring in a mix of wireline and wireless technologies to build a cost-effective solution for rural communities. This way the current digital divide can be bridged and e-government programmes and services can be brought closer to the rural communities.

ICT industries and services

Currently there are five telecommunication operators with a PSTN line capacity of 145,729 lines of which 91,289 lines are in use. A fibre optic STM-16 with a capacity of 2.5G connects all 17 provinces. There are 338 public telephones and 793 rural telephones covering 118 of 141 districts.

Twelve companies, including five telecom operators, have been granted Internet Service Provider (ISP) licenses. However, only six companies are able to provide the service, as the rest have problems setting up their own Internet infrastructure and marketing plan. There are about 300 Internet cafés in the country; about half of these are in Vientiane and about a sixth are in the Luang Prabang province. Around 20,000 Laotian use the Internet regularly; most of them are students. The main purpose of Internet use appears to be chatting as well as other forms of online entertainment. Internet use for education and research is limited as students connect to the Internet via Internet cafés and Internet use in universities and colleges is reserved for the administration only. Government and project staff who access the Internet in their offices comprise the next biggest group of Internet users.

There is generally no problem finding computer equipment and software in Vientiane, where there are around 67 computer companies (resellers) of which half are joint ventures with foreigners or local distributors of international brand names. The same cannot be said of the provinces however. In fact, three provinces in the north and two provinces in the south do not have any computer service companies.

A relatively large number of small businesses import computer parts and assemble computer systems according to customer specifications. There is a 5 per cent import tax on office supplies, computers, photocopiers and peripherals, which in many countries would be marginal but which is high in Lao PDR, since the purchasing power is so constrained. If it is intended for resale or retail, the import tariff is 10 per cent for office equipment. For telecom equipment, the import tariffs are 3 per cent for office use and 10 per cent if intended for resale/retail. There is also a 3–5 per cent profit tax, bringing the total tax to 18–20 per cent. There is an ongoing debate on whether to lower the tariffs to stimulate the sector. Specialized items, larger servers, telecom equipment and the like are generally ordered internationally. There is an important second-hand market serving the large segment that finds new computers and equipment too costly. A new PC costs around USD 1,200 in Lao PDR.

Most ICT-related consulting services, including system administration, development, integration and programming, can be found in Vientiane. However, perhaps due to limited capacity or the low quality of services, large ICT projects tend to be awarded to foreign companies or companies with foreign consultants.

There is a limited understanding of intellectual property (IP) rights. Software piracy is common, due mostly to inability to pay for proprietary software. There is great interest in open source software. However, Linux platforms have not yet taken off.

There is hardly any investment—whether foreign or local—in IT production in Lao PDR. Yet the government has been successful in attracting some investment (both foreign and domestic) into the telecom sector as well as for IT human resources generation. The investment policy encourages private investment in telecommunication and in IT training. Though there is no explicit policy for attracting investment in ICT, the Decree of the Prime Minister (No. 46/PM) states that hardware consultancy, software consultancy and supply, data processing, maintenance and repair of office, accounting and computing machinery, and areas relating to human capital formation like general education, technical and vocational secondary education, and adult and other education, are open to foreign (as well as domestic) investment without any restrictions.

Investments in the country's ICT sector have so far been confined mostly to ICT services. There is a need to focus on promoting investment in the production of ICT goods. This is because in an economy like Lao PDR, investment in ICT production might act as a catalyst in sustaining the use of ICT. Moreover, the production of information technology goods offers an opportunity for the country to diversify its export basket. Given the landlocked nature of the country, engaging in the production of any transport-intensive goods may not be advisable. However, it might be possible to enter into the production of certain ICT goods with a low level of technology like passive components, electromechanical components and other intermediate goods for the ICT industry and less skill-intensive IT-enabled services.

Digital content

The National Data Center established under the Lao PDR-India cooperation on ICT was inaugurated on 28 May 2006. This state-of-the-art centre is designed to be the nerve centre of all e-governance activities in Lao PDR. It has a storage area network (SAN) with a capacity of 1 Terabyte backup from a Tape Library, multiple Web servers, application servers and database servers.

The national portal (www.laopdr.gov.la) was launched at the same time as the National Data Center. It is intended to be the gateway to a variety of information and services provided by different government departments. It shall also provide base infrastructure for various government entities to launch their own websites and other e-governance initiatives on the Internet. Considered the digital face of the Government of Lao PDR, the portal is based on an Open eNRICH v4.0 Community Software Solution framework developed by the National Informatics Centre, the Government of India in collaboration with UNESCO, and OneWorld International Foundation.

In addition, the Government of Lao PDR has identified as core activities of the national and provincial government back-office computerization with a citizen interface and the development of Web-based content in the Lao language in the areas of government reporting, culture and heritage. To be developed by STEA in the next two years are systems for e-reporting, e-documentation, e-procurement, e-registration, e-project management, e-maps, e-taxation, decision support, SMS and teleconference.

Other ongoing initiatives with donor assistance are:

- The e-tax system of the Department of Tax with assistance from the Swedish International Development Agency (SIDA).
- A database of government employees being developed by the Public Administration and Civil Servant Agency with assistance from the United Nations Development Programme (UNDP).
- A Social Insurance System being developed by the Ministry of Labour and Social Welfare with assistance from the International Labour Office (ILO).
- A driver's license and transport database being developed by the Ministry of Communication, Transport, Post and Construction with assistance from the Australia International Development Centre.
- An education database being developed by the Ministry of Education with assistance from the French Development Agency.

However, there is still very limited online content in the Lao language. This is due to the lack of Lao language versions of most software applications, as well as low Internet and computer penetration especially in the rural areas.

There are some online newspapers in Lao, such as the *Pasaxon Newspaper* (http://www.pasaxon.org.la), the *Lao News Agency* (http://www.kpl.net.la) and the *Vientiane Mai*

Newspaper (http://www.vientianemai.net/indexder.php). Lao National Television (http://www.lntv.gov.la) has a website, as does Lao National Radio (http://www.lnr.org.la).

Other local websites are those of the:

- Lao National Tourist Information Service (http://www.tourismlaos.gov.la)
- *Vientiane Times Newspaper* (http://www.vientianetimes.org.la)
- *Lerenovateur* (http://www.lerenovateur.org.la), a newspaper in French
- Lao Trade Promotion Center (http://www.laotrade.org.la)
- National Statistics Center (http://www.nsc.gov.la)
- *Update Magazine* (http://www.laoupdate.com)
- Lao Localization Development Effort (http://www.laol10n.infor.la)
- Inlao.net contains stories, jokes and entertainment in both Lao and English, while mahasan.com is an information resource on Laos.

Online services are provided by the Planet Internet Service Provider (http://www.planet.com.la), Lanexang Internet Service Provider (http://www.lanexangnet.com/), Lao Airline Service Company (http://www.laoairlines.com), Lao Telecom Company Ltd. (http://www.laotel.com), Enterprise Telecom of Laos (http://www.etllao.com) and Lao National Internet Exchange (http://www.laonix.net.la).

ICT policies and programmes

National ICT policies

The overarching national goal is to advance beyond LDC (least developed country) status by the year 2020 through sustainable and equitable development. To realize this vision, the Government of Lao PDR aims to bring the country into the information age by increasing general access to ICT through the provision of modern telecommunications infrastructure and computer networks, by fostering enterprise and industry, by promoting research and development in ICT and by developing the necessary human resources and institutional capacities. Nine priority policy areas have been identified by the government: Infrastructure and Access, Enterprise and Industry, Research and Development, Applications, Human Resource Development, Legal Framework, Awareness, Poverty Alleviation and Standardization and Localization.

The National Policy on ICT aims to ensure that the necessary institutional, human capacity, sectoral conditions and legal frameworks are in place for leveraging and applying ICT to meet the needs of the country. A National ICT Board (NICTB) will be formed to carry out the tasks outlined in this policy document.

Telecommunications policies

Following a review of the state of the telecommunication sector and the future telecommunication requirements of its citizens, the government has formulated a new telecommunications sector policy framework. The policy objectives are to develop the national telecommunication infrastructure especially in regional and remote areas of the country; to establish a financially viable telecommunications sector through sustainable investment in telecommunications infrastructure by the private and public sectors as well as aid agencies; to improve the efficiency and effectiveness of telecommunications service delivery to end users; to achieve cost-effectiveness in meeting end-user demand for telecommunications services at affordable prices; and to strengthen regulatory capability and skill sets within government so as to ensure a high standard of sector governance and oversight of market participants.

The government has adopted the following major policy initiatives which are collectively designed to achieve the telecommunications sector policy objectives mentioned previously:

- Issuance of telecommunication regulations based on the existing Telecommunication Act 2001 for the effective governance of the telecommunications sector.
- Overhaul of sector regulation to ensure efficient and effective supervision and management of the sector. Specifically, a new Telecommunications Regulatory Authority within the Ministry of Communication, Transport, Post and Construction will be created.
- Requiring all operators to secure the relevant licenses under the Licensing Regulation. Licensing fees will be paid into the Consolidated Revenue and/or the Telecommunications Development Fund (as originally created pursuant to Article 5 of Telecommunication Act 2001).
- Encouraging network interconnections to ensure non-discrimination between operators and to ensure good communication quality.
- A moratorium on the issuance of new major network facilities or service licenses to any new entity pending a review of sector policy in 2007.
- Promotion of fair competition, facility sharing and collocating of equipment.
- Appropriate management of telecommunication resources such as radio frequencies, numbering, rights of way and Internet domain names.

- Monitoring and control of the cross-subsidization and accounting separation of the significant market power operator.

e-Government action plan

The vision for implementing e-governance in the country is to adopt ICT tools across various tiers of administration at the ministries, departments, provinces, districts and villages of Lao PDR to bring about SMART Government and offer appropriate interfaces to the people (in cities as well as villages) through electronic delivery channels. The e-government objectives are: anywhere, anytime access to government information to bring about transparency, efficiency and empowerment of citizens; e-delivery of government services to citizens through Web and Integrated Citizen Service Centres, which are expected to be particularly beneficial to poor communities in remote areas; increased internal efficiency and prompt delivery of citizen services; and strengthening of communitization.

ICT for development

The first multi-purpose telecentre in Laos

The developmental opportunities arising from strategic use of ICT are to be demonstrated with the establishment of a multi-purpose telecentre in the provincial capital in Luang Prabang and eventually in rural communities. The project started in 2003 with support from the International Development Research Centre (IDRC) of Canada. The ultimate goal of the project is to 'help create opportunities and reduce poverty by utilizing the benefits of connecting rural communities and integrating ICT in their daily activities'.

The telecentre at Luang Prabang is running at full capacity. Nicknamed 'e-Way', it is open seven days a week. There is a long queue of people who have signed up to attend the courses being given at the e-Way Centre. The centre is actively supporting training to increase the number of people in the community with basic computer skills and English language competence. In the last three years, the centre has trained over a thousand trainees, which contributed to the ICT development of the province. For example, the number of computer training schools in the area increased to more than 10 and a third of the school owners and at least one teacher from each training school were trained at the e-Way Centre. Moreover, there are now more than 50 Internet cafés in the area, with many of the Internet café owners trained at the e-Way Centre.

However, the e-Way has not achieved all of its objectives. In particular, it has not become a place for information access and dissemination for the people of Luang Prabang because of the lack of Internet connectivity, which in turn is due to the centre's inability to afford the fees (telephone charges). Potentially, the telecentre can sustain itself, but there is a difference of opinion between local government and project personnel regarding the management of the telecentre. The management was revamped toward the end of 2004 and the telecentre now offers telephone and fax facilities mostly to small business people. The services have not yet been extended to the rural areas as envisaged.

The National Internet Exchange Point

The project aims to connect the six different ISPs in Laos through an Internet Exchange Point (IXP). Currently, each ISP has its own leased line to a foreign country where they connect to the global Internet. If the ISPs could be connected through an IXP in Laos, there will be both economic and technical gains, since the ISPs could have a reduced traffic load on their leased lines and therefore pay less. Laonix, as the IXP is called, will make Laos less dependent on the narrow interconnection to the international Internet as it will be possible to keep local traffic local by interconnecting all commercial Lao ISPs and the academic network. End users will benefit from a faster network when they access local content within Laos. The project was initiated by LANIC, which hopes that with the support of SIDA, IXP will create an environment suitable for local ICT development within Laos.

A policy framework and digital standardization for information exchange in Lao

This ICT4D Project supported by UNDP aims to harness the potential of ICT for enhancing progress towards national development goals, by strengthening the capacities of STEA to establish and manage national ICT standards. The project consists of two components: (a) strengthening the government's strategic framework for ICT policy development; and (b) improving the utilization of the Lao language in electronic communication through the development of a standard Lao character set.

The first component builds on a UNDP sub-regional project that helped develop a national ICT policy which sets the foundation for an ICT Master Plan. This draft plan outlines the priority development initiatives involving ICT, and the required coordination arrangements and funding. The project aims to help raise awareness of and to disseminate the approved national ICT policy.

Through the second component, the project provides initial assistance to STEA in formulating and implementing standards for entering, storing and processing digital information in the Lao language. The ability to communicate electronically using the Lao language is currently hampered by the absence of a standard Lao character set, that is, the standards to enter, store, display, print and exchange Lao text on computers. Standardized use and application of the Lao language in digital data exchange will enable use of the Lao language on the Internet, enhance public and private sector performance and help preserve Lao culture. Activities in this regard will be developed under the guidance of a broad-based network of stakeholders led by STEA.

Legal and regulatory environment for ICT

The Ministry of Communication, Transport, Post and Construction (MCTPC) is responsible for national telecommunication policy and regulation. The Department of Post and Telecommunications is the functional unit within MCTPC whose tasks include telecom and post policy, frequency management, long-term development strategy, licensing and regulation. The MCTPC issued Telecommunications Act No. 02/NA in 2001.

The Lao National Internet Committee (LANIC) is responsible for national Internet policy and regulation, including the approval of ISP licenses and Internet café licenses. On 15 April 2000, LANIC issued Regulation No. 141/pmo on the implementation, service and usage of Internet systems in Laos.

The Science Technology and Environment Agency (STEA) is responsible for national information technology policy and regulation, including approval of computer-related business licenses. STEA is currently developing cyber law and e-government regulation.

However, a major problem is that the areas of responsibility of MCTPC, LANIC and STEA have not been properly defined. Lack of coordination and overlapping responsibilities have contributed to a situation where these organizations are engaged in drafting competing ICT policy and regulation papers to be presented before the legislature for selection and approval. In sum, although significant progress has been made particularly in fibre infrastructure networks, ICT4D efforts in Laos have generally been rather slow, fragmented and uncoordinated. In this connection, there are discussions about establishing a Ministry of Information Communication Technology that would combine the ICT-related activities of MCPTC, LANIC and STEA. There are also talks of establishing an independent ICT regulatory body.

Open source

The Lao government supports open source software because it is free and anyone can develop it. It saves a lot of money when compared to other software applications, according to the Department of Science and Technology. However, there is no concrete policy and strategy for promoting open source software. In the national ICT policy paper, there are few statements related to open source. For example, in the Human Resources Development section it states that the Government of Lao PDR shall focus on world-class curriculum development for the Bachelor and Master's degrees in Computer Science/ Engineering and degrees related to ICT, and that the government shall promote the integration and teaching of free and open source software (FOSS) in the computer science/engineering curricula. The section on Standardization and Localization stipulates that the government shall establish a network of national and international experts from the academe, government and the private sector to give advice on all issues relating to the localisation of ICTs, including open source and proprietary software.

The UNDP-supported ICT4D project being managed by Anousak Souphavanh, the author of LaoNux, has made much progress on the localization of open source software into the Lao language. Specific achievements include:

- Phetsarath Fonts—With the assistance of a calligrapher and graphic designers, a new aesthetically pleasing font has been developed. Called Phetsarath OT, it will be released for public use soon. The font is UNICODE Open Type, and therefore based on open standards.
- An open source e-mail client called 'Salika' is being developed as an option for Lao users to replace Microsoft Outlook. It is also UNICODE-enabled, with Lao language menus.
- A Lao version of Open Office (named Xangdao) is being developed to lower the threshold for low-income Lao people with limited English skills to benefit from computers. The product is a multi-platform productivity application suite that can be installed on Windows or any other computer operating system.

Research and development

There are hardly any publications available that would indicate that there is genuine research going on. There is also no research institute in Lao PDR. Independent researchers customize foreign software applications for resale in the local market. The significant research initiative can be seen only at the Information Technology Center of STEA, which has established the open source laboratory and R&D division for software development.

The Information Technology Center is implementing the Lao Localization Project under the PAN Localization Project, the regional initiative to develop local language computing capacity in Asia supported by IDRC of Canada. The Natural Language Processing (NLP) Group of Lao PDR composed of an architecture team, a linguistics team and a programmer, was established in 2003 to carry out the research on Lao localization. During the first phase, from 2003 to 2006, the NLP completed the development of a Lao character set, Lao fonts, Lao keyboard, Lao ASCII to Unicode converter, Lao syllabification and line breaking utility, and Lao collation and sorting utility. In the second phase, from 2007 to 2009, the NLP will conduct the research work on translation of gTLDs and ccTLDs in Lao, Corpus Translation Phonetic Module for Lao TTS, Lao Speech Corpus, English-Lao Parallel and Aligned Tagged Corpus 100k words, Lao TTS and Lao Diphone Database. This phase will also include the localization of Open Office and the creation of e-content.

Education

As of now, computer and IT-related education in Lao PDR is provided not only by the National University of Lao PDR (NUOL) but also by the private sector. The computer science programme of the NUOL began in 1998 under the Faculty of Science, Department of Mathematics. Every year since 2002, the programme has produced around 30 graduates.

The Faculty of Engineering and Architecture (FEA) is considered to be the best equipped with IT facilities in the NUOL system. The main component of these facilities is the Lao-Japan Technical Training Center (LJTTC), which offers a combination of general application courses, computer-aided engineering courses and a course on network software. In 2002, with assistance from the Japan International Development Agency (JICA), the faculty implemented an IT bridging course, a two-year programme leading to a Bachelor of Information Technology Application after completion of a Bachelor's degree in mathematics, electronics, engineering or management. Forty ICT specialists have graduated from this programme, which has a capacity of 20 students per year. The target is to graduate 100 graduates and to establish a Master's level phase.

The vacuum in IT education facilities in the public sector is filled, at least partly, by the private sector. The following private colleges are currently providing IT education, albeit at a very preliminary level: Vientiane College, a private institution with the academic and financial support of Monash University in Australia; Lao American College which has established working relations with the NUOL, City University of Washington State, USA, Ohio University, USA and Bangkok University, Thailand; and Quest Colleges which has a joint venture with Vaasa Polytechnic in Finland. In addition to these educational institutions with foreign investment, there are local initiatives like Rattana Business Administration College, Com Centre Collage, Sengsavanh Collage and Sousaka Collage. There are also a number of computer dealers providing short-term training in computer operations.

Through the Laos-India ICT Bilateral Cooperation, the National Informatics Centre is setting up an ICT Training Laboratory with 25 computers, software tools, training accessories and the required infrastructure at the Information Technology Center of Laos. In addition, a four-course computer training module (on Office Productivity Tools, Database, Web Development and Network System) for 150 government officials has been conducted.

Under the e-Government Action Plan, by 2007 STEA will be implementing an ICT human resource development programme with the following components:

1. Construction of three computer training rooms at STEA;
2. A three-month training scholarship at the ICT engineer level for 30 members of staff of various government organizations at Alcatel Shanghai Bell University;
3. Training courses at engineer level at the Training Center in Vientiane for 260 staff members (142 staff members from districts, 18 staff members from provincial offices and 100 staff members from government ministries);
4. Onsite computer training in the provinces for at least two weeks for 30 participants;
5. Development of the training curriculum and training materials for STEA to carry out further training. The curriculum for engineers aims to reach international accredited standards in the following areas: office automation, database, networking, Internet, security, decision support system, e-government system and computer programming; and
6. Development of e-learning applications for the training curriculum.

An adequate supply of IT human resources is needed not only for effective use of the Internet and the generation of Internet content, but also for the country to enter the rapidly growing area of IT-enabled services. To meet this need, the Ministry of Education has adopted a 'top-to-bottom' approach in which the Ministry first develops ICT access and skills within the Ministry

> **Education over IP**
>
> With the installation of a fibre network and ongoing development of new applications, in particular IP telephony and video conference facilities built by a team of KTH students (Royal Technology Institute of Sweden), the National University of Lao PDR (NUOL) is now able to conduct distance education, mainly between the main campus and the remote campuses. Distance education will help solve problems arising from a shortage of qualified teachers.
>
> In addition, university staff can now make internal telephone calls for free within the university using new IP phones or with special software installed on their computers.
>
> NUOL is expanding its data network. By upgrading the wireless backbone to fibre, it has all of the opportunities to implement new applications for students and staff. The project was supported by SIDA.

itself, and then at the university level, and then finally within the schools. Given that the younger generation considers 'Internet and English as their lifeline', it may be advisable for the Ministry to consider a 'bottom-up' approach in which computer access and computer education shall be provided in primary and/or secondary schools.

Conclusion

Lao PDR is a country with a small population and limited financial resources. It faces many challenges, including an undeveloped road network, limited water and electricity supplies and undeveloped educational and health systems.

The ICT situation is no exception: telephony and data connectivity is scarce and unevenly distributed, although it is improving due to an aggressive deployment of a national fibre network and expansion of mobile telephony. There is a lack of both basic and advanced ICT skills. The negative ICT situation in general, with the limited ICT capacity and low computer penetration, is exacerbated by the fact that English proficiency is low, which in turn represents another training obstacle since most literature and training materials are in English. This problem is amplified by the fact that Lao people are as yet unable to study and work on computers in their native language. Last but not least, most people cannot afford a computer and, even if they could, only major urban areas have access to electricity and telephony services.

Several major technical remedies are underway, while the task of accelerating human capacity building remains. Several initiatives in this area are ongoing, but not yet at a scale that brings broad impact.

References

Government of Lao PDR. (2005). *National statistics yearbook 2005*.

Japan Telecommunications Engineering and Consulting Service (JTEC). (n.d.). *Study report on the broadband access network development project with optical fibre in Vientiane*.

Lao National Internet Committee. (2005). *Report on Internet development in Lao PDR*.

Ministry of Communication, Transport, Post and Construction. (2005). *Report on telecommunication policy and development in Lao PDR*.

Science Technology and Environment Agency, Government of Lao PDR. (2005). *E-government action plan and report of ICT development in Lao PDR*.

Soderberg, B. and Pehrson, B. (2004). *ICT-related challenges and opportunities for the government of Laos*. Swedish International Development Agency. Retrieved from http://www.spintrack.com/itadvice/reports/SPINTRACK%20Laos%20ICT%20report%202004-09,%20Sida,%20Final%20-%20PUBLIC%20VERSION.pdf

United Nations Development Programme. (2006). *Report on ICT4D projects in Lao PDR*.

United Nations Economic and Social Commission for Asia and Pacific. (n.d.). *Report on the development of enabling policies for trade and investment in the IT sector of the GMS countries. Chapter 4: Lao PDR*. Retrieved from http://www.unescap.org/tid/projects/gmsti_chap4.pdf

.my

Malaysia

Musa **Abu Hassan** and Mohd Safar **Hasim**

Total population	26.64 million (2nd Qtr 2006)
GDP per capita	USD 4,904 or MYR 18,145 (USD 1 = MYR 3.70)
Key economic sectors	Manufacturing, Services and Agriculture
Computers per 100 inhabitants	21.8 (October 2005)
Fixed-line telephones per 100 inhabitants	16.3 (2nd Qtr 2006)
Mobile phone subscribers per 100 inhabitants	80.8 (2nd Qtr 2006)
Internet users per 100 inhabitants	14.0 (dial-up), 2.5 (broadband) (2nd Qtr 2006)
Domain names registered under .my	83,709 (January 1995–December 2006)
Broadband subscribers per 100 inhabitants	2.5 (2nd Qtr 2006)
Internet domestic bandwidth	2 Mbps (October 2006)
Internet international bandwidth	2 Mbps (October 2006)

Introduction

Since achieving independence in 1957, the development programmes of Malaysia are conducted through Five-Year Development Plans. The year 2006 marks the beginning of the country's Ninth Development Plan. Besides the Five-Year Plan, Malaysia has Vision 2020, a plan to achieve developed country status for Malaysia by the year 2020 that was envisioned and launched by ex-Prime Minister Tun Mahathir Mohamad in 1991.

Vision 2020 contains nine central strategic challenges, the sixth of which is the challenge of establishing a scientific and progressive society, a society that is innovative and forward-looking, and one that is not only a consumer of technology but also a contributor to the scientific and technological civilization of the future (Government of Malaysia 2006). Key to meeting this challenge is the adoption and use of information and communication technology (ICT) by all sectors of government and society.

The deployment of ICT as an important component of development started in the Seventh Malaysian Plan 1996–2000. During this period the National Information Technology Council (NITC) was established as a think-tank and advisor to the government on IT development. The NITC initiated the formulation of a national IT plan and the identification of key programmes for the transformation of Malaysian society into a knowledge-based society (Malaysia 1996).

The government also launched the Multimedia Super Corridor (MSC) during this development period. Conceptualized in 1966, the MSC has since grown into a thriving and dynamic ICT hub, hosting multinationals, and foreign-owned and home-grown Malaysian companies involved in multimedia and communication products, solutions, services, and research and development (Multimedia Development Corporation 2006). The MSC has seven flagship applications to enhance the socio-economic development of Malaysia: Electronic Government, Multipurpose Card, Smart School, Tele-health, Research and Development Clusters, E-Business and Technopreneur Development.

Another crucial component of ICT development in Malaysia is the National Information Technology Agenda (NITA) that was developed to spearhead and guide ICT development in the country. Its three main interrelated components are people, applications and infostructure. Hashim (2000) noted that NITA grants equal opportunity to every citizen to access the infostructure in order to transform Malaysia into the value-based knowledge society envisioned in Vision 2020.

This chapter presents the Malaysian scenario of ICT for development. For further information, readers are encouraged to refer to the websites listed at the end of the chapter.

Technology infrastructure

Since 1987, Malaysia has been actively involved in reforming and restructuring the telecommunications sector (Kementerian Tenaga, Air dan Komunikasi 2006). The participation of the private sector has ensured the development of the necessary information infrastructure. For instance, trunk fibre networks now crisscross peninsular Malaysia and extend across the South China Sea to connect Sabah and Sarawak in the eastern part of the country. With Malaysia's own satellite (MEASAT I, MEASAT II and MEASAT III), completion of the Malaysian infrastructure superhighway is within reach.

Six licensed telecommunication companies provide telephony services: Telekom Malaysia (TM), Celcom, Maxis Communication, Time Telekom, DIGI Telekom and Prismanet. Two of these companies, TM and Maxis, are also Internet service providers (ISP). Another ISP currently operating in Malaysia is Jaring by MIMOS.

At of the end of 2005, there were 2,839,000 fixed-line residential telephone subscribers (49.5 per 100 households) and 1,527,000 fixed-line business subscribers. The total of 4.366 million fixed-line subscribers represents a penetration rate of 16.6 per 100 inhabitants in Malaysia (Malaysian Communication and Multimedia Commission 2006).

The cellular phone penetration rate was higher at 74.1 per 100 inhabitants at the end of 2005. There are 19.545 million cellular phone users in the country, with a ratio of eight prepaid users to one postpaid user.

The penetration rate of dial-up Internet in the same period was 13.9 per 100 inhabitants. The penetration rate for broadband Internet was 1.9 per 100 inhabitants. The personal computer penetration rate in 2005 was recorded at 21.8 per 100 inhabitants (Malaysia 2006).

The Malaysian government is pursuing an overall technology infrastructure development strategy called the MyICMS 886 Strategy, which is short for Malaysian Information, Communication and Multimedia Services 886 Strategy. The numbers 886 refer to eight new services to catalyze and promote the development of eight essential infrastructure that are to generate growth in six areas that have been identified as key for the consumers and business in Malaysia (Ministry of Energy, Water and Communication n.d.). The eight new services are high-speed broadband, 3G and beyond, mobile TV, digital multimedia broadcasting, digital homes, short-range communication, VoIP/Internet telephony, and universal service provision (USP). The eight essential infrastructures are multi-service convergence networks, 3G cellular networks, satellite networks, next-generation Internet protocol (IPv6), home Internet adoption, information and network security, competence development, and product design and manufacturing. The six growth areas are content development, ICT education hub, digital multimedia receivers, communication devices, embedded components and devices, and foreign venture.

Specific to broadband service, in 2006 TMnet started providing Streamyx at 384 Kbps and above. Broadband services are available via fixed/cable (ADSL and fibre), satellite (VSAT and DTH) and wireless.

The growth of the communications industry in Malaysia is also underpinned by demand for new services arising from the convergence of information technologies in the field of switching and transmission. For example, ATM, ISDN and SDH have created new services like VOD, video conferencing and many other multimedia applications on the Web, such as graphics, audio and animation, video and virtual reality, to mention a few. The key to this trend is liberalization, whereby government has allowed private participation in this sector. The thrust of the competition policy as envisaged in the Equal Access Policy is that it must lead to improvement in the quality of service as well as bring down prices as a result of improvements in operation efficiency. In keeping with this principle, 3G services have been awarded to four private companies.

Key institutions dealing with ICTs

All of Malaysia's 27 ministries and the Prime Minister's Department are urged to utilize ICT to the fullest in delivering services to their target sectors. The following government organizations are specifically tasked with ICT development in Malaysia:

- Ministry of Energy, Water and Communication—responsible for communication infrastructure, policy formulation and service regulation.
- Ministry of Science, Technology and Innovation (MOSTI)—tasked with creating an environment that is conducive to the advancement of science and technology and providing efficient technical and management support services to ICT projects and programmes.
- Ministry of Rural and Regional Development—in-charge of community access, telecentres, rural information programmes, bridging the digital divide and Infodesa projects for computer training skills and computer literacy.
- Ministry of Information—responsible for Smart Community projects, community access and bridging the digital divide.
- State government—tasked with providing community access through the State Information Technology Advancement Unit (KIT) for Electronic Government System, Education Net, Electronic Community and Electronic Commerce.
- Local Authorities or City Councils—must provide community access and computer training skills and computer literacy.
- Malaysian Communications and Multimedia Commission (MCMC)—issues licenses, implements regulations and facilitates universal service provision. This agency is under the Ministry of Energy, Water and Communication.

Private companies such as Maxis are expected to target ICT exposure, Internet access, basic computer training skills and computer literacy. Cybercafé entrepreneurs are expected to provide access to the Internet and the digital experience to various communities.

R&D in microelectronics and IT is the responsibility of government-linked companies such as MIMOS Berhad. The Multimedia Development Corporation (MDeC) is responsible for the implementation of the MSC.

Education content is provided through the education portal Utusan Melayu.

ICT industries

The MSC was set up in 1996 to build a competitive cluster of local ICT companies and a sustainable ICT industry in a 750 sq km area south of Kuala Lumpur. Five cities and cyber centre ecosystems have been developed within the area: Cyberjaya, Kuala Lumpur City Centre (KLCC), KL Tower, Technology Park Malaysia and UPM-MTDC incubator centre. By the end of 2005, a total of 1,421 companies were awarded MSC status. More than half of these companies are engaged in software development for general enterprise solutions and data warehousing, as well as high-end specialized applications and e-commerce (Malaysia 2006). More than 500 companies were set up in 1996–2003, the first phase. Some 22,000 high-value jobs were created and some MYR 6 billion (USD 2.2 billion) in revenue was generated.

By August 2006, the number of MSC-status companies had increased to 1,556, of which 1,485 are MSC Malaysia technology companies, 52 are institutions of higher learning and 19 are incubator companies. Sixty-seven companies are international world-class companies, such as Nokia, Siemens, Motorola Multimedia Lotus, Lucent Oracle, Intel, Fujitsu and Unisys.

Under phase one there is only one corridor, the MSC. Under phase two, called Next Leap (2003–10), there will be a web of corridors with Penang-Kulim in the north, Johor in the south, the east coast, and Sabah and Sarawak. Under phase three, (2010–20), MSC benefits will be extended to the rest of the country, thereby transforming Malaysia into a value-based knowledge society and a one-nation Multimedia Super Corridor. To further speed up the development from August 2005, the states of Perak, Melaka, Pahang, Johor, Negeri Sembilan and Kuala Lumpur are to receive MSC benefits.

Enabling policies and programmes

When Malaysia chose ICT to drive its economic and social development in the early 1990s, it was in ready mode. By the late 1980s, Malaysia had privatized its telecommunication and broadcasting industries (Hasim 2000). Equally important, the Microelectronic Systems of Malaysia (MIMOS) was established on 1 January 1985.

In 1994, Malaysia set up its National IT Council with then Prime Minister Mahathir Mohamad as chairperson. The Council was tasked with policy formulation, setting the strategic direction, policy coordination and evaluation and technology assessment and adoption, as well as industry promotion. The Council unveiled Malaysia's IT Agenda (NITA) in 1996. Rooted in Vision 2020, NITA identified ICT as the means to help Malaysia leapfrog from being an industrial society to a post-industrial society. Entitled *Turning Ripples into Tidal Waves*, the document formed the basis for the informatization of Malaysian society—that is, the use of ICT in all walks of life to improve productivity and enhance quality of life. The focus was ICT for development (ICT4D) to address the issue of equitable development. Thus, the NITA document outlined a balanced and people-centred approach to ICT development.

The K-Malaysia migration strategy is part of the NITA document. Its vision is to evolve a values-based knowledge society in the Malaysian mould, where the society is rich in information, empowered by knowledge, infused with a distinctive value-system and is self-governing (John 2002). The strategy follows three stages of development: information society by 2005, knowledge society by 2010, and values-based knowledge society by 2020. The phases coincide with the three phases of physical development, where the first phase is development within the MSC; the second phase extends the corridor to the north, the east coast, the south, and Sabah and Sarawak; and the third phase will cover the whole country.

Under the Ninth Malaysia Plan, Malaysia will enhance its position as a global and multimedia hub, expanding the communication network to ensure more equitable access to information and services and bridging the digital divide. MSC phase II will be rolled out, expanding multimedia applications, identifying new sources of growth in ICT, developing a skilled ICT workforce, accelerating e-learning acculturation and enhancing information security.

Legal and regulatory environment

Malaysia's approach to ICT development is two-pronged: providing the hard infrastructure in terms of physical development and laying out the soft infrastructure in terms of laws and regulations. At the start of Malaysia's push for using ICT, six cyber laws were enacted, namely: Communications and Multimedia Act 1998, Malaysian Communication and Multimedia Commission 1998, Computer Crime Act 1997, Digital Signature Act 1997, Telemedicine Act 1997 and Copyright (Amendment) Act 1997 (Hasim 2002). Together with the Bill of Guarantee, these laws became the soft infrastructure for the development of ICT. Each is discussed briefly below.

Communications and Multimedia Act 1998

This is the cornerstone of all cyber laws in Malaysia. It is the basic document for the MSC and Malaysia's efforts for future digital development in Malaysia. One of the most important elements of the Act is absence of censorship of the Internet. The Act repealed the Telecommunications Act 1950 and Broadcasting Act 1988.

Malaysian Communication and Multimedia Commission 1998

This law enables the Malaysian Communication and Multimedia Commission to be set up as a new regulator for the communication and multimedia industry in Malaysia. The Act is based on the principles of transparency and clarity, more competition and less regulation, more emphasis on process rather than content, administrative and sector transparency and industry self-regulation. The Act covers only networked services and activities.

Computer Crime Act 1997

The law covers six crimes related to misuse of computers: unauthorized access, unauthorized access to commit fraud or dishonesty, modification of computer content, communication of a password to persons other than the person for whom the password is intended, abetting a computer offence, and custody or control of a computer program or data without authorization.

Digital Signature Act 1997

This Act, which legitimizes and provides for the use of digital signatures, aims to encourage electronic commerce. On 1 November 2001, the Malaysian Communications and Multimedia Commission was appointed as the Certifying Agency for digital signatures.

Telemedicine Act 1997

The Act provides for the use of multimedia in telemedicine. It provides a legal framework for the practice of telemedicine.

Copyright (Amendment) Act 1997

This is an amendment to the Copyright Act to include the Internet.

Security issues

The extent of security threats through the Internet in Malaysia can be gauged by the reported cases recorded by Niser, a body set up by the Malaysian government to monitor such incidents. Niser classifies security threats as mail bomb, spam, harassment, forgery, hacking, virus, malicious code, denial of service, destruction and intrusion. MyCert, which was set up before Niser and is now part of it, began collecting data on these security threats in 1997. In 2005, a new category called malicious code (such as W32.Brontok worm and W32.Nyxem worm) was included.

During the past 10 years (August 1997–2006) Niser has recorded 30,803 cases of security threats: 23,330 cases of spam, 2,890 cases of virus, 1,775 cases of intrusion, 1,675 cases of hack threats, 339 cases of harassment, 198 cases of fraud, 137 cases of malicious code, 104 cases of mail bomb, 99 cases denial of service and 27 cases of destruction. It must be noted that these figures represent only incidents reported to Niser; the actual number of incidents could be much higher.

Police have taken action on many of the reported cases under Computer Crime Act 1997 and Communication and Multimedia Act 1998. Police action under the Communication and Multimedia Act 1998 resulted in 282 cases prosecuted, with damages totalling about MYR 1.17 million (USD 307,000). Most of the cases are related to licensing (Hasim 2005). Since 2000, when the Computer Crime Act was enforced, some 4,846 cases have been prosecuted with a total value loss of MYR 10.5 million (USD 2.76 million). Except for 2001 and 2002, the number of cases prosecuted under the Computer Crime Act appears to be dwindling. It is not clear whether the decline in number of cases is a result of fear of police action or inability of the police to take action. The police have recently taken in many IT graduates to beef up its computer crime section.

Education and capacity building

One of the flagship MSC projects is the Smart School project, which aims to: (a) prepare school leavers for the Information Age; (b) bring about a systematic change in education, from an exam-dominated culture to a thinking and creative knowledge culture; (c) re-emphasize science and technology education with a focus on creativity and innovation; (d) equip students with IT competence; and (e) inculcate Malaysian values among students and produce a generation of caring, peace-loving and environmentally concerned citizens (Curriculum Development Centre 1997). The Ministry of Education is implementing the SchoolNet Project in collaboration with the Ministry of Energy, Water and Communication towards achieving the Smart School initiative. The initial target was connecting 10,000 rural and urban schools and some educational institutions. However, in the Ninth Malaysian Plan the government decided on 'Making All Schools Smart.' The Ministry of Education is also developing

MySchoolNet, an educational portal for students, schoolteachers and school administrators (Kementerian Pendidikan Malaysia 2002).

The Malaysian Grid for Learning (MyGfL) is another national initiative for e-learning. It is being undertaken by MIMOS Berhad to: (a) provide e-learning systems and tools to enable and support e-learning activities and processes for life-long learning; (b) bring together all relevant players in the e-learning ecosystem (learners, enablers and providers) to participate in the overall e-learning value chain and be part of the national learning grid; (c) develop e-learning standards to ensure conformance and adoption of best practices in e-learning content and systems; and (d) encourage sharing and development of local/indigenous content, thus stimulating the content industry (MIMOS Berhad 2005). An example of projects under MyGfl is Cikgu.net, maintained by Jaring, a subsidiary of MIMOS Berhad.

The private sector also plays a role in providing educational resources to students and the public at large. For example, the Utusan Melayu Berhad has set up an educational portal called Smart Utusan Education Portal. This educational gateway attracts an average of six million visitors every month. It carries learning and teaching materials for all school levels, including pre-school and pre-university (Portal Pendidikan Utusan 2005). Within the portal, students can get access to the past year's national examination questions, and teachers can view teaching plans and mathematics and science teaching scripts which provide ideas or activities to make teaching and learning more interesting. There is also an interactive section for visitors to share their thoughts through an online forum and cyber chat, try the e-laboratory or play games. Under the 'Community' heading is *Sekolahku* (Myschool) where schools can publish their Web pages.

Online services

ICT has brought about changes in the way Malaysians conduct their work, social activities and leisure, especially since the Internet was made available in 1992 (Abu Hassan 2002). Many Malaysians now get their news, transact with banks, request services from or file complaints to the Local Authority, interact with the government officials, and communicate with friends and family online.

From the Malaysian government official portal, citizens and residents of Malaysia can get information and services related to education, employment, health, social welfare and community, property and investment, legal matters, public amenities and utilities, security and safety, taxation and collections, and travel and transport. There are also links to all government offices, government tenders, job vacancies in government offices, and many more. Thus, it can be said that it is a comprehensive portal. In August 2006, the portal received an average of 42,000 visits monthly (MAMPU n.d.).

To ensure that the public has access to e-services provided by government, telecentres have been set up in rural and urban areas by the Ministry of Rural and Regional Development, the Ministry of Energy, Water and Communication, the Ministry of Information, the State Government and the Local Authority. The Ministry of Rural and Regional Development has the Infodesa Project (see 'Digital Content' section). The Ministry of Energy, Water and Communication is collaborating with Pos Malaysia, Telekom Malaysia, and the MSD group of companies (Kementerian Tenaga, Air dan Komunikasi) in setting up Rural Internet Centres (RIC). The RIC project aims to bridge the digital divide between rural and urban communities. In 2006, there were 42 RICs throughout the country, all located at the post office where people congregate to pay utility bills and conduct other business. Members of the RIC can use the computer and Internet facilities for free. Each RIC has a website containing information on topics that are of interest to the local communities, such as health, agriculture, fisheries, tourism, arts and culture, security, entrepreneurship, local and national news, and local history. The RICs also conduct regular workshops to train the local community in the use of computers and the Internet. One RIC has conducted a workshop on starting a blog.

The Maxis Bridging Community (MBC) project is an example of a private initiative to help bridge the digital divide in Malaysia. Maxis has trained 1,343 teachers and 2,722 school children from 719 schools in the Cyberkids Camp started in 2005. Using the 'Train the Trainer' concept, the camp includes classroom and outdoor activities designed to enable participating teachers and students to effectively use computers and the Internet and subsequently train others in their respective communities. The programme targets rural primary schools (Maxis Communication Berhad 2006).

Open source initiatives

The Malaysian Administrative Modernization and Management Planning Unit (MAMPU) is mandated by the government to take the lead in implementing the open source software (OSS) initiative for the public sector (Malaysian Public Sector 2006a). The four main components of the OSS Master Plan are: (a) framework and strategic thrusts, (b) technical implementation plan and roadmap, (c) open source competency centre and (d) OSS policies and guidelines.

Subsequently, the government formulated six objectives for the OSS implementation plan, namely: (a) reduce total cost of ownership, (b) increase freedom of choice of software usage, (c) increase interoperability among systems, (d) stimulate the growth of the ICT industry, (e) promote the growth of the OSS user and developer community and (f) reduce the digital divide (Malaysian Public Sector 2006b). Since the launching of the OSS initiative, several workshops and seminars to create public awareness have been conducted regularly, mainly by the Open Source Competency Centre (OSCC) set up by the government.

The 2005 human resource targets of the OSS initiative were for all Chief Information Officers (CIOs) and Information Technology Personnel (ITP) to be OSS literate, 60 per cent of ITP to be trained in OSS and 10 per cent of ITP to be certified in the use of OSS. Other targets for 2005 were for 60 per cent of all new servers procured to run on open source operating systems, 60 per cent of Web servers (software) to use OSS, 30 per cent of office infrastructure (e-mail, DNS, Proxy) to use OSS and 30 per cent of desktop solutions to use OSS (Malaysian Public Sector 2006c).

Digital content initiatives

The NITA document identifies content development as one of the strategies to be pursued as part of the implementation of the seven MSC flagship applications. An important aim is to develop culture-appropriate content.

A key programme under content development is the Demonstrator Application Grant Scheme (DAGS). Its objectives include: (a) acculturating Malaysians to ICT, enabling them to maximize the benefits of ICT applications at work and at home, (b) building an integrated network of electronic communities using multimedia technology and (c) enhancing closer cooperation and collaboration between public agencies, private corporations and non-profit organizations through joint ventures and institutional linkages (NITC 2002). The DAGS funds Demonstrator Application programmes (DAs), which are small, focused and short-term projects that seek to create, develop and promote new ICT-based applications that create new content value for community development within specific contexts. To date, 75 projects have been started (some have been completed) under the DAGS strategic priority areas of Social Digital Inclusion, Economic Competitiveness and e-Public Services. Examples of DAGS projects are e-Bario, e-Homemakers, ICT in Masjid as a Neighbourhood Centre, MyBiz, Penang e-Doctor and Smart Taxipreneur (Demonstrator Application Grants Scheme 2003).

Another content development programme is Infodesa by the Ministry of Rural and Regional Development. This content exchange platform intended for rural communities has two main components: *Medan Infodesa* or MID (Infodesa Centre) and *Titian Digital* (Digital Bridge). MID is a physical entity that is built or housed in the rural areas and equipped with ICT infostructure and functions; it serves as the district training centre and as a catalyst for local entrepreneurs. Digital Bridge is a communication gateway for the local community to interact through ICT with service providers and those involved in rural development activities. At present, there are 30 MIDs throughout the country. Examples of MID are e-Bayangan in Sabah, e-Gulang in Selangor and e-Bujang in Kedah (Kementerian Kemajuan Luar Bandar dan Wilayah 2005).

Research and development

MIMOS Berhad has been designated to lead R&D activities related to ICT in Malaysia. Its approach to R&D has been collaboration and smart partnerships with universities, industry, research institutions and the government (MIMOS Berhad 2004). Four technology thrust areas are being emphasized by MIMOS: Pervasive Computing, Cyberspace Security, Microelectronics, and Grid Computing and Bioinformatics.

To support research activities, the Malaysian government has introduced several research grants. For instance, the Ministry of Science, Technology and Innovation (MOSTI) has the task of coordinating the ScienceFund, TechnoFund and InnoFund (Ministry of Science, Technology and Innovation 2004). There are few research grants from the private sector so far.

A number of studies regarding various aspects of ICT have been conducted by several universities in Malaysia. For example, academic staff and undergraduate and graduate students of the Department of Communication, Universiti Putra Malaysia have completed more than 50 ICT-related studies since the early 1990s. The research respondents include the general public, family members, youth, women, public and private sector staff, cybercafé users, members of urban and rural communities, as well as cyber communities. Among the aspects of ICT that have been studied are Internet addiction, techno stress, computer anxiety, telecommuting, IT in agriculture, the contents of community Web pages and societal readiness to accept IT.

At Universiti Kebangsaan Malaysia, an e-community research centre was set up in 2000. Among others, the centre monitors and assesses the impact of ICT on society, especially on changes in quality of life. It is hoped that systematic monitoring of e-community programmes would lead to better strategic planning, policy formulation, as well as theory-building. Membership in the e-community research centre is open to researchers from other academic and government institutions.

There is a need for information sharing among researchers from various universities and other research institutions regarding what ICT studies are being undertaken. In this way, new areas of research can be identified, the findings of previous research can be utilized, duplication can be reduced and research resources can be optimized.

Challenges

Given Malaysia's ICT capability, it would be good to get as many Malaysians as possible online. Making citizens ICT users is one of the major challenges in the Malaysia ICT scene. Equally challenging is increasing Internet penetration throughout Malaysia. The way forward is to lower cost in terms of tariff and maintenance, so that more people can use the Internet. The government announced recently its intention to equip 1,500 schools with cybercafé facilities as part of the bridging the digital divide initiative. These facilities should also be made accessible to the public, especially in marginalized and rural areas.

Another challenge is keeping cyber laws up-to-date. It has been about a decade since the current laws were passed. A review of these laws would make them more relevant and attuned to changing technologies and online applications. For instance, the US government has come up with a law disallowing credit card companies from honouring payments for gambling through the Internet.

Above all, Malaysians should strive to ensure that Malaysia is ready to realize Vision 2020—that is, Malaysia as a developed nation and a value-based knowledge society.

References

Abu Hassan, M. (2002). Internet in Malaysia. In Rao, S. and Klopfenstein, B. (eds), *Cyberpath to development: Issues and challenges* (pp. 137–57). Westport, CT.: Praeger.

Curriculum Development Centre. (1997). *Concept*. Retrieved 28 September 2006 from http://www.ppk.kpm.my/smartschool/concept.html

Demonstrator Application Grant Scheme. (2003). *Demonstrator application grant scheme (DAGS)*. Retrieved 29 August 2006 from http://www.dags.net.my

Government of Malaysia. (2006). *Malaysia as a fully developed country—One definition*. Retrieved 9 September 2006 from http://www.pmo.gov.my/website/webdb.nsf/

Hashim, R. (2000). Memasyarakatkan teknologi komunikasi maklumat (TKM) menerusi projek e-komuniti: Yakin boleh? [Socialization of information and communication technology (ICT): Have the confidence?] Paper presented at Seminar on Socialization of ICT, 10–11 May 2000. UPM Serdang.

Hasim, M.S. (2000). South East Asian countries' response to the advent of the information era: With emphasis on Malaysia. *Jurnal komunikasi*, 16, 1–32.

Hasim, M.S. (2002). *Introduction to media and cyber laws*. Kuala Lumpur: Utusan Publications and Distributors Sdn. Bhd.

Hasim, M.S. (2005). Opportunities and challenges of the Internet: Coping with legislations in cyber space. Paper presented at IAMCR Conference, 27 July 2005. Taipei, Taiwan.

John, K.J. (2002). Towards a K-Malaysia: A vision and strategy. Paper presented at the International Conference Partnership Networks as Tools to Enhance Information Society Development and Knowledge Economy, 9 December 2002, Moscow, Russia. Retrieved 4 February 2007 from http://www.globalknowledge.ru

Kementerian Kemajuan Luar Bandar dan Wilayah. (2005). *Infodesa*. Retrieved 26 September 2006 from http://www.rurallink.gov.my/Pk_Default.php?

Kementerian Pendidikan Malaysia. (2002). *MYSchoolNet*. Retrieved 28 September 2006 from http://www.myschoolnet.ppk.kpm.my/main/perihal.htm

Kementerian Tenaga, Air dan Komunikasi. (2006). *The industry*. Retrieved 11 September 2006 from http://www.ktkm.gov.my/template.asp

Kementerian Tenaga, Air dan Komunikasi. (n.d.). *Portal komuniti desa* [Rural community portal]. Retrieved 30 August 2006 from http://idesa.net.my

Malaysia. (1996). *Seventh Malaysian Plan 1996–2000*. Kuala Lumpur: Percetakan Nasional Malaysia Berhad.

Malaysia. (2006). *Ninth Malaysian Plan 2006–2010*. Kuala Lumpur: Percetakan Nasional Malaysia Berhad.

Malaysian Communication and Multimedia Commission. (2006). *Facts and figures*. Retrieved 15 September 2006 from http://www.mcmc.gov.my/facts_figures/stats/index.asp

Malaysian Public Sector. (2006a). *Government decision on OSS implementation*. Retrieved 18 September 2006 from http://opensource.mampu.gov.my/index.php

Malaysian Public Sector. (2006b). *Objectives of OSS implementation*. Retrieved 18 September 2006 from http://opensource.mampu.gov.my/index.php

Malaysian Public Sector. (2006c). *Master plan—Target to achieved*. Retrieved 18 September 2006 from http://opensource.mampu.gov.my/index.php

MAMPU. (n.d.). *Citizen and resident of Malaysia*. Retrieved 26 September 2006 from http://www.gov.my/MyGov/BI/Directory/Citizen/

Maxis Communication Berhad. (2006). *Giving back to society*. Retrieved 2 August 2006 from http://www.maxis.com.my/personal/about_us/profile/giving.asp

MIMOS Berhad. (2004). *The organization*. Retrieved 16 August 2006 from http://www.mimos.my/about2.html

MIMOS Berhad. (2005). *About MyGfL*. Retrieved 28 September 2006 from http://www.mygfl.net.my/mygfl/index_php3?page=About&Lang=571&ms=main

Ministry of Energy, Water and Communication. (n.d.). *MyICMS886 strategy*. Putrajaya.

Ministry of Science, Technology and Innovation. (2004). *MOSTI grants*. Retrieved 18 September 2006 from http://www.mosti.gov.my/MasterPortal/website/index.jsp

Multimedia Development Corporation. (2006). *The multimedia super corridor*. Retrieved 9 August 2006 from http://www.msc.com.my/msc/msc.asp

NITC. (2002). *DAGS' pride: A collection of DAGS projects*. Kuala Lumpur: NITC Malaysia.

Portal Pendidikan Utusan. (2005). *Tentang kami* [About us]. Retrieved 28 September 2006 from http://www.tutor.com.my/tutor/info.asp?pg=about.htm

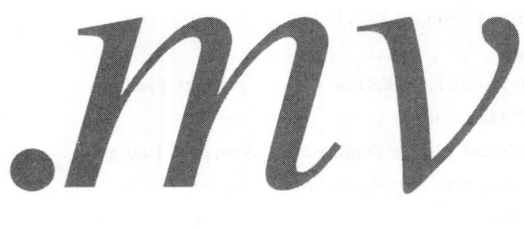

Maldives

Malika **Ibrahim** and Ilyas **Ahmed**

Total population	298,842 (2006)
GDP per capita (as of 2005)	USD 2,271 (USD 1 = MVR 12.75)
Key economic sectors	Tourism, Fishing
Fixed-line telephones per 100 inhabitants	11
Mobile phone subscribers per 100 inhabitants	91
Internet users per 100 inhabitants	25
Broadband subscribers per 100 inhabitants	2
Domain names registered under .mv	1,210

Notes: All indicators are based on 2006 year-end statistics, unless otherwise specified. Internet users are estimated by adding subscription customers and casual users (consisting of users of open access dial-up, users who share home or office connections, and cybercafé users).

Overview

Like most other countries in Asia Pacific, the Maldives experienced some successes in the telecommunications sector, especially in terms of mobile coverage. At present it enjoys a 100 per cent geographical coverage and over 90 per cent telephone penetration. Mobile phones form a very basic foundation for communication for the communities in the 200 inhabited islands of the Maldives.

The importance of ICTs was seen especially after the tsunami of 26 December 2004. This was the worst disaster ever to hit the Maldives, an archipelago about a metre from sea-level. Waves of about 1–3 metres were reported all over the country. These waves destroyed the infrastructure and livelihood of affected islands. It was reported that 82 perished, about 12,000 people were displaced and 26 went missing and were presumed dead. Tourism, the largest contributor to GDP, suffered with the closing down of 19 of the 87 resorts in the country. The tsunami also affected a significant part of the telecommunications network, causing interruptions in all telecommunications services in a major part of the country. Telecom services in the Maldives run primarily on a microwave network, extending from the capital Malé towards the northern and southern parts of the country. During the tsunami, five major nodes of the network were badly damaged, disrupting communications to 13 of the country's 20 atolls. The radio equipment and shelters housing the equipment were also damaged beyond repair. The simultaneous failure of the main power on the islands added to the difficulties in maintaining communication services, even on islands where the damage to the main telecom equipment was not as severe as in the worst hit areas.

With coordinated action by Telecommunications Authority Maldives (TAM) and the telecom company Dhiraagu, the telecom network and services were restored using available means and equipment. TAM, in collaboration with its operators and other international aid organizations, is now aiming to improve the resilience and diversity of the telecommunications network.

In August 2005, with the introduction of the second mobile phone operator, telecom services began to expand rapidly. Mobile technology is being used not only for voice communication but also for the Internet and as a substitute for fixed-line telephones. Once alien, the Internet is now a household term even in the outer islands. Under the e-government project, all government offices are being connected through a wide area network. This network will extend government service online to citizens. The e-government project will commence service during the last quarter of 2007.

Technology infrastructure

Like other developing countries, the Maldives is striving to catch up with modern technology and to adapt to the rapid changes in the ICT world. Since 2001, both government and the private sector have exerted significant efforts to develop the ICT sector and to strengthen related institutions in order to modernize the country. Specifically, the government of the Maldives adopted an accelerated ICT development policy when it launched its first Telecommunications Policy in 2001. The positive results of the development of telecommunications benefit all social strata.

Comprehensive communications services, including telephone on demand and ADSL broadband Internet, are now available in Malé and the major population centres. These areas cover 13 of the 200 inhabited islands, corresponding to about 42 per cent of the population. All inhabited islands have access to fixed-line telephones and cellular telephone services are available throughout the country. Teledensity as of end 2006 was 11 per cent for fixed telephones and 89 per cent for mobile telephones.

Aside from ADSL, broadband Internet is available via cable TV networks (CATV) in Malé and a few other islands. Forty-five per cent of the population thus have access to broadband Internet services. In the islands that do not yet have broadband access, the Internet is accessed primarily via small telecentres and through the use of mobile phones. The two cellular phone networks have EDGE technology countrywide.

The telecommunication backbone is a digital microwave network that runs from north to south, connecting all inhabited islands as well as industrial islands. Either fixed-line switches or remote local switches are installed in major population centres.

The current international gateway is established via satellite. However, plans are underway to connect the Maldives to the international optical submarine cable network. To this end, two submarine cable projects were undertaken in 2005 and were partly functional by the end of 2006. One of the cable projects

---

Connecting the Maldives to the international submarine cable network

Like most small countries, the Maldives has been relying on satellite technology to connect to the outside world. The main reason for resorting to this technology is the cost-effectiveness of satellite for the level of international tele-traffic that a small country like the Maldives generates.

Having submarine optical fibre connectivity instead of satellite has been a dream for the Maldivians for quite some time. While global submarine optical fibre cable networks like SE-ME-WE (South East Asia-Middle East-West Europe) have gone around the Maldives to offer opportunities to tap into them, the high cost of joining these cable consortiums prevented the country from reaping the technical benefits of optical fibre technology. When considering a cable system for Maldives, the typical argument is: Satellite is adequate for low traffic and the cost of a wide fibre bandwidth is neither required nor justified.

So, should the Maldives live with satellite forever? Recall the story of cars and roads. The typical argument against building extra roads is that for the number of cars that exist, the roads are enough. But consider what happens when new roads and highways are built: people who did not previously own cars, buy cars and existing cars are put to more use. Furthermore, types of vehicles that did not exist before, such as heavy freight trucks, start making use of the roads. In no time, the new highways are congested. And the cycle goes on. Can this analogy be applied to connecting the Maldives to an international fibre system? Satellite technology was sufficient in the past when voice telephony was the driver of international communications. Now, voice is no longer the major consumer of international bandwidth, and the bandwidth consumed by data applications such as the Internet has surpassed the bandwidth usage of voice.

In 2004, the Government of the Maldives took the initiative of revisiting the feasibility study for installing an international optical fibre system in the country. In 2005, the bold decision was made to connect the Maldives to the rest of the world using an optical fibre system. A consortium was established among the telecom service providers, Wataniya Telecom Maldives, Focus Infocom Maldives and Reliance Infocom of India, to proceed with connecting Maldives to a fibre system. The company established by the consortium, WARF Telecom International, brought the first fibre into the country in October 2006. It connects the Maldives to Falcon Network at a node in Trivandrum, India. A few days later, in early November, Dhiraagu brought in a cable connecting the Maldives to Colombo, Sri Lanka. As of this writing, work is underway to test both cables. The Dhiraagu cable was inaugurated in December 2006 and the WARF cable in March 2007.

To use the analogy of roads and cars once again, with the cables in place, the 'roads' are now established. Will traffic increase correspondingly? Voice and data traffic alone justify the costs. Equally important, industries that were not possible in the past, due to the limited connectivity are expected to develop. One example is call centres. Data-centric activities, such as data warehousing, is another potentially lucrative industry. Other sectors and industries not directly related to ICT are also expected to benefit from the optical fibre systems.

was undertaken by Dhiraagu while the other was undertaken by a company formed by the other telecom service providers. The cable systems are to become fully functional by the end of the first quarter of 2007, relieving the Maldives from the inherent limitations of satellite technology in the international gateway.

To reap the benefits of ICT, the government has embarked on a project to connect government institutions via a comprehensive computer network. This e-government project has two major components: developing the physical network, and building the applications that would run on the network. It is expected that citizens will start using e-government services by the last quarter of 2007.

Key ICT institutions

The Ministry of Transport and Communication is the line ministry responsible for policymaking with respect to ICT. Under the Ministry are two institutions directly involved with the ICT sector. One of these is TAM, which is responsible for the development and regulation of telecommunications. The second is the National Centre for Information Technology (NCIT), which looks after IT development and the establishment and operation of the government network.

Within the telecommunications sector, there are currently three licensed operators. The national telecom service provider Dhiraagu provides all telecommunication services, including mobile phone and Internet services. Focus Infocom is the second Internet service provider, while Wataniya Telecom Maldives provides mobile services as the second mobile telephone operator. Competition in the telecommunications sector is still in the early stages, with the new players working hard to gain a market share. Focus Infocom and Wataniya have become significant in the market, but Dhiraagu still has the largest share in the Internet and mobile services market.

Digital content initiatives

Like most developing countries, the Maldives lacks available local digital content. The National Centre for Linguistic and Historical Research (NCLHR) has attempted to develop digital content in the Maldivian language, Dhivehi. One of their initiatives is the compilation of the 'Basfoiy', a CD-ROM of Dhivehi words and some common phrases used in everyday communication. It is used in government offices where the administrative language is Dhivehi. It is also widely used in academic institutions by students studying the Dhivehi language.

Another digital content project is the Digitization of Local Content Initiative of the Ministry of Transport and Communication (formerly the Ministry of Communication, Science and Technology) and the United Nations Development Programme. Under this initiative, practical information, such as those in pamphlets, booklets and guidebooks, was collected and then digitized on a CD for distribution to island communities that lack easy access to this kind of information. Initially the project team collected content that is freely available from the organizations involved. Later the project team met with some key organizations, such as the Ministry of Health, Ministry of Fisheries and Agriculture and Ministry of Atolls, that contributed useful information in printed and electronic formats. Some information was also taken from the organizations' websites. One of the challenges for the project team was converting some printed material into a readable format that would then be browsable and searchable on the CD. The CD has been distributed to all atoll schools, atoll capitals, schools in Malé, and ministries, government departments and NGOs.

There has also been an increase in the content available on the Internet. Local newspapers have websites with content in English and Dhivehi. Their biggest audience are Maldivians studying or living abroad. The online newspapers are also finding an audience in islands where it is difficult to circulate the print version on a daily basis. Most government agencies and major private agencies have websites. The websites of resorts are accessed mostly by would-be tourists and resort guests. The Maldives Tourism Promotion Board also provides links to these websites. Government websites are accessed mostly by government employees and citizens living in Malé. However, some websites offering online services are also accessed by people living in the islands.

Another project that aims to provide local content especially to the island communities is called Digital Empowerment of Island Communities. The biggest component of this project is the establishment of two Web portals for the island communities. The original plan was to complete the project by early 2005. However, due to the change in government structure in mid-2005, project ownership was transferred from the former Ministry of Communication, Science and Technology to the National Centre for Information Technology, which caused delays in providing online services through the portals. The preliminary work on the portals has been completed, such as the system (hardware and software) and the training necessary for hosting the portals.

Online services

There has been a significant increase in online services in the Maldives since 2002. The private sector, especially the tourism sector, took the initiative of developing online services when it started offering online booking for tourists. The telecommunications operators also provide online services to their customer base, such as downloading application forms and checking bills online. A favourite service among customers is the Web-based short message service (SMS) to mobile phones provided by Dhiraagu. Dhiraagu also provides directory inquiry services on its Web portal. Dhiraagu customers can use the directory service for name searches as well as for number searches (reverse directory). The latter is another customer favourite.

The health sector also provides basic online information to the public. The Ministry of Health offers information on nutrition, diseases and national health indicators on its website. The website also has e-books on baseline health studies that can be downloaded for free. The website of the Department of Public Health provides information on various diseases, food and safety, and maternal and child health care. IGM Hospital and ADK Hospital, the two major hospitals in Malé, also provide information on doctors' schedules and other services on their websites.

ADK Hospital is also the first hospital in the Maldives to offer telemedicine services to its patients. The hospital is affiliated with a foreign medical partner to which it sends medical records, such as medical investigations and laboratory reports, for a more accurate diagnosis. This way, the doctors at ADK Hospital can make well-informed decisions in treating their patients.

Other health-related public organizations that provide information online include the Maldives Nursing Council, the Board of Health Sciences and the National Thalassaemia Centre. Two major NGOs, the Society for Health Education and Care Society, also provide online information through their websites.

Most of the government websites give information on obtaining services. Almost all websites offer various forms that can be downloaded, filled in and mailed or faxed to the relevant organization. For example, people can download the passport application form from the Department of Immigration and Emigration website and report lost passports, which is a highly useful feature for Maldivians who are travelling. The Maldives Police Services has a similar feature on its website for reporting stolen or lost National Identification Cards and passports. The Ministry of Higher Education, Employment and Social Security website has information on scholarships, employment, expatriate employment, employment agencies, employer guidelines and employment contracts, as well as downloadable scholarship application forms.

The Ministry of Fisheries, Agriculture and Marine Resources publishes market prices for different species of fish and local agricultural produce, as well as fish-catch by different fishing regions in the Maldives, which is important information for local fishermen. The Ministry of Planning and National Development publishes the *Statistical Yearbook* on its website. The Ministry of Construction and Public Infrastructure has an online database called Harbour Permit and Mooring Database, which can be used to obtain information about the types of permits given to vessels using the inner harbour of Malé and Villingili.

ICT and ICT-related industries

In 2004, a concept paper to develop the key initiatives that need to be undertaken to facilitate the development of what was then called 'The Maldives IT Village' was drafted. The National Centre for Information Technology (NCIT) decided to pursue the concept further and develop the IT industry to help diversify the economic base, which in turn would attract foreign IT companies to invest in the country. It is hoped that the IT industry will generate employment opportunities, especially for the youth (NCIT 2006).

NCIT states the objectives of the IT Industry Plan as follows:

1. To achieve significant growth in terms of employment and financial turnover;
2. To be export-focused and make a positive contribution to trade performance;
3. To underpin innovation and growth in other key industry sectors;
4. To foster a reasonable number of small, dynamic and rapidly growing firms able to work alone as well as in strategic alliances; and
5. To attract and land one or two key technology multinational companies to act as anchor tenants.

The ICT industry in the Maldives consists of computer hardware and software dealers, software developers, Web developers and network solution providers. A number of vendors are selling hardware and software. One of the factors contributing to the bigger presence of hardware vendors could be the low duty (only 5 per cent) levied on imported computer hardware and consumer electronics equipment.

Despite the many software vendors, the number of software developers is relatively low. One of the reasons for this could be the lack of expertise to provide continuous support for software applications. Because there is a lack of local personnel trained in software development, software companies often have to recruit staff from neighbouring countries like India and Sri Lanka.

A mobile payment and banking system in the Maldives

The Maldives Monetary Authority, the country's Central Bank, is planning to introduce a 'Mobile Telephone Payments and Banking System'. The system will be available in the whole country and all mobile phone companies and banks will participate. This is achievable because more than 95 per cent of the Maldives receives mobile telephone service and mobile telephone ownership is close to 90 per cent.

There will be two components to the system:

(a) an Electronic Funds Transfer Exchange, which would be both a software platform and a physical business unit to provide a clearing system for payments (similar to a cheque clearing system), answer customer queries, maintain the software, and sign up and maintain banking agents; and
(b) a network of banking agents comprised of shops and similar entities around the country to operate as cash handling points.

Initially, the Exchange would be housed as a separate department at the Maldives Monetary Authority. Later it will be spun off as a separate legal entity owned by banks, mobile telephone companies and the public.

Implementation of the system would require an extensive awareness campaign and education of the users and banking agents. It would also require proper legislation to ensure that customer rights are protected. The mobile telephone companies already have the technological capability to provide the service with minimum investments.

This technology has been tested in many countries, including Indonesia, the Philippines and South Africa. With this system, anyone in any part of the country can open a bank account, subscribe to a mobile phone, and buy insurance; settle utility bills; pay for groceries, cinema tickets, taxi fares, and café bills; repay loans; pay government license fees, taxes, and other fees; pay salaries and make welfare payments and accept deposits. In sum, the mobile phone becomes an electronic wallet—a wallet that would not lose the funds in it even if the mobile phone is lost or damaged.

According to the Maldives Monetary Authority, the technology would save costs associated with printing and supplying cash, free up resources currently used by banks to deliver banking services, automate the banking and clearing system, reduce the transaction costs of banking and payments settlements, deliver banking services to the whole country, increase national savings, and revolutionize the financial industry. The system is also expected to contribute to national efforts to create a knowledge workforce. It is hoped that with the introduction of the e-payment gateway through the e-government project, more e-commerce services will be introduced in the Maldives.

However, as the number of institutes and training centres offering computer courses have increased, there might be enough human capital to sustain a software industry in the future. The few software developers in the market have been quite successful in developing software for the hospitality industry. They have even managed to market the software to other countries. Point-of-sale systems and hospitality management software for tourist resorts are the most commonly developed software in the Maldives.

There are also a few network solution providers integrating hardware, software and services. The widespread use of mobile services has given rise to a boom in mobile phone shops within the last three years. The shops not only sell the latest mobile phones and gadgets, but also provide comprehensive after-sales service.

Enabling policies and programmes

ICT programmes in the Maldives are driven mostly by the Telecommunications Policy and the National IT Policy. Telecommunications Policy 2001–05 sought to reduce telecom charges, open up the telecommunications market, improve accessibility and strengthen the regulator. Most of its provisions and action plans were implemented. Telecommunications Policy 2006–10 places emphasis on providing telecom services with non-discriminatory charges to all islands, further developing the telecom infrastructure and providing broadband services. It also aims to make the regulatory authority an autonomous body with clearly defined powers. Telecommunications development is paramount, with special emphasis on providing extended services using mobile communications technology (m-services). Competition will be stimulated further and new services are to be opened for competition during the policy's tenure.

The National IT Policy has been on the drawing board for quite some time now. The project to formulate the National IT Policy was a collaborative effort between UNDP Maldives and the National Centre for Information Technology of the Ministry of Transport and Communication. The project ran into delays due to the unavailability of local consultants to work on the project as well as the change in government structure in July 2005, which dissolved the former Ministry of Communication, Science and Technology and transferred the communications functions to the newly formed Ministry of Transport and Communication.

Legal and regulatory environment for ICTs

Currently there is no comprehensive telecommunication law. The key legal instrument governing the telecommunications sector is Maldives Telecommunication Regulation 2003 issued under a Presidential decree. Work is underway to formulate an Act covering telecommunications services in the Maldives. The draft, which has been completed, will be submitted in the form of a bill to the Parliament of the Maldives.

There are likewise no specific cyber laws as yet. However, the new telecom policy being drafted for the period 2006–10 calls for controlling cyber crime and articulates the need for cyber laws. Their introduction is among the policy's thrusts.

Open source initiatives

Open source software (OSS) is not very common in the general consumer market in the Maldives. The reason perhaps is the ready availability of proprietary software. However, the service providers use a number of applications and backend software based on open source systems. In ISP services, both Dhiraagu and Focus Infocom use Linux/UNIX-based applications.

Other than the ISPs, there are small groups and individuals working to promote open source applications. They carry out this activity mostly as a hobby when they have time. They occasionally produce small applications such as instant messaging in the local language and add-ons for existing applications. Given the right environment and facilities, these highly talented individuals could be motivated to produce high-quality applications comparable with pricey proprietary applications.

Capacity building and R&D

Within the last five years, there has been a major increase in the number of training centres and institutes offering ICT training courses. Some of these are locally developed while others are offered through international affiliations. The courses vary from basic computer skills to professional programs in networking and Web development. In 2005, the Faculty of Management and Computing of the Maldives College of Higher Education started offering degree programmes in IT.

As for research in ICT, there are currently few research activities within the ICT sector. However, there are efforts to promote R&D initiatives. Incubator facilities are being developed by the NCIT to help small players to establish themselves. The new telecom policy calls for the development of new services based on mobile technology. Actions under this policy include the establishment of a think tank to foster the development of modern and innovative applications using wireless and mobile technology. In establishing the think tank, which is expected to commence in 2007, it is important to create the right work environment so that free thinking and creativity is given the highest priority and bureaucratic and government formalities are avoided as much as possible. Appropriate remuneration should be given to members to sustain participation.

Challenges

As a small island developing state that is geographically dispersed with more than 1,900 islands, the Maldives faces numerous development challenges. One of the main challenges is the difference in population densities in different atolls and islands. While some islands have a population of more than 5,000 people, other islands have a population of fewer than 500 people. The government faces difficulties in replicating basic services, such as health and education, in these islands. The sparse population in these islands translates to lack of economies of scale for certain initiatives, which makes it extremely difficult to attract private parties to invest in these islands.

Not surprisingly, there are disparities between these islands and the capital Malé in terms of ICT use. Extending ICT services to these islands is often time-consuming and expensive. However, despite these difficulties, the Maldives boasts a 100 per cent geographical coverage of its telecommunications network. The network itself consists of different technologies, as one technology is not enough to provide a total solution to these low-lying, flat islands scattered over an area of 90 square kilometres.

Telecom operators and government agencies like NCIT are working continuously to extend the ICT infrastructure to the outer islands. By the end of 2007 or sooner, the Maldives will be connected to the rest of the world by an optical fibre system. All of the atoll capitals will be connected via a network under the e-government project and the e-Government Service Platform will offer services online to the citizens of the Maldives.

References

Dhiraagu Private Limited. (2007). *Press releases 2007*. Retrieved 1 January 2007 from http://www.dhiraagu.com.mv/newsdesk/index.php?type=pressreleases

Focus Infocom Private Limited Homepage. Available at http://www.rol.net.mv/

Government of Maldives. (2003). *Maldives telecommunication regulation 2003*.

Government of Maldives. (2006). *Maldives telecommunications policy 2006–2010*.

Ministry of Planning and National Development. (2006a). *Statistical yearbook of Maldives 2006*. Retrieved 2 October 2006 from http://www.planning.gov.mv/publications/yrb2006/intro.htm

Ministry of Planning and National Development. (2006b). *Maldives key indicators 2006*. Retrieved 1 December 2006 from http://www.planning.gov.mv/publications/KeyIndicators/KeyIndicators2006.pdf

National Centre for Information Technology. (2006). *Projects*. Retrieved 20 September 2006 from http://www.ncit.gov.mv/projects/

Republic of Maldives. (2006). *Vision 2020*. Retrieved 15 October 2006 from http://www.planning.gov.mv/contents/misc/vision2020.php

Telecommunications Authority of Maldives Homepage. Available at http://www.tam.gov.mv

Wataniya Telecom Maldives Homepage. Available at http://www.wataniya.com.mv

World Bank, Asian Development Bank, and the UN System. (2006). *Tsunami: Impact and recovery joint needs assessment World Bank–Asian Development Bank–UN System*. Retrieved 1 October 2006 from http://www.mv.undp.org/Images/joint_needs_assessment.pdf

.mn

Mongolia

Lkhagvasuren **Ariunaa** and Batpurev **Batchuluun**

Total population	2,646,000 (2006)
GDP per capita	USD 1,093 (USD 1 = MNT 1,165 as of March 2007)
Key economic sectors	Agriculture, Mining, Animal Husbandry, Animal Production and Processing (Meat, Milk, Cashmere, Wool), Tourism
Computers per 100 inhabitants	5
Fixed-line telephones per 100 inhabitants	6
Mobile phone subscribers per 100 inhabitants	33
Internet users per 100 inhabitants	0.42
Domain names registered under .mn	1,887
Broadband subscribers per 100 inhabitants	0.21

Overview

Since 2005, the information and communication technology (ICT) sector in Mongolia has developed with the support of a favourable policy and regulatory framework, institutional setup, and commitment from private software, hardware and infrastructure development companies extending ICT services to citizens.

The establishment of the Information and Communication Technology Authority (ICTA) under the direct supervision of the Prime Minister of Mongolia spurred ICT development, beginning with the development and implementation of the e-Mongolia national programme, changes to the legal and regulatory framework for ICT, and cooperation and coordination with companies and local and international organizations and donors, such as the World Bank, Asian Development Bank (ADB), United States Agency for International Development (USAID) and the International Development Research Centre (IDRC) of Canada.

ICT infrastructure has changed with the decision of the Government of Mongolia to separate services from networks. The Mongolian Telecom Company (MTC) has been divided into two companies: the service provider (ServCo) and the network maintenance company (NetCo). ServCo is mandated to provide services over the telecommunications network on the same terms as other service providers while NetCo is tasked with managing the telecommunications network backbone. This has opened up opportunities for service companies to use the network infrastructure on a competitive basis.

Two more mobile operators have commenced operations. The Unitel Company started providing services in June 2006 using GSM 1800 technology. G-Mobile has been awarded a license to provide ICT services in rural areas.

The software development companies have redirected their targets towards introducing and implementing software outsourcing. The hardware companies have set up the Mongolian Association of Computer Suppliers' Companies (MASCO). The ISPs have formed the Mongolian Association of Internet Service Providers (MISPA) to protect the rights of ISPs and to collaborate on addressing issues common to service providers, such as cost of bandwidth and outreach to the rural areas.

Nevertheless, there remain big challenges for the ICT sector in Mongolia. First, the draft laws on ICT are still under discussion. Second, there is a need to integrate ICT with public sector reform. Third, the ICT capabilities of the country's human resources need to be strengthened through curriculum change at the secondary and tertiary education levels. Fourth, there is a significant digital divide between the rural and urban populations. Although majority of the population lives in the rural areas, ICT penetration in rural areas is much lower than in urban areas. For example, the 7,726 rural users of fixed-line telephones represent only 6.4 per cent of the total users of telecommunications services (ICTA 2006) and 86 per cent of ICT businesses are concentrated in the major cities of Ulaanbaatar, Darkhan and Erdenet.

Technology infrastructure

The base line of the Mongolian telecommunications network consists of about 5,200 km of digital system radio relay line, over 7,400 km of fibre optic network, around 27,000 km of air lines connecting *aimag*[1] centres with *soum*[2] centres, 23 VSAT stations with 10 Mbps of bandwidth and international Intelsat satellite stations providing services to over 332 communication stations with a capacity of 135.2 thousand telephones.

There are 342 telecommunications branches nationwide. Of these, urban telecommunications branches have 272 small-capacity telephone stations, 210 branches have radio connections and all branches are able to receive national TV broadcasting. There are 313 communication lines between *aimags* and *soums*, and 288 communication lines between *soums* and *baghs*[3]; 120 *soums* are connected to the central energy system and 105 operate using a solar energy source.

The fibre optic network has been deployed since 2000. It was recently extended to reach over 5,000 km in length, covering north to south along the railway, the South Gobi close to the mining areas, the east from Ulaanbaatar to Dornod, the farthest eastern *aimag*, the northwest from Khuvsugul *aimag* through to Zavkhan *aimag* and Uvs *aimag*, and all the way to the westernmost *aimag*, Bayan-Ulgii.

Five companies provide telecommunications services: Mongolia Telecom, Railcom, Incomnet, Mobicom and Skytel. There are two fixed-line telephone service companies: Mongolia Telecom and Mongolian Railway Authority. There are three mobile telephone operators providing services in all *aimag* centres and 62 *soums*. There are over 774,900 mobile phone users, which means that one in four citizens has a mobile phone.

There are five companies providing VSAT services: MTC, Civil Aviation Authority, Incomnet, Orbitnet and the Meteorology Office of the Ministry of Environment.

Wireless Local Loop (WLL) services were introduced in Mongolia in 1999. Five companies have licenses to operate WLL services: two are licensed to operate in Ulaanbaatar and the rest are obliged to cover rural Mongolia. Currently, they provide services in over 25 locations. As of the end of 2006, there are over 44,200 users of WLL services in Mongolia.

There are 14 ISPs, the majority providing services in Ulaanbaatar. Internet connections include dial-up, ADSL, leased line, Wi-Fi, CATV modem, fibre optic link and VSAT. The major service is through ADSL and dial-up connections.

Key institutions dealing with ICTs

The major institutions dealing with ICT are the Information and Communication Technology Authority (ICTA) and the Communications Regulatory Commission (CRC).

ICTA was established in November 2004 under the Office of the Prime Minister of Mongolia. Its primary responsibilities are to coordinate ICT policy, coordinate and implement ICT-related programmes and projects, and cooperate with national, international and donor organizations.

CRC, which was established in 2004 as an independent institution, handles the regulatory aspects of ICT. Its objective is to establish fair and effective competitive conditions in the IT market and to ensure the provision of high-quality services through the most advanced technologies. It issues licenses and permits, regulates tariffs, sets regulatory service fees, allocates radio frequencies and presides over the settlement of disputes.

In addition, the National Information Technology Park was set up with support from the Government of Korea to provide incubator services for newly established software development companies.

A number of NGOs are also active in the ICT sector of Mongolia. The list is headed by the Mongolian Information Development Association/Mongolian Information Technology Association (MIDAS/MONITA), which was established in 2001 as part of a joint project of the United Nations Development Programme (UNDP) and the Mongolian Foundation for Open Society (Soros Foundation) (MFOS). MIDAS/MONITA has actively promoted ICT in Mongolia through software and hardware exhibitions and projects undertaken with UNDP, the World Bank and other donor organizations. Two other ICT-focused NGOs were established more recently—the Mongolian Association of Computer Suppliers' Companies (MASCO) and the Mongolian Internet Service Providers' Association (MISPA).

The Japan-Mongolian Information Technology Association (JMITA), established in 2002, has been actively engaged in introducing Japanese software engineering examinations for Mongolians since 2005. The second round of training started in September 2006 and will be followed by the first round of examinations to be held in Mongolia. It is hoped that Mongolian software developers will gain the qualifications required by the Japanese software development industry, which would enable software development companies in Mongolia to do outsourcing jobs from Japan.

The Mongolia Development Gateway (MnDG) was established in 2002 with support from the Development Gateway Foundation to harness the use of ICT in sustainable development and poverty reduction activities and strengthen partnerships for development in Mongolia. At present MnDG runs an ICT basic skills training centre and a portal site (www.gateway.mn).

The software development companies have formed the Mongolian Software Industry Association (MOSA) with the aim of protecting their legal rights and of transforming the software industry into one of Mongolia's leading industries with the capacity to penetrate the international market.

Digital content initiatives

The number of websites developed and maintained in the Mongolian language is growing, compared to 3–5 years ago when extensive attempts were taken to support the development of Mongolian-language websites. According to a study conducted by Intec, there are over 2,000 such websites which are hosted either in Mongolia or in other countries. Of these, 62.9 per cent use the .mn domain name, 20.6 per cent use .com, and 16.5 per cent use other domain names, such as .net, .org and .tk. It was observed that 23 per cent regularly update their website contents and 70 per cent have a dynamic structure. Of all websites, 62 per cent are in the Mongolian language and 36 per cent are in two languages (mostly English and Mongolian). Around 50.4 per cent belong to private companies, 12.1 per cent have education and discovery-related content, 11.6 per cent are websites of state and government organizations, 11.2 per cent are information and news websites, and the rest are NGO and personal websites.

The popular websites are information websites such as www.olloo.mn and www.mongolmedia.com, public discussion websites such as www.open-government.mn and www.forum.mn and portal sites such as www.pmis.gov.mn and www.gateway.mn.

The development of local language content in CD-ROMs has also been growing lately. There are learning materials like the e-learning CD-ROM package of the Microsoft Office Suite in the Mongolian language. The 'Innovating ICT for Rural Education of Mongolia' project of the ADB, Ministry of Education, Culture and Science (MOECS), and the Japanese Fund for Information and Communications Technology (JFICT) has supported the development of CD-ROM-based materials for teachers as tools for integrating ICT in teaching practices. Eleven CD-ROMs on using ICT in teaching English, Mathematics, Physics, Chemistry, Biology, Labour, History and other subjects have been developed by teams of software and applications developers and teachers and educators. The CD-ROMs, which are all in the Mongolian language, have been distributed to all of the 600 schools in Mongolia to be used by teachers in the classroom.

Online services

Online services are offered mostly by banks. The Golomt Bank and the Trade and Development Bank of Mongolia have websites through which citizens and individuals can check account balances, transfer money and conduct other online transactions.

One of the thrusts of the e-Mongolia programme is to encourage organizations to develop their own websites and to encourage governmental organizations to make information for the public available via their websites. The websites of the Mongolian Taxation Authority (MTA) and the Ulaanbaatar Mayor's Office are examples of openness and transparency. The MTA website contains an extensive array of information, laws and regulations related to taxes, and over 50 downloadable taxation forms commonly used by businesses and individuals. The website of the Ulaanbaatar Mayor's Office contains all of the orders and decrees issued by the Mayor of Ulaanbaatar City.

In contrast, few businesses in Mongolia are introducing online services. The following provide e-commerce services: www.rose.mn, www.asuult.net, www.banjig.net and www.call2mongolia.com. The first online insurance system in Mongolia was introduced by Practical Daatgal Co. Ltd. (www.practical.mn).

In addition, online distance diagnosis and training services for the rural population are now available through the project 'ICTs for Health Services in Rural Mongolia' implemented at the Health Sciences University of Mongolia (HSUM) and supported by IDRC in 2003–05. The 'Doctor system' (www.pi-hsum.mn/dd/) for distance diagnosis works even with low bandwidth. At present, it is used to transmit patient information from remote areas to the capital city for diagnosis and treatment advice. An e-learning system for medical professionals who live and work in the rural areas is also part of the project. This system (www.pi-hsum.mn/de/) allows users to take online courses, quizzes and a final examination. If they qualify, they receive a credit certificate issued by the Postgraduate Institute of HSUM. It is also possible to take paid courses by using a prepaid scratch card.

ICT and ICT-related industries

The number of mobile service providers has increased with Unitel Company starting operations in June 2006. Unitel offers GSM 1800 technology-based services, mainly in the Ulaanbaatar metropolitan area. A fourth operator, G-Mobile, has been awarded a license to provide mobile services based on GSM 450 technology to rural parts of Mongolia. G-mobile is expected to provide services to 125 *soums* initially and then to an additional 80 *soums* within two years.

Outsourcing is a new ICT industry in Mongolia. Several software development companies have been providing outsourcing services for other companies in Mongolia and in developed countries, such as Japan, the USA and the UK. In this connection, the Mongolian government has redirected the support provided by the National IT Park to companies that are able to provide support to outsourcing bridge companies. Currently, there are about half-a-dozen companies developing software and applications for Japanese and Mongolian companies with staff working in Japan.

In addition, wireless technologies have been penetrating the Mongolian market, with several companies offering wireless solutions. Following the trend, ICTA has set up four free access hotspots in Ulaanbaatar—at the airport, railway station, National IT Park building and Sukhbaatar Square. Hotels, restaurants and other businesses are integrating wireless applications in their businesses. This new trend finds support in the Last Mile Initiative (LMI) of USAID, which aims to pilot test Wi-Fi networks in the rural areas of Mongolia (*see the next section*).

Enabling policies and programmes

The e-Mongolia national programme developed by ICTA and approved by the Government of Mongolia has outlined 16 objectives which are indicated in a plan of action and which are being pursued through projects like 'Computers for All' through which Intel will provide quality computers at affordable prices to the citizens of Mongolia. Six months after the programme's implementation in July 2005, over 8,000 computers had been sold (ICTA 2006), which is a significantly higher number than the average monthly sales of computers without this project. Supplementing the Computers for All project is the effort to get ISPs to reduce the cost of Internet services and to lower telecommunications costs associated with access to the Internet. As a result, low-cost Internet options are now being offered to families and individuals, with telecommunications charges for dial-up connections drastically reduced from MNT 40 per minute to MNT 1 per minute. Moreover, as of November 2006 the Skytel Company, one of the VSAT service providers, has been offering ADSL services.

Legal and regulatory environment for ICTs

Following its establishment and the introduction of the e-Mongolia programme, one of the first activities carried out by ICTA was to develop proposals for a favourable legal and regulatory framework for ICT development in Mongolia. Aside from existing ICT-related laws, such as the Law on Telecommunications, Law on Radio Frequency and Law on Post, there are several laws and regulations that make reference to ICT. These include the Law on Education (approved in 2002), Law to Protect Intellectual Property, Law on Author's Rights, Patent Law, Law on Technology Transfer and Law on Science & Technology. Each of these laws contains clauses on the introduction and integration of ICT in the sector. For example, the Law on Education has a clause about reforming the curriculum of informatics subjects in the secondary school and developing a world-class curriculum for ICT specialists.

Although extensive work has been undertaken in drafting a general law on ICT and laws on e-government, digital signatures and e-commerce, they need to be revised in light of the latest developments in the ICT sector. The World Bank's Information and Communications Infrastructure Development Project has components to review the existing legal environment for

The last mile initiative

USAID's Last Mile Initiative (LMI) is a global programme to expand the rural poor's access to communications. Launched in April 2004, LMI intends to spur increases in productivity and transform the development prospects of farmers, small business, new start-ups, and other organizations in rural areas presently underserved by the world's major voice and data telecommunications networks. Six countries were selected to participate in LMI in its inaugural year. Fifteen additional countries were selected in 2005, Mongolia being one of them.

The goal of the Mongolia LMI Project is to install low-cost community-centric telecommunication services—primarily voice—in four *soum* centres as a way of providing those living in these rural communities with telephony access that would link them to the other cities in Mongolia and to the world. During the initial assessment, a Wi-Fi-based pilot project was suggested as the best means of connecting rural Mongolia in an effective and affordable way. The project carried out tests of Voice-over-Wireless Fidelity (VoWi-Fi) phone networks in rural Mongolia and developed a more detailed project plan. Advanced VoIP/Wi-Fi technologies are expected to deliver Internet-based telephony to rural communities at a substantially reduced cost compared to other technologies.

Currently, the LMI project is being piloted in Saikhan *soum* of Bulgan *aimag* and Tsengel *soum* of Bayan-Ulgii *aimag*.

the ICT sector and to provide recommendations to improve public–private partnerships for e-government. At the same time, the European Bank for Reconstruction and Development is funding the revision of the existing telecommunications law by the CRC. A working group composed of representatives of ICTA, the private sector and NGOs has been established to coordinate the efforts of various stakeholders and to develop a draft package law on ICT which would include a general IT law and draft laws on communications, e-government, digital signatures, e-transactions, information security and freedom of information.

Education and capacity building

The major state institutions offering ICT-related courses are the Computer Science and Management School (CSMS) of the Mongolian University of Science and Technology (MUST), the School of Mathematics and Computers of the National University of Mongolia (NUM), the School of Information Technology of NUM and the School of Computer Science and Technology of the Mongolian State University of Education (MSUE). There are also private ICT schools, such as the Khuree Institute and the Ulaanbaatar Institute.

Microsoft-certified and CISCO Academy training programmes have likewise been available in Mongolia in the last 3–6 years. The Intec Company recently introduced the Aptech Certified Computer Professionals Programme from Aptech, India. This initiative is expected to give Mongolians the opportunity to receive certified computer education from well-known and highly reputable providers in Southeast Asia.

The Asian Development Bank has been working extensively with the Ministry of Education, Culture and Science (MOECS) on the development of the education sector, with a focus on the development of ICT-related education. One of the components of the Second Education Development Programme (SEDP) is introducing ICT in secondary complex schools through provision of computers and hardware and basic computer skills training. In addition, technical assistance is being provided to introduce ICT in rural *soum* schools, including teacher training in computer and technology-related skills and in integrating ICT in the classroom.

Open source and open content initiatives

There is currently no government policy paper for open source software and open content-related activities. Not even the e-Mongolia programme mentions open source. However, members of the Mongolian ICT community appreciate the value of open source applications and content.

ICTA has launched, within the framework of the e-Mongolia programme, the 'Web and E-mail' sub-project which involves the customization and adoption of the Joomla open source content management system for government organizations. A recent InfoCon study shows that nine out of 15 ISPs use OSS in their operating, networking, Web serving, mail serving, or database serving systems (for example, Redhat, FreeBSD, BIND, Samba, Apache, Tomcat, MySql, PostgreSql, Sendmail and Mailman).

Two NGOs, OpenMN and the Mongolian UNIX Users Group (MUG), are advocating the promotion, utilization and development of open source software in Mongolia. OpenMN (www.openmn.org), which was founded in 2003, has released the Mongolian version of a UNIX-like operating system named Soyombo 1 as a modular distribution based on Morphix with live CD support (bootable and operable from a CD). At present, it is translating KDE and GNOME into the Mongolian language. MUG, which was established in 2004, successfully organized the first Mongolian Linux Festival and SysAdmin Summit in 2006.

Also in 2006, the Ministry of Education, Culture and Science developed the Mongolian e-library (www.elibrary.mn) to meet the information needs of Mongolians.

Other major sources for online open content in the Mongolian language are www.gateway.mn operated by MnDG, www.forum.mn by Open Society Forum NGO, and www.montsame.mn by the Mongolian National News Agency. The goal of the MnDG portal is to provide a variety of quality open and inclusive content. The portal system intends to link existing knowledge networks and bring together the best available information about development issues. The site uses a 'deferred publishing' approach, whereby content suggested by users is reviewed prior to publication on the site. The Open Society Forum is a new initiative of the Mongolian Foundation for Open Society. Its goal is to provide broad access to information resources about policies, laws and regulations and to provide a venue for public engagement in the policy formulation and implementation monitoring process. At present, it is focusing on three broad themes: Governance, Economic and Social Policies.

The Information Technology Thesaurus Project (http://www.itdic.edu.mn/term/) is an online dictionary designed by IT specialists and teachers to standardize the use of terminology in the IT sector. A team of editors has been appointed to produce a print version that will be made available to the public.

The Evaluation and Adaptation of OSS for Distance Learning in Asia project (http://www.infocon.mn/eng/index.php?inf=projects#9) aims to evaluate existing open source distance learning software and identify suitable software that can be customized to meet specific needs of educational institutions in the Asian region. The project is supported by IDRC.

Research and development

In May 2006, the Information and Communications Infrastructure Development Project, a USD 10 million technical assistance package of the World Bank, was approved by the Government of Mongolia. The project has three major components: (a) universal access, (b) monitoring the radio spectrum and (c) e-government. The first component was first implemented in 2004, when the feasibility study was conducted among herders of three *soums* of Arkhangai and three *soums* of Bayankhongor *aimag*. The study showed that herders are willing to spend at least MNT 5,000 per month (about USD 4.29) for telephone services as long as the service will be provided in locations closer to them. Based on the study results, the universal access component of the project aims to provide telecom services in *soum* centres and *bagh* centres. Such services are now available in centres in Tariat, Undur-Ulaan and Khangai *soum* of the Arkhangai *aimag* and in at least 10 *bagh* centres in each of these *soums*. Mobicom and Incomnet are to provide the services. On the drawing board is a study of demand for telecommunications among herders in the eastern, western, northern and southern *aimag* to be undertaken in May 2007.

Also worth mentioning is the Naraa Foundation ICT4D Scholarship funded by IDRC. Awards are primarily for students of IT schools of government universities. Students from other faculties (such as health, agriculture and education) who are incorporating ICTs in their research are also eligible. At present about 30 students are receiving the scholarship. The project aims to establish an ICT4D scholarship fund to support Mongolian tertiary-level students who are completing their Bachelor and Master's degree courses in ICTs.

Challenges

First, more work is required for the further development of ICT-related laws and regulations in Mongolia. In light of the rapid changes in the ICT field worldwide, the draft laws on ICT, including the general law on ICT and the law on e-commerce, digital signatures and e-government, need to be reviewed and adjusted if necessary to take into account new developments such as those related to e-security. The draft laws must also be approved by the Mongolian Parliament.

Second, e-government needs to be developed. Initial steps in this direction are: introducing public servants to e-mail, using an internal filing system and integrating ICT in the everyday work of public administration officials.

The third major challenge is in the area of human resource development. There is a need to amend the curriculum of tertiary institutions and secondary schools, including the curriculum of ICT-related subjects, courses and degrees. The Aptech Centre in Mongolia is expected to play a significant role in integrating the latest ICT-related curriculum in the curriculum of tertiary institutions as well as in integrating ICT in the secondary schools.

Finally, the biggest challenge concerns the education and training of the general public, especially individuals who are not working in ICT fields. Education and training programmes in basic ICT literacy and in the use of ICT-related services such as online payments and online registration need to be organized. Equally necessary is the development of local content responding to the needs and demands of users, given that there are currently few online resources and services available to Mongolians.

Notes

1. An *aimag* is an administrative unit equivalent to a state or a province. An *aimag* consists of *soums* and *baghs*.
2. A *soum* is an administrative unit equivalent to a county. There are over 300 *soums* in Mongolia. Each *soum* consists of 3–5 *baghs*.
3. A *bagh* is the smallest administrative unit of Mongolia. There are over 1,500 *baghs* spread throughout Mongolia.

References

Communications Regulatory Commission. (2004). *Handbook*. Ulaanbaatar.

ICTA, CRC, Intec and MIDAS/MONITA. (2006). *White paper on ICT development in Mongolia—2006*. Ulaanbaatar.

Infocon. (2006). *Mongolian web sites—2006*. Ulaanbaatar.

Information and Communications Technology Authority. (2005). *Concept for developing common network of Mongolia*. Ulaanbaatar.

Intec Company. (2005) *Study of Mongolian-language websites*. Ulaanbaatar.

Mongolian Taxation Authority Homepage. Available at http:www.mta.mn

.mo
Macau
Geoff Long

Population	508,473 (2006)
GDP per capita	USD 24,274 (2005)
Key economic sectors	Gaming, Tourism, Construction
Fixed telephone lines	174,400 (2005)
Mobile phone penetration	111 per cent (2005)
Internet subscribers	88,694 (2005)
Internet household penetration	66 per cent (2005)
Annual Internet usage	79.2 million hours (2005)
Internet international bandwidth	1,125 Mbps (2004)

Source: *WTO Trade Policy Review*, Macau Government.

Overview

Macau has transformed itself over the past two decades from a manufacturing-based economy to one that is now dominated by the services sector. In particular, the liberalization of the gambling industry in 2002 has led to massive investments that have helped boost activity in numerous areas of the economy, particularly construction, the hotel sector and tourism.

Since it was returned to China from Portuguese administration in 1999, the Macau Special Administrative Region (SAR) has recovered from the Asian financial crisis and SARS, and is now one of the most vibrant economies in the world, with an average annual growth of 14.5 per cent recorded between 2001 and 2005. This growth is expected to continue over the next few years, as a number of high-profile projects are completed and foreign investment shows no signs of waning.

Macau is situated at the Pearl River Delta on the southeast coast of Mainland China, only 60 km southwest of Hong Kong. Home to 500,000 residents, Macau's roughly 28 sq km land mass consists of the Macao peninsula and the islands of Taipa and Coloane. The land area has actually been enlarged by more than 50 per cent since 1912 through land reclamation projects, which are continuing.

Prior to 2002, all of Macau's casinos were controlled by the STDM syndicate. In March 2002, the government awarded three new casino concessions, one to an STDM subsidiary and the other two to Wynn Resorts of Las Vegas and Galaxy Casino Company of Hong Kong, which has since given a sub-concession to the Venetian Group of Las Vegas. This in turn stimulated private investment of around USD 1.55 billion in various casino projects, hotel resorts and other tourist infrastructure, while several billion dollars of investment are planned for between 2006 and 2010.

Today, Macau's services sector accounts for around 90 per cent of GDP, while manufacturing has fallen to just 4.3 per cent of GDP in 2005. Gaming and related tourism services constitute the core of the economy, with gross gaming receipts accounting for 49 per cent of GDP in 2005 and direct taxes on gaming providing 75 per cent of government revenue in 2006. Because of these revenues, the government can afford relatively low taxes, which also attracts private investment. Traditionally, governments in Macau have allowed the private sector to be the prime economic driver, including for information technology development.

Despite the success of the gaming sector and related tourism, the government has recognized that relying too heavily on any one industry can be a dangerous strategy—as demonstrated by the devastating effect of SARS on tourism in 2003. As a backup, it has started to undertake measures to encourage the diversification of the economy, such as incentive schemes for the industrial and commercial sectors that have not benefited from the growth in tourism and gaming. Initiatives include loan schemes for small and medium-sized enterprises (SMEs) and the construction of the Macau-Zhuhai Transborder Industrial Park, which aims to take advantage of the Closer Economic Partnership Agreement (CEPA) concessions signed in 2003 to access the Mainland Chinese market. Preference in the allocation of sites has been given to projects offering new technology, high value added, or similar benefits to Macau's industrial development.

The government is also looking at the promotion of Macau as a destination for meetings, incentives, conventions and exhibitions (MICE).

ICT infrastructure

Most telecommunication services in the past were provided under a monopoly concession by Companhia de Telecomunicacoes de Macao (CTM). That concession was revamped in 1999 in the lead up to partial liberalization of the telecom sector. Liberalization has been successful in broadening communications infrastructure, particularly mobile and Internet services, which were opened to competition in 2000 and 2002, respectively.

CTM still has a monopoly concession over local fixed and international services and leased lines until 2011. The company is a private joint venture and its shareholders are Cable and Wireless (51 per cent), the Portugal Telecom Group (28 per cent), CITIC Pacific (20 per cent), and the Macau government (1 per cent). The Macau government has resisted World Trade Organization (WTO) recommendations to open up the remainder of its telecommunication services to competition and/or foreign players until after 2011.

The Internet sector was the first to be opened to competition and there are now 10 Internet service providers (ISPs) compared to just two at the end of 1999. There are also eight Internet content providers. There were a total of 88,694 Internet subscribers at the end of 2005, against just 17,169 in 1999, according to information from government authorities. The Internet household penetration rate is 66 per cent.

Macau now has five mobile phone operators. In July 2002, the first three GSM licences went to CTM and two offshoots of Hong Kong operators: Hutchison Telephone and SmarTone Mobile. Kong Seng Paging was also given a licence to operate as a mobile virtual network operator (MVNO), which means it offers mobile services to consumers using leased capacity from other mobile operators. There were 532,758 mobile subscribers at the end of 2005, giving a penetration rate of more than a hundred per cent. In 1999, there were only 118,101 mobile subscribers.

In 2006, Macau Unicom, a subsidiary of Mainland operator China Unicom, was granted a licence to operate CDMA2000 1X service. It initially offered a roaming service for visitors from Mainland China, but also plans to offer local service. It was also the first of the three 3G licences awarded by the government. The other two 3G licences were for W-CDMA networks and were awarded to CTM and Hutchison Telephone. Both SmartTone and Kong Seng Paging had also bid for 3G licences but were unsuccessful. The first 3G services are expected to begin commercial operation in 2007/2008.

Macau operates two satellite earth stations connected through an Intelsat satellite and has high-capacity fibre optic links to Hong Kong and Mainland China through Zhuhai in Guangdong province. It is also connected to the Sea-Me-We3 submarine cable system and the China-US cable for international bandwidth. International Internet bandwidth was 1,125 Mbps at the end of 2004, compared to just 19 Mbps in 1999.

In terms of broadcasting, there are five television stations and two radio stations in Macau. There is additional strong competition from neighbouring Hong Kong, whose broadcasting and media services are popular with Macau residents.

The government is also undertaking a study on the feasibility of establishing a second fixed or wireless network for the provision of broadband services, but no framework or licences have been put forward.

Enabling policies

The following are enabling policies that will affect ICT developments in Macau, some of them indirectly and others more directly.

Science and Technology Development Fund

The Science and Technology Development Fund was set up in 2004 to promote new technologies and enhance scientific knowledge and capacity. It provides funding for science-related education, research and project development.

Productivity and Technology Transfer Centre

The Macau government is aiming to promote the concept of lifelong learning through vocational training and retraining within the community. The Macau Productivity and Technology Transfer Centre provides advanced training in languages, information technology and textiles.

The Closer Economic Partnership Agreement (CEPA)

CEPA is one of the more important policies for Macau, although its coverage is much wider than ICT. It is designed to liberalize trade in all goods and services and to facilitate trade and investment between Macau SAR and Mainland China.

Ties with Portugal

A Framework Agreement on Cooperation exists between Macau and Portugal based on their historical ties. It is intended to promote cooperation in economics, finance and science, as well as serve as a platform for trade between China and Portuguese-speaking countries. The past two Forums on Economic and Trade Cooperation between China and the seven-member Community of Portuguese-speaking Countries were held in Macau; the third will be also be held in Macau in 2009.

Macau-Zhuhai transborder industrial park

Opened in December 2006, the Macau side of the Macau-Zhuhai industrial park is the second such facility after the Concordia Industrial Park. Information technology is one of a number of sectors that will be catered to in the park, which is expected to facilitate cross-border trade in goods and services.

Key national initiatives

e-Macao

e-Macao is a project to build a foundation for electronic government in Macau through readiness assessment, software research and development, and capacity building for the government workforce.

The e-Macao project started in July 2004. By June 2006, it had completed a comprehensive survey of 44 government agencies, developed three prototype systems to deliver representative public services online, trained close to 200 government staff in relevant technical and management skills, and organized 22 seminars and workshops to raise the level of awareness about

---

ICT and the gaming industry

With all the noise and excitement associated with a typical casino, most people probably do not stop to think of the technology that is involved. In fact, technology and communications infrastructure are cornerstones of the gaming industry—and will be so increasingly in the future.

Take the Wynn Macau, the hotel and casino complex that opened last year and which follows in the steps of the famous Las Vegas Wynn establishment. It has a staff of 50 IT specialists to handle the requirements of the casino and hotel complex. Each of its 200 gaming tables is loaded with technology, while the cages that dispense cash and receive deposited chips have been described as a bank branch where funds and assets can be moved from one place to another.

In the hotel there is Wi-Fi coverage in all rooms as well as wireless controls in some suites for lighting, drapes and AV equipment that rides on the Wi-Fi network. In the future, the Wi-Fi network will cover the entire property and allow for the use of wireless POS and wireless gaming machines.

Staff members are fitted with an employee ID that can track their movement within the complex as well as restrict access to secure areas or provide intelligence by sending data back to a data warehouse. For example, the data can help to plan staff meal times so that they maximize cafeteria operations. Staff attendance is automatically tracked for various purposes. There is an automated uniform-distribution system where employees drop uniforms into a slot at the end of their shifts, after which the uniforms are laundered and replaced on the racks. When the employee returned for the next shift, a touch of their ID card would bring that unique uniform for collection via the automatic rack.

The Wynn Macau has also developed its own technology that has since been used by the sister establishment in Las Vegas. For example, it developed an automated system to mark down results in a game of baccarat. Tables were modified to incorporate result displays and produce cards pre-printed with the history of the game in progress to hand out to newly-arrived patrons. The result has increased interest in the game because when people see a banker run or a player run on the result displays, it entices them to sit down.

Casinos in Macau are also big users of Radio Frequency ID (RFID) technology. At the Wynn Macau, RFID chips are embedded inside the gaming chips, which allow them to be uniquely identified. Even if someone could copy an RFID and somehow figure out the encryption keys used, it would not be registered in the database and hence would not be recognized. RFID also allows the casino to track the lifecycle of a chip across the gaming floor, see how often it is exchanged at a cage and generally know how and where the chips are being circulated.

Source: Summarized from an interview with Andre Ong, CIO of Wynn Macau, by Stefan Hammond, the editor of *Computerworld Hong Kong*. Full article at http://www.cw.com.hk/computerworldhk/article/articleDetail.jsp?id=402472

electronic governance. Project activities in 2006 concentrated on building prototype software infrastructure for e-government and supporting rapid development and run-time execution of various Electronic Public Services (EPS).

The first phase of the project was officially completed on 30 June 2006. Funding for the project was from the Macau government via the Macau Foundation. The United Nations University International Institute for Software Technology (UNU-IIST) was the lead partner. The second phase of the project has been approved by the government and extended for three more years until the end of 2009. Besides managing and executing its own portfolio, UNU-IIST's role is to provide advice to the e-Macao Project on e-government development in Macau and chair its technical committee.

More information can be found at the project portal at www.emacao.gov.mo.

UNeGOV.net

The UNeGov.net—Community of Practice for Electronic Governance initiative, was established in order to transfer the experience gained through the e-Macao Project to other parts of the world, particularly to developing countries. The aim is to build a global community of practice composed of experts and practitioners interested in developing, sharing and applying concrete solutions for e-governance.

The initiative was announced at the World Summit for the Information Society in Tunis in 2005. Additional network-building workshops have since been held in Bethlehem (Palestine), San Luis (Argentina), Bahia Blanca (Argentina), Kathmandu (Nepal), Abuja (Nigeria) and Ulaanbaatar (Mongolia). Some workshops were followed by a three-day school on the foundations of e-governance. More workshops and schools are planned with an international conference on e-government to be hosted by UNU-IIST in Macau in 2007.

In 2006, UNeGov.net developed four thematic areas resulting in new research, development and organizational transformation projects: Strategic IT Planning for Public Organizations, a collaboration with the Macau Institute for Tourism Studies; Semantic Interoperability for Electronic Government, a collaboration with Microsoft; Software Infrastructure Development for e-Government, a collaboration with the Macau government; and Software for Communities of Practice.

Most of the projects are also part of the UNU-IIST project portfolio for the new e-Macao Program. More information can be found from the community portal at www.unegov.net.

Regulatory environment

In March 2006, a new independent regulator was created, the Bureau of Telecommunication Regulation, which took over from the Office for Development of Telecommunications and Information Technology. Its duties include regulation, supervision and promotion of telecommunications services and ensuring fair competition in the telecom sector. Other responsibilities are the granting of new mobile and Internet licences, promoting the competitiveness of the telecommunications market and safeguarding the rights of users.

Regulations and licensing requirements are specified under the Basic Telecommunications Law. All mobile licenses are valid for a period of eight years, while qualified companies can apply for a license at any time to operate Internet services for renewable five-year periods. Internet licensees are not allowed to operate online gaming businesses. Companhia de Telecomunicacoes de Macao (CTM) has a monopoly on fixed-lines until 2011.

One of the stated goals of the government is to ensure a competitive environment. For example, ISPs are free to establish new pricing schemes unless they are found to be anti-competitive or against the public interest. According to Article 8 of the Basic Telecommunications Law, all forms of cross-subsidization or other practices that go against competition or user freedom are prohibited.

The Macao Network Information Center (MONIC) administers the registration of '.mo' country-code domain names (ccTLD). MONIC has been operated by the University of Macau since 1992.

Content and services

The Library of the University of Macau (http://library.umac.mo/lib.html) subscribes to over 24,000 e-journals and maintains a collection of more than 20,000 e-books. It plans to increase its online databases and full-text databases.

Many customs declarations are now made through an electronic data interchange system. The government's stated goal is for all customs clearance procedures to be paperless in the near future. Most license applications and approvals can now be done electronically.

The Macau SAR Government Portal (www.gov.mo) is provided in Cantonese, English and Portuguese. It offers government news, city information and e-services and has areas for citizens, tourists and merchants. It is also a useful gateway to the public sector, with links to almost all government agencies and departments.

The Macau Computer Association (MCA) was established in 1983 for people who are interested in computer science and applications. It has organized a number of major ICT events and was also a co-organizer of the sixth Asia Pacific ICT Awards (APICTA) 2006, which was held in Macau SAR for the first time.

Teledifusao de Macau (www.tdm.com.mo), a free-to-air television and radio broadcaster, also offers programmes in streaming video formats from its website, mostly in Cantonese and Portuguese, although there are also English news broadcasts online. Audio files from Radio Macau are also available.

Open source community

A promising open source initiative known as the Global Desktop Project unfortunately stalled in January 2007. The project ran for two years and was a major effort to increase the number of open source software programmers in developing countries.

The Global Desktop Project was brought to the UNU-IIST in Macau by Scott McNeil in 2005. However, funding after 2007 was conditional on outside funding that did not come through. UNU-IIST said that despite initial encouragement and considerable effort, including meetings with the chief executive in Macau and the Minister of Science and Technology in China, funds were not found and McNeil left UNU-IIST to seek another home for the project.

UNU-IIST says that it remains committed to open source software development in developing countries. In future, it plans to develop courses related to open source and put them online. It also has an ongoing project with Microsoft on interoperability between open source and proprietary software.

Intellectual property rights

In a report to the WTO in May 2007, the Macau government noted that there have been important developments in the enforcement of intellectual property rights laws over the past five years. Its copyright and industrial property laws were overhauled in 1999, and since 2004, the examining entity for patent applications has been the State Intellectual Property Office of the People's Republic of China. As a result, invention patents granted in the mainland can now be extended to Macau SAR.

The number of trademark applications increased from 1,696 in 2001 to 4,651 in 2005, and applications for patents and models and designs increased in proportion, although from much lower levels. The system for the registration of rights is essentially the same as in the EU, having been based on that of Portugal.

The Macau Customs Service, created in November 2001, has strong powers to prevent and punish violations of intellectual property rights. The government reports that more than 30 illegal producers have been closed down and 25 production lines seized. The sale of optical discs is also strictly controlled, with retailers needing to obtain advance authorization and discs required to carry an individual Source Identification Code. The Macau Customs Service was awarded the Sixth Annual Global Anti-Counterfeiting Award by the Global Anti-Counterfeiting Group in 2003.

Future trends

By most estimates, Macau's economy will continue to expand, with many new projects planned for the coming years. While it does not have ambitions to become an ICT hub in the way other economies in the region do, notably Malaysia, Singapore and Hong Kong, there will still be a need for greater ICT infrastructure and skills in Macau. In fact, it would be easy to underestimate the level of ICT infrastructure and skills needed in the gaming industry. The increase in number of hotels, tourists and casinos is also driving demand for communication services. Many of the major ICT vendors have set up a direct presence in Macau in the last year because of growing demand. These vendors which previously worked through third parties, often shipping in experts from Hong Kong, include Cisco, Microsoft and HP.

However, according to many in the ICT sector, there is already a skills shortage and the situation could get worse, particularly when some of the major new hotels and casinos open. This has been a problem in the past too, as casino workers are often offered higher salaries, making it hard to recruit people not just for IT but also for the government and banking sectors.

The Macau government has acknowledged the need to diversify away from the gaming and tourism sectors, but that may not be so easy given the amount of investment coming in and the recruitment difficulties. Often touted is the promotion of Macau as a destination for meetings, incentives, conventions and exhibitions (MICE). MACAU has been relatively successful in attracting major conventions and events, including the 2006 Asia Pacific ICT Awards (APICTA), which aims to stimulate ICT innovation and facilitate technology transfer. APICTA has 15 member economies: Australia, Brunei, Hong Kong, India, Indonesia, Korea, Macau, Malaysia, Myanmar, Philippines, Singapore, Sri Lanka, Thailand, Vietnam and China.

Macau faces extensive competition from neighbouring countries, particularly Hong Kong, Mainland China and Singapore. However, Macau's advantage is that it is visa-free for around

64 countries and has an Individual Travellers Scheme that allows residents from 44 Chinese cities to travel to Macau. As a result, the immigration process is easier for MICE attendees. In addition, most of the major hotel and casino projects are also promoting business conventions.

Thus, the prospects look bright for Macau SAR. If it can manage to diversify further and encourage the development of ICT skills, Macau's prospects would look even better.

References

Hammond, S. (2006a). Macau: expansion drive. *Computerworld Hong Kong*, 1 May 2006. Retrieved from http://www.cw.com.hk/computerworldhk/article/articleDetail.jsp?id=323191

Hammond, S. (2006b). Jetfoil to Fisherman's Wharf. *Computerworld Hong Kong*, 1 August 2006. Retrieved from http://www.cw.com.hk/computerworldhk/article/articleDetail.jsp?id=362218

Hammond, S. (2007). Bold steps in Macau. *Computerworld Hong Kong*, 1 February 2007. Retrieved from http://www.cw.com.hk/computerworldhk/article/articleDetail.jsp?id=402472

Reed, M., George, C., and Hoi, W. (2007). *UNU-IIST annual report 2006*. Retrieved from http://www.iist.unu.edu/newrh/III/1/docs/techreports/report351.pdf

Report by Macao, China. (March 2007). *WTO trade policy review*. Retrieved from http://www.wto.org/english/tratop_e/tpr_e/g181_e.doc

Report by the Secretariat, Macao, China. (March 2007). *WTO trade policy review*. Retrieved from http://www.wto.org/english/tratop_e/tpr_e/s181-00_e.doc

.mm

Myanmar

Thein **Oo** and Myint Myint **Than**

Overview

There are indications that the State Peace and Development Council, the governing body of Myanmar, realizes that ICT can help to improve social and economic conditions. There are dedicated and concrete efforts at ICT development. Thus, although Myanmar is at an early stage of ICT development, there is a clear potential for developing a viable ICT industry and effective use of ICT to make the country more productive and competitive in the international market.

ICT Parks have been established in Yangon and Mandalay, the two former capital cities. Moreover, construction of a new Yadanabon cyber city is progressing speedily. It is intended to be a self-contained city with a teleport, an incubation centre, local and foreign software and hardware companies, residential areas and a shopping centre, among others.

There are efforts to formulate national ICT policies. The two ICT Master Plans, the ICT Master Plan Framework for 2001–05 prepared by the Myanmar Computer Federation and the Myanmar ICT Development Master Plan for 2006–10 prepared by the e-National Task Force, provide general guidelines for ICT development efforts. Unfortunately, the plans have not been officially adopted and their implementation is now being evaluated.

There are noticeable improvements in the ICT infrastructure even if it is still very weak at present. The construction of the national backbone linking the major cities by fibre is progressing satisfactorily. The grid will cover more than 15,000 kilometres. A satellite link is also being established to cover the remote areas that the fibre link cannot reach. The international bandwidth is still too low to meet the demand but there is a good effort to improve the situation. The big challenge is improving the last mile link to enable universal access. Reliability and affordability are the two key issues for future development.

The legal framework for ICT development needs to be improved considerably. Only the Computer Science (ICT) Development Law and the Electronic Transaction Law have been promulgated. The Telecommunication Law and IPR laws are at the draft stage. Myanmar may need to work closely with the international community in the preparation of its cyber laws.

The ICT industry is at the infancy stage and its contribution to national GDP and export is still negligible. e-Government has been initiated through efforts to establish an e-government network and data centres. However, there is little effort to develop the appropriate applications. Moreover, while there was a big plan and initial efforts for e-education development, the results were not satisfactory. Two computer universities and 24 government computer colleges have been set up throughout the country with the goal of producing 1,197,350 professionals. But shortage of faculty members, facilities and quality standards undermine the effort. Other key issues are training for trainers, courseware development, operation and maintenance.

ICT infrastructure

Building and upgrading the ICT infrastructure is the most basic and most important component of ICT development. The ICT infrastructure is the essential prerequisite of ICT applications and ICT industry development, e-commerce, e-government and e-learning. As ICT infrastructure requires high-tech equipment and personnel, it also has a critical impact on economic development.

However, Myanmar's current ICT infrastructure is very poor even among developing countries, as the following will suggest:

- Teledensity, including mobile telephony, is only 1.23 per cent, among the lowest in ASEAN countries.
- The number of Internet users is too small.
- The telephone supply does not meet the increasing demand.
- Most of the switching systems are manual, with a small capacity.
- The transmission systems consist mostly of microwave systems.
- The Myanmar Posts and Telecommunications (MPT) buries cables underground instead of installing ducts or poles, which implies that maintenance and repair cost may be high in the future.
- The Ministry of Communications, Posts and Telegraphs (MCPT) provides all telecommunications services, including fixed and mobile access, and local, national and international calls and leased lines.
- There is no plan to separate the business part from the administration in the near future.
- Telecommunications personnel are unskilled.
- There is no billing system in place.

On the other hand, the demand for telephone services is very high and has a great potential. Also, the average revenue per user (ARPU) is increasing, which implies that there are network externalities for returns on investment to increase in the future.

The development of the infrastructure for fixed-line telephones has been delayed for various reasons. One important reason is lack of funds due to insufficient foreign currency and the current tariff rate structure for telecommunication use.

Until recently, the access line of fixed-line subscribers had been copper-wire cable. Bagan Cyber Tech (BCT) is now providing Wireless Local Loop (WLL).

MPT is the only organization providing mobile telephone service. An Advanced Mobile Phone System (AMPS) and Digital-AMPS (D-AMPS) are currently in use. CDMA and GSM, the second generation mobiles (2G), were introduced in 1997 and 2002, respectively. MPT plans to install a GSM system in about 20 cities in border areas, as well as in coastal areas.

The telecommunication network system consists of the following:

Exchange
Auto Exchange	137
Manual Exchange	640
Transit Exchange	4
International Exchange	2
Mobile Exchange	6
Packet Exchange	1
Internet Gateway	2

Microwave Stations
Digital	181
Analogue	79

Earth Stations
International Stations	1
Domestic Stations	416

International Submarine Cable Landing Point   1

A Data Communication System has been in use in Yangon and Mandalay since 1997.

Myanmar carries out overseas communication services through satellite system and the Asia-Europe underwater cable system. The country has been linked to the Internet with 57 lines since 1999. There are two Internet Service Providers (ISP). MPT is the only gateway to foreign countries. The current bandwidth is 64 Mbps. As of August 2006, there were about 88,500 Internet users using dial-up (74 per cent of users), ADSL (13 per cent of users), WLL (7 per cent) and satellite terminal (6 per cent). Cybercafés, which emerged in 2002, increased the number of Internet users among the general public. There are 17 cybercafés in Yangon and three in Mandalay.

The number of personal computers per thousand inhabitants remains low: 5.1, 5.6, 6.6 and 7.3 in 2002, 2003, 2004 and 2005, respectively.

Yangon, the former national capital, is the centre of the nationwide network. MCPT is constructing optical fibre backbone routes within Myanmar. The fibre link between Yangon in lower Myanmar and Mandalay in upper Myanmar, and linking major cities, has already been completed.

The WLL service provided by BCT has a transmission speed of 128 Kbps to 2 Mbps. The system was installed first in business areas and has been expanded to cover the whole municipal area of Yangon. The biggest advantage is its affordability: the price is estimated to be one-fifth of the current system. BCT also introduced Broadband Satellite Data Service using iPStar by means of Thaicom 4 Satellite in 2002. The system is primarily intended to provide network services in rural areas.

A VSAT system, called DOMSAT (DOMestic SATellite system), is under operation as a telephone network to connect Yangon and 10–20 remote cities. However, some parts of the system fail and the usage rate is rather low.

Cross-border fibre links between China and Myanmar, India and Myanmar, and Thailand and Myanmar have been set up. International telecommunication is provided through the Sea-Me-We3 submarine cable system and INTELSAT earth station, directly linking Myanmar to 34 countries. MCPT plans to build another international switch and earth station in Mandalay, which will increase the reliability of the international telephone network.

While the building of the national backbone and international links is progressing satisfactorily, improvement of the 'Last Mile Link', which links subscribers to the backbone, seems to be very slow. BCT is working on introducing ADSL to improve the last mile link. Both, MPT and BCT plan to expand the IP-based VSAT and iPstar broadband network to the villages. It is necessary to formulate policies to increase ICT accessibility in the rural areas with concrete implementation measures.

ISDN (Integrated Services Digital Network) has been provided in some parts of Myanmar. In the future, however, the whole telecommunication network will be more and more IP-based. At the moment, the introduction of IP-related technologies is way behind in Myanmar. This is due not only to the technological or hardware problems, but also to the immaturity of policies and institutions to promote and administer the spread of these new technologies. The conventional telecommunication technologies were developed with the involvement of the central government. However, the new IP-related technologies have been developed in an atmosphere of open discussion among academic groups and the private sector. It is necessary to develop this environment of academic and private groups. The IP-based network will be developed as a common infrastructure for the transmission and processing of all kinds of information. From this point of view, Myanmar has to study the development of Internet-related technologies and related institutions in other parts of the world and begin to develop an adequate institutional structure for an IP-based telecommunications network.

Moreover, the telecommunication infrastructure in industrial zones needs to be improved considerably, as both teledensity and Internet access are insufficient. The development of the telecommunication infrastructure in the industrial zone to support ICT applications development should correspond with industrial development policy and priorities.

The cost of Internet access in Myanmar is the highest among ASEAN countries. It is necessary to reduce the Internet access cost considerably. Strategic plans for an affordable pricing mechanism for telecommunications and Internet access services, especially for priority groups such as educational and health institutions and rural areas, need to be put in place.

Key institutions dealing with ICTs

Myanmar introduced ICT quite early. The first computer centre, Universities Computer Center (UCC), was established in 1971. In the mid-1980s there were efforts to introduce e-government mainly for administrative purposes via the Computing Development Project (CDP), a UNDP project. But there was no national ICT development policy.

The organizations responsible for ICT development in Myanmar are shown in Figure 1.

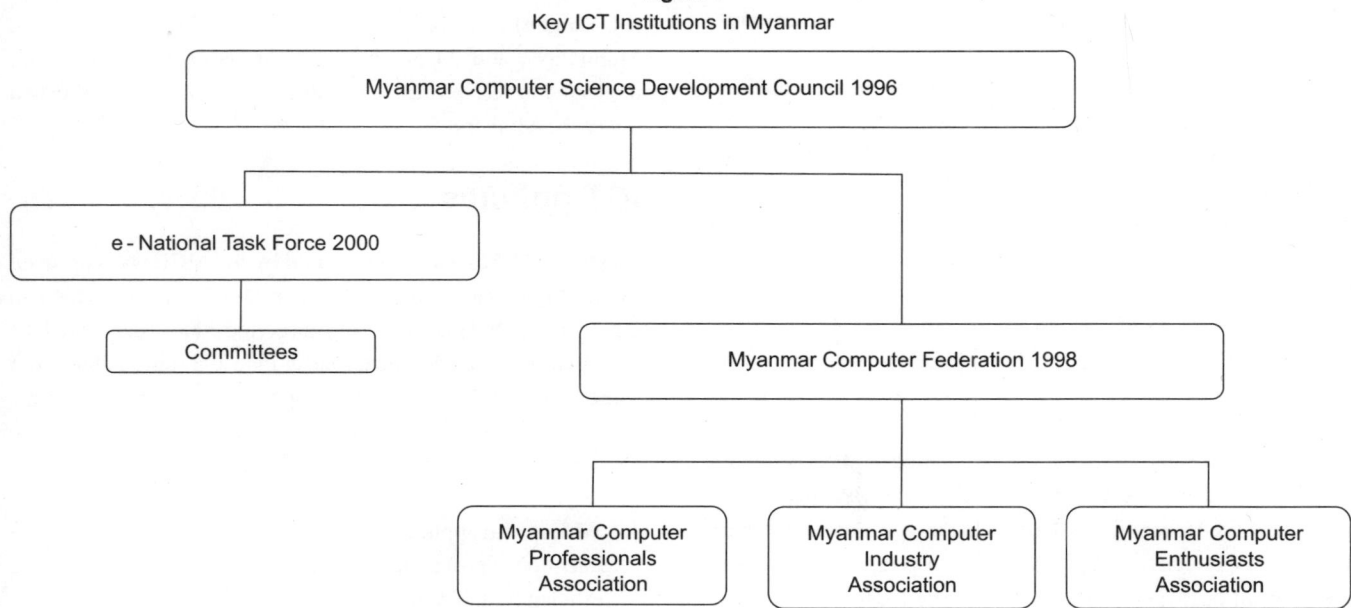

Figure 1
Key ICT Institutions in Myanmar

Myanmar Computer Science Development Council

The Myanmar Computer Science Development Council was established following the promulgation of the Myanmar Computer Science Development Law on 20 September 1996. The Council is the highest ICT policymaking and implementing body. It is headed by the Secretary of the State Peace and Development Council (SPDC). Its members include the heads of various Ministries, including Home Affairs; Communications, Posts and Telegraphs; Cooperatives; Information; Industry; Education; Science and Technology; Immigration and Population; Culture; Foreign Affairs; and Defence. Other members are the Director of the Directorate of Signals, the Director General of the Department of Higher Education, the rectors of the University of Computer Studies, Yangon and the University of Computer Studies, Mandalay, and the Director General of the Advanced Science & Technological Research Department.

e-National Task Force (e-NTF)

The e-National Task Force, which was formed on 30 October 2000, is responsible for the implementation of the e-ASEAN Framework Agreement. It is headed by the Minister of Communications, Posts and Telegraphs and its members are from the public and private sectors.

The functions of the e-NTF are as follows:

1. Give policy recommendations for the building of a National Information and Communication Infrastructure;
2. Develop regulatory and legislative frameworks for e-commerce development in accordance with international standards and practice;
3. Coordinate with government agencies in the implementation of the e-ASEAN Framework Agreement;
4. Prepare an implementation plan for ICT applications development;
5. Provide estimates of resources needed for the implementation of projects for improving e-readiness; and
6. Evaluate ICT projects.

Myanmar Computer Federation (MCF)

The Myanmar Computer Federation, an NGO formed in 1998, consists of representatives of three associations—the Myanmar Computer Professionals Association (MCPA), the Myanmar Computer Industry Association (MCIA), and the Myanmar Computer Enthusiasts Association (MCEA)—as well as representatives of government agencies dealing with IT. The federation and the three associations under it do a good job of raising awareness of ICT.

Specifically, the MCF is tasked with promoting the utilization of computers in different areas of work through training courses, lectures, competitions and study tours; developing computer science curricula for computer training schools; developing standards and an accreditation system for computer training schools; giving assistance to manufacturers to enhance the quality of computer hardware and computer software; conducting and supporting computer science research; liaising with international computer organizations; developing computing in the local language; compiling, publishing and distributing publications on computers; developing computer literacy especially among young people and helping develop outstanding computer scientists and inventors; and recommending to the Council honorary titles and awards for outstanding computer scientists and inventors.

The three associations under the MCF may be briefly described as follows: the MCPA is an NGO composed of ICT professionals, the MCIA is an NGO composed of ICT companies, and the MCEA consists of individual computer users.

The e-National Task Force, which is a public institution, and the MCF and the associations under it from the private sector side, work closely under the guidance of the Council.

Ministry of Science and Technology (MOST)

The Ministry of Science and Technology, created in 1997, is responsible for ICT human resource development. Two computer universities, one each in Yangon (University of Computer Studies, Yangon) and Mandalay (University of Computer Studies, Mandalay), and 24 government computer colleges (GCC) dedicated to ICT professional education have been established under the Ministry.

ICT policies

The first ICT Master Plan prepared by the MCF was approved by the Myanmar Computer Science Development Council in 2001. The second plan, covering the period 2006–10, is called the Myanmar ICT Development Master Plan and Action Plan. Both Master Plans guide all ICT development efforts in Myanmar.

The Master Plan specifies the following broad mission elements:

1. Widespread application of IT in state management with the intention of providing better services to the public, improving efficiency, and reducing costs;

2. Widespread application of IT in business organizations to improve productivity and render better services;
3. Utilization of IT as a low-cost communication infrastructure for the smooth operation of socio-economic organizations;
4. Utilization of IT as a vehicle for business organizations penetrating the international market;
5. Widespread application of IT to improve the educational level of the whole population;
6. Development of the IT Industry to become one of the main economic sectors;
7. Development of IT human resources;
8. Creation of an IT-intelligent society;
9. Facilitation of the growth of e-commerce; and
10. Reduction of the digital divide.

ICT Industry

According to a survey of 250 companies conducted in 2005, 27.46 per cent of the companies surveyed are engaged in training, 27.14 per cent are in hardware sales, 11.59 per cent are in system integration, 8.55 per cent are in network solutions and 24.70 per cent are in software development and others. Most are members of the MCIA. About 75 companies have offices in the ICT Parks established by the Myanmar Info-Tech Corporation. The IT market is estimated to grow between 10 and 20 per cent per year, largely due to Internet-related IT services.

The commercial market consists mainly of customized applications such as accounting packages, computer-based training (CBT), Web-enabled applications and multimedia design. Most of the end-users are banks.

The government is the largest consumer of information technology. However, there is as yet no government-wide information system and there is no effort at standardization of telecommunications, hardware and software. Standardization and compatibility of data models and structure and telecommunications protocols, need to be addressed.

The IT Industry Association was established to focus on marketing and promotion, and to foster networking and collaboration among ICT companies. This consortium of 50 ICT companies was formed in 2001 by the Myanmar Info-Tech Corporation in order to develop the Myanmar ICT Park with international facilities in Yangon and Mandalay. Other goals of the consortium are to develop a home-grown solution for e-government, provide ICT-related services, provide infrastructure for ICT companies, tap local and overseas markets, and develop human resources.

To improve the capabilities of ICT companies, the Myanmar Info-Tech Corporation and MCF conducts seminars and workshops, training courses, an on-site training programme (jointly with CICC, Japan), and an on-the-job training programme (jointly with AOTS, Japan). The corporation and MCF are also collaborating on the development of a Software Quality Certification (CMMI) Programme with Software Park, Thailand. A business matching programme, trade shows and exhibitions, and the National ICT Awards programme are also arranged by the associations under MCF.

In 2002, a consortium of private companies established the Myanmar ICT Park, a special zone where adequate facilities and support are provided for ICT companies.

There are many efforts by MCF for international cooperation. However, cooperation through ASEAN is not yet fruitful.

e-Government

Myanmar is at the very low end of the UN's e-Government Readiness Index: 123rd among 191 countries. Its Web index rank is 100th; its telecom index rank is 182nd (weakest point); and its human index rank is 122nd.

The current ICT Master Plan states that e-government should be given priority. Widespread application of IT in state management is envisioned to provide better services to the public, improve efficiency and reduce costs. The Master Plan also states that as the biggest buyer of IT products and services, the state should act as the main demand force for domestic IT application growth, and that it should establish demonstration projects to show the benefits of IT applications to motivate the public and private sectors and the whole population to use IT extensively. However, except for these broad statements, there is no specific e-government plan.

At present, the e-NTF is the focal point for computerization of government agencies. It is tasked with preparing an e-Government Master and Action Plan, training government employees, and building infrastructure. Under the direction of the e-NTF, all government agencies at ministry and department or enterprise level have already appointed a Chief Information Officer (CIO).

The Yangon City Development Council (YCDC) and Mandalay City Development Council (MCDC) are active in computerization. They have developed and maintained their own websites and some operations, such as tax collection and building registration, are already computerized. On the other hand, there is no information that other local government agencies have any IT plan. The IT development effort at state/division level is still very low. Most agencies do not have any e-government project, although some ministries, departments and enterprises have formed computerization steering committees.

PC penetration (the number of PCs per employee) is still very low in government agencies. The agency with the highest PC penetration has 0.3 PCs per employee, which means that three employees share one PC. The lowest rate is 0.002. There are still some government agencies that do not have computers. Five agencies have one or more information systems but only 11 per cent provides public services online.

Governmental IT System Development consists of infrastructure building and development of application systems particular to each government agency. Currently underway is the Myanmar Basic e-Government System project with a loan of more than USD 10 million from the Daewoo International Corporation & KCOMS Co. Ltd., Korea. It includes a Basic Database Management System, including common applications and data exchange.

Legal and regulatory environment for ICTs

By enacting or amending ICT laws based on a concrete roadmap, the Myanmar government will be able to not only overcome a stagnant telecom market but also drive economic growth. An ICT legal framework is a very important part of ICT development. Whenever there is a paradigm shift in telecom policies, the ICT legal framework has played an important role in achieving the goal of providing consumers with better services at a lower price.

To facilitate a regulatory environment that is supportive of National ICT Development, the following ICT related laws have been promulgated:

1. *Myanmar Computer Science Development Law* (State Law and Order Restoration Council Law No. 10/96): Promulgated in 1996, the law specifies the formation of institutions that will be responsible for ICT development in Myanmar, as well as their responsibilities and scope of authority.
2. *Electronic Transactions Law* (State Peace and Development Council Law No. 5/2004): Promulgated in 2004, this law aims to use electronic transactions technology in building a modern, developed nation; to obtain more opportunities for all-around development of sectors, including human resources, economic and the social and educational sectors, by electronic transactions technologies; to recognize the authenticity and integrity of electronic records and electronic data and give legal protection thereof in matters of internal and external transactions making use of computer networks; to enable transmitting, receiving and storing of local and foreign information simultaneously, making use of electronic transactions technologies; and to enable speedy and effective communication and cooperation with international organizations, regional organizations, local and foreign government departments and organizations, and private organizations and persons.

The drafts of the new Telecommunications Law and Intellectual Property Rights Law have already been completed. The Telecommunications Law can lay the foundation for better services at lower prices by promoting market competition in the telecom market.

Education

There are good efforts at IT application in education. Most of the schools have a multimedia classroom and there is an educational intranet. But there is a lack of well-planned and coordinated efforts for effective use of the facilities. There are 1,712 high schools, 3,099 middle schools and 36,004 primary schools in Myanmar. A total of 991 schools have been designated as Multimedia Schools. Of these, 238 have been upgraded as Electronic Learning Centres. More schools will be upgraded in the coming years. In the meantime, schools are encouraged to study the possibility of acquiring technology through their own initiative.

Computer universities and colleges, as well as certification programmes, have been established. But there is a need to improve quality. Yangon University and Dagon University under the Ministry of Education also provide Bachelor, Diploma and Master's degree courses in ICT.

The ICT industrial base is very weak in Myanmar. Consequently, the employment share of ICT occupations is still very low. Among ICT occupations, computer programmers account for the lion's share. Among a sample of 3,343 workers, 17 per cent are computer programmers, 14 per cent are computer support specialists, 11 per cent are computer operators, 10 per cent are computer software engineers, 8 per cent are computer hardware engineers, 8 per cent are computer repair staff, 8 per cent are computer and information systems managers, 7 per cent are network or computer systems administrators, 4 per cent are data entry and information processing workers, 4 per cent are systems analysts, 4 per cent are desktop publishers, 3 per cent are Web masters, 2 per cent are database administrators, 1 per cent are computer scientists, and 1 per cent are computer-control programmers and operators. According to an MCF ICT survey conducted in 2005, the pattern of future demand for ICT workers is likely to be the same as the existing occupational pattern. The most needed occupations include computer programmers, computer and information systems managers, and computer software engineers.

ICT occupations pay higher wages than other occupations in Myanmar. ICT fields are popular among students and it is relatively easy to attract top domestic talents to ICT fields. In terms of labour cost, Myanmar's ICT sector has an advantage over equivalent sectors in neighbouring countries. For example, the monthly wage of a computer programmer in Myanmar is only about a third of the monthly wage of a computer programmer in India.

On the other hand, the number of IT researchers/engineers in Myanmar is negligible. The computer universities are offering PhD courses, the students of which can be considered as IT researchers. As of November 2006, there were about 90 who had completed these PhD courses and about 74 doing their thesis.

MCF established a small ICT Research Centre in 2004. Most of its research work is on Natural Language Processing. MCF and its associations also provide ICT professional training. There are around 90 private computer schools, including 70 in Yangon. Altogether, these institutions produce about 900 ICT professionals annually. Most of the private training centres offer Basic Computer Skills Courses such as MS Office, desktop publishing, e-mail and Web design, graphics design, Introduction to the Internet, AutoCAD (Engineering Drawing), ACCPAC Plus (Computerized Accounting) and Multimedia Creation. Some offer professional courses such as Software Engineering, Software Engineering in Database & e-Commerce, Network Engineering, Specialized Programming Courses, Networking with Linux and Practical Network (Server-Based), as well as Oracle, CISCO and Microsoft Certification Examination Courses.

Informal certification programmes were also initiated recently. The MCPA certification programme has been successfully implemented and the Japanese Information Technology Examination Center (JITEC) and MCF have entered into an agreement to implement a cross-certification programme. The Japanese Information Technology Engineer Examination (JITEE) has been conducted twice a year since 2002. These programmes provide IT training opportunities to those who cannot afford or who do not want to join formal education programmes. Bilateral, regional and international certification programmes such as JITEE should be implemented effectively and systematically promoted.

An e-Learning Centre was established in 2001 with support from Japan. The centre is currently providing training courses for JITEE. The e-Learning Centre should be strengthened to become a full-fledged centre providing all of the facilities needed to support informal ICT human resource development.

It should be recognized that the quality of ICT professionals in Myanmar still needs to be improved to meet international standards. This gap should be closed quickly through effective human resource development programmes.

Towards an ICT society

Most of the current efforts in Myanmar are in improving awareness. IT literacy is still very low for the whole society. The Myanmar Computer Federation and the three associations under it have conducted an 'IT Caravan' to the rural areas, with support from MCPT and the Center of the International Cooperation for Computerization (CICC), Japan.

There are good initiatives to narrow the digital divide between urban and rural areas, such as the establishment of Public Access Centers (PAC). However, the weak telecommunication infrastructure and high costs continue to widen the digital divide. There is a need for specific measures to liberalize investment, production and distribution of IT products and services. There is also a need to give business organizations special incentives for using IT.

References

Information and Public Relations Department. (2002). Transport, construction and communications. In A. Myo Thant, B. Maung Hlaing, C. Soe Myint and D. Tin Tin Win (eds). *Myanmar facts and figures 2002*. Yangon: Ministry of Information.

Information and Public Relations Department. (2003). Endeavours made on economic development of the State. *Magnificent Myanmar (1988–2003)* (1st ed.). Yangon: Ministry of Information.

.np
Nepal

Sushil **Pandey** and Basanta **Shrestha**

Total population	25,886,736 (Central Bureau of Statistics)
GDP per capita	PPP USD 1,500 (as of January 2007) (CIA World Fact Book 2007)
Key economic sectors	Agriculture, Tourism
Fixed-line telephones per 100 inhabitants	2.46
Mobile phone subscribers per 100 inhabitants	4.03
Internet users per 100 inhabitants	0.19
Domain names registered under .np	10,062 (as of February 2007) (Mercantile Communications)
Internet international bandwidth	37 Mbps uplink and 90 Mbps downlink (as of February 2007)

Overview

Political and social instability in Nepal slowed down the development of ICT in the country in 2006–07. Nevertheless, a landmark event in this period has been the localization of computing. Nepalese whose access to computing was restricted by language barriers can now use computers and the Internet in the Nepali language. They also have a choice of two operating systems in Nepali—MS Windows and Linux.

Recent developments in nationwide connectivity through optical backbone cabling and incremental growth in teledensity likewise bode well for the development of the ICT sector in Nepal. The liberalization of the telecom sector is expediting growth and improvement in the quality of ICT services and resulting in some price reduction. Nepal also stands to benefit from optical cable linking with its neighbouring countries. To provide the rural areas with basic telecommunication facilities is the rural VSAT project and the establishment of telecentres through multi-stakeholder partnerships. The recent de-licensing of Wi-Fi bands for public use is also expected to facilitate the deployment of wireless networking to the rural areas.

However, there is an urgent need to revise Nepal's IT Policy, the blueprint for attracting foreign investments in ICT. It is also necessary to address emerging issues such as use of open source software, human resource development, e-commerce, ICT for development, cyber security, VoIP, and the like.

Technology infrastructure

Lack of connectivity still prevents many Nepalese from reaping the benefits of ICT. However, some recent developments in nationwide connectivity and infrastructure are encouraging.

The optical fibre network and the Asian superhighway

The government, through Nepal Telecom (NT), has completed the first phase of the East-West Highway Optical Fibre Project that will build the optical fibre backbone for telecom services. This has substantially upgraded the reliability and quality of long-distance calls within Nepal and between Nepal and India. Phase I of the Project covers a distance of 850 km from Bhadrapur to Lamahi and from Kholapur to Nepalgunj. Phase II aims to cover 900 km from Birtamod to Kakarvitta and from Lamahi to Mahendranagar. Completion of both phases means that Nepal will be connected to India through fibre links via different connecting points and the country will have an optical fibre backbone from east to west. Several points have recently been hooked up and aligned with India's leading service provider. The completion of this ambitious project will also make Nepal part of the Asian Information Superhighway, a cheaper and more reliable alternative to existing satellite communications.

Funding for the fibre optic project comes mostly from the Indian government. But Nepal Telecom is receiving grant assistance from the Chinese government for the optical fibre project along the 115 km long Arniko Highway linking Kathmandu to Khasa, which borders China. China is already laying optical fibre cables from Beijing to Lhasa and there are plans to extend the cables from Lhasa to Khasa. When the Arniko Highway project is completed, Nepal will be connected to the information superhighway between Nepal and China. Since Beijing is already connected to Hong Kong via cable, then Nepal shall also be connected to Hong Kong, one of Asia's communication gateways to the rest of the world. This gives Nepal an alternative route for international communication links, besides the existing satellite.

Efforts should be made to link the Arniko Highway Project and the East-West Highway Optical Fibre Project, in order to link the software cities of India like Bangalore and Hyderabad, to the hardware centres of China like Beijing and Shanghai. Nepal will thus become a transit state for trade and communication between China and India, two giant economies. Nepal stands to reap huge revenues from this development, as Nepal's policymakers must have realized when they undertook these projects. At the very least, with the completion of the Arniko Project, Nepal Telecom should be able to provide 24-hour high-speed information access to Nepal's first and only Information Technology Park in Banepa, which links Bangalore and Beijing.

Nepal rural VSAT project

The extension of infrastructure services to rural areas where the majority of the poor reside is an important goal. A key effort in this regard is the rural VSAT project that aims to connect 1,000 Village Development Committees (VDC) in the mountainous regions where other modes of telecommunications are not possible, to the national telephone network. Each VDC will have two telephone lines via solar-powered VSAT technology. Started in 2002, the project shall be completed in 2007. The VSAT phone lines are used as a public call office (PCO) or phone shop and effectively link remote rural mountain people to the lowlands.

There is also a special rural operator, STM Telecom Sanchar (P) Ltd., licensed to provide telecom facilities to 534 villages in the Eastern Development Region using VSAT technology.

Expansion of fixed-line services nationwide with CDMA

Nepal Telecom aims to provide on-demand telecom connections throughout the country through the addition of one million fixed-lines in five stages. This is in accordance with the aim of improving the fixed-line penetration to 4 per cent by 2007 and 20 per cent within the next six years. Using a wireless facility based on CDMA technology, Nepal Telecom recently launched a Limited Mobility service in Kathmandu. This will be gradually extended throughout the country. With its average data throughput of 120 Kbps, data users should find CDMA-based services attractive.

Enabling policies and programmes

ICT development in Nepal continues to be based on two policies, the IT Policy of 2000 and Telecommunication Policy 2004. Both have been extensively discussed in the 2003–04 and 2005–06 editions of the *Digital Review of Asia Pacific*. At this point, suffice it to say that a more realistic policy needs to be developed.

The High Level Commission on IT (HLCIT) has produced a draft IT policy with the following vision:

> By the year 2015, Nepal will have transformed itself into a knowledge-based society by becoming fully capable of harnessing information and communication technologies and through this means, achieving the goals of good governance, poverty reduction and social and economic development.

Given the dynamic developments in ICTs and the new opportunities they bring for the country's overall development, there is an urgent need for the draft IT policy to be finalised, endorsed and implemented.

The adoption of Telecommunication Policy 2004 was a strategic move to liberalize the telecommunications sector by promoting private sector participation and competition in all market segments for the purpose of broadening access to telecommunications facilities in the country. At present there are several operators in important market segments—fixed-line and wireless local loop, mobile, rural connectivity and value added services. The policy of promoting competition is bearing fruit. For example, getting a telephone connection is not as difficult as it used to be. The telecommunication sector kept the country going despite political instability and incessant conflicts, although it was not spared attacks on infrastructure by rebel groups and disruptions in service by the previous government. The cost for users has come down due to competition among telecommunication service providers. However, more needs to be done to ensure that telephones and telecommunications in general become inclusive tools for the common people rather than a luxury for the rich.

Legal and regulatory environment

The Nepal Telecommunications Authority (NTA) oversees the development of telecommunications and the Internet. It plays a key role in effective interconnection regulation, which is necessary to ensure the twin goals of maximizing the productivity and efficiency of rapidly growing networks, and extending these networks to those who cannot enjoy them and thus bridging the gap between technology haves and have-nots.

Nepal has recently moved a step further with the promulgation of the much discussed Electronic Transaction Act, also widely known as the Cyber Law, which legalizes all electronic transactions and digital signatures. Computer and cyber crimes, such as hacking, piracy and fraudulent activities, have also been defined and penalties set accordingly. The new law should facilitate business processes and it is important that it is put into practice.

VoIP has not yet been legalized in Nepal. Recent moves to deregulate the frequency bands used by Wi-Fi could help make VoIP free and legal in the country. IP-based networks are increasingly being used as alternatives to traditional telephone networks in other countries, and VoIP should be seriously considered for the growth of the country's ICT sector.

Security

In Nepal as in other countries, Internet service providers (ISPs), business process outsourcing (BPOs) and the banking industry are the most sensitive to ICT security issues, as security breaches can cause them severe damage. Majority of Internet users are dependent on the security of ISPs. Corporations have their own security system, mainly in the form of a firewall against intrusion, hacking, viruses and spam. In government offices, the only means of protection is antivirus software, which is not enough.

Corporations, banks and ISPs also understand security issues from the point of disaster recovery. Because Nepal is in an earthquake-prone belt, all ICT assets need to be protected from natural disasters. Thus, the bigger business houses and banks have security mechanisms ranging from backing up of data on tape/disk to database replication and having a dual system, such as a mirror site or an offshore backup centre.

Security is another important area for IT policymakers and government to think about. The security and safety of various ICT platforms in government offices must be given priority consideration before any e-governance base is made fully functional. This assumes the adoption and use of security measures, including empowering law enforcers and the judiciary with knowledge of cyber forensics and digital evidencing.

ICT industries

Hardware and software

There is still very little hardware production in Nepal. Mercantile and Beltronix are the only two companies offering branded computers under their name. Both are ISO 9001 certified. There are however, plenty of local companies assembling computers from imported parts.

Some software companies are producing good quality software and even catering to the international market. Still, the software industry in Nepal is not growing as rapidly as envisaged.

Experts are urging the Computer Association of Nepal (CAN) to establish a separate cell under it to more effectively address the problems confronting the software industry. Software companies are in need of basic policy guidance, support for starting software businesses and nurturing entrepreneurs, R&D and marketing, promotion of software developed in country, international cooperation and software certification.

IT-enabled services

Some IT-enabled services are finally transforming Nepal into a global outsourcing centre. They are not only contributing to the country's foreign currency earnings but also providing jobs to the youth and unemployed.

Serving Minds and D2Hawkeye are two local companies that have broken into the BPO scene, proving that the Nepalese can compete with better-known businesses in other countries in serving clients in a globalized economy. Serving Minds is a call centre located in Kathmandu. It is now one of the largest employers of Nepali graduates. Starting with only a few employees in 2003, it hopes to have 1,000 staff members by 2007. D2Hawkeye Services is an offshore centre for D2Hawkeye, a premier medical data-mining company that provides decision support, builds fully integrated medical and financial databases for its clients, and leverages these databases with applications that are relevant to clients. In three years, the company grew from three to over 50 computer engineers. It is the largest software company in Nepal.

There are other IT-enabled companies in Nepal that are offering services to clients in the USA and Japan. They are working in the areas of medical transcription, digitization of maps and call centres, like the newly established Link Tree Pvt. Ltd. This sector needs to be developed further with appropriate policies, as it not only brings in foreign exchange for Nepal but also provides employment on a mass scale.

Telecom sector

Fixed-line telecom service is growing more slowly than mobile telecom service. There are currently two operators providing fixed-line service—Nepal Telecom (NT) and United Telecom Limited (UTL). NT, with a market share of 88 per cent, provides PSTN as well as wireless local loop (WLL) services, while UTL provides only WLL services. The fixed-line penetration rate as of February 2007 is 2.46 per cent.

NT and Spice Nepal Pvt. Ltd. are licensed to operate mobile GSM telephone services. NT's mobile phone subscriber base has reached 690,369. Spice Nepal has a subscriber base of 351,450 as of February 2007. The mobile customer base is growing rapidly, thanks to the prepaid mobile scheme which has already crossed the 933,052 mark and with the demand both inside and outside Kathmandu Valley still growing. Spice Nepal Pvt. Ltd. is extending its service to other districts as well. NT already has mobile service presence in 40 of 75 districts in Nepal.

NT is exploring use of CDMA technology to improve the quality and capacity of its mobile phone services. The company will be able to provide one million lines of mobile phones with the full implementation of CDMA. NT has already introduced multimedia messaging (MMS) services as well as Global Positioning Radio System (GPRS) services.

As of February 2007, teledensity in Nepal was around 6.48 (2.46 for fixed-line, including WLL and Limited Mobility service, and 4.03 for mobile). Mobile teledensity in 2002 and 2004 was 0.09 and 0.42, respectively, and fixed-line teledensity in the same period was 1.42 and 1.64, respectively. Lately, mobile teledensity has overtaken fixed-line teledensity.

Internet service providers

There are currently 39 licensed ISPs, of which 32 are operating in Kathmandu Valley. It is hoped that Internet services will be expanded to areas outside of Kathmandu Valley in the near future. There are around 50,000 active Internet subscribers in the country. However, the number of Internet users could be higher since many people access the Internet in offices and cybercafés.

Thirteen ISPs are currently connected to NPIX (Nepal Internet Exchange), which is owned by ISPAN (Internet Service Providers Association of Nepal). The international bandwidth used as of February 2007 for uplink and downlink is 37 and 90 Mbps, respectively.

Key ICT4D institutions

The key government institution coordinating ICT developments, including ICT for development (ICT4D), is the High Level Commission for Information Technology (HLCIT). It provides crucial strategic direction and helps formulate appropriate policy responses for the development of the ICT sector.

There are few institutions focusing fully on ICT4D in Nepal. Many seem to be counting on telecentres as a medium for development in the rural areas.

The Forum for Information Technology Nepal (FIT Nepal) (http://www.fitnepal.org.np/) is an NGO founded by a group of IT enthusiasts to promote ICT as a tool for development. FIT Nepal strives to bring the benefits of ICT to rural and marginalized communities by establishing community centres and through capacity building programmes.

Swaabhimaan is an alliance between government, civil society organizations, and public and private agencies. Its mission is to create a knowledge revolution in Nepal through some 1,500 telecentres. Swaabhimaan is an offshoot of Mission 2007 in India, a civil society-led nationwide initiative launched in July 2004 with the motto, 'Every Village a Knowledge Center'.

E-Networking Research and Development (ENRD) is an NGO operating from Kathmandu. It has been working closely with the Nepal Wireless Networking Project since 2005, helping to de-license Wi-Fi bands in Nepal. ENRD organizes basic computer education and hardware training programmes for villagers.

Nepal's Rural Education and Development (READ), an NGO based in Kathmandu, is providing no-cost public access to computers and the Internet and is committed to promoting information and literacy.

International and regional organizations like the International Centre for Integrated Mountain Development (ICIMOD), South Asia Partnership—Nepal (SAP), PANOS South Asia, BELLANET, Mountain Forum and the One World through Open Knowledge Network (OKN) initiative are also engaged in ICT4D in Nepal through knowledge sharing, diversification of knowledge delivery, including multimedia and radio, and creation of online communities.

Online services

e-commerce

Until recently the development of e-commerce was hampered by the absence of an Electronic Transaction Act. Moreover, most Nepalese do not have credit cards, connectivity in Nepal is low and few people have ready access to the Internet. Nevertheless, a number of businesses, such as munchahouse.com.np and nepalshop.com, provide e-commerce services by accepting and delivering orders through shops in Nepal, supported with offline payment. There are also full-fledged e-commerce-enabled websites where the payment function is based outside of Nepal and only the delivery of product is managed within the country.

e-governance

The government portal (http://www.nepal.gov.np) provides information on government activities and services. However, there is a need to go beyond providing information to enabling online transactions such as payment of bills and e-procurement.

The government has been preparing an e-Governance Master Plan. E-governance is envisioned to reduce costs and to improve public service delivery. However, government officials will need to be reoriented and trained in e-governance. In line with the government's thrusts, the Asian Development Bank (ADB) is programming a loan assistance package for electronic public service delivery particularly in rural and semi-urban areas.

Digital content initiatives

Much of the digital content produced in Nepal is Web-based and English is the dominant language. The topics are usually tourism, news and media, culture, arts, entertainment, business, government, banking and finance, health, education and information technology. There are online directories and yellow pages. The tourism industry, a major economic sector in Nepal, has a strong Web presence.

Many government ministries and departments are putting up their own websites. Many of these are in English and, as mentioned previously, most are limited to providing information. Moreover, the URL for the government portal, which is managed by the National Information Technology Centre (NITC), is a cumbersome www.nepal.gov.np, instead of a more straightforward http://gov.np.

Most English language newspapers in Nepal have an online presence. News and current affairs are still the most sought after content. However, content in specialized areas is also surfacing. For example, Nepali journals will soon be available online. The International Network for the Availability of Scientific Publications (INASP) is working with the Central Library of Tribhuban University to establish an online platform for Nepal journals. Also, a network for sharing of programme content is helping Nepal's FM stations connect better. FM radios outside Kathmandu Valley are using the Internet, and wireless and other new media technology, to share six hours a week of audio software produced from a central hub. ICIMOD via menris.icimod.net is facilitating access to digital spatial content through Web mapping and using well-defined metadata, thereby reinforcing cooperation among digital content stakeholders. Moreover, the Nepal Bar Association makes Supreme Court decisions available to appellate and district courts through a website and periodic publication of case compendia on CD-ROM.

A big gap in digital content production is in local language publishing. Few newspapers in Nepali languages are online. The problem stems partly from the lack of support for local language computing. The recent landmark breakthroughs in Nepali computing, including support for Nepali Unicode (see the section 'Research and development in ICT'), should stimulate development of Nepali content.

Open source and open content

The development of NepaLinux was a significant contribution to the promotion of free and open source software (FOSS) and computing in Nepali and other native languages in the country. NepaLinux seems to be user-friendly. But because Linux is still relatively unknown to most Nepali desktop users, there is a need to popularize its use. A policy promoting open source software or content is also needed.

It is also time to discuss and experiment with the open content approach to fostering open and collaborative content development and dissemination.

Education and capacity building

The majority of ICT personnel (69 per cent) are in Kathmandu Valley. According to a CAN survey conducted in October 2005, females comprise 18 per cent of the total IT workforce. Almost 44 per cent of the workforce has qualifications in IT-related fields. Still, most ministries and government agencies lack IT human resources to meet technical capacity. There are few ICT-related positions in the civil servant career path, making it difficult to attract technicians to the government sector.

There are high-quality educational programmes in ICT that can turn out well-trained graduates. But the number of such graduates needs to be substantial, in order to sustain the ICT industry.

Nepal has four universities—Tribhuwan Univeristy, Kathmandu University, Pokhara University and Purbanchal University—offering several IT-related courses through their affiliate colleges. These include the Bachelor of Computer Engineering (BE), Bachelor of Engineering in Information Technology (BEIT), Bachelor in Computer Application (BCA), Bachelor in Computer and Information Systems (BCIS), Bachelor in Information Technology (BIT) and Bachelor in Information Management (BIM). Each of the courses has its own objective and intake stream. Some courses are more oriented to engineering (for example, BE, BEIT), some are a mixture of management and ICT (for example, BCIS, BIM), while others are focused on ICT applications (for example, BCA, BIT). The important characteristics of these courses are that they are imparting current trends in ICT, targeting a range of students and preparing students for a rewarding career in ICT. The total intake is about 6,000 students per year, of which 50 per cent graduate.

There are a few advanced courses in ICT, such as the Master in Computer Applications (MCA). Students seeking Master-level courses in IT can also study abroad. But the requisite ICT training is available within Nepal itself and at a lower cost.

Another 1,000 students complete their ICT studies in neighbouring countries, in particular India, Bangladesh and the Philippines, and then return to Nepal. Around 4,000 ICT graduates are produced each year. However, this number exceeds the number of ICT jobs available. Although there are job opportunities in the international market for ICT professionals, the discrepancy between supply and demand with respect to ICT graduates and jobs in the domestic market needs to be addressed by both IT policy and education policy.

Apart from the formal ICT courses offered by the universities, computer learning centres and training institutes, which have mushroomed all over the country, offer short courses on a wide variety of computer software use and applications, including certifications of internationally recognized courses like Microsoft Certified Professional (MCP), Microsoft Certified Systems Engineer (MCSE) and Cisco Certified Network Associate (CCNA). These are significant value additions to the development of human resources in ICT in Nepal. According to the CAN survey in October 2005, 28 per cent of the total ICT workforce work as trainers and instructors in these training institutions.

Research and development in ICT

A landmark development in the ICT scene in Nepal in recent years has been in the area of localization. Nepalese who were unable to use computers due to language barriers can now use computers not just for word processing but also for database, spreadsheet, layout, Internet and e-mail. They also have two alternatives to choose from—Windows XP or Linux. Microsoft, in collaboration with Unlimited NuMedia, released the Nepali Language Interface Pack for WindowsXP and Microsoft Office 2003. Almost at the same time, Madan Puraskar Pustakalaya (MPP), a not-for-profit NGO that maintains the principal archive of books and periodicals in the Nepali language, unveiled the all-Nepali Linux, called NepaLinux.

NepaLinux is a complete operating system in Nepali developed over three years under the PAN Localization project with the support of IDRC of Canada. A Nepali spell checker and thesaurus are also integrated in the OpenOffice suite and is available in the NepaLinux package.

MPP further proved its commitment to expanding use of NepaLinux by coming out with a new release in November 2006. Microsoft for its part has decided not to charge any license fees for the use of the localized editions of the new product, which has been under development for the last one and half years.

Another significant R&D output is the development of Dobhase, a Web-based machine translator that translates text in English to Nepali. Dobhase, which literally means interpreter, is the product of an 18-month project of the Information and Language Processing Research Lab of Kathmandu University in partnership with MPP. The project was funded by the PAN (Pan Asia Networking) ICT R&D grants programme. More needs to be done to achieve accurate translations and the two partners are working to refine the current software by expanding the lexicon size and incorporating a facility for translating text in Nepali to English.

Conclusion

The historic signing of a peace agreement between the Nepalese government and the Maoists could finally lead the country to peace and stability. It is hoped that this will also lead to the desired development of the ICT sector. There is no dearth of specific recommendations in reports and workshops, including this review. What is needed above all is the willingness to implement these recommendations based on a commitment to ensure the growth of the ICT sector within the changed political reality of a new Nepal.

References

Computer Association of Nepal (CAN) Website. Available at http://www.can.org.np

Dobhase English-Nepali Translator Project Website. Available at http://nlp.ku.edu.np/cgi-bin/dobhase

High Level Commission on IT (HLCIT) Website. Available at http://www.hlcit.gov.np

Internet Service Provider Association of Nepal (ISPAN) Website. Available at http://www.ispan.net.np

IT Professional Forum Website. Available at http://www.itpfnepal.org

Madan Puraskar Pustakalaya Website. (2003). Available at http://www.mpp.org.np

Ministry of Information and Communication Website. Available at http://www.moic.gov.np

Ministry of Science and Technology Website. Available at http://www.most.gov.np/

Nepal Telecom (NT) Website. Available at http://www.ntc.net.np

Nepal Telecommunications Authority (NTA) Website. Available at http://www.nta.gov.np

Nepal Telecommunications Authority (2007). *MIS report*, February.

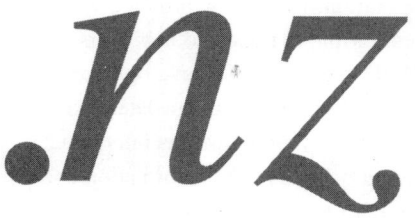

.nz New Zealand

Andy **Williamson**

GDP per capita	USD 24,498 (USD 1 = NZD 1.44)
Computers per 100 inhabitants	62
Fixed-line telephones per 100 inhabitants	96
Mobile phone subscribers per 100 inhabitants	74.2
Internet users per 100 inhabitants	60.5
Domain names registered under .nz	261,283
Broadband subscribers per 100 inhabitants	11.5
Internet domestic bandwidth	10 Gbps (2006)
Internet international bandwidth:	300 Gbps (2006) (~500 Gbps potential)

Introduction

New Zealand (Aotearoa) is a country of over four million people located to the east of Australia and at the southwestern edge of the Pacific Ocean. First settled by Polynesian seafarers around 800 years ago, New Zealand became a focus for European colonization after the signing of the Treaty of Waitangi (te Tiriti o Waitangi) by Māori chiefs and the British Crown in 1840. The Treaty establishes a unique model of biculturalism where the rights of indigenous Māori are protected alongside those of the Crown. New Zealand has been an independent state within the Commonwealth since 1947.

Post independence, New Zealand was heralded as a leader in the development of social welfare systems. During the 1980s, it was at the forefront of neo-liberal economic transformations, dramatically reducing the role of the State and the size of the public sector through a policy of privatizing State Owned Enterprises. New Zealand today is experiencing a moderate swing back to greater government intervention with the repurchase of public assets such as railway tracks and the introduction of legislation that recognizes the national interest in areas such as telecommunications and aviation. The country is also becoming more ethnically diverse, the result of changes to immigration policy dating back to 1987. New Zealand today is 79 per cent European, 15 per cent Māori, 7 per cent Pacific peoples and 7 per cent Asian. However, the relative size of the European population is falling and by 2021 the Asian population will reach 15 per cent. One in five New Zealanders—one in three in Auckland—was born overseas (Statistics New Zealand 2002b, 2003).

New Zealand is well positioned in terms of the uptake and use of ICT. It is one of the highest per capita users of the Internet. However, it lags behind the OECD average in terms of broadband adoption and telecommunications infrastructure investment. Where New Zealand has made significant progress is through the Digital Strategy released in 2005, which takes a 'whole of government' approach rather than isolating ICT within a single Ministry. This strategy brings together existing work and identifies new legislation and funding opportunities to make New Zealand 'a world leader in using information and technology to realize its economic, social, environmental and cultural goals, to the benefit of all its people' (New Zealand Government 2005, p. 4). It is overseen by a core group of 'Digital Ministers' within the Cabinet and operationalized through a Digital Strategy Steering Group, external Advisory Group and Secretariat that works across all government ministries, departments and agencies. The Digital Strategy is derived in part from the World Summit on the Information Society (WSIS) model targeting the benefit of ICT for government, business and civil society, and it is achieved through three areas—content, confidence and connection.

Digital content initiatives

New Zealand has a wealth of digital content producers, ranging from world-renowned Weta Digital, made famous by movies such as 'Lord of the Rings' and 'King Kong', to small local companies, such as video and interactive designer Oktobor and Web developers Springload who designed the 2006 Webby Award-winning New Zealand Festival website (www.nzfestival.telecom.co.nz). New Zealand Tourism's international travel website (www.newzealand.com), built around the Shado customer management system from New Zealand-based Straker Interactive, is another Webby-winning example of innovative digital content. NZLive.com was launched in September 2006. Managed by the Ministry of Culture and Heritage, it links existing content and events in the creative sector.

> **Hector the Protector (www.netsafe.org.nz)**
>
> Hector the Protector is New Zealand's Internet safety ambassador. Developed by The Internet Safety Group with support from Microsoft New Zealand and produced by Inkspot Digital, Hector is an easy way to educate children and keep them safe online. Named after New Zealand's rare Hectors Dolphin, Hector is downloaded and floats—or swims—in the corner of the screen. If something upsetting or inappropriate appears on the screen, the child can click on Hector and the screen is blanked out with an underwater scene and a reassuring message. At this point, Hector encourages the child to seek some help from an adult. The Hector button is supported by a set of Web-based resources and a recently released short animation for children, with more episodes in development.

Public digital content is being brought together through the National Digital Forum. The Digital Content Strategy, a subset of the Digital Strategy, is currently under development, led by the National Library. The Digital Content Strategy defines content in its broadest sense and aims to address economic, social and cultural challenges and opportunities. It targets two outcomes:

- New Zealanders will create, access, share, use, preserve and protect a broad range of quality content that supports the push to transform the economy and strengthen national identity and community; and
- Businesses will be digital content savvy and New Zealand will have a strong and internationally competitive digital content industry.

The Digital Content Strategy acknowledges formal content, such as that originating from research, libraries and museums, informal content such as that from community groups and blogs and commercial content created for profit. It is based around a framework that recognizes the need for content to be created and, where appropriate, commercialized; the need for appropriate models governing access and sharing of content, including creative commons; and the need for enhanced understanding of digital content and guidance on the protection and preservation of content.

The outcomes are to be achieved through a range of initiatives (at this stage, 16 are defined). These include 'Digital New Zealand', which is a programme to increase the responsiveness of the community, business and government sectors to digital content opportunities and challenges, an annual showcase of digital talent, funding and business development programmes, skill enhancement through targeted and general ICT access programmes, the development of standards and a framework for content repositories, and a national content portal.

Online services

New Zealanders have historically been quick adopters of new technology, as demonstrated in their high usage of electronic point-of-sale transactions and electronic and Internet banking. In 2004, 96 per cent of adults had a telephone at home and 71 per cent had access to a mobile telephone (ITU 2004). Over 65 per cent of adults had Internet access, the 8th highest Internet penetration rate in the OECD and up from 37 per cent in 2000. This figure rises to 70 per cent in the 18–64 age group (Ministry of Social Development 2006). Among the New Zealand population groups, Pacific people were least likely to have Internet access (39.5 per cent), followed by Māori (45 per cent). Māori also experienced the lowest growth in Internet usage over the fours years to 2004. And despite overall high levels of Internet use, New Zealand ranks poorly in terms of the transition to broadband, with only 11.5 per cent of the population having access to a broadband connection (OECD 2005).

All mainstream media are available online and news aggregation, portal and shopping sites are popular. Demonstrating the changing demographic of New Zealand are sites such as SkyKiwi (www.skykiwi.com), a website for Chinese-speaking New Zealanders. Local auction website 'Trade Me' is consistently New Zealand's top-ranking website and has over one-third of the population as registered users and over 700,000 auctions live at any one time. It was recently sold to Australian media conglomerate Fairfax for NZD 700 million. Telecom established an online retail presence last year with the launch of Ferrit, a site that aggregates new products from over 125 retailers. Other popular websites include NZDating, the major banks, and news portals such as Stuff, XtraMSN and the *New Zealand Herald*.

Research suggests that proximity of Internet access in part determines usage and value, such that those with access at home or work report being more satisfied with the Internet as a tool and more reliant on it than those without immediate access (Williamson and Dekkers 2005). There is a strong correlation between income and access to ICT in New Zealand (as elsewhere). The urban poor, those living in rural locations, and the elderly are more likely to lack Internet access at home (Craig 2003).

According to the 2001 census, 50 per cent of those owning their own home had Internet access as opposed to only 11 per cent of those living in State or local authority rental housing (Statistics New Zealand 2002a). Only 50 per cent of single-parent families have Internet access at home, compared to 78 per cent of two-parent families (Ministry of Social Development 2006). As a result, many community ICT initiatives are focused on providing ICT access within the local community and developing ICT literacy skills (Craig and Williamson 2005).

The New Zealand e-Government Strategy states that by June 2007 ICT will be an integral part of the delivery of government information, services and processes, and that by mid-2010, government operations will have been transformed through the use of the Internet (State Services Commission 2001). The State Service Commission's 2004 review shows that many agencies now have downloadable versions of forms and a significant number are offering fully online services. This could provide significant cost savings to end-users since a survey of interactions with government agencies shows that obtaining forms was the primary activity of respondents (Curtis et al. 2004). While the security and authentication of users of online government services is currently an issue for both end-users and agencies, a project is underway to identify and implement solutions across all government agencies (New Zealand Government 2004).

Two government agencies that stand out in terms of their current level of online service delivery are the New Zealand Customs Service (www.customs.govt.nz) and the Inland Revenue Department (www.ird.govt.nz).

> **The Inland Revenue Department online (www.ird.govt.nz)**
>
> The Inland Revenue Department (IRD), a leader in government use of the Internet, has an extensive range of online service and payment options for individual and company tax payers. The IRD site allows visitors to complete most standard taxation forms electronically, to register for key services, such as Goods and Services Tax, and to communicate via a secure e-mail system. There is integration with New Zealand Post's online 'change of address system' and payment interfaces that provide secure payment from the online banking system of all major banks.

ICT industries

ICT is one of the three focus areas for the government's Growth and Innovation Framework (GIF). The key objective is to increase the sector's contribution to GDP from 4.3 per cent to 10 per cent by 2012. This will occur by developing programmes that target globally competitive businesses, sustaining (and growing) a skilled ICT workforce, building a culture of entrepreneurship and targeting government regulatory changes to support the industry.

New Zealand is a market of overseas vendors and local ICT companies the majority of which appear focused primarily on the domestic market. It is estimated that the industry is made up of around 2,000 businesses employing two or more people. It generated NZD 14.3 billion in sales of ICT-related goods and services in the 2005 financial year, of which NZD 1.1 billion was exported (Griffin and Muller 2005; Statistics New Zealand 2006). Of the total sales, 55 per cent was for ICT services, 34 per cent for telecommunications and programme distribution services, and 5 per cent for software. Major employers in the sector include Microsoft, IBM, Vodafone and local companies Datacom, Gen-I and Eagle. Christchurch-based companies Tait Electronics, Jade and UK-owned Allied Telesis, as well as Auckland-based (but US-owned) Navman, are internationally recognized examples of New Zealand's innovative ICT industry.

According to the Ministry of Economic Development (2004), small and medium-sized enterprises (SMEs) account for 97 per cent of New Zealand's businesses and 38 per cent of total economic output. This demographic is reflected in the ICT sector where 84 per cent of businesses were classified as SMEs. However, the 'large enterprises' that made up only 6 per cent of the ICT sector accounted for 79 per cent of total sales.

Enabling policies and programmes

The value of ICT for community development was first recognized at a policy level through the Connecting Communities Strategy (Community Employment Group 2002) and latterly in the Digital Strategy. Defining a 'connected community' as one that uses ICT as an enabler to reach its goals and needs effectively and efficiently, Connecting Communities aimed to improve access to and effective use of ICT amongst communities. More recently, the Draft Digital Strategy (New Zealand Government 2004) was published, followed a year later by the full strategy (New Zealand Government 2005). This strategy provides 'an integrated framework for existing and future initiatives to encourage the uptake and effective use of ICT for economic, social and cultural gain' (p. 2) and sets out to create the conditions necessary for this through three interrelated areas (p. 3):

1. Content: Information made available via digital networks
2. Confidence and capability: The necessary skills to use ICT effectively
3. Connection: Affordable access to ICT infrastructure

The Digital Strategy is significant in that it signals a realization that a whole government approach to ICT is required and that, even in a 'developed' country such as New Zealand, ubiquity and sustainability of ICT, or innovation through ICT, cannot be assumed. The model used in the strategy is internationally significant because it emerged through the 1st World Summit on the Information Society (WSIS 2004) and adopts the WSIS tri-sectoral model of government, business and civil society (or community). The Digital Strategy sets out a platform for ICT up to 2010 that, in order to be made operational, must then be aligned with the government's current key policy platforms of economic transformation, families and national identity. It also links closely with the government's Growth and Innovation Framework and the Sustainable Development plan.

The strategy commits up to NZD 400 million of funding to a wide range of digital initiatives delivered by government, business and the NGO sector. While some of these are existing work programmes, there is approximately NZD 60 million of new funding aligned with the Growth and Innovation Framework, which includes two competitive funds: the Broadband Challenge (NZD 24 million) focused on broadband infrastructure partnerships in key urban centres, and the Community Partnership Fund (NZD 21 million) which provides matched funding for grassroots ICT initiatives. The strategy encompasses projects such as PROBE (provincial broadband extension), a regional broadband initiative that has been extended to address broadband availability in remote and underserved communities, and a number of projects within the Ministry of Education. The latter includes the innovative Digital Opportunities project, which funds partnerships between schools, ICT organizations, and the Ministry of Education in order to 'improve learning through the innovative use of leading edge technologies' (Ministry of Education 2006, p. 1).

Within a local government context, there is increasing but slow recognition of the role of ICT. Some city and district councils have been notable in their recognition of ICT within a service delivery and community development framework – for example, Porirua, Waitakere for the development of a joint community and council 'Digital city strategy' (Williamson and Edwards 2005), and Wellington, which recently launched an ICT policy that supports community ICT and eDemocracy initiatives in the city (Wellington City Council 2006).

Legal and regulatory environment

New Zealand is often considered to be lightly regulated in contrast to many other developed countries. However, this has been changing over the last three Labour-led governments and increasingly there is a focus on developing legislation that protects the national interest over allowing market forces to solely determine outcomes.

The government has made a number of legislative changes to address the increasing importance of ICT within business and society. These include the Electronic Transactions Act 2003, which is based on the United Nations Commission on International Trade Law's 'Model Law on Electronic Commerce' and which clarifies the legal position of electronic commerce, and the Unsolicited Electronic Messages Act 2007, which deals with spam. Amendments have been made to existing statutes, such as the Crimes Act 1961, extending definitions to encompass crimes involving computers and communication interception offences. Changes have also been proposed to the Copyright Act 1994 to encompass new digital technologies.

The most significant examples of increased regulation have been in telecommunications, particularly in relation to mobile and broadband Internet. The Telecommunications Act Amendment Bill seeks to regulate the day-to-day management of Telecom New Zealand's physical network to encourage other ISPs to deploy their own broadband equipment. At present, New Zealand permits only bitstream unbundling, which is seen as largely ineffectual in encouraging competition since Telecom still owns and sets the cost and performance levels for the majority of Internet customers (via its ADSL service). The new legislation will fully unbundle the local loop, allowing access to network cabinets and exchanges, and create the provision for naked-DSL, meaning that retail customers will no longer need a Telecom telephone line to receive ADSL. There is also provision in the legislation for the regulatory division of Telecom's wholesale and retail services, following the model adopted by British Telecom in the UK. These changes are significant because New Zealand lags behind other OECD countries in terms of broadband uptake (22nd) and investment in telecommunications infrastructure (also 22nd) (OECD 2005).

Other regulatory activities relate to the operation of the Telecommunications Commissioner to regulate competition in the industry and the government's plan not to automatically renew radio spectrum allocations around 2010, given the emergence of new technologies such as WiMAX and the rapidly changing digital landscape.

Open source

Although organizations such as the New Zealand Open Source Society (www.nzoss.org.nz) and Openz (www.openz.org) continue to promote the use of open source in New Zealand, the local open source community remains relatively small and appears to be more focused on promoting global open source tools and standards than on innovating locally. Open source is clearly established in the New Zealand market, as demonstrated by the purchase of local open source pioneer, Asterisk, by commercial heavyweight Gen-I (now itself part of Telecom). Its use is sanctioned at government level where government agencies are encouraged to include open source alternatives in their evaluation, where they exist (State Services Commission 2003). One example of open source adoption is the Ministry of Education's contract with Novell to provide 2600 schools with Linux-based servers and PCs. Examples of locally created open source software include the NZ Open Source Virtual Learning Environment (VLE) Project (eduforge.org/projects/nzvle/) and the Greenstone digital library collections software produced by the New Zealand Digital Library Project at the University of Waikato (www.greenstone.org).

Research and development

New Zealand universities are undertaking considerable research in the area of ICT. Projects of interest include the Human Interface Technology Laboratory New Zealand (HIT Lab NZ), a human–computer interface research centre at the University of Canterbury and a partner of University of Washington-based HIT Lab US (www.hitlab.org.nz), and MediaLab, an academic and commercial partnership focused on innovative uses of ICT (www.medialab.co.nz).

There is limited but increasing research on ICT4D and community usage of technologies. Waikato University has been funded for some time by the Foundation for Research Science and Technology to undertake research into the economic, strategic and structural impact of ICT in relation to disadvantaged groups. Researchers at Victoria University and Massey University in Wellington have undertaken research into the use of ICT in the community and the voluntary sector. Government-funded research has identified barriers to the uptake and effective use of ICT in the community and the voluntary sector. These include lack of funding and over reliance on volunteers. The research shows that support for ICT is often ad hoc, training is rarely planned and structured, or it is privately funded, if funding is at all available. While there is general recognition that planning is valuable, there appeared to be a dearth of skills in this area and therefore few organizations were effectively planning for the use of ICT (Craig and Williamson 2005). The Families Commission has recently commissioned research to look at how ICT (in particular, the Internet) affects family life, identifying issues such as the ongoing digital divide, cyber safety, the potential to harness children's interest in gaming and the tangible benefits of the Internet and mobile communications to family communication (Weatherall and Ramsay 2006).

Security issues

A strategic approach to security and confidence in ICT is being developed by the Ministry of Economic Development to 'ensure that there is a clear overall vision to underpin and guide planning and to harmonize actions by government and other participants' (Ministry of Economic Development 2006, p. 7). This draft framework encompasses government, business and civil society and ranges from protecting critical technology infrastructure to data protection, online crimes, spam, Internet safety and education. It follows from an e-Government-centric assessment of Internet-based threats undertaken by the State Services Commission (e-Government Unit 2004) and is intended to raise awareness of security issues in the design and use of ICT. It recognizes that a broad-based culture of security and safety is a prerequisite for building and maintaining confidence in ICT (confidence being one of the three areas within the Digital Strategy).

The Internet Safety Group's most recent survey of home computer users reported that 16 per cent had no (or were not aware of) anti-virus software installed on their computers. This rises to 35 per cent for security software, such as a firewall. Only 52 per cent were proactively ensuring that their operating system was up to date. Of those that did have anti-virus software installed, 76 per cent were updating it regularly (Internet Safety Group 2005b). It appears that small and medium-sized business are more security-conscious than the general public, with 95 per cent having anti-virus software installed and 84 per cent using a firewall. Almost three quarters of businesses reported an ICT-related security incident in the last year. Thirty-six percent identified inappropriate content on a business computer but only 7 per cent had disciplined an employee and 72 per cent did not know where to report computer-related security breaches (Internet Safety Group 2005a).

Key ICT4D institutions

The Ministry of Education continues to fund community computing programmes, although changes to the funding categories mean that these have been reduced. The Ministry also continues to support projects such as Computers in Homes (www.computersinhomes.org.nz), which is run by the 20/20 Communications Trust (www.2020.org.nz) and provides computers to low-income households with school-age children.

The Digital Strategy's Community Partnerships Fund (CPF) is a contestable fund of NZD 21 million managed by the Department of Internal Affairs. The first funding round recently allocated NZD 6.5 million across 53 projects to support grassroots ICT initiatives that build capacity in or provide resources to communities. Examples of funded projects include an online clearing house for research undertaken on the community and voluntary sector, capturing of community content, community-based digital broadcasting and the establishment of an e-Rider programme (www.eRiders.net) in New Zealand. The fund is designed so that the emphasis is on projects that are scalable and that can become sustainable after the initial government funding ends. Unfortunately, awareness of this contestable fund in the community and voluntary sector appears to be relatively low, with a recent survey reporting only 27 per cent of NGOs aware of it and, of that group, 84 per cent not understanding the eligibility criteria and scope of the fund (NZFVWO 2006).

Also part of the Digital Strategy, the Broadband Challenge fund has recently allocated approximately NZD 2.5 million in funding to partnerships to develop broadband projects that provide access in rural or otherwise underserved communities.

Local non-government programmes also exist, such as Clubhouse 274 (www.clubhouse274.org.nz) which is part of the international Computer Clubhouse model providing a computer-oriented after-school learning environment for young people from underserved communities in South Auckland. Numerous local councils have funded computer facilities and training within their libraries and notable examples include Christchurch, Porirua, Manukau and Waitakere.

Educational programmes

The Digital Strategy contains considerable reference to ICT initiatives in the compulsory and tertiary education sectors, with an investment of at least NZD 69 million between 2006 and 2010 (New Zealand Government 2005). The Ministry of Education's strategic framework for ICT identifies 65 proposed or existing projects which range from directly supporting student learning to resource development and digital content creation, open-standards network infrastructure deployment and professional development to upskill teachers. (Ministry of Education 2005).

Numerous community ICT training schemes exist in New Zealand. Many of these are by tertiary education providers and are funded through the Ministry of Education's community education budget. Microsoft's Unlimited Potential training was introduced in New Zealand in 2005 and is jointly funded by Microsoft and the Department of Internal Affairs. In New Zealand, the project's UPLIFT programme targets building ICT skills in existing community organizations with a focus on training community workers to train others in their own communities. The programme is operated by Whitireia Community Polytechnic.

The library sector, which in New Zealand is operated by local councils, has been a significant promoter of online access and training. Numerous libraries now provide computer and Internet access and some provide one-on-one or group training. This is likely to be significantly enhanced as a result of the Digital Content Strategy. One of the strategic goals outlined in the Public Libraries Strategic Framework (2006–16) is to enable access such that 'local communities and individuals have access to the digital world and the skills to participate in an informed way free from unnecessary restrictions or charges' (LGNZ et al. 2006, p. 57).

Technology infrastructure

There is a need to overcome New Zealanders' resistance to broadband and, as discussed earlier, current legislative changes address this from a regulatory perspective. New Zealand has

Waitakere City's Community Area Network

Developed as a partnership between Waitakere City Council, a local community trust and a private company, the Waitakere City Community Area Network project is one of the five remote and underserved community broadband projects to receive funding from the first round of the government's Broadband Challenge Fund. The project focuses on providing wireless broadband to small communities on the western edge of Auckland that are beyond the reach of ADSL and often have poor mobile coverage. These small-scale, community-based networks use open-access wireless standards to provide a relatively low-cost broadband solution for communities that would otherwise miss out. The network uses a meshing technology to optimize performance and is connected directly to fibre via a long-distance radio link. The project has been designed in a modular way so that it can easily scale up or down depending on demand and be packaged so that it can be used in other isolated communities.

100 per cent broadband coverage through ADSL, wireless and satellite services. ADSL is the primary technology for broadband delivery, servicing 90 per cent of broadband users. Wireless is increasingly seen as a viable alternative with companies such as Call Plus and Woosh investing in WiMAX technologies. There is also activity around public access Wi-Fi, particularly in the main urban centres, and examples include CafeNet in Wellington (www.cafenet.co.nz).

The ADSL network is controlled by former State-owned monopoly Telecom New Zealand, which resells the service to different ISPs, the largest of which, Xtra, it owns. It has been argued that this has led to poor consumer choice and a level of service below that in other comparable countries. Ironically, it has also led to New Zealand broadband pricing being relatively low (albeit for a lower quality service). A recent study ranks New Zealand poorly at 22nd out of 26 OECD countries in terms of broadband service quality and price. The research observes that New Zealand consumers receive, on average, poorer quality broadband and are significantly more likely to be affected by low data usage caps (Williamson 2006).

Government activity in terms of infrastructure primarily relates to funding through the Broadband Challenge (urban fibre) and PROBE Extension (remote and under-served communities) components of the Digital Strategy. In addition, TelstraClear has been awarded a government contract to build and manage the Research and Education Advanced Network, New Zealand (REANNZ), which will provide high speed connections between universities, crown research institutions, related companies and overseas institutions at a cost of NZD 43 million.

In the commercial sector, mobile Internet continues to grow with the availability of 3G services from both Telecom Mobile and Vodafone, and the planning of limited new networks by Econet and TelstraClear. The latter recently also announced a project to deliver wireless telecommunications and Internet service in the north island city of Tauranga.

Conclusion

New Zealand is at an interesting point with regard to digital maturity. Internet usage is high but there remains a demonstrable failure to engage with broadband. Government has embarked on regulating the monopoly telecommunications network with a view to increasing opportunities for competition in the market and increasing broadband uptake. Recognition of ICT is increasing within local and central government and also at a community level. However, disparities still remain in terms of access and skills and these are largely a reflection of broader socio-economic and educational disparities.

The innovative Digital Strategy takes a whole of government approach to ICT, delivering funding for alternative network infrastructure extending broadband coverage and for community-focused ICT initiatives. Altogether, the strategy encompasses over NZD 400 million in funding that also includes significant investment in the education sector and the development of a Digital Content initiative. The only cloud on the horizon would appear to be the need to overcome a lack of awareness of the Digital Strategy in the community and voluntary sector.

References

Community Employment Group. (2002). *Connecting communities: A strategy for government support of community access to Information and Communications Technology*. Wellington, NZ: Department of Labour.

Craig, B. (15–16 September 2003). Social consequences of connecting low-income communities: The New Zealand experience. Paper presented at the Electronically Enabling Communities for an Information Society: A colloquium, Prato, Italy.

Craig, B. and Williamson, A. (2005). *Survey of New Zealand community ICT organisations and projects*. Wellington, NZ: Victoria University of Wellington/Department of Labour.

Curtis, C., Vowles, J. and Curtis, B. (2004). *Channel-surfing: How New Zealanders access government*. Wellington, NZ: State Services Commission.

e-Government Unit. (2004). *Trust and security on the Internet*. Wellington, NZ: State Services Commission.

Griffin, J. and Muller, G. (2005). *The HiGrowth project: New Zealand ICT sector profile*. Wellington, NZ: IDC.

Internet Safety Group. (2005a). *Cybersafety and security for small to medium enterprises in New Zealand*. Auckland, NZ: Internet Safety Group.

Internet Safety Group. (2005b). *Netsafe home computer security survey*. Auckland, NZ: Internet Safety Group.

ITU. (2004). *Asia-Pacific telecommunication indicators*. Geneva: International Telecommunications Union.

LGNZ, LIANZA, and National Library of New Zealand. (2006). *Public libraries of New Zealand: A strategic framework 2006 to 2016*. Wellington, NZ: Local Government New Zealand, Library & Information Association of New Zealand Aotearoa and the National Library of New Zealand.

Ministry of Economic Development. (2004). *SMEs in New Zealand: Structure and Dynamics—2004*. Retrieved 7 April 2004 from www.med.govt.nz/irdev/ind_dev/smes/2004/index.html

Ministry of Economic Development. (2006). *A strategic consideration of ICT security and confidence in New Zealand: Discussion paper for key agencies/organisations*. Wellington, NZ: Ministry of Economic Development.

Ministry of Education. (2005). *Enabling the 21st century learner*. Wellington, NZ: Ministry of Education.

Ministry of Education. (2006). *Welcome to DigiOps*, from www.digiops.org.nz

Ministry of Social Development. (2006). *The social report 2006: Indicators of social wellbeing in New Zealand*. Wellington, NZ: Ministry of Social Development.

New Zealand Government. (2004). *Digital strategy: A draft New Zealand strategy for consultation*. Wellington, NZ: New Zealand Government.

New Zealand Government. (2005). *The digital strategy: Creating our digital future*. Wellington, NZ: New Zealand Government.

NZFVWO. (2006). *Digital Strategy survey*. Wellington, NZ: New Zealand Federation of Voluntary Welfare Organizations.

OECD. (2005). *OECD broadband statistics*. Paris, France: Organization for Economic Co-operation and Development.

State Services Commission. (2001). *New Zealand e-government strategy*. Wellington, NZ: State Services Commission.

State Services Commission. (2003). *Open source software: Briefing to the Minister of State Services*. Wellington, NZ: State Services Commission.

Statistics New Zealand. (2002a). *Census 2001*. Retrieved 13 May 2003 from www.stats.govt.nz/census.htm

Statistics New Zealand. (2002b). *Census snapshot: Cultural diversity*. Retrieved 25 January 2005 from www.stats.govt.nz/products-and-services/Articles/census-snpsht-cult-diversity-Mar02.htm

Statistics New Zealand. (2003). *National Asian population projections (2001–2021)*. Retrieved 25 January 2005 from www2.stats.govt.nz/domino/external/pasfull/pasfull.nsf/web/Reference+Reports+2001+Census:+Asian+People+2001

Statistics New Zealand. (2006). *Information and communication technologies supply survey*. Statistics New Zealand.

Weatherall, A. and Ramsay, A. (2006). *New communication technologies and family life*. Wellington, NZ: Families Commission.

Wellington City Council. (2006). *Information and communications technology policy*. Wellington, NZ: Wellington City Council.

Williamson, A. and Dekkers, J. (2005). ICT as an enabler in the community and voluntary sector in New Zealand. In G. Irwin, W. Taylor, A. Bytheway and C. Strümpfer (eds), *Community Informatics Research Conference (CIRN) 2005* (pp. 408–29). Cape Town, South Africa: Community Informatics Research Network.

Williamson, A. and Edwards, R. (2005). *Draft digital city strategy*. Waitakere City, NZ: Wairua Consulting, Waitakere City Council & Work Raft Trust.

WSIS. (2004). *World summit on the information society*. Retrieved 21 August 2004 from www.itu.int/wsis

North Korea

Kyungmin **Ko**, Seungkwon **Jang** and Heejin **Lee**

Total population	23,113,019 (July 2006 est.)
GDP per capita	USD 1,800 (2006 est.)
Key economic sectors	Military Products, Machine building, Electric power, Chemicals, Mining (coal, iron ore, magnesite, graphite, copper, zinc, lead and precious metals), Metallurgy, Textiles, Food processing, Tourism
Fixed-line telephones	980,000 (2003)
Mobile phone subscribers	20,000 (2004) (*Yonhap News* 2005, p. 234)

Source: CIA. (2006). *The world factbook*. Retrieved from https://www.cia.gov/cia/publications/factbook/geos/kn.html

Overview

North Korea often appears in the media due to its disputes with the international community regarding its development and testing of nuclear weapons. Little else is reported and known about the country. ICT development is an example of an area in North Korean development in which there is much activity but about which little is known outside. Since the mid-1990s, ICT has been emphasized as a strategic industry in North Korean economic development.

In this chapter, we report the current status of ICT development in North Korea. It is not easy to retrieve reliable information on ICT in North Korea, and the information that is available is not always accurate. Even statistics published by international organizations like ITU are not updated regularly. Therefore we have constructed a data set that is as consistent as possible with some policy reports (for example, KIPA 2000; Lee, J.-W. 2003) and newspaper articles from North and South Korea.

Infrastructure

North Korea's telecommunication infrastructure lags behind the telecom infrastructure of many other countries. According to the *World Telecommunication Development Report* (ITU 2001), for example, the number of fixed-line telephone subscribers in 2000 was only 1.1 million out of a total population of about 23 million. This figure is higher than the *The World Factbook* (CIA 2005) estimate of 0.98 million as of 2003. At present, mobile telephones in North Korea are only for the privileged. The first mobile service was launched in Pyongyang and Rasun in November 2002. The number of subscribers at the time was 3,000. This grew to over 20,000 by December 2003.

Since the 1990s, the telecommunication infrastructure has been built up with some notable achievements. In 1994, a fibre optic cable factory was built in Pyongyang and a cable network was completed between Pyongyang and Hamhung. In 1998, a fibre optic cable network was laid out from Pyongyang to Sinuiju (400 km) with aid from UNDP. Fibre optic cable networks connect over 200 districts (called *kun*) nationwide and are being extended to the smallest administrative units (called *li*) (Lee, J.-W. 2003). Mobile communications has been extended to major cities. For the completion of the nationwide mobile communication network by 2007, infrastructure, including mobile base stations, is being built (Yonhap News 2005).

A nationwide network or Intranet, called Kwangmyong (meaning 'bright light'), was built in November 2002. The Central Information Agency for Science and Technology (CIAST), which is responsible for building and managing the Intranet, runs a portal site that is also called Kwangmyong. The portal offers a variety of services, such as database search, e-mail, website search, electronic library, real-time chatting, electronic commerce, and entertainment (Kwon 2002). This domestic network is open to the public in Pyongyang, the capital city. They call it the '170 Network', after its modem connection number. The modem speed is 56 Kbps. The network is also connected via fibre optic leased lines. It is used mainly for e-mail (Lee, S. 2004).

In April 2004, an Internet café was opened through which ordinary people can use the Internet. However, access is limited to the domestic network; the real Internet is closed to the general public.

There are no commercial ISPs in North Korea. The servers of North Korean websites are located in Japan, China and

Germany (Akutsu 1999; Ko 2004). There are very few websites with North Korea's domain name .kp, and those that exist have appeared only recently. Examples are the Academy of Sciences' websites for science and technology (http://www.stic.ac.kp) and for medical science (http://www.icms.he.kp) (*Junja Shinmun* 28 July 2003). However, these sites are available only on the Intranet and are not accessible from the Internet. It seems that these are test sites for the opening of the Internet in the future.

ICT industries

The hardware industry in North Korea is not developed due to limitations such as lack of financial resources and technological capability and, most of all, the 1996 Wassenaar Arrangement which bans countries that endorsed it from exporting to North Korea high-technology materials and technologies that can be converted to military use.

Nonetheless, with the support of the government, there have been some developments in the North Korean hardware sector. For instance, the Ministry of Electronic Industry was established in 1999 to foster computer hardware technology. The Morning-Panda Joint Venture Computer Company, which produces personal computers, was founded in 2002 by the Electronic Products Development Company and Nanjing Panda Electronics Company, a Chinese company (*Chosun Shinbo* 15 March 2001). Since April 2004, the Korea Institute of Industrial Technology and the Samilpo Information Technology Center have jointly developed and are working on the export of Hana 21, a personal digital assistant (Yonhap News 2005).

Due to the limitations mentioned earlier, more attention is being given to software development, which is viewed as strategic to the growth of the North Korean economy (Park 2003). North Korean organizations have developed a number of software products (Hayes 2002), some of which have received positive evaluation (Lee, J.-W. 2003; Ko 2004). For example, Silver Baduk, the computer version of the Asian chess game Baduk which was developed in the Korea Computer Center (KCC), was awarded the first prize at the world computer Baduk competition in 1998, 1999, 2004, 2005 and 2006 (Shin 2004).

Like other developing countries that are motivated by India's success story, North Korea is keen on exporting software, including offshore services. For example, KCC recently invited a Dutch IT consultant specializing in offshore software development projects (Tjia 2006). Some of the mobile games made by KCC are already popular in Japan. KCC also provides offshore services for clients in China, South Korea and Japan. In addition, some companies in Pyongyang are involved in entry-level business process outsourcing activities. For example, they offer back-office services to Western companies engaged in doing business with Japan and conduct data-entry work for international organizations like the UN and the International Red Cross. According to Tjia (2006), the level of IT expertise in North Korea is high, quality is assured through adherence to ISO 9001, CMMI and Six Sigma standards, and the IT sector is dynamic, with new firms and ventures being established.

To promote and support the development of ICT in general and the software industry in particular, two significant laws have been passed—the Computer Software Protection Law in June 2003 and the Software Industry Law in June 2004. They are significant because they signal to foreign companies that North Korea will abide by business practices common in the West and that the country is interested in attracting foreign investment in its software industry. The Computer Software Protection Law aims to protect software copyright, promote adherence to copyright provisions of international treaties and promote international cooperation (North Korean Law Society 2005). The Software Industry Law defines principles of software industry development, software writing and distribution, infrastructure building and international cooperation (North Korean Law Society 2005). These two laws are unprecedented in other industries and they clearly indicate that the North Korean authorities consider the development of the software industry a strategic goal (Lee, C. 2003; Lee, T. 2003).

In addition, other laws and regulations such as the Foreign Investment Law, Contractual Joint Ventures Law, Joint Ventures Law and Foreign Enterprise Law (Park 2004) contribute to software industry development by allowing tax exemptions and tax reductions for foreign investors. Further, legislation enacted in 2002 for the development of special economic areas such as Sinuiji, Mount Kumgang and Kaesong are specifically designed to boost foreign direct investment in high-technology industries, including ICT (Park 2004).

In 2001, North Korea signed international agreements on intellectual property rights, namely, the World Intellectual Property Organization (WIPO), the Paris Convention for the Protection of Industrial Property and the Patent Cooperation Treaty (PCT) (Park 2004). However, we have yet to see whether these laws work in practice, as declared.

The Single Leap strategy and key institutions

The North Korean government has sought to boost the ICT industry since the mid-1990s (Lee and Hwang 2004). In 1998, Kim Jong-Il adopted a policy emphasizing science and technology. In 2000, major newspapers, in a joint New Year editorial, noted science as one of the three major pillars for the construction of Kangsung Taeguk ('a strong and big nation'), along with ideology and arms (Bae 2001). Kim Jong-Il also emphasized the need to develop ICT (Lee and Hwang 2004). ICT is identified as a strategic industry in policies emphasizing the importance of science and technology.

North Korea aims to maintain the socialist system while leapfrogging to the information era (Bae 2001; Seo 2001). This development strategy, called Single Leap, aims not just for catch-up but for breakthrough. North Korean leaders view ICT as playing a key role in the 'Single Leap' strategy (Hahn 2003). Thus, after his May 2000 secret visit to Zhongguancun, the Silicon Valley of China, Kim Jong-Il ordered his government to promote ICT (Bae 2001).

In North Korea, the institutions in the ICT sector are steered by the central government. The Ministry of Posts and Telecommunications is responsible for telecommunication policies and the Ministry of Electronic Industry for IT industries. However, to achieve centralized control over information and communications, the Korean Workers' Party (KWP) and intelligence agencies like the Department of National Security and the Ministry of People's Security help formulate telecommunication policies (Ko 2004).

Since the development of ICT industries was identified as one of the strategic goals for national development in 2001, the Party has taken the lead in ICT development. In December 2001, the Party created a new unit, the Bureau 21, to direct ICT development. The bureau is so named because of Kim Jong-Il's idea that 'the 21st century is the age of information technology industries' (North Korean Central Broadcasting Agency 19 April 2001).

Guided by the Ministries and the Party, six key players are working in the ICT sector: the Academy of Sciences, Pyongyang Informatics Center (PIC), Korea Computer Center (KCC), Kim Il Sung University, Kim Chaek University of Technology and Pyongyang University of Computer Technology (Hayes 2002).

The Academy of Sciences set up the Department of Computer Science in 1983, to focus on basic research in ICT. Software products developed by the Department are exported by the Baeksong Trading Company, which is the commercial arm of the Academy.

PIC was founded in 1986 with support from the Pro-North Korean General Association of Korean Residents in Japan and UNDP. It aims to become the centre of development and education for computer systems (Hayes 2002). PIC sells some of the products it has developed at its Singapore office. PIC has a partnership with the Osaka Information Center (OIC) and runs a training centre for IT professionals (Lee, J.-W. 2003).

Founded in 1990, KCC is the largest IT company in North Korea with more than 1,000 employees (Tjia 2006). It aims to accomplish computerization in a variety of sectors, to develop technology for programming and to promote technology exchanges in computing areas. KCC has a company for international businesses, which deals with joint development, subcontract development, software export and technology and expert exchange programmes with foreign companies. KCC also has an office in Beijing specializing in export (KCC 2003).

The three other prestigious universities are discussed in the next section.

ICT education

Until 1985, education and training in most universities and computer-related institutions had focused on hardware (Song and Choi 2005). Software became the centre of ICT education in North Korea after Kim Jong-Il emphasized the importance of software development and software specialist education at the 8th National Computer Program Contest and Exhibition in December 1997. Kim Jong-Il also ordered the inclusion of a computer subject in the Year 11 curriculum at the National Contest and Conference of Voice Recognition Program held in February 1998 (KDI North Korea Team 2001).

Some top universities play a major role in ICT education. They have been offering programming courses since 1999.

At the Kim Il-Sung University, the most prominent university in North Korea, the Faculties of Automation and Physics have been restructured to form the Faculty of Computing. The university also established the Information Center for software research and development in 1999.

Kim Chaek University of Technology is one of the leading universities in science and technology. As early as 1983, it set up the Computing Research Center and the Faculty of Computer Engineering. In November 2001, it created the School of Information Science and Technology (Song and Choi 2005).

The Pyongyang University of Computer Technology was established in 1985 to develop both hardware and software experts. Its reputation in IT education was formed in 1997 when its graduates swept prizes in many national computer programming contests. The university plays an important role in educating computing teachers as well as in training IT specialists for industry.

One of the distinctive characteristics of North Korea's education system is the focus on gifted students. Since the publication of his paper on education systems in July 1984, Kim Jong-Il has emphasized the importance of educating gifted young people for the development of high-technology industries. Special classes in computers, science and mathematics were established for talented students in Kim Chaek University of Technology in 1999. In secondary schools, there are special classes for computer prodigies who then move on to prestigious universities like Kim Il-Sung University and Kim Chaek University of Technology (*Rodong Shinmun* 29 April 2001). This specially trained cadre of computer specialists forms the core of the ICT workforce in North Korea (*Chosun Shinbo* 7 May 2001).

Aside from focusing on training highly gifted students, the North Korean government also seeks to promote the importance of ICT to the whole of society by organizing festivals, exhibitions, lectures, contests and conferences. These activities aim to raise public awareness of science and technology development and encourage researchers and technologists to improve their productivity and share their experiences and knowledge with other researchers and with the rest of society.

Digital content and online services

The Internet in North Korea is under the strict control of the State. Despite the recent opening of an Internet café in Pyongyang, Internet access is restricted. Therefore there is little digital content for public consumption. Most content is produced for political purposes. For example, news websites deliver online news contents from government-operated offline media like Chosun Tongsin (Central News Agency of Democratic People's Republic of Korea) and *Pyongyang Times*.

However, some commercial websites have appeared recently (Ko 2003; Song 2003). To attract tourists to North Korea, a website called Arirang (www.arirang.dprkorea.com) was launched in April 2002; it also promotes the national mass game that is also called Arirang. Three months earlier, in January 2002, the DPRK National Tourism Administration opened a website called Chosun Tour (www.dprknta.com) in Japanese with the assistance of the Japan National Tourist Organization. Since then a few other commercial websites have been put up. A website in English was developed for e-commerce (www.chollima-group.com) and the North Korean Embassy in Austria opened a website (www.dprkorea-trade.com) to sell North Korean specialties online.

Since the economic reform in July 2002, some online activities have emerged. The Advanced Technology Service Center in Pyongyang runs an online shopping site. Authentication systems are required for electronic transactions. The 626 Technology Service Center has been appointed as the national authentication certificate issuer, indicating that more shopping sites will be put up in the future (Lee, S. 2004).

Although the Internet is recognized in North Korea as increasingly important, only the privileged are allowed to use it. North Korean leaders are concerned about the impact of the Internet on the stability of the regime. In an interview with *Chosun Ilbo* (3 October 2003), a North Korean newspaper, the president of Chosun Posts and Telecommunications Corporation, Mr Hwang Chol-Poong said that 'North Korea will not open the Internet due to the possibility of threats to the national security.' However, the North Korean government recognizes the potential economic value of the Internet. Thus, it faces the dilemma of whether to court political instability in exchange for economic gains by opening up the Internet.

Open source initiatives

KCC has developed Red Star Linux version 1.2 as its official operating system for nationwide use (*Junja Shinmun* 19 March 2002). Linux Arirang 2.0, another operating system developed at the Kim Chaek University of Technology, was awarded the silver medal at the 14th National Computer Program Contest and Exhibition in 2003 (Ministry of Unification 2003).

Linux-related R&D is growing. Kim Chaek University of Technology has completed a series of projects for the development and upgrade of Linux operating systems. It was thought that R&D in Linux would be the first step toward an indigenous operating system in North Korea (*Hankyoreh* 31 October 2001). At KCC, about 20 per cent of programmers use Linux, and over 70 programmers have been working on the development of North Korea's own operating system (Yoon 2004; Jung 2006).

Kim Jong-Il has also underlined the significance of North Korea having its own programming tool, which explains the North Korean authorities' eagerness to use open source software, including Linux. The leaders of the North Korean computer institutions believe that tools borrowed from abroad allow only a limited range of development, and that making their own tools, such as programming languages, will enable North Koreans to achieve more (*Yonhap News* 21 June 2006). Using Linux as the foundation of the indigenous Korean operating system is expected to foster independence from US-made software (Yoon 2004).

However, the campaign for independence in the software industry using open source software is fraught with many obstacles. For example, it is said that efforts to register a North Korean version of the Mozilla Web browser as a formal project (ko-KP) and requests for information regarding how to support the source code to use the North Korean encoding protocol (euc-KP) in Mozilla, were stymied by the US rules forbidding the export of encryption algorithms to countries like North Korea, which the US government considers to be a terrorist threat (Yoon 2004).

Cooperation between North and South

Since the historic summit between North and South Korean leaders in 2000, there have been many activities promoting exchanges and cooperation in ICT between the two states. ICT is considered a promising area where North and South could realize mutual benefits. Although the early enthusiasm has dimmed partly due to the worsening of political relations as a result of North Korea's nuclear programmes, the ICT sector is still one of the most viable options for economic cooperation between North and South Korea. There are many exchanges in software in particular.

The Hana Program Center in Dandong, a Chinese city on the border between China and North Korea, is a joint North-South software company. It has successfully completed many projects from China and South Korea by hiring programmers from North Korea. The projects include the development of CAD (computer-aided design) with Dooson Cadcam Engineering and of network software with Dasan CNS Inc. The company is also developing Linux operating systems jointly with MontaVista Software Korea, a subsidiary of a US Linux development company (*Junja Shinmun* 2 February 2004).

Furthermore, there have been many North-South joint projects in digital content development (Shin 2004). Dinga, the first 3D animation jointly produced by North and South Korean software developers, was rated highly for its quality. It got an agent contract for distribution in Japan with Media International Corporation, NHK's (Japan Broadcasting Corporation) subsidiary in entertainment. Similarly, the Little Penguin Pororo, which was broadcast on South Korea's EBS TV (Education Broadcasting System) in November 2003, was produced jointly by South Korea's EBS and North Korea's Samchonri General Company. Also, there are several mobile games in service on LG Telecom, such as Pro Beach Volleyball from Samchonri General Company. South Korea's KISTI (Korea Institute of Science and Technology Information), in collaboration with North Korea's CIAST (Central Information Agency for Science and Technology), has produced a CD-ROM titled 'Nature in Mount Baekdu'. In April 2002, Hoonnet, a South Korean company, came out with a website jointly operated with DPR Korea Lotto (Ko 2004), which was founded by North Korea's Jangsang Trade and the Pan-Pacific Economic Development Association of Korean Nationals. This joint venture is especially significant as it is a case of economic cooperation between North and South Korea using the Internet.

Challenges

Despite North Korea's enthusiasm, there are internal and external factors that constrain ICT development in the country (Lee and Hwang 2004). First, telecommunication and telephone networks, which are the basic infrastructure of ICT industry, remain underdeveloped. Although the situation is improving, this aspect of ICT development in North Korea is still far behind the level of global competition. Second, there is a severe lack of basic equipment. Hardware, such as computers and modems, is in short supply. Third, public Internet access is severely restricted. Fourth, advanced technology industries require a large amount of capital, which North Korea cannot afford. The necessary capital must come from foreign investors, but there is as yet no sufficient incentive for this. For example, many South Korean ICT companies are interested in North Korea for its supposedly cheap but good quality ICT workforce. However, large companies are still cautious about investing in North Korea due to political risks. The sixth constraint to ICT development in North Korea is the adverse impact of international politics such as the barring of ICT-related exports to North Korea by the Wassenaar Arrangement. Most of all, North Korea's nuclear test in October 2006 has destabilised North Korea's relations with South Korea, the USA, Japan and China, with devastating impact on its economic plans.

In sum, ICT development in North Korea faces many obstacles. It is uncertain how North Korea can cope with these internal and external challenges which are interrelated. To overcome the internal obstacles, North Korea needs significant investments for capacity building of ICT professionals and widespread provision of basic telecommunication services. Since it cannot afford to make such investments, North Korea should seek external assistance. Within the foreseeable future, however, it is unlikely for North Korea to receive such aid from the international community given the current international political climate. Unless the stalemate caused by North Korea's recent nuclear tests is broken, the challenges facing North Korea are not likely to be resolved.[1]

Note

1. As of proofreading this article, international relations surrounding the Peninsula (for example, USA–North Korea and North–South) are rapidly improving in favour of ICT development in North Korea. For example, the second summit between North and South Korea has been announced for early October 2007. However, we have yet to see how those relations will develop both in the short and long term.

References

Akutsu, H. (1999). *Is the Internet on the side of 'rogue states'? A lesson from the North Korean case*. Retrieved 3 August 2002 from http://ifrm.glocom.ac.jp/gii/akutsu19990621en.html

Bae, S. (2001). North Korea's policy shift toward the IT industry and inter-Korean cooperation. *East Asian Review*, 13(4), 59–78.

CIA. (2005). *The world factbook*. Retrieved from https://cia.gov/cia/publications/factbook/geos/kn.html

Hahn, J.S. (2003). *ICT in education: Democratic People's Republic of Korea*. Retrieved from http://www.unescobkk.org/fileadmin/user_upload/ict/Metasurvey/DKOREA.PDF

Hayes, P. (2002). DPRK information strategy? Does it exist? Paper presented at the Conference on IT Revolution and National Security in Korea, Asia Pacific Center for Strategic Studies, Honolulu (8–10 October). Retrieved from http://www.nautilus.org/archives/pub/ftp/Phayes/DPRKInformationStrategyPubVersionOct11-021.htm

ITU. (2001). *World telecommunication development report*. Geneva: ITU.

Jung, C. (2006) Independent OS development completed for the national IT industry. *Minjog* 21, 1 August. Retrieved from http://www.minjog21.com/. (In Korean)

KCC. (2003). *Introduction to the Korea Computer Center*. Retrieved from http://www.stepi.re.kr/upload/cyber_research/fileboard/%C1%B6%BC%B1%C4%DE%C7%BB%C5%CD%BC%BE%C5%CD%282003%29.hwp. (In Korean)

KDI North Korea Team. (2001). Development of computer software in North Korea. *KDI review of North Korea economy*, January. (In Korean)

KIPA (Korea IT Industry Promotion Agency). (2000). *The current status of North Korea's IT and suggestions for cooperation*. (In Korean)

Ko, K. (2003). A study of method and strategy to build e-government in North Korea. *North Korean studies review*, 7(2), 299–329. (In Korean)

Ko, K. (2004). *IT strategy in North Korea*. Seoul: Communication Books. (In Korean)

Kwon, J.-J. (2002). Science and technological information service system, Kwangmyong. Paper presented at the 42th Conference of Science and Technology Association of Korean Residents in Japan and the Symposium on Science and Technology for Unification, Japan, 5–6 October. (In Korean)

Lee, C. (2003). North Korea's open policy and necessity of inter-Korean cooperation. In Chonghee Lee (ed.), *ICT in North Korea* (pp. 15–40). Seoul: Thinking Tree. (In Korean)

Lee, H. and Hwang, J. (2004). ICT development in North Korea: Changes and challenges. *Information technologies and international development*, 2(1), 75–87.

Lee, J.-W. (2003). The information and communications technology sector. In Choong Yong Ahn (eds), *North Korea development report 2002/03* (pp. 188–217). Seoul: Korea Institute for International Economic policy. Retrieved from www.kiep.go.kr/inc/download_pub.asp?fnm=(6E531D1B)%20NK1.pdf

Lee, S. (2004). The present situation of internet diffusion in North Korea. *Science & technology policy*, 14(4), 79–91. (In Korean)

Lee, T. (2003). North Korea's ICT industry. In Chonghee Lee (ed.), *ICT in North Korea* (pp. 83–116). Seoul: Thinking Tree. (In Korean)

Ministry of Unification (Republic of Korea). (2003). *Jugan Bukhan Donghyang*, no. 675, 26 December. Retrieved from http://www.unikorea.go.kr/kr/KNU/KNU0103R.jsp. (In Korean)

North Korean Law Society. (2005). *Collection of North Korean laws and regulations*. Seoul: North Korean Law Society. (In Korean)

Park, C. (2003). North Korea's software technologies. In Chonghee Lee (ed.), *ICT in North Korea* (pp. 117–61). Seoul: Thinking Tree. (In Korean)

Park, J.W. (2004). *A study of North Korean laws relating to IT and the institutionalization of IT exchange and cooperation of inter-Korea*. Korea Legislation Research Institute, Research Paper No. 2004–18. (In Korean)

Seo, J. (2001). *North Korea: From economic crisis to the IT industry*. Seoul: Jishik Madang. (In Korean)

Shin, S. (2004). Digital contents are surging from North Korea. *Digital contents*, July. Retrieved from http://www.dbguide.net/know/know103001.jsp?divcateno=3&divcateno_=3&idx=2227&mode=view. (In Korean)

Song, K.J. and Choi, H.K. (2005). A study of North Korea's IT education. *North Korean science and technology review*, 3, 65–104. (In Korean)

Song, S.-S. (2003). Present situation on foreign Internet websites of North Korea. *North Korean science and technology review*, 1, 103–30. (In Korean)

Tjia, P. (2006). *North Korea: An upcoming software destination*. Retrieved 19 December 2006 from http://www.gpic.nl/IT_in_NKorea.pdf

Yonhap News. (2005). *2005 Yonhap yearbook*. Seoul, Korea: Yonhap News. (In Korean)

Yoon, S.C. (2004). *Is there open source software in North Korea?* Retrieved 27 December 2004 from http://zdnet.co.kr/news/column/scyoon/0,39025737,39132512,00.htm. (In Korean)

Newspaper articles (In Korean)

Chosun Ilbo. (2003). North Korea, impossible to open the Internet due to regime maintenance. 3 October. Retrieved from http://nkchosun.com/news/news.html?ACT=detail&cat_id=4&res_id=39323&page=2

Chosun Shinbo. (2001). Our goal is 'The World First': Computer prodigies studying with kid artists. 7 May. Retrieved from http://www.korea-np.co.jp/news/ViewArticle.aspx?ArticleID=7094

Chosun Shinbo. (2003). The first home made PC, 'Morning-Panda': A joint venture with a Chinese firm for PC production. 15 March. Retrieved from http://www.stepi.re.kr/research/cyber/fileboard/view.asp?No=231&page=1&block=0&form=list

Chosun Central Broadcasting Agency. (2001). *The 21st century is the age of IT industries*. 19 April.

Hankyoreh. (2001). North Korea universities develop Linux applications. 31 October. Retrieved from http://www.hani.com/section-003100000/2001/10/003100000200110312246005.html

Junja Shinmun. (2002). North Korea's policy change in software. 19 March. Retrieved from http://www.etnews.co.kr/news/detail.html?id=200203180270

Junja Shinmun. (2003). North Korean homepages with 'kp' domain name open. 28 July. Retrieved from http://www.etnews.co.kr/news/detail.html?id=200307270031

Junja Shinmun. (2004). South-North joint IT company, Hana Program Center conducts new IT projects. 2 February. Retrieved from http://www.etnews.co.kr/news/detail.html?id=200401300088)

Rodong Shinmun. (2001). Superiority of 'our' socialism in IT industry. 29 April.

Yonhap News. (2006). North Korea's 'Our Own Way' in operating systems. 21 June. Retrieved from http://www3.yonhapnews.co.kr/cgi-bin/naver/getnews_new?14200606210190016564+20060621+1654

Pacific Island Countries

Esther Batiri **Williams**

Overview

In the Pacific, the ICT sector continues to be very complex and diverse. There are 22 countries[1] in this region and each represents a different set of unique environmental conditions and stage of development. There are challenges imposed by the small scale and scattered nature of the Pacific island economies, difficult environments and the lack of supporting infrastructures such as electricity. Geography, population and the availability of content and appropriate platforms add to issues like language, culture and politics. However, in the past 12 months, great strides have been taken by the Forum[2] to put together a regional digital strategy[3] to carve a roadmap for the application of ICT for Pacific island countries at national and regional levels. Hence, based on the rate of ICT development and proliferation, a classification of the countries can be made. No reliable data is available at this time to further support this classification.[4]

The first group comprises countries that have ICTs as a priority and are committed to moving ahead with their development plans for this sector. Some of the countries that fall in this category are Fiji, Samoa, Tonga, Niue, Vanuatu, Republic of the Marshall Islands, Papua New Guinea and the Federated States of Micronesia. The second group includes those countries that acknowledge the importance of ICT for development, have plans to push forward developments and are able to achieve some gains, but whose government and other leaders are not providing the required recognition or funds. The countries that can be included in this category are the Cook Islands, Solomon Islands, New Caledonia, Palau, French Polynesia and Tuvalu. The third group comprises countries that are still not putting ICT in any prominent position overall and lack leadership in and commitment to advance ICT development. These countries include Kiribati, Nauru, Tokelau, and Wallis and Futuna.

Within these categories there are some common elements. More people are now accepting the use and impact of computers and ICTs in development in all of the Pacific countries and soon there may be only one category. In areas where infrastructure is good, ICT services are satisfactory, computers and related equipment are available, and user demand is high. Computer prices are dropping and becoming affordable. A large part of the population of the Pacific region is ICT-savvy. They are aware of what ICTs can do and know that ICTs can link them to the global world. A large part of the population of the Pacific Island States also has access to the Internet and broadband, and they use wireless technology, 3GSM phones and state-of-the-art technologies at work. Because of the isolation of the islands and their distance from major business and trading centres, Pacific islanders are forced to find better means of communicating all the time.

Within the three groups, there are major common challenges in the implementation of the regional digital strategy. These include policy and regulatory issues, human resources development and research, technological infrastructure development, building awareness and support for ICT at various levels, and sourcing resources to support these initiatives. There is also a growing digital divide between and within countries.

The regulatory and policy challenges can be very frustrating especially in countries which have the technologies in place but which lose opportunities because of exclusive licenses in telecommunications, data and voice transmission, as in Fiji. In such contexts, there is limited freedom of access to low-cost bandwidth, quality multimedia and voice transmission, and providers cannot offer the best and most affordable service to users. Moreover, a great deal of time is taken, for instance in negotiating deals and preparing memoranda of agreement with the providers to make education and research opportunities more universally accessible, equitable and affordable. Theoretically, in a deregulated market, and using the technologies available, the sky is the limit. However, the current situation resembles a technological communications straitjacket.

In the regional digital roadmap, one of the major activities is to encourage countries to develop their own national ICT policy. But while the countries in the first category, such as Fiji Islands, Tonga, Cook Islands, Vanuatu and Samoa, have national ICT policies in place, their implementation has been slow due to limited financial and skilled human resources to fulfil the aims and objectives of these policies. Human resources development is a crucial area for the sustainability of ICT projects and programmes that the countries of the region should endeavour to promote and develop.

In terms of political support, making leaders aware of the opportunities and realities of ICT applications is a challenge that needs to be addressed in various forums every year. While many Pacific island nations have indicated support for ICT, such support has generally been peripheral. The only country that has identified ICT as a priority and allocated substantial funds to the sector is Fiji. In its 2007 budget, Fiji has allocated over USD 6 million for ICT development focusing on e-government. In addition, it allocated USD 1 million for general ICT development and has supported the establishment of a regional ICT centre worth over USD 15 million at the University of the South Pacific through a bilateral aid programme with the Government of Japan.

At a meeting of Forum Communications Ministers held in Wellington in March 2006, it was clear that a crucial factor in getting ICT development moving in the region is availability of funds which many countries do not have. Leaders in the region recognize that millions of dollars are needed to build technical infrastructures, train people and put in place the appropriate applications. For individual countries, this level of financial support would be difficult to secure. Thus, calls were made to approach this challenge as a region. For instance, satellite telecommunications costs are very high but bulk buying for the region could be considered and the costs shared between the countries based on volume needed. In addition, while different donors are currently supporting different projects in the different countries, harmonizing donor support for various initiatives, programmes and projects at both national and regional levels will minimize overlap and maximize output.

Furthermore, it is important that developments in ICT in the Pacific region be more inclusive, proactive and responsive. There needs to be greater awareness of the opportunities offered by ICTs to meet development needs in various sectors of the communities. Constraints will need to be identified and solutions found. And work towards improvement should be tackled on a co-operative and collaborative basis between governments, regional and national organizations, non-government organizations—particularly women's organizations, and the communities. These actions and identified initiatives are included in the Pacific Plan Digital Strategy, discussed below.

Infrastructure and development

The Pacific islands face a multitude of infrastructure and technological constraints coupled with high costs of satellite communications, wireless technology, bandwidth charges and equipment and related facilities. While a reduction in equipment costs has been experienced in recent months, the high cost of telecommunications is historic. For many years and in many countries, tariffs have remained relatively unchanged. There have been attempts to rebalance these costs as well as open up the market in countries where monopolies exist, such as in Fiji, Solomon Islands, Palau and Vanuatu, but little progress has been achieved in this front.

In Fiji, the Commerce Commission ruled on the lowering of telecommunications tariffs for three of the major providers, Telecom Fiji Limited, Vodafone Fiji and Fiji International Telecommunications. In the first year (October 2005 to October 2006) costs were reduced by 30–80 per cent for different services. This will be followed by another review a year later with the ultimate aim of opening up the market. While accepting this determination, the providers also submitted a proposal to government to put in place a legal framework for deregulation that will prevent new entrants from exploiting only profitable areas while neglecting unprofitable services. A new Telecommunications Bill was prepared and after extensive consultations, this was passed by the Fiji government in October 2006 and will come into effect in 2007.

In Tonga, mobile costs have remained relatively low as a result of having a deregulated communications market for sometime now.

Aside from the high costs of telecommunications in most of the region, there are large differentials in access rates between urban and rural and remote areas. In many countries, this gap is increasing with the provision of broadband and mobile services in the urban areas. Specific policies are being examined to improve access for rural and remote communities, as in Fiji, where Vodafone Fiji has undergone a major services upgrade to reach remote islands and rural areas. The impact has already been felt in many communities, especially those in a number of scattered islands where there has been no communications at all.

Local online content

Local online content continues to grow exponentially with increased industry growth and demand for such services. Aside from academic institutions developing a range of local content material for courses delivered online and face-to-face, media organizations, industry, tourism outlets and businesses are all developing their own local online materials for promotion as well as for awareness and information. Non-government organizations are also developing their own local online materials and linking these to other related websites globally.

With broadband now available in the region, use and creation of local online content are expected to grow. Local newspapers are already available online. Bookings for travel, hotel accommodations and other services in the region can now be accessed online. These are changing the way business is done and the way work is managed and operated. Costs are still on the high side but the use of the Internet for local content will continue to grow in the future.

Online services and industries

In the past 12 months, with the opening up of the telecommunications industry in Fiji, licences have been approved for a number of ISPs. This has led to the setting up of a number of Internet cafés in all of the urban cities and towns in Fiji. Furthermore, with the reduced costs of telecommunications effective in 2005, the availability of high speed broadband at cheap rates, and the low cost of laptops and computers, new information companies and related industries have been established. Examples are public relations and advertising companies where multimedia graphics and the Internet are basic tools. These are small companies but their number is growing.

Also remarkable is the increase in the use of ICTs in small and medium-sized enterprises (SMEs) in various fields, including the knowledge industry, clothing, agriculture, music and film, tourism and travel, and marketing.

The potential to develop ICT further for economic and social development will vary for the different countries and will depend largely on size and the economic base. The larger islands may have more capacity to undertake developments as their markets are larger and therefore draw investors. These would include the countries in the first category. The smaller states may not be in a position to take advantage of the economic environment and may lack resources to develop ICT businesses and the ICT sector. These are largely countries in the second and third categories. In such situations, a regional approach to industrial development in ICT may be worth considering for the small Pacific countries.

ICT policy

Because infrastructure, applications, services, user knowledge and skills, costs, the regulatory environment and potentials are different in the different countries, each one needs to craft a national policy with a strong sense of opportunity and reality. Such a policy will need to be integrated with other national plans and programmes. In Fiji and Samoa, the establishment of the ICT Advisory Committee to the Minister to assist the Minister to develop policies has proved to be important to both policy formulation and application.

Apart from infrastructure issues, a national policy should cover tariffs, market structure of operators, plans for networks in rural areas, and new services and human resources development. In addition, every country will also have to deal with the impact of convergence of media, telecommunications and data. New technologies will continue to be introduced into the market and there will be a continued need to evaluate the impact of these on development.

At the regional level, there is a need to create a mechanism to review the status of ICT in the region and to put plans together. Meetings of Ministers of Communications have endorsed plans that have not been fulfilled due to the lack of resources. While regional organizations have continued to develop their own ICT programmes and agenda, there is a clear need for these organizations to work together to advance aims in this sector. Some areas where work could proceed and opportunities could be exploited include building regional ICT infrastructures, ICT applications and training, e-government, e-commerce, security, banking, development of content and policies. Also needed is a regular forum that can manage discussions of regional ICT issues and lead these further to development stages where necessary and appropriate.

Donors continue to play an active role in ICT developments in the region. Assistance has largely been in the areas of training and capacity building, infrastructure development, equipment, and capital to implement projects. But while regional and national agencies have received aid from donors and other sources, there is growing recognition that these should be coordinated.

The Pacific Plan Digital Strategy

The Pacific Plan Digital Strategy is a key initiative put together by the ICT Council of Regional Organizations in the Pacific (CROP) Working Group and approved by the Forum Leaders meeting in Papua New Guinea in October 2005. Its main purpose is to assist member countries to develop their ICT infrastructure and capacities in recognition of the positive impact that ICTs can have on economic and social development. This strategy has seen rough sailing through to its present status, but having now been endorsed, it will be the responsibility of all stakeholders to make the strategy work.

To assist with the implementation of the Digital Strategy and to enhance the delivery of desired results, the Digital Strategy Roadmap was developed proposing initiatives for the short and medium terms on the strategy's three pillars: the national pillar, the regional pillar and the global pillar. These priority initiatives are:

- Improving access to communications technologies
- Reducing ICT costs
- Providing higher bandwidth to the global ICT backbone
- Establishing the appropriate regulatory environment for business and economic development
- Strengthening ICT skills

The Pacific Plan Digital Strategy implementation plan in Appendix 1 lists the agreed initiatives, resources required and organizations responsible for each initiative. An initiative for implementation in the first year is the assessment of the e-readiness of the Forum Island countries to determine whether they are ready for a networked world. Appendix 2 provides a list of key actions and allocated responsibilities for all CROP members, donors and other agencies.

In addition, the Digital Strategy aims to promote the full and equal participation of women and other marginalized groups by mainstreaming gender and equity perspectives in ICT policy and programmes. In the implementation of the plan it is important that women are considered not only as recipients of ICT-based programmes but also as equal partners in the development and management of ICT initiatives. femLINKPACIFIC continues to play a critical role in ensuring that this objective is attained.

In October 2006 the Task Force on Regional Approaches to ICTs in the Pacific met and agreed to a number of issues. These included the establishment of a regional resources centre which could facilitate regional policy and regulatory and technological information sharing, the development of a roadmap that includes initiatives and their implementation, and the conduct of a review of Forum member countries focusing on ICT regulatory framework and e-readiness of the Pacific Island States. Australia and New Zealand offered to perform a supporting role in the area of policy formulation, information sharing and technical assistance.

A second ministerial meeting will be convened in mid-2007 to review the progress made in the implementation of the strategy in the first 12 months of 2006.

Capacity building

An increasing number of national and regional institutions and organizations offer training in ICT at different levels, from short-term training courses to formal degrees.

The University of the South Pacific (USP) is perceived as taking a coordinating and 'lead' role in ICT education, research and training for improved economic and social development. Thus, USP's programmes in computing science have been reviewed and a new degree structure has been recommended to include new topics and areas of study. A professional certificate course in Red Hat Linux has been offered in the region once a year since 2004, while non-credit computing courses are offered throughout the region all year round through the Community and Continuing Education Programme, using university computer laboratories and USPNet when necessary. A graduate course called CS493: Advanced Information Engineering for Small Business with Open Source Software continues to be offered with the University of Electro Communications, Japan via IP-based satellite together with two other courses. In addition, ICT is being integrated in all courses in science and engineering, economics, tourism, governance, creative media and statistics.

USP also has a number of databases on education, economics, labour statistics, population, and law and governance. More emphasis is being given to linking job-related courses and linkages with the industry. And the university is attempting to incubate new and small industries for short periods.

In distance learning, the following activities are continuing: multimedia database installation, streaming of video programmes, online course development in Computer Science and specialist training.

USP also maintains a lead role in research in ICT. Under a recent project in capacity building funded by the Japan International Cooperation Agency (JICA), the university completed the following studies:

- Evaluation of the Computer Science Curriculum in Fiji Secondary Schools
- Maximizing the Benefits of ICT/Multimedia in the South Pacific: Cultural Pedagogy and Usability Factors
- GIS—Community-based Tourism Development Project
- ICTs, Sustainable Rural Development and Poverty Reduction in the Solomon Islands: PFNet case
- Redefining Telecommunication Legislation and the Regulatory Environment in Fiji for Improved Economic Growth and Social Development
- Economic Impact of e-Commerce Strategies for Marketing Small and Micro Tourism Businesses

- ICT and the Promotion of Indigenous Identity
- Open Source in the Region

A number of research projects are in progress, including a CROP study of the e-readiness of Pacific island countries, ICT and small and micro-enterprises, e-governance and e-business.

The open source movement

In 2006, USP became a Centre of Excellence for Free and Open Source Software (FOSS) in the Pacific region. The Centre is conceptualized as a consortium between USP and regional organizations, including the Pacific Islands Applied Geoscience Commission (SOPAC), the Pacific Islands Chapter of the Internet Society (PICISOC) and the Secretariat of the Pacific Community (SPC). The vision is that developing countries in Asia Pacific can achieve sustained economic growth and social development by using affordable FOSS solutions.

The International Open Source Network Pacific Islands Chapter (IOSN PIC) has actively supported USP in prototyping a FOSS course management system, Moodle, with the establishment of a Moodle server. Over the past semester, USP has been able to make available on the Moodle server an increasing number of courses, especially courses in computing studies. In addition, IOSN PIC has identified a need for border control systems (immigration information systems) for the small island states of the Pacific.

The work plan for the next 12 months includes research projects, training and education, user awareness and education, and publications. All components are critical to the functioning of the Centre. In training and education, the training of trainers in Red Hat course certification is the most active, with sessions held in Suva, Fiji and Apia, Samoa. IOSN PIC is also collaborating with USP to develop and offer undergraduate and postgraduate programmes in science.

However, the centre needs to identify more skilled people. It is hoped that in a year or two, USP will be graduating students who have gained expertise in FOSS development. It is abundantly clear that FOSS is very young and new in the region. Apart from USP and SOPAC, no other institutions have voiced their interest in using FOSS.

Networks

There are a number of networks in the region that operate and are used for various development matters. The most well known is the USPNet, a USP-dedicated VSAT telecommunications network that is used for education and research purposes. The network was developed in 1978 and has undergone two major upgrades (in 2000 and 2006) using satellite communications as well as cable. The 2000 upgrade was made possible with

---

ICT in the regional university

The University of the South Pacific (USP) perceives ICT as a means of achieving equitable access to education and of reducing the cost of education for all citizens in the University's 12 member countries. ICTs are also considered a means of alleviating poverty through the development of human resources with the skills to create new businesses and take advantage of new employment opportunities and markets, improve health care in remote areas, and participate in global communication networks, and thereby help raise standards of living.

Thus, in addition to curricular reforms to integrate more ICT-related and ICT-supported courses, USP is setting up the Japan-Pacific ICT Centre which will operate a research, business and entrepreneurial incubator development programme. Through this programme, the University will provide a comprehensive and integrated support package for the R&D work of its partners, including physical space, student expert skills, research partners, basic office equipment and secretariat assistance.

USP also has a tax-free IT Park in the Statham Campus in Suva, Fiji which has attracted a number of software industries. The industries are expected to employ the University's graduates.

Moreover, USP has embarked on a programme to upgrade and expand libraries in the region to enable them to provide telecentre services that will include a mix of library, Internet, communication, education, health and community services.

There is a vibrant ICT group at USP and it will continue to forge ahead with work in new areas, including ICT and cultural heritage, ICT and special education, ICT and biotechnology, and ICT and SMEs.

assistance from the governments of Australia, New Zealand and Japan which implemented a new cost-effective platform for regional educational and telecommunication needs.

The most recent enhancement of USPNet (in 2006) moved the network from Intelsat to a new satellite service provider, New Skies. This increased the satellite spectrum by 36 per cent and reduced spectrum costs significantly. In addition, the new satellite provider allows for future expansion. The enhancement supports higher data rates and efficient use of bandwidth. This new satellite ground Gilat equipment supports Internet-based (IP) traffic. The move to IP-based technology provides USPNet with a technology platform that will receive increasing capacities and capabilities. The system has provided USP with faster, more flexible more expandable facilities that can link directly and converge with other existing information networks such as the Australian Academic Research Network (AARNET).

The new enhancement offers the possibility for USP to be part of the new Global Development Learning Network (GDLN) of the World Bank and to establish a Pacific GDLN. A link up to such a network will allow Pacific connectivity for distance learning and enable linkages with other academic and learning institutions in the world, especially those in the Asia Pacific region, including the People's Republic of China, Japan, Indonesia and Vietnam.

A very useful legal information network is the Pacific Legal Information Institute (PACLII) based at the USP Law School. PACLII, through the commendable work of the World Legal Information Institute, gives lawyers access to digital copies of laws, judgements, cases and trials in all Pacific Island States. It is hoped that this service will change the way lawyers conduct their work and that rural and low-income communities can benefit from access to legal information. At the same time, the network highlights the increasingly global nature of the practice of law, with lawyers representing clients with global interests and businesses. The Internet and other technologies have the potential not only to make information accessible but also to improve existing legal structures.

There are other regional networks in areas such as agriculture, marine studies, business, governance, community services, environment and health that are managed and serviced by other regional organizations and academic institutions in collaboration with USP and the Pacific Community.

Trends

The Pacific region is facing difficult times politically and economically. With recent political upheavals in the Solomon Islands, Tonga, French Polynesia and Fiji, development in these countries will not be easy, to say the least. These upheavals will no doubt affect the development of the ICT sector as well as other sectors. What may be useful to consider now is how ICTs can assist the Pacific island countries in strengthening governance, expanding distance learning, making people aware of the global community that we are all part of and instilling the rule of law in every citizen.

For the benefits of using ICTs as tools for development to be realized, it is necessary for countries to determine their own priorities. Unfortunately this is not happening too often as countries are often influenced by outside forces. For instance, the outcomes of the ministerial meeting in Wellington in March 2006 seemed to have been heavily influenced by considerations that were not discussed during the pre-ministerial meeting. Instead of the ICT needs identified by the countries during the pre-ministerial meeting, the final communiqué contained recommendations bringing in new players, such as the Asian Development Bank (ADB), Commonwealth Secretariat and Internet Corporation for Assigned Names and Numbers (ICANN), to review, undertake feasibility studies of and take a central role in ICT developments in the region. The fact is that there is no dearth of studies already undertaken by regional people and organizations in areas such as networking, satellite communications, distance learning, mobile communications and training.

It has also been argued that for the Pacific countries, state-of-the-art communications is not commercially viable and is impractical. However, different countries, especially developing countries, have shown that when communications is made possible and costs brought down, the use of that means of communication grows exponentially. In Fiji and Niue, for instance, wireless and broadband have taken off, to the surprise of many.

Regarding deregulation, governments and a number of providers in the Pacific do not seem serious enough and willing to go along with deregulation. If they understood the positive impact of communication economically and socially, they would use everything in their power to find a roadmap to deregulation.

In the end, it is fair to say that customers may not care too much about the legal and technical side of things, as the most important thing to them is affordable access and reliable and efficient communications services. If telecommunications providers in the region can meet these demands, then we might begin to see the positive impact of ICTs on economic and social development in the Pacific Island States.

Notes

1. These include American Samoa, Cook Islands, Federated States of Micronesia, Fiji, French Polynesia, Guam, Kiribati, Nauru, New Caledonia, Niue, Northern Mariana Islands, Palau, Papua New Guinea, Pitcairn Islands, Republic of the Marshall Islands, Samoa, Solomon Islands, Tokelau, Tonga, Tuvalu, Vanuatu and Wallis and Futuna.
2. Forum countries and territories include Australia, Cook Islands, Federated States of Micronesia, Fiji, Kiribati, Nauru, New Zealand, Niue, Palau, Papua New Guinea, Samoa, Solomon Islands, Tokelau, Tonga, Tuvalu and Vanuatu. While Australia and New Zealand are Forum members, this article does not include them.
3. The Pacific Plan Digital Strategy was agreed to and approved by Forum leaders at a meeting of leaders in Papua New Guinea in 2005.
4. An e-readiness survey is currently being undertaken by CROP agencies and results would provide up to-date statistics. The expected date of completion is July 2007.

References

Pacific Islands Forum. (2006a). *Pacific Plan Digital Strategy Roadmap*.

Pacific Islands Forum. (2006b). *Wellington Declaration*. Forum on Information and Communications Technologies. Ministerial Meeting in Wellington, New Zealand.

Pacific Islands Forum Meeting of the Task Force on Regional Approaches to ICTs in the Pacific, Nadi, Fiji. (7 October 2006). *Agreed Outcomes*.

Williams, E. (2006a). Change the digital divide society to digital opportunity society. Paper presented at the International Cooperation Forum on Telecommunications and Broadcasting held on 6 October 2006 at Chiyoda Hoso Kaikan.

Williams, E. (2006b). *Pacific Plan and the Digital Strategy: USP's role, input and contribution to its implementation*. Suva: USP.

Williams, E. (2006c). The possibility of distance education and challenging issues in developing countries: A case study of the University of the South Pacific's distance education and ICT policy in the Pacific. Paper presented at a conference of the Institute for International Cooperation, JICA.

Appendix 1

Pacific Plan Digital Strategy implementation plan regional organizations—USP input

Initiatives for the first three years (2006–08)	Milestones	Implement	Agree in principle	Further analysis required	Resource requirements (USP involvement is noted but others are speculative)	CROP partners
Support to leaders for the establishment of an ICT Steering Committee to develop policy, plans, programmes of activities ensuring the integration of ICT in all sectors and plans. Representation on these issues and user groups and national committees ensure gender equity.	• Countries are at different stages of progress in this objective. Fiji established a government ICT committee two years ago but is now reviewing membership and the TOR. Samoa has also established an ICT Steering Committee and is active in priority areas. Niue, Cook Islands, Tonga, PNG, and the French Territories are in various stages of development.	✓			National activity but closer working relations and collaboration with PIF where possible. PIF, SOPAC and USP can assist countries in drawing up and establishing such a committee.	PIFS, USP, SOPAC, SPC, PITA
Development and drawing up of policies.	• This is an important process and countries should be assisted in establishing ICT Steering Committees.					
Development of measures in ICT and impact targets to be agreed to.	• This is an area that needs to receive support. Efficient planning is not possible if reliable data and statistics are not available. Many studies have been undertaken but there is no clearing house for this data. A desk study could be initiated and updated regularly. This study should be undertaken as soon as possible. • This should be a PIF project with USP researchers to assist under its ICT project and research plan work.		✓		World Bank could be approached to fund the study to be undertaken by CROP members.	PIFS, USP, others
Database of ICT statistics, papers, good practice, reports, etc.	• A database that maintains up-to-date ICT statistics and reports is part of the USP ICT Centre project. Such a database including e-governance work is already being developed. • Digital index		✓ ✓		JICA, World Bank, UNESCO and ITU	USP, SPC and others SPC, USP
Universal Service Obligation—telecommunication services to rural and remote areas to be pursued	• The Fiji Ministry of Information and Media and industry are working closely together to develop rural services. A USF proposal is being looked at as well as a new Telecommunication Bill. Niue provides services to the whole country and to all citizens free of charge. Tonga and Samoa also provide services but more information is needed on this. Other countries still have to decide on a process.		✓		Proposals need to be drawn up. National government responsibility.	Countries and industry. PIF to coordinate. USP can provide advisory services.
Telehealth Services—national and regional	• A number of countries have pursued work in this area. Sustainability has been a problem. Fiji, Cook Islands, Micronesian countries and Solomon Islands involved in telehealth projects in the past three years.			✓	NZAID, WHO could be approached to fund. Approach UH to utilize their communication system.	SPC, FSM, USP, SOPAC

Area	Description		Partners
	There is a need to get greater funding and ensure that capacities continue to support such services. The University of Hawaii is active in this area and can be approached to partner the countries to develop such services.		
Education and human resources—national level	This is an important area and needs range widely from advising and developing ICT curriculum in schools, to vocational ICT training to more training at higher levels. Under the ICT capacity building project at the USP, the needs for training and development were recognized as critical. Under this aid project a number of activities were launched and have been continued by USP. Others have not been sustained and countries need to be assisted.	✓	UNESCO, AUSAid to assist in drawing up the curriculum, vocational training support.
	An ICT in School curriculum needs to be developed at national level.	✓	USP, SPC, SOPAC
	Short-term training is undertaken by various training outlets in the different countries. The list of providers is available.	✓	
Promote competition and liberalize the telecommunication market	Support for this is needed by regional governments and can only be taken at government level. The justification necessary is that it promotes all principles and objectives of the Digital Strategy and it is crucial to the success of the Digital Strategy.	✓	AUSAid, NZAID, Commerce Commission (Fiji) could act as advisor; also PIAS-DG, USP.
	Establishment of a regional commerce commission styled against the Fiji Commerce Commission	✓	PIF, USP, PITA, SOPAC
ICT leadership in specific areas (regional)	All CROP partners undertake various activities in ICT. USP plays an active role in:		
	o education, ICT curriculum for schools, vocational training, professional qualifications	✓	JICA—Pacific Centre for ICT (pending)
	o distance and flexible learning	✓	JICA—ICT for Human Security (pending)
	o ICT and industry collaboration—incubator centre	✓	Australia—Library in a box and digitization project (implemented)
	o research	✓	Telecentres and HIV/AIDS (pending)
	o software development—linking with ICT Park at Statham Campus and is a business venture for the university		ICT Park—funded by industry in partnership
	o library in a box		USP
	o telecentres and HIV/AIDS	✓	
	o statistics	✓	

(Appendix 1 continued)

(Appendix 1 continued)

Initiatives for the first three years (2006–08)	Milestones	Implement	Agree in principle	Further analysis required	Resource requirements (USP involvement is noted but others are speculative)	CROP partners
	o e-commerce					
	o e-governance	✓				
	o multimedia databases	✓				
	o networking and satellite communications	✓				
	o technical and specialized training—RedHat, etc.	✓				
	o telecommunications					
	o digitization of course materials, archives, records					
	Many of the short-term courses are organized and run by the university in collaboration with other regional organizations. All courses have been reviewed with new programmes offered in collaboration with UEC, Tokyo through the university network.	✓				
	Infrastructure (USPNet, computer labs and equipment) has been expanded with assistance from JICA, AusAID, and NZAID.					
Promote software development for education, social, economic, and political development	• USP ICT Park focusing on software development for industry, government and SMEs	✓			Self-financed	USP
	• FOSS development and plans	✓	✓		APDIP, UNDP	USP
	• FOSS Pacific node		✓			SOPAC, SPC, PICISOC
Broadcasting in the region	• Under USP's Pacific Centre for ICT project, the role of radio, in particular digital radio, in small island states and remote areas will be re-examined. The DFL mode of delivery through radio will be expanded.	✓			COL, UNESCO	SPC, USP
	• More content development promoted; alternative energy source for radio station; converging radio, Internet and television	✓				

Appendix 2
Pacific Plan Digital Strategy key actions and allocated responsibilities

Issue/aspect	Facet	Short-term (1–2 years)	Medium-term (2–5 years)	Long-term (beyond 3 years)	Prime responsibility	Supporting agencies	Specific activities
Country assessments and status monitoring		Commence and complete	Continuing	Continuing	Country Minister	UNDP, USPSPC, PIFS	UNDP and CROP Agencies will jointly compile a survey of the current status of ICTs in the countries in 2006. SPC will support the collection and presentation of basic ICT data under PRISM. USP is developing a data base of ICT reports, analyses, statistics and benchmarks as part of the USP ICT Center Project (JICA).
National policies and ICT steering committees		Establish	Continuing	Continuing	Country Minister	Forum, USP, SOPAC, SPC	Existing policies will be reviewed and all countries supported in developing actionable ICT Policies. WB will sponsor, with Forum and PITA support, a workshop on convergence issues as they affect telecom regulators in April 2006. Forum and USP will assist countries with policy formulation.
Network infrastructure	Rural and remote access to ICTs	Identify the scale; Include intent in ICT policy	Country policies on Universal Service Obligation; Establishment of Access Funds and administration	Roll-out completed	Country Minister	SOPAC, DFC, PITA, USP	SOPAC will propose the establishment of a rural access expert to assist countries with technology selection, project proposal, and programme management of 'talanoa centres'.
Network infrastructure	Regional satellites and cables	Seek buy-in and feasibility studies	Feasibility study	Construction	PITA Forum	Countries	Donors will be approached to fund a regional study of the future demand for international capacity and the optimum immediate term approach, and develop a plan for implementation.
ICT applications	e-Government	Survey	Capacity building	Maintenance Potential regional approach	Country	SOPAC	SOPAC to coordinate and investigate potential common portal arrangements.
ICT applications	e-Commerce	Survey	Legislation and Regulation	Potential harmonized regional legislation and regulation	Country	PIPSO	SPTO development of the tourism-related websites, content management systems, e-learning for tourism SMEs, and ICT for tourism-related crisis management.
ICT applications	Telehealth					FSM, USP, SPC	SPC, USP, FSM

(Appendix 2 continued)

(Appendix 2 continued)

Issue/aspect	Facet	Short-term (1–2 years)	Medium-term (2–5 years)	Long-term (beyond 3 years)	Prime responsibility	Supporting agencies	Specific activities
ICT application	Education				Country	SPC, USP	USP engagement in curricular development and training. SPC support to applications of ICT in the sector. SPC to increase capacity in ICT HRD capacity monitoring.
ICT application	Disaster management	Include ICTs in planning	Appropriate applications. Capacity building	Early warning systems	Country	SOPAC	Early warning systems, development of disaster management plans.
ICT application	GIS and remote sensing	Expand basic platforms and data	Develop regional capacity	Mobile ground receiving station	SOPAC	EU	

Pakistan

Jamshed **Masood** and Salman **Malik**

Total population	160 million (estimate)
GDP per capita	USD 847 (equivalent PKR 50,820; USD1 = PKR 60)
Fixed-line telephones per 100 inhabitants (including WLL)	3.5
Mobile phone subscribers per 100 inhabitants	30
Internet users per 100 inhabitants	7.5
Internet subscribers per 100 inhabitants	1.5
Broadband subscribers per 100 inhabitants	0.03125
Total Internet bandwidth consumption	800 Mbps

Sources: Pakistan Telecommunication Authority (2006, 2007a, 2007b); Ministry of Finance (2006); and Population Census Organization (n.d.).

Technology infrastructure

Despite the impact of negative shocks to the economy, including the tragic earthquake of October 2005, macroeconomic problems resulting from high international oil prices, the geopolitical scenario and other issues, 2006 was yet another landmark year of developments in the telecommunications field in Pakistan.

The telecommunications sector remained a high priority area for investments, spurring the development of the much needed technology infrastructure to support Pakistan's vision of becoming an Information Society. The sector attracted approximately USD 2 billion in foreign and local investments, translating into 54 per cent of the total foreign direct investment in the country (Pakistan Telecommunication Authority 2006). Thus, the cellular phone subscriber base went up to 48.5 million (Pakistan Telecommunication Authority 2007) and the fixed-line/Wireless Local Loop (WLL) subscriber base went up to about 5.6 million lines, or 3.5 per cent of the population.

The inland and international submarine fibre optic infrastructure for transport and transmission of telecommunication traffic was also a focus of attention. Two new Terabit international submarine fibre optic cable ventures came into operation, connecting Pakistan to the rest of the world. The first project is TW1, a private sector venture with a capacity of 1.28 Tbps. It is connected with FLAG, a global submarine fibre optic cable venture. The second project is the Sea-Me-We 4 project, a 20,000 km long submarine fibre optic cable managed by a consortium of 16 leading operators including PTCL, the dominant telecom operator.

Pakistan has an extensive fibre optic backbone infrastructure operated by PTCL. The aggregate length of the domestic fibre optic network exceeds 11,000 kilometres. Three consortia—Multinet, which is a wholly owned subsidiary of Telecom Malaysia; LinkDirect, a subsidiary of Mobilink, the largest cellular company in Pakistan; and Wateen, a subsidiary of the Dhabi Group which has cellular operations in Pakistan and some other countries in Southeast Asia and Africa—are building domestic nationwide fibre optic networks to meet the growing demand for voice and value-added data service traffic. Wateen is also in the process of rolling out a nationwide wireless broadband voice and data WiMAX network in Pakistan.

However, in terms of basic and broadband access to the Internet, the fruits of the deregulation of the telecom sector have not been fully realized. The Internet subscriber population is 2.4 million (or 1.5 per cent of the total population), with less than 30,000 DSL subscribers across the country. The entire broadband population is less than 50,000, including subscribers to wireless or cable broadband connections. The total Internet bandwidth consumption in the country has reached 800 Mbps.

Pakistan has seen large-scale deployment of CDMA Wireless Local Loop (WLL) networks, which are data access-enabled. One 3G CDMA cellular network is expected to go live before the end of 2007. Existing GSM cellular operators have introduced EDGE services. These initiatives are expected to address access problems faced by potential users and to generate demand for broadband Internet services throughout the country.

ICT and ICT-related industries

The telecom sector of Pakistan has experienced positive developments and hyper growth since deregulation. Investments in telecom infrastructure development have increased and the telecom industry is beginning to make a significant contribution to GDP. But one cannot say the same of the IT sector. The reasons behind this lag are complex and varied.

The IT industry in Pakistan is characterized by low barriers to entry, a preponderance of small units, intense competition from industry power houses in neighbouring countries, limited skill sets and a highly competitive global marketplace. The IT industry is a mix of specialized businesses and entities, including more than 100 international call centres/business process outsourcing (BPO) companies, 50 medical transcription/data entry companies, five engineering outsourcing companies, 200 custom software companies, and several other businesses specializing in animation, mobile content, retail banking, industry-specific ERP (enterprise resource planning), and document management. There is also a sizeable population of businesses catering to hardware and local office automation applications.

The IT industry currently employs more than 4,000 call centre/BPO agents, 2,000 in the medical transcription/data entry business, 500 in the engineering outsourcing field, about 5,000–7,000 software professionals in export-related work, and undetermined numbers in other specialized areas.

Pakistan's IT industry is still in its infancy, with export revenues of under USD 72 million (Pakistan Software Export Board 2006). But government and private entities are trying their best to ensure that Pakistan becomes a competitive destination for local and international IT services and manufacturing and related industries.

Key institutions dealing with ICT

The Ministry of Information Technology (MoIT) is the Government of Pakistan's policy and implementation arm for planning, overseeing and implementing activities to promote information technology and telecommunications programmes and projects aimed at economic development. MoIT oversees a number of departments. In addition to the Federal Government, there are departments within provincial governments with the specific task of ICT promotion initiatives.

Also under MoIT is the Pakistan Software Export Board (PSEB), which has the sole mandate of promoting Pakistan's IT industry. It has undertaken the following programmes:

- Provision of appropriate IT-enabled office space for IT companies in PSEB-designated IT Parks;
- Quality certification programmes under which 100 IT companies have received International Standards Organization (ISO) certification;
- Human resource development initiatives, including an internship programme that has placed 2,500 interns within the IT industry;
- Promotion of Pakistan's IT industry in the international media; and
- Subsidized participation of Pakistan-based IT companies in international trade shows.

PSEB's Board of Directors consists of representatives of government, industry associations and the private sector.

The Electronic Government Directorate was set up in 2002 to drive the e-government initiative envisioned under the IT Policy and Action Plan adopted in 2001. The Directorate is responsible for the planning, preparation and implementation of e-government projects and for providing technical advice and guidelines for the implementation of e-government projects at the federal, provincial and district levels.

Enabling policies and programmes

Cognizant of the importance of developing the IT sector in Pakistan, the Government of Pakistan has undertaken measures to generate demand for ICT services and applications and to create a critical mass of companies and expertise in the ICT field. The Federal Government has been actively pursuing an enabling policy framework since 2001, with the aim of attracting investments into the ICT sector. The incentives offered include corporate tax exemptions on export earnings, 100 per cent repatriation of profits, tax holidays for IT venture capital funds and provision of a subsidized enabling infrastructure.

The e-Government Plan and the resulting ICT initiatives serve as the key demand drivers for the local ICT industry. The private sector also actively contributes to the growth of ICTs in the country.

To ensure that the fruits of technological advancements, starting with basic access to telecommunications, reach poor and underprivileged communities across Pakistan, the government has adopted a Universal Service Policy with the following goals:

- to make voice telephony affordable and Internet access available to progressively greater proportions of the Pakistan's population;
- to foster conducive conditions and an enabling environment in which teledensity can grow; and
- to jumpstart the broadband and ICT markets to facilitate e-services.

In terms of proliferation of ICT services, the following targets have been set:

- 85 per cent of the population should have telecommunication coverage and therefore access to e-services if desired;
- 5 per cent teledensity in the rural areas;
- 1 per cent broadband penetration; and
- preferably one telecentre for every 5,000 people, or at least one telecentre for every 10,000 people in Universal Service Fund (USF) contract areas (MoIT 2006a).

Telecommunication operators are required to contribute to the USF. In turn, they can tap into this fund for telecom service projects in underserved areas. Currently, the Fund has over USD 49 million in its account. An additional contribution of over USD 15 million annually will be used to improve rural access.

e-Government services

The following are the broad areas in which initiatives have been undertaken under the e-government plan:

1. Infrastructure development to provide government departments with computers, intra-Ministerial networking (LAN/WAN), office productivity tools and electronic communication to facilitate their work.
2. e-Office applications, which include internal communication, human resource, budgeting/finance, project management, document/file management and collaboration modules. An application suite is being implemented in test sites and will be replicated across all Federal Government departments in phases.
3. Agency-specific applications and e-services for citizens. High-impact core services and processes are being identified and automated at all government departments.

The following are representative e-government projects:

1. *Online Recruitment Management System for the Federal Public Service Commission (FPSC)*. The FPSC is responsible for the hiring and induction of government employees. Millions of candidates are processed for induction to government service annually. As part of the e-government initiative, an online job portal has been set up where job openings are posted, candidates can file online applications and examination and interview results can be accessed.
2. *Applications for the Dissemination of the Proceedings of Parliament*. Specifically, this project includes automation of the flow of legislative processes within the Secretariats of the two Houses of Parliament, provision of a platform for cross-referencing of information, online interface for members of the National Assembly and Senate, and public access to information on Parliamentary proceedings.
3. *Online Access to Case Laws at District Bar Associations*. Online access to databases of statutory and case laws has been made available to 70 District Bar Associations, the Supreme Court Bar Association and 11 High Court Bar Associations to improve the quality of legal decision-making in remote areas and to ensure the ready availability of judgements in comparable cases in electronic form. Aside from unlimited 24 × 7 access to statutes and case law in all locations, there are computer operators to help users at each location.
4. *Online Processing of Hajj Applications*. Each year about 150,000 Pakistanis undertake a pilgrimage to Saudi Arabia to perform the Hajj, one of the pillars of Muslim faith. To facilitate the process for pilgrims, a suite of applications has been deployed with the following features: online Hajj application submission, balloting of Hajj applications, travel management, passport printing, pilgrim tracking (in Pakistan as well as in Saudi Arabia) and a private tour operator management system.

In addition, the e-government plan envisions the automation of services in government hospitals and civil service agencies, police and security services, training and human resource development, licensing and registration processes, and filing and processing of patents. A national data centre for hosting Internet and intranet applications of the Federal Government is also being deployed.

Education and capacity-building programmes

The Government of Pakistan is committed to promoting and supporting ICT initiatives to enhance Pakistan's capacity to develop and produce ICT products and services. This includes the development of a critical mass of ICT experts. The Higher Education Commission (HEC) was set up in 2002 with the objective of improving the quality of higher education and meeting the requirement for trained human resources that the country needs to meet the challenges and opportunities posed by the rapid evolution of ICTs.

As part of its mandate of human resource development, HEC has sponsored programmes to establish ICT infrastructure in universities across the country. The objective is to lay the platform for the delivery of a range of ICT-based educational services, including a world-class digital library and conferencing facilities. HEC is setting up video conferencing lecture rooms and has initiated the 'Online Lecturing and Video Conferencing System' in all public universities. The conferencing system, which will come with collaborative tools to enhance student–teacher interaction, is in aid of distance learning, which is envisioned as a means of addressing the shortage of faculty members in universities located in far-flung areas and, ultimately, of uplifting the standard of education in Pakistan.

As part of its ICT promotion activities and to complement its effort to improve the quality of higher education programmes and establish new universities and institutions, HEC is conducting ICT awareness programmes in smaller cities and towns and providing opportunities for talented individuals from these areas to join the mainstream ICT industry in due course. A recent initiative in this respect is the Outreach Scholarship Programme (OSP) funded through the National ICT R&D Fund of the Government of Pakistan. The programme reaches out to talented young people who cannot afford higher education in top-tier universities in Pakistan and provides them with scholarships in ICT disciplines in these universities. The programme not only benefits these individuals but also develops a more diverse human resource pool for Pakistan's ICT industry. During the first phase of the programme, which was conducted in higher secondary schools, 250 secondary schoolteachers were trained to help students hone their skills and participate in the competitive process to join mainstream universities. More than 3,000 secondary school students participated competitively in the university studies scheme during the first phase of the programme.

There are also government-funded human resource development programmes in software development, including internships, apprenticeships and specialized training. At present about 20,000 IT graduates are being added to the workforce annually. Government-sponsored internship programmes have placed about 2,500 interns in the industry during the last four years, with almost 80 per cent of the interns finding permanent employment. In the next four years, the government plans to place 10,000 more graduates in the internship programme. In future, the apprenticeship programme will assist 1,000 new hires to undergo year-long industry-specific training.

Research and development

A National ICT R&D Fund has been established by the government to provide an incentive for local development of ICTs that are vital from commercial and development perspectives. The projects that are qualified for funding are based on an innovative value chain concept through deployment of integrated and interactive thematic strategies. The following areas have been prioritized for the first policy cycle: HR Capacity Building, National Productivity Enhancement, ICT Product Development, ICT Market Development and Multi-Sectoral Support Programme.

Telecommunication operators are the primary contributors to the ICT R&D Fund, which is managed by the Ministry of Information Technology. Currently, the Fund exceeds USD 40 million.

Open source initiatives

One of the key deterrents to growth and widespread use of software and hardware in Pakistan is affordability. Computers are expensive for average citizens and numerous copyright and cost issues hamper the widespread use of software. Recognizing that open source software (OSS) and applications could jumpstart overall ICT development, the Government of Pakistan considers OSS adoption as a key ICT policy objective.

Some level of success has been achieved in the public, academic and private sectors with respect to developing a nucleus of users and developers of open source applications. And with interest in OSS spreading to educational institutions and the corporate sector, it is likely that OSS will enter the mainstream in the next few years.

In the public sector, the Government of Pakistan is the key sponsor of OSS promotion activities in terms of human resource development initiatives in Linux, extending support to interested companies for migration to and adoption of OSS applications, and development of open source applications for the automation of small and medium-sized enterprises. In addition, the government is contributing to programmes for creating awareness of OSS. The Open Source Resource Centre (www.osrc.org.pk) was established by the Government of Pakistan in 2005 to bring together expertise and resources on open source technologies under one umbrella. The Resource Centre has also brought together established technology vendors, open source community members, and enterprise IT users/customers to jointly explore new opportunities for OSS deployment in the local context. To date, the Resource Centre has helped train 4,000 end-users and 800 system administrators in OSS applications, created a network of open source developers, conducted numerous workshops and seminars, and assisted numerous government and private-sector organizations in adopting OSS.

Business organizations have successfully reduced their costs by eliminating proprietary software licensing through the adoption of OSS. This has helped in the creation of a critical mass of vendors and a support industry actively providing and supporting solutions for desktop, server and middleware stacks, including enterprise-level technical support.

Civil society organizations have also been actively pursuing the OSS promotion agenda. The Linux Pakistan User Group is composed of 3,000 professionals. The Free and Open Source Software Foundation of Pakistan (FOSSFP), which is dedicated to promoting the overall adoption, development and usage of OSS, has over 850 Ubuntu-Linux user group members and over 4,900 registered certified users. FOSSFP claims to have trained and certified over 5,000 Ubuntu-Linux users in Pakistan and is also leading the initiative to develop a localized version of Ubuntu-Linux in Urdu, to be followed by other local and regional languages.

Conclusion

Developing IT and IT-enabled services in Pakistan along global standards is not only a possibility; it is also a pressing need. The correct set of policies and the necessary fundamentals have been put in place, and bold and strategic initiatives are being undertaken. It is now up to the direct and indirect stakeholders of the IT industry to grab the opportunity, not only for their own benefit but also for achieving the ultimate objective of sustainable economic development in Pakistan.

References

Higher Education Commission Website. Available at http://www.hec.gov.pk

Ministry of Finance. (2006). *Economic survey 2005–06*. Government of Pakistan. Retrieved 11 January 2007 from http://www.finance.gov.pk/survey/home.htm

Ministry of Information Technology. (2006a). *USF policy framework*. Islamabad.

Ministry of Information Technology. (2006b). *R&D Fund policy framework*. Islamabad.

Pakistan Software Export Board Website. Available at http://www.pseb.org.pk

Pakistan Software Export Board. (2006). *Industry overview*. Retrieved 2 January 2007 from http://www.pseb.org.pk/page.php?pid=2

Pakistan Telecommunication Authority. (2006). *Annual report 2006*. Islamabad.

Pakistan Telecommunication Authority. (2007a). *Cellphone users jump by 145pc*. Press Release, 17 January. Retrieved 17 January 2007 from http://thenews.jang.com.pk/print1.asp?id=39259

Pakistan Telecommunication Authority. (2007b). *Pakistan has 12 Mln Internet users*. Press Release, 1 January. Retrieved 3 January 2007 from http://www.app.com.pk/en/index.php?option=com_content&task=view&id=1386&Itemid=2

Population Census Organization. (n.d.). *Pakistan at a glance*. Retrieved 13 January 2007 from http://www.statpak.gov.pk/depts/pco/index.html

PTCL R&D Fund Secretariat. (2005). *Annual report 2005*. Islamabad.

Philippines

Lorraine Carlos **Salazar**, Shelah **Lardizabal-Vallarino** and Zorayda Ruth **Andam**

Total population	86,972,500 (2006)
GDP per capita in USD	1,168 (2005)
Computers per 100 inhabitants	4.46 (2005)
Fixed-line subscribers per 100 inhabitants	4.18 (2006)
Mobile phone subscribers per 100 inhabitants	49.29 (2006)
Internet users per 100 inhabitants	5.32 (2005)
International internet bandwidth	3,214.5 Mbps (2005)

Introduction

According to the National Telecommunications Commission (NTC), the Philippines' telecom regulator, in 2005 Filipinos sent an average of 500 million text messages per day in 2006, double the number of text messages that they sent daily in 2005 (Amojelar 2007). Indeed, the Philippines is the undisputed 'SMS capital of the world'. Ninety-five per cent of the 42.8 million mobile phone subscribers in the Philippines use their phones for text messaging. In a country where computer and Internet penetration remains very low, text messaging is the equivalent of e-mail and instant messaging. It has become an indispensable communication tool for social relations and corporate and government transactions.

Indeed, the number of mobile phone owners in the Philippines has grown by leaps and bounds in the past five years alone. From a teledensity of less than one per 100 people between 1970 and 1990, the Philippines achieved a fixed-line teledensity of 4.18 per 100 people and a mobile phone density of 49.29 in 2006. This came about from rapid technological advances and intense competition following a liberalization drive that transformed a monopoly into one of the most profitable and highly competitive economic sectors.

However, while telecommunications infrastructure is one of the success stories in information and communication technologies (ICTs) in the Philippines, much more needs to be done in terms of ICT access, opportunity and utilization. Today, there is an increasing consensus that ICTs have the potential to bring about social, economic and political change and development when people have access to them. ICTs can be utilized to help reduce poverty and socio-economic disparities by bringing the traditionally marginalized within reach. Development practitioners generally agree that ICTs need to be integrated systematically into poverty reduction strategies. There is a need to go beyond small pilot projects into larger national or regional implementation of ICT programmes. And multi-stakeholder partnerships between government, civil society and the private sector must be created to share specific competencies and resources (Weigel and Waldburger 2004).

This chapter documents the state of ICTs in the Philippines. The first section reviews the state of ICT infrastructure in the country, focusing on the telecoms infrastructure, the success of mobile phone diffusion, the very low diffusion of personal computers and its usage, and the problem of ICT data availability. The next section reviews the policy and regulatory environment as well as the institutions involved in regulating ICTs in the Philippines. The third section examines the various digital content initiatives undertaken by government, the private sector and civil society, focusing mainly on efforts to bridge the digital divide. Finally, the last section reviews the challenges faced by the Philippines in its efforts to ensure that ICTs will have a positive impact on the lives of its citizens.

ICT infrastructure

After over a decade of liberalization, the Philippine telecommunications sector has produced a highly competitive environment with 74 local exchange carriers, 14 inter-carrier carrier services, 11 international gateway facilities and seven cellular telephony providers.

While the erstwhile virtual monopolist, PLDT, continues to dominate the fixed-line sector, and its wireless subsidiary Smart

Communications Inc. is the leading cellular phone company, the presence of two major competitors makes mobile telephony the most competitive and fastest growing sector. In 2005, the Philippines had a total of 42.8 million mobile telephone subscribers, the third highest mobile phone density in ASEAN, next to Singapore and Malaysia. In stark contrast, fixed-line teledensity in the Philippines (4.18 fixed-line telephones per 100 people) is the fourth lowest among ASEAN countries, ranking only higher than Cambodia, Laos and Myanmar.[1] Intense competition among the mobile phone companies has resulted in affordable mobile services and innovative packages, including prepaid services. The latter, introduced by Globe Telecommunications, has made mobile phones affordable to lower income groups.

The first mobile phones introduced in the 1990s used analogue technologies. Their uptake was slow because of the high cost of handset and service, as well as poor billing and cloning problems. Philippine telcos then shifted to 2G technologies. By 1999, GSM had become the dominant technological standard in the country. As of the end of 2005, three 3G licenses had been issued. The top two companies, Smart Communications Inc. and Globe Telecommunications, started offering 3G services in early 2006.

As mobile teledensity increased, it became clear that Filipinos were not using their mobile phones for voice calls. Instead they were using their phones for SMS (short messaging service) or text messages. Analysts estimate that 'texting' (SMS) in the Philippines exceeds voice traffic by a factor of 10 to 1. Mobile phone companies earn about half of their revenues from these non-voice services. For instance, in its 2005 financial report, Smart earned PhP 36.8 billion (USD 707 million) from data services whereas its revenues from voice services totalled PhP 34.3 billion (USD 659 million) (PLDT 2005).

Corporate data are supported by consumer surveys. A June 2003 survey found that 94 per cent of mobile telephone subscribers use their phones for text messaging. Of these, 70 per cent send about 10 messages per day and about 14 per cent send between 10 and 20 messages per day. Given the average of 250 million text messages a day, at two US cents per message, telcos in the Philippines earn a hefty USD 5 million per day on text messaging alone! In 2006, the volume of text messages increased significantly more than the three-million increase in mobile phone subscribers. A key factor here was the entry of a third niche player, Sun Cellular, which began operations in 2004.

The decision to open the market to Sun Cellular despite protests from the major players, Smart and Globe, can be credited to NTC. Often criticized for lack of independence, the regulatory body surprised many when it allowed Sun Cellular to offer discounted prices for its services. The other mobile phone companies had no choice but to offer lower tariffs and unlimited texting packages, which in turn led to subscriber gains.

The liberalization of the telecommunications industry has led to economic and social gains. Both corporate and individual users are benefiting from the increased competition among providers. Market competition and new technologies are driving developments in the telecommunications sector, making it a growth engine for the country's development. In 2005, the sector accounted for 4 per cent of GDP and 6 per cent of total tax receipts. Telecommunications companies are also the most profitable and actively traded companies in the Philippine Stock Exchange. In terms of its contribution to social development, the liberalized telecommunications sector is providing connectivity to millions of overseas Filipino workers, enabling them to keep in constant touch with their families.

While the picture is rosy for telecommunications, the same is not true for personal computers (PC) and Internet penetration and utilization. The International Telecommunications Union (ITU) estimated about 3.5 PCs per 100 inhabitants in 2003 and 4.5 per 100 inhabitants in 2004. Unpublished data from the 2003 Family Income and Expenditures Survey (FIES) of the National Statistics Office (NSO) indicate that only about 4.4 per cent of households own a PC, whereas 52.7 per cent own a television, 75.2 per cent own a radio, and 14.1 per cent own a fixed-line or cellular phone (Astrologo 2006).

With regard to the Internet, while broadband and broadband wireless services have been introduced in the past couple of years, dial-up Internet is still the most widely used type of Internet access. NTC reported a total of 1.44 million Internet subscribers for the 177 registered Internet service providers (ISPs) in 2005. The ITU estimate for 2005 was 4.4 million Internet users or 5.32 users per 100 Filipinos. This number is disappointingly lower than Iran's (10.07 per 100 population) or Zimbabwe's (10.08 per 100 population).

While the available data seem to point to low utilization, it is also true that there is a gap in the official data on Internet use. The most recent official information comes from the NSO's Functional Literacy, Education and Mass Media Survey (FLEMMS) conducted in 2003. The FLEMMS study found that the Internet is a source of knowledge and information for about 13.8 million (20 per cent of the population) Filipinos aged six years and above. Of these, only 7.4 per cent access the Internet for information daily. On the other hand, there has been a marked increase in the number of Internet cafés, telecentres and other public access points.

Indeed, one of the biggest challenges in understanding how ICT has changed, is changing and will change the daily life of Filipinos is collecting accurate data on ICT utilization. The government needs to ensure that its policy decisions are based on real and up-to-date figures. However, there is a lack of funding for ICT statistical collection (Astrologo 2006). In 2000, the National Statistical Coordination Board (NSCB) created a Task Force on the Measurement of e-Commerce to develop a framework and to identify methodologies and strategies for the generation of data and other indicators of electronic commerce. The Task Force found that official statistics on ICT utilization are sorely lacking. In 2002, the NSO, in collaboration with the Information Technology and E-Commerce Council (ITECC, now the Commission on ICT), conducted the Survey on Information and Communication Technology of Philippine Business and Industry, the first attempt to collect and generate benchmark information on the availability, distribution and utilization of ICTs by business establishments in the country. However, due to funding constraints, the survey has not been repeated. And no study of public ICT utilization has been undertaken.

ICT laws and regulatory bodies

Legal provisions

Republic Act (RA) 7925 (and its implementing rules and regulations) is the primary law governing the telecommunications sector. It consolidates a number of policies and practices contemplated under Executive Order (EO) 59 (compulsory interconnection among authorized telecoms carriers) and EO 109 (provision of local exchange service policy). RA 7925 effectively liberalized telecommunications services. Its salient provisions include: the responsibilities of the regulatory authority (NTC) and the Department of Transportation and Communications (DOTC), the classification of telecommunications entities, the management and allocation of the radio frequency spectrum, the need to obtain a legislative franchise, interconnection rules, NTC's mandate to establish rates and tariffs, access charges and revenue sharing, the rights of telecommunications users, and ownership of telecommunications entities.

On 23 August 2005, the NTC issued Memorandum Circular No. 05-08, also known as the VoIP (Voice over Internet Protocol) Rules, which declared VoIP as a Value-Added Service (VAS). A VAS provider does not have to secure a legislative franchise to be able to provide VoIP.[2] The Rules recognize VoIP 'as an application that digitizes and transmits voice communications in packets via the Internet, enhances or improves upon traditionally telephony that is conducted through circuit switched connections by allowing the convergence of voice with other data applications, and by providing economic benefits in the form of greater efficiencies and lower costs'.[3]

To further promote competition in the Philippine telecommunications market, the NTC published a consultative paper in December 2005 that discusses the merits of introducing four pro-competition policies, notably: (a) imposing obligations on carriers with significant market power (SMP), (b) mandating local loop bundling, (c) requiring carriers to allow for resale of their services, and (d) changing the basis of price regulation from *ex ante* to *ex post*.

On 24 August 2006, the NTC released a consultative document on the proposal to impose obligations on carriers with SMP. The document discusses the rationale behind the imposition of SMP obligations and the critical processes for implementing SMP obligations, namely: (a) defining markets to be used as basis for regulatory intervention, (b) determining whether one or several operators in the defined markets have the degree of market power that merits regulatory intervention, (c) identifying appropriate SMP obligations to achieve policy objectives, and (d) determining conditions that justify withdrawal of regulation.

With regard to e-commerce, Republic Act No. 8972 (E-Commerce Act of 2000) provides for the legal recognition of electronic documents and electronic signatures as functional equivalents of their paper-based forms. The law likewise provides a temporary solution to the blurring of distinctions and foreseen regulatory inconsistencies in the cable industries due to digital convergence. It also attempts to solve the constitutional limitations on ownership of broadcast and mass media enterprises by distinguishing between physical infrastructure and programming and content, as well as the management thereof.[4] Under the law, the physical infrastructure of cable and wireless systems for cable TV and broadcast shall be considered to be within the activity of telecommunications for purposes of e-commerce and convergence. This means that foreign ownership of the network or its physical infrastructure is allowed up to 40 per cent of total capital stock. Programming, content and management, on the other hand, do not fall under the 'physical infrastructure' of cable TV and broadcast.

EO 205 governs the cable TV industry in the Philippines while EO 467 stipulates policy guidelines and regulations on the operation and use of satellite communications facilities and services.

As regards data protection, the Philippines has no existing comprehensive legislation on personal data protection and privacy of information. In general, privacy rights hinge on the due process clause of the 1987 Philippine Constitution. In particular, the right to privacy is connected with the Constitutional guarantees for the privacy of communications and correspondence, and against unreasonable searches and seizures. These Constitutional guarantees protect private citizens from intrusive actions by the State and do not in any way serve as limitations on private transactions and activities of non-State entities in relation to private persons.

Articles 26 and 32 of the Philippine Civil Code provide a cause of action for damages and other equitable relief in favour of an individual whose right to privacy has been violated. The Philippine courts have yet to interpret the Civil Code provisions on privacy with respect to personal data. The Philippine E-Commerce Act (RA 8972), however, imposes an obligation of confidentiality on any person who gains lawful access to information covered by and/or contained in an electronic data message or electronic document. In fact, the law imposes a penalty of PhP 1 million (about USD 20,000) or imprisonment of six years on an individual who violates the obligation of confidentiality. Sections 33(a) and 33(b) of the law likewise penalize hacking or cracking and piracy in electronic commerce. Moreover, EO No. 269 mandates the Commission on Information and Communications Technology (CICT) to define policies that will preserve the rights of individuals to privacy and confidentiality of personal information.

While the use of personal data for direct marketing purposes is not prohibited in the Philippines, NTC has adopted a regulatory position in response to consumer/subscriber complaints concerning marketing advertisements transmitted and received via broadcast messaging services, through the issuance of NTC Memorandum Circular No. 03-03-2005. The Circular regulates all commercial and promotional advertisements and surveys sent via broadcast and push messaging services, and covers all public telecommunications entities and content providers operating within the country. The NTC may impose administrative sanctions for any violation of these guidelines, including the revocation or cancellation of a permit or authority to operate.[5]

At present, there are efforts towards the drafting and enactment of a data protection legislation patterned after the European Union Data Protection Directive. A Cybercrime Prevention Bill, which penalizes the transmission of unsolicited commercial communication or spam mail, among others, is likewise being deliberated on by a Congressional Technical Working Group.

Regulatory bodies

The Department of Transportation and Communication (DOTC) was created in 1987 to oversee communications in the Philippines. It formulates and recommends national policy guidelines, establishes and administers comprehensive and integrated programmes for transportation and communication, provides direction to transportation research and development, and administers all laws and regulations pertaining to transportation and communication.

The NTC is an attached agency of the DOTC. It has the sole authority to issue a Certificate of Public Convenience and Necessity (CPCN) for the installation, operation and maintenance of communications facilities and services, radio communications systems, telephone services and telegraph systems. It also has the authority to determine the areas of operations of applicants for telecommunications services.

The NTC has been given the mandate to be the principal administrator of RA 7925. It is mandated to undertake the necessary measures for the implementation of policies and objectives with respect to: facilitation of the entry of qualified service providers and adoption of an appropriate pricing policy; assurance of quality, safety, reliability, security, compatibility and interoperability of telecoms facilities and services; fair and reasonable interconnection of facilities of authorized public network operators and other providers of telecoms services; the protection of telecoms entities from unfair trade practices of other carriers; and the promotion of consumers' welfare.

Hence, the DOTC is the policymaking body for the promotion, development and regulation of coordinated networks of transportation and communications systems, while the NTC is the regulating authority.

CICT was created under EO No. 269 (2004) to address the urgent need to make the country's approach to ICT development coherent and efficient, and to play an active role in streamlining, managing, coordinating and implementing the various ICT-related plans and policies of government. Its Five strategic directions are:

1. provision of affordable Internet access to all segments of the population;
2. development of an ICT-enabled workforce;
3. creation of an enabling legal and regulatory environment;
4. online provision of government services to stakeholders; and
5. development of the country as a world-class ICT services provider.

CICT took the place of the Information Technology and E-Commerce Council (ITECC). It also assumed the DOTC's functions of providing policy directions in the area of information and communications. Currently attached to the Office of the President, it is a transitory body towards the creation of a proposed Department of Information and Communication Technology (DICT).[6] It has direct supervision and control over the National Computer Center (NCC), the Telecommunications Office (TELOF), and other operating units of the DOTC. The Philippine Postal Corporation is also attached to CICT. The NTC receives policy guidelines from CICT but remains independent with respect to its quasi-judicial functions.[7]

ICT plans and initiatives[8]

At present, CICT is finalizing the Philippine ICT Roadmap, which lays down the government's strategies and programmes in the use and development of ICTs for better governance, corporate performance and overall economic development. The Roadmap was drafted through workshops and multi-stakeholder consultations and was to be released before the end of 2006. It is guided by the following principles:

- Commitment to a people-centred, inclusive and development-centred information society;
- Government providing an enabling policy, legal and regulatory environment;
- A multi-stakeholder approach to ICT for development;
- Using ICTs as tools for sustainable development, with an eye to accessibility, availability, security and accountability, interoperability and sustainability of ICT programmes;
- Promoting the development of digital content that is relevant and meaningful to Filipinos;
- Creating a safe, trustworthy online environment for all Filipinos; and
- Establishing a strong CICT organization to facilitate ICT development and ICT for development.

The Roadmap has five key strategic programmes and initiatives:

1. Ensuring Universal Access and bridging the digital divide through the:
   - Establishment of Community e-Centres, Internet in schools (iSchools), eLGUs (e-Local Government Units), and Regional ICT Centres, in partnership with the private sector, local government and civil society stakeholders;
   - Provision of Low-Cost Computing through the *PC ng Bayan* initiative and the distribution of Free and Open Source Software (FOSS) as an alternative to commercial software;
   - National Broadband Policy which provides for nationwide broadband connectivity and public access points; and
   - Last Mile Initiative.

2. Developing Human Capital for Sustainable Development by:
   - Creating an ICT Competency and Standards Development and an ICT Competency Assurance Body;
   - Establishing ICT for Education (ICT4E) programmes.

3. Using ICT to promote efficiency and transparency in government through:
   - Financing, via the e-government fund, frontline government ICT projects, such as the Bureau of Internal Revenue's computerization project, the Bureau of Custom's Web-based application system, and the NCC's eLGU Real Property Tax System;
   - Developing common applications for national government agencies;
   - Creating a government communication network;
   - Enhancing ICT training for government; and
   - Revising the Government Information Systems Plan (GISP).

4. Strategic Business Development to Enhance Competitiveness in the Global ICT Market through:
   - The Workforce Mobilization Programme with the Commission of Higher Education (CHED) and the Technical Education and Skills Development Authority (TESDA) to improve English competence, industry certifications and career advocacy programmes.
   - Launching the Philippine CyberServices Corridor, an ICT belt stretching over 600 miles from Baguio City in northern Philippines to Zamboanga in the south, to be wired by a USD 10 billion high-bandwidth fibre backbone and designed to provide a variety of cyberservices at par with global standards. The Corridor aims to support government's priorities for job creation, expansion of the middle class and regional development.
   - Strengthening the ICT capacities of small and medium-sized enterprises.

5. Legal and Policy Agenda for the Philippine ICT Sector, through the passage of the following bills:
   - Creating a Department of Information and Communication Technology that will ensure effective coordination and implementation of the national ICT agenda.
   - NTC Reorganization to give the country's telecom regulator fiscal autonomy and political independence.
   - Convergence Bill
   - e-government Bill
   - Privacy and Data Protection Act
   - Cybercrime Bill
   - Freedom of Information Law

While the ICT Roadmap is being finalized, CICT is focusing its energies on the development of the country's Cyber Corridor Plan. In July 2006, CICT Chair Ramon Sales was designated as the Plan's champion, and the business process outsourcing (BPO) sector was named the driver of economic growth and a solution to unemployment. The Department of Trade and Industry (DTI) is leading efforts to increase the competitiveness of the BPO sector and to promote the Philippines as a hub for IT-enabled services. The government is also allocating funding to improve the English proficiency of graduates and the English, science and mathematics skills of school teachers. Congress is also seriously considering a bill mandating the use of English as the medium of instruction in place of Filipino.

The Philippines: The next BPO hub?

In September 2006, Dell announced that it was opening a second call centre office in the Philippines. The hardware giant is one of a growing number of Fortune 500 companies like IBM, Intel, Motorola, Citigroup, and Barnes & Noble that are setting up offices in the country—proof that the business process outsourcing (BPO) sector in the Philippines is booming. In 2000, the BPO sector employed a mere 2,400 workers and earned USD 24 million. Five years later, it accounted for 112,000 workers and USD 2.2 billion in revenues. By 2010, the government expects the sector to employ over one million workers, generate USD 12.8 billion in revenues, and contribute 10 per cent to the GDP.

For developing countries like the Philippines, investment in the BPO sector is desirable as it provides a means of reducing unemployment, generating foreign exchange and diversifying the economy.

The Pros: What makes the Philippines attractive as a BPO location site? First, it is the third-largest English-speaking country in the world, with a long-standing familiarity with the US market's accents and idioms. Second, the Philippines produces 400,000 college graduates annually. Of this total, 110,000 graduate with commerce and business skills, 80,000 with IT expertise and 30,000 with medical degrees. Third, investors praise Filipino workers for their work ethic, loyalty and ability to adapt. Their cultural emphasis on service and their familiarity with Western business practices reduce training time. Fourth, the cost of labour is hard to beat: a Filipino university graduate is paid only about USD 7 a day. Finally, due to reforms in the telecommunications sector, telecom prices are dropping and infrastructure is improving. The country has one of the cheapest leased line rates in Southeast Asia, costing about USD 2,500–3,000, cheaper than India's USD 6,000–8,000.

The Cons: The Philippines has been dogged by fears of political instability, fuelled by rumours of coups and fairly frequent mass protests. However, the economy continues to grow, buoyed by massive inflows of foreign exchange from overseas workers, fiscal reforms and the growth of sunrise sectors such as agriculture, mining, ICT, tourism, health care and BPOs—prompting talk of a 'firewall' between the country's economic and political systems. Thus, perhaps more challenging than the Philippines' political situation is the variable supply of skilled human resources. Despite its ranks of skilled workers, the country is having problems matching its supply of graduates to emerging needs. For one, the country loses many of its skilled workers to overseas markets. The government predicts a total shortfall of 314,800 workers from 2006 to 2010. Second, despite the Philippines' reputation as an English-speaking country, industry leaders have voiced concern about the English proficiency of recent graduates. At present, only 5–6 people are recruited out of every 100 applicants, prompting employers to recruit outside urban areas and set up intensive training programmes for 'near-hires'.

Government Responses: The government is championing the BPO sector as the new economic growth driver and a solution to its unemployment problem. In her State of the Nation Address in July 2006, President Arroyo announced the plan to develop a CyberServices Corridor of IT-enabled services, call centres and other BPOs to run from north to south of the archipelago. While 425 of the 555 BPO firms are currently in and around Manila, growing numbers are being established around Cebu, a centre for the electronics industry. In addition, in August 2006 President Gloria M. Arroyo directed DTI to lead a public-private sector task force in drafting a BPO Master Plan to increase the country's competitiveness in the sector and to promote the country as a hub of IT-enabled services.

The government has also announced several measures to improve the English language proficiency of Filipino graduates, as follows:

- In June 2006, PhP 600 million (about USD 12 million) was allocated to upgrade the English language skills of English, Science and Mathematics teachers.

- In May 2006, a PhP 500 million (about USD 10 million) 'Training for Work' scholarship fund was announced for 100,000 'near-hires' to bring their language skills to the required standard.
- As early as 2004, a bill was filed in Congress mandating the use of English as the medium of instruction, in place of Filipino.

Indian Recognition: On 3–7 September 2006, a high-level Indian delegation composed of the Secretary of the Ministry of IT and Communications, the President of the National Association of Software and Service Companies (NASSCOM) and representatives of tier-one BPO firms operating in India visited the Philippines. The visit, which was hosted by the Business Process Association of the Philippines (BPAP), led to the signing of an agreement between BPAP and NASSCOM to promote outsourcing in the two countries. The two industry associations agreed to cooperate in strategic communications, geographic risk mitigation, workforce development and infrastructure improvement. They also agreed to share best practices and adhere to international standards in data security and privacy. The agreement suggests recognition of the Philippines as a serious rival by the Indian BPO sector. Indeed, the Philippines is increasingly becoming the second most preferred location for BPO operations and is well poised to take a bigger slice of this market.

On another front, the tight fiscal situation of the government has led to scrutiny of its expenses, including its annual expenditure for communication. In 2005, the national government spent PhP 3.48 billion (about USD 632.7 million) in communication expenses. This huge expenditure has led to calls for the government to start using VoIP (Cruz 2006).

ICT4D projects

In January 2004, the Department of Science and Technology, in partnership with the Canadian International Development Research Centre (IDRC), launched the main initiative for the use of ICTs to foster development and alleviate poverty through community projects (see its website at www.ict4d.ph). As of August 2006, there were 449 documented ICT4D projects under the following categories: e-Agriculture (24 projects), e-Business (41 projects), e-Employment (14 projects), e-Environment (28 projects), e-Government (237 projects), e-Health (16 projects) and e-Learning (89 projects).

As the breakdown shows, the government is leading the way in ICT use, improving services and its accessibility to citizens and investors. As proof of the government's commitment to ICT4D, a four billion peso e-government fund was created in 2003, to which was added PhP 1 billion in 2005, to finance high-impact and mission-critical ICT projects of government agencies, under the management of CICT. As of June 2006, there were 31 ongoing projects financed by the e-government fund, amounting to about PhP 3.46 billion. Examples of these projects are:

- The Bureau of Customs' ASYCUDA World Project (e-Customs), which aims to facilitate trade exchanges and generate government revenues by creating a Web portal that upgrades the Bureau's operational facilities, streamlines processes and encourages transparency in transactions.
- The Bureau of Internal Revenue's PhP 678.51 million Integrated Computerization Project envisioned to promote transparency and efficiency in government transactions, contribute to stronger tax administration, facilitate increased revenue collection, minimize graft and corruption in government, and provide more convenient frontline services to taxpayers. If successfully implemented, the project would contribute significantly to improving tax collection, expected to total 4.1 trillion pesos from 2006 to 2010. The project includes, among others, systems to facilitate e-filing, a tax compliance verification drive using mobile technology, a Computer-Assisted Audit Programme (CAAP), and Automated Excise Data Management System-Phase II (AEDMS).
- The eLGU project of the National Computer Center which assists local government units (LGUs) in their computerization efforts to enable better and faster delivery of government services. The project consists of four major components, namely: development of e-Government applications, LGU capability building, advocacy and promotions, and empowering local communities in ICT through Community e-Centres (CECs).
- The eLibrary project of the Department of Science and Technology (DOST), the University of the Philippines, the Department of Agriculture, the National Library and the Commission on Higher Education creating a Web portal that now contains digitized text of 800,000 bibliographic records consisting of 25 million pages of local and international materials, 29,000 full-text copies of foreign journals and 15,000

theses and dissertations. The eLibrary provides member agencies free access to content, subject to a fair use policy.

There is also the Anti-Money Laundering project and the NCC's online payment portal for government-related services.

In 2005 the priorities of the 2005 e-government fund were the following:

- CICT's Community e-Centres (CeC) project, which involves the establishment of single access points for online delivery of national and local e-government services to smaller communities. The CeCs will provide voice and data services, Internet access, PC rental, business services and community-based services such as agricultural price monitoring, trading and local content development. Also to be provided are special services for overseas Filipino workers (for example, VoIP, e-mail, job search), commercial services (remittance services, real estate tax payments, business permits and licenses), and national government services such as online applications for birth certificates, payment of social security and government insurance contributions, and passport renewal. At present, there are already 54 operational CeCs in the Visayas and in Mindanao.
- DTI's Business Name Registration System
- The local government information portal and the public safety information portal of the Department of Interior and Local Government

Since 2000, private foundations have sought to provide Internet connectivity packages to secondary schools. The most recent programme is a multi-sector initiative called GILAS (Gearing Up Internet Literary and Access for Students). Launched in 2005, GILAS aims to provide Internet access and basic Internet literacy programmes in all of the 5,433 public secondary schools in the Philippines within five years time. The project is led by a consortium of top private corporations and civic organizations in cooperation with the Department of Education.

Establishing a Web presence

As of mid-2006, 348 out of 375 national government agencies (NGAs) had established a presence on the Web. In terms of stages of e-government following the UN-ASPA (American Society for Public Administration) standards (Digital Philippines), 10 NGAs are at Stage 4 (transactional Web presence), 142 are at Stage 3 (interactive Web presence), 131 are at Stage 2 (enhanced Web presence) and 65 are at Stage 1 (static Web presence).

A December 2004 report by the National Computer Center showed that all but 15 of the 1,709 local government units have a Web presence. A hundred of these are at Stage 2 of the UN-ASPA standards and 40 are at Stage 3. Among the 79 provincial governments, 42 have static websites, 26 have enhanced websites and 11 have interactive websites. In addition, out of 115 cities, 68 have static websites, 32 have enhanced websites and 15 have interactive websites.

As for state colleges and universities (SCUs), 60 of the 111 SCUs had websites as of the third quarter of 2006. Of these, 19 websites are static (Stage 1), 21 are enhanced (Stage 2) and 21 are interactive (Stage 3) (National Computer Center Website).

Education and capacity building

The Philippines Open University is a leader in online learning. From only two online courses in 2001, it now delivers online tutorials for all of its more than 200 courses using an open source learning management system. Online learning has also caught on in a number of private and public Philippine universities.

At the secondary school level, the Philippines is part of a regional programme called ASEAN SchoolNet, which is funded by a grant from the World Bank-managed Japan Social Development Fund awarded to World Links, an international NGO with extensive experience in developing local capacity to integrate ICTs in education. The Foundation for IT Education and Development (FIT-ED) is responsible for the programme's implementation in the Philippines. The other target countries in the region are Cambodia, Lao PDR, Vietnam and Indonesia. The programme aims to promote appropriate and sustainable use of computers and Internet technology for teaching and learning in the public secondary school system through a teacher professional development programme, as well as training in technology planning for school administrators. As of 2006, 31 public and private secondary schools and 17 teacher education institutions are participating in the Philippine project.

Another ICT for education project in the Philippines is the Intel® Teach to the Future (ITTF) Program, a global initiative of Intel Corporation promoting inquiry-based and project-based learning and the integration of computers in the school curriculum. The Philippine programme was launched in 2000 by Intel Technology Philippines, Inc. in partnership with the Department of Education (DepEd), the Department of Science and Technology-Science Education Institute (DOST-SEI), the University of the Philippines National Institute for Science and Mathematics Education Development (UP NISMED), EduQuest Inc., and FIT-ED. It serves as a major teacher training component of the government's computerisation initiatives for basic education, such as the PCs for Public Schools project. Since

2000, the ITTF Programme has trained more 50,000 schoolteachers in effective technology use in the classroom.

FIT-ED also successfully co-organized with the Department of Education and CICT the First and Second National ICTs in Basic Education Congress in December 2004 and September 2006, respectively. The Congress brought together teachers, school administrators, teacher trainers, educational technologists and other practitioners and stakeholders from the public, private and non-profit sectors.

CICT is spearheading the development of ICT Competency Standards to be applied in education and training, and to help professionalize ICT personnel in government and the private sector through the design, formulation and administration of competency-based certification examinations. To date, three draft standards have been formulated: the National ICT Standards (NICS)–Basic, NICS–Advanced and NICS–Teachers. An ICT Competency Assurance Body is to be established to oversee accreditation, certification and coordination with concerned stakeholders.

Moreover, CICT has drafted an ICT in Education Master Plan for all levels, including a National Roadmap for Faculty Development on ICT in Education. In 2005, CICT assisted the Department of Education and FIT-ED in formulating the draft National Framework Plan for ICTs in Basic Education (2005–10).

With respect to content development, CICT has the Open Content in Education Initiative (OCEI) which aims to convert Department of Education curricula into interactive multimedia content, develop computer applications for schools, and sponsor student and teacher competitions to promote the development of education-related Web content. The iSchool WebBoard aims to develop and share online learning materials, and to facilitate immediate access to useful references and interactive facilities on the Internet. Another project, PhEdNet, is a 'walled garden' hosting teaching and learning materials and applications for use by Filipino students, teachers and parents. All public high schools will be part of this network, which would ensure that only DepEd-approved multimedia applications, materials and mirrored Internet sites will be accessible from school PCs.

eSkwela is a CICT project that aims to establish Community e-Learning Centers for out-of-school youth (OSY), providing them with ICT-enhanced alternative education opportunities. At the tertiary level, CICT has the eQuality Programme in partnership with State colleges and universities.

The Digital Media Arts Programme aims to develop digital media skills for government using open source technologies. The ICT Skills Strategic Plan adopts an inter-agency approach to the identification of strategic, policy and programme/project recommendations to address the ICT skills demand-supply gap.

Open source initiatives

The Philippines has an active community of open source advocates. In October 2006, the University of the Philippines, in cooperation with the International Open Source Initiative, conducted the Linux Training for Trainers series to enhance awareness and knowledge of free and open source software (FOSS). Also in the pipeline is a training programme for service providers and small to medium-scale enterprises looking for alternatives for building their own IT infrastructure. These initiatives aim to develop personnel who are knowledgeable about open source applications development and administration in order to promote its use.

Already, a bill on the use of FOSS is being contemplated in the Philippine Congress, with the goal of mandating the use of FOSS by government agencies and encouraging its use and development in the public and private sectors. CICT supports this move, given that its ongoing projects like eLGU, eGovernance Center of Excellence, iSchools and eSkwela are already using FOSS.

Conclusion

For a developing country, the Philippines is doing a lot in terms of utilizing ICTs as a developmental tool. The main challenge is really to ensure that its ICT4D programmes have a meaningful impact on the daily lives of people. The first step towards this is to understand that ICT adoption is a social process with the potential to exacerbate existing inequalities, as much as it holds the promise of bridging these same inequalities. Careful planning is needed to ensure that ICT4D goals are realized.

The institutionalization of a separate ICT department will result in effective coordination and implementation of the country's ICT agenda. Currently the Philippines is the only country among ASEAN-6 that does not have a dedicated executive department dealing with ICTs. It is hoped that the new department will see the light of day in the near future.

Notes

1. In fact, in 2005 installed fixed-lines grew by 1 per cent while subscribed lines declined by 2 per cent due to disconnections.
2. The VoIP Rules explicitly define VoIP service as the 'provision of voice communication using Internet Protocol (IP) technology, instead of traditional circuit-switched technology'. See VoIP Rules, Section 2(f).

3. See VoIP Rules, 10th WHEREAS Clause.
4. Republic Act No. 8792, Section 28, par. 3.
5. See Section 5.1 of the NTC Memorandum Circular No. 03-03-2005.
6. The proposed DICT Bill consolidated by the Congressional Technical Working Group has not yet been passed into law.
7. EO 263, Section 5.
8. For additional ICT policies, see www.ncc.gov.ph/default.php?a1=2&a2=3&a3=0

References

Amojelar, D. (2007). Text messages hit 250M per day. *Manila Times*, 3 August.

Astrologo, C. (2006) *Show me the data*. Retrieved 14 August 2006 from www.nscb.gov.ph

Commission on Information and Communications Technology Website. Available at www.cict.gov.ph

Cruz, M. (2006). Govt officials spend billions on phone calls. *Manila Times*, 2 October 2006.

GILAS Website. Available at www.gilas.org

Government of the Philippines. (1987). *The 1987 Philippine Constitution*.

Information and Communications Technology for Development in the Philippines (ICT4D) Website. (2004). Available at www.ict4d.ph

Librero, F. (2004). Digital learning environment in the Philippines: Perspective from the UP Open University. Paper presented at the Symposium on Digital Learning, Keio University, Japan, 11–13 December 2004.

National Computer Center Website. Available at www.ncc.gov.ph

National Telecommunications Commission Website. Available at http://portal.ntc.gov.ph/wps/portal

National Telecommunications Commission. (2005a). *Memorandum Circular No. 05-08*.

National Telecommunications Commission. (2005b). *Memorandum Circular No. 03-03-2005*.

Philippine Long Distance Telephone. (2005). *Annual report 2005: Broadbanding the future*. Manila, Philippines.

Republic of the Philippines. (1993a). *Executive Order 59*.

Republic of the Philippines. (1993b). *Executive Order 109*.

Republic of the Philippines. (1995). *Republic Act 7925 (Telecommunications Act of the Philippines)*.

Republic of the Philippines. (2000). *Republic Act No. 8972 (E-Commerce Act)*.

Weigel, G. and Waldburger, D. (eds). (2004). *ICT4D—Connecting people for a better world: Lessons, innovations and perspectives of information and communication technologies in development*. Switzerland: Swiss Agency for Development and Cooperation and Global Knowledge Partnership.

Singapore

Goh Seow Hiong

Total population	4,483,900 (2006)[a]
GDP per capita	SGD 44,765 or USD 29,091.23 (1 USD = 1.53878 SGD)
Key economic sectors	Manufacturing, Construction, Utilities, Wholesale and retail trade, Hotels and restaurants, Transport and communications, Financial services, Business services
Computers per 100 inhabitants	64.48 (2006)
Household computer penetration	78 per cent household with at least one computer (40 per cent with 1 computer, 38 per cent with 2 or more computers) (2006)
Fixed-line telephones per 100 inhabitants	41.2 per cent (November 2006)
Mobile phone subscribers per 100 inhabitants	101.5 per cent (November 2006)
Internet users per 100 inhabitants	34.1 per cent assuming dial-up users only (November 2006); 50.77 per cent if dial-up subscriptions + broadband subscriptions/population
Domain names registered under ".sg"	898,762 (July 2006)
Broadband subscribers per 100 inhabitants	16.7 per cent (*Household penetration*: 61.1 per cent)
Internet international bandwidth	28 Tbps submarine cable capacity (end 2005); 20 Gbps direct international Internet connectivity (June 2005)

Sources: Statistics Singapore 2006; World Bank 2001; Infocomm Development Authority of Singapore 2006; Internet Systems Consortium, Pyramid Research.

Note: [a]The total population comprises all citizens and permanent residents with local residence and foreigners staying in Singapore for one year or more. Singapore residents refer to citizens and permanent residents with local residence.

Overview

Indicators

Singapore's key economic sectors continue to be in manufacturing, construction, utilities, wholesale and retail trade, hotels and restaurants, transport and communications, financial services, and business services. The country's GDP was SGD 194,359.8 million in 2005, with a per capita GDP of SGD 44,666 (SingStat 2006).

In the ICT sector, the number of households with access to a computer at home was stable at about 74 per cent from 2003 to 2005, up from 68 per cent in 2002. The proportion of households with two or more computers was at 28 per cent, up from 23 per cent in 2003. The proportion of households with access to the Internet at home was also stable at 65–66 per cent in 2003–05, up from 59 per cent in 2002. Sixty-five percent of Singapore's resident population aged 15 and above were computer users, up from 57 per cent in 2003. Of the same population 61 per cent were Internet users, up from 53 per cent in 2003. Twenty-seven percent of Internet users have made purchases online (up from 17 per cent in 2003), averaging about SGD 1,068 of Internet purchases in 2005 (IDA 2006a).

As of May 2006, the fixed telephone line population penetration is 42.4 per cent, and fixed-line household penetration is 97.8 per cent. Mobile phone penetration is 99.3 per cent, and household broadband penetration is 55.6 per cent (IDA 2006b).

Since the downturn in 2002–03 as a result of the SARS outbreak and the global economic slowdown, the Singapore economy has made significant strides in restoring the strength and competitiveness of the economy. New initiatives have been announced even amidst efforts to contain rising costs and to streamline for greater efficiency. The ICT or infocomm sector, composed of international and local players in a vibrant and competitive market, remains strong.

Technology infrastructure

It has been 25 years since Singapore embarked on its infocomm journey in 1981 with the creation of the then National Computer Board. In 1997, new cellular and paging operators were introduced in the market to compete with the incumbent operator. In 1998, Singapore ONE, the nation's broadband network with a core fibre and ATM backbone and access through cable and ADSL, was made commercially available nationwide. In 2000, after the creation of the Infocomm Development Authority of Singapore (IDA), the country's telecommunications market was completely liberalized, with competition in both the mobile and fixed-line segments and the lifting of equity restrictions on foreign operators. By 2002, submarine cable capacity was at least 21 Tbps and there was direct Internet connectivity to over 30 countries, and at least 2,250 Mbps of international Internet connectivity. In 2005, 3G services were commercially rolled out by mobile operators. By 2006, total submarine cable capacity was 28 Tbps and direct international Internet connectivity was 20 Gbps.

Under the iN2015 Masterplan (*discussed later*), the Singapore government intends to put in place a Next Generation National Infocomm Infrastructure by 2012. The aim is to have pervasive connectivity around the country, comprising complementary wired and wireless networks to provide seamless connectivity. The envisioned wired infrastructure will deliver symmetric broadband access speeds of up to 1 Gbps, and will be Internet Protocol Version 6 (IPv6) compliant. The ultra high-speed network will connect all homes, schools and businesses, and it is expected to enable new broadband-enabled services applications, such as immersive learning experiences, telemedicine, high definition TV, video conferencing and grid computing (IDA 2006c).

To achieve this vision, the government has issued a Request-for-Concept to solicit proposals from local and international telecommunications providers and hardware and software vendors. The submissions reflect the need for an infrastructure with open access, with fibre commonly proposed to provide ultra high-speed broadband access. Innovative business models were also proposed to provide competitive end-user pricing to consumers (IDA 2006d).

Following the Call for Collaboration on the wireless network, the government awarded three service providers contracts to provide free basic 512 Kbps wireless at public places in key catchment areas for at least two years, with a second tier of higher bandwidth access speeds for paid subscribers. While Wireless Fidelity (Wi-Fi) will be deployed, the SGD 100 million project will also trigger enhancements to existing deployment of WiMAX (Worldwide Interoperability for Microwave Access) and High-Speed Downlink Packet Access (HSDPA). The project, dubbed Wireless@SG, will complement and extend to public places broadband access currently available in homes, offices and schools, and will enable access to Internet-based services like email, instant messaging, online games and Voice over Internet Protocol (VoIP) calls (IDA 2006e).

Online services

e-Payment

A collaboration between IDA, the Land Transport Authority (LTA) and industry players aims to enable seamless electronic payments (e-payments) for a range of daily needs through the new Singapore Standard for Contactless ePurse Application, or SS 518 CEPAS. The standard is aimed at creating a nationwide interoperable micro-payment platform for use across different sectors, including transit and retail. Establishing a standard aims to encourage more card issuers, such as banks and merchants, to participate in e-payment, allowing existing multi-purpose stored value schemes to interoperate and giving consumers the convenience of using a single card rather than multiple cards for different purposes. This will reduce the number of paper-based transactions and double the annual value of electronic transactions. Enterprises will be encouraged to use the standard. Its development is a pioneering effort as there are no international standards in this area yet (IDA 2006f).

e-Government

After the successful implementation of two previous e-Government Action Plans, the Singapore Government has announced iGov2010, a new SGD 2 billion five-year Masterplan developed in consultation with the public and private sectors.

Through the earlier plans, all government services that can be placed online are already available on the Internet. A survey in March 2006 indicated that nearly nine out of 10 customers had transacted with the government electronically at least once in the last 12 months. The new Masterplan aims to transform backend processing to achieve front-end efficiency and effectiveness. There would be greater emphasis on transcending organizational structures, changing rules and procedures and integrating government services around customer and citizen's needs.

The iGov2010 Masterplan has four main thrusts:

- *Increasing Reach and Richness of e-Services*—Government e-services will be made more accessible to a larger population and opportunities will be created for more innovative services. Given the high mobile penetration rate, use of the mobile channel will be increased. More proactive, responsive, user-friendly and integrated e-services will be delivered. For those without Internet access or who need help in transacting online, the CitizenConnect Centres at Community Clubs will be expanded from five to 25 centres.
- *Increasing Citizens' Mindshare in e-Engagement*—The government will actively engage citizens in the policymaking process, and strengthen its relationship with citizens through the use of infocomm technologies. The Government Online Consultation Portal will be enhanced to meet the needs of different groups more effectively.
- *Enhancing Capacity and Synergy in Government*—The government will take further steps to transform its business and operating model to make the government more agile and efficient, capable of achieving a higher level of service delivery. Agencies will streamline common functions and integrate backend processes (for example, in human resource and finance). The use of unique identifiers across different agencies for companies, businesses, societies and non-profit organizations will create greater efficiencies.
- *Enhancing National Competitive Advantage*—The government will continue to collaborate and partner with the private sector through various infocomm projects to allow the

co-creation, development and export of iGov solutions. The government will allow intellectual property ownership to be retained by companies to enhance their business and solution export opportunities (IDA 2006g).

An e-Government Leadership Centre jointly set up by IDA, the National University of Singapore's Institute of Systems Science, and the Lee Kuan Yew School of Public Policy was launched in June 2006 to build on the country's leadership in e-government developments and to share knowledge with foreign governments keen on leveraging e-government to further their countries' developmental goals. Topics related to public policy, ICT policy and management, and innovative case studies in e-government in Singapore will be covered to meet the varying needs of different organizations (Computerworld 2006).

A showcase known as the 'Government Executives in the New Information and Knowledge Era' (GENIE Showcase) demonstrated how government would use new infocomm technologies through partnerships with industry. Solutions expected in the near future include authentication through biometrics; a rich integrated user interface that is self-learning, adaptive and predictive of the content required; integrated communication services; and self-recovery and diagnostics for computers (IDA 2006n).

Web services

IDA launched the WEAVE Programme (Web Services Add Value to Enterprise) in 2003 to promote Web services to industry. A Web Services Chapter was also set up under the Singapore infocomm Technology Federation (SiTF) to accelerate the adoption of Web applications and services. Various initiatives under the Chapter are designed to realize business opportunities and enhance interoperability. As of August 2006, an industry rate adoption of over 28 per cent has been achieved. In terms of capability development, there are now more than 3,000 professionals trained in Web services, with a third attaining certification through IDA-endorsed courses. Altogether, about 65 industry projects have been supported by IDA, among them well-known companies like Singapore Airlines, PSA (formerly Port of Singapore Authority) and United Premas. The total value of investments in the industry is estimated to be SGD 246 million.

Government has also made significant headway in the use of Web services, making it more convenient for businesses and the public to transact with government. For example, four licensing agencies (the Health Sciences Authority, Media Development Authority, National Environment Agency and Public Utilities Board) have implemented 40 strategic Web services that enable real-time integration of services in the Online Business Licensing Service portal. In many cases, this effectively reduces the turnaround time for businesses to obtain their licenses from a few weeks to a few working days, or even within the same working day.

The government will be extending Web services to the private sector. Through the Government Web Services Exchange, businesses will be able to leverage common e-government services to save time and effort. For a start, the Web services currently available include the business-related services offered by the Accounting and Corporate Regulatory Authority, the carpark-related services offered by the Housing Development Board and Urban Redevelopment Authority, the library book catalogue services offered by the National Library Board under the Ministry of Information, Communications and the Arts (MICA), and National Servicemen (NSmen)-related services offered by the Ministry of Defence (MICA 2006).

e-Lifestyle

A survey has indicated that in addition to the Internet being a tool for work, six in 10 Internet users aged 15 years and above also use the Internet for leisure, including playing computer games, downloading or uploading digital photos, listening to online music, reading publications and watching films over the Internet. Almost six in 10 Internet users made an online transaction with the government within a 12-month period and four out of 10 Internet users have made other online transactions. The most popular online application was online banking. (IDA 2006i)

Industries

The Singapore infocomm industry's revenue grew by 8.9 per cent in 2005, reaching SGD 37.89 billion (about USD 24.75 billion). External demand grew by 11 per cent. The export market drove most of the industry's growth in 2005.

There was little change in industry composition from 2003 to 2005. More than half (51 per cent) of the total revenue in 2005 was attributed to hardware, 19 per cent to telecommunication services, 14 per cent to software, 9 per cent to IT services and 7 per cent to content services. Growth in software was 5.5 per cent in 2005, compared to –14.9 per cent in 2003. Growth in IT services in 2005 was –3.4 per cent, compared to –25.3 per cent in 2003. Growth in content services was 0.9 per cent compared to 19.0 per cent in 2003.

The export market contributed more to the total revenue compared to the domestic market (58–42 per cent) in 2005. In the domestic market, 29 per cent of the revenue was attributed

> ## Online gaming
>
> Online gaming, with its high potential for growth, fits well into Singapore's objective of being a hub to host and manage regional games. The interest in this sector started in 2001 from the government's e-Celebrations campaign to promote infocomm adoption and e-lifestyle. It has since grown to be a rapidly emerging sector, with the market for Asian gamers expected to hit SGD 23 billion (about USD 15 billion) by 2009.
>
> Surveys indicate other benefits to online gaming, including broadening a person's knowledge of various subjects, such as history, military strategy, economics and sports; stimulating creativity; and developing interest in computers and software skills.
>
> Singapore has established the Games Exchange Alliance (GXA) to localize and deliver games to Asian gamers. With 25 members, the GXA provides a comprehensive spectrum of resources to take games from the development stage to full deployment, including 'last-mile' commercialization to 13 countries in Asia. Some successes include the creation of a cross-platform MMORPG (massively multi-player online role-playing game) that is playable on both computers and mobile phones in real time. GXA's bundling of complementary games services among its members allows for an integrated value-chain approach and greater economies of scale through the wider spectrum of services.
>
> Government initiatives such as Games Bazaar (a scalable hosting platform for games companies, publishers and distributors to test and deploy games regionally), Games Market Access Programme (a suite of services of regional deployment, hosting and distribution, community building and marketing support for a fixed monthly price), and GXA aim to attract more game publishers and developers to base themselves in Singapore. Benefits to companies include shortening time-to-market of the games they develop (IDA 2004, 2006h).

to hardware, while 41 per cent was attributed to telecommunication services. In the export market, 66 per cent of the revenue was attributed to hardware, with 3 per cent attributed to telecommunication services. In software, the domestic market contribution was only 4 per cent compared to 22 per cent in the export market. In IT services, it was 16 per cent of the domestic market and 4 per cent of the export market. A tenth (10 per cent) of the domestic market was attributed to content services compared to 5 per cent in the export market. The domestic market rebounded from its negative growth in 2004 to register a positive growth in 2005 (IDA 2005a).

Key national initiatives

Singapore iN2015 Masterplan

Launched in June 2006, the new 10-year Infocomm Masterplan named iN2015 aims to achieve by 2015 the vision of Singapore as an Intelligent Nation and Global City powered by infocomm. The government cited innovation, integration and internationalization as the basis of the Masterplan which was formulated with inputs from the government, the public, as well as specific industry sectors such as education, health care, manufacturing and logistics, finance, tourism and retail, and digital media.

Among the specific targets for 2015 are:

- for Singapore to become a global leader in harnessing infocomm to add value to the economy and society;
- a two-fold increase in value-added of the infocomm industry to SGD 26 billion;
- a three-fold increase in infocomm export revenue to SGD 60 billion;
- 80,000 additional jobs;
- at least 90 per cent of homes using broadband; and
- 100 per cent computer ownership for all homes with school-going children.

The plan not only addresses national economic competitiveness; a digital inclusive angle also ensures that the elderly, the less-privileged and those with disabilities can benefit and have opportunities for development.

Four key strategies are outlined in the Masterplan:

- Spearhead the transformation of key economic sectors, government and society through more sophisticated and innovative use of infocomm;
- Establish an ultra-high speed, pervasive, intelligent and trusted infocomm infrastructure;

- Develop a globally competitive infocomm industry; and
- Develop an infocomm-savvy workforce and globally competitive personnel in infocomm industries.

Initiatives to strengthen domain and technology capabilities within the industry, and to nurture local companies to expand and grow internationally, are expected. The infocomm competencies of the general workforce will also be raised. Techno-strategists who have both the technical and business expertise will be groomed to achieve business and organizational goals through the strategic and innovative use of infocomm. Top-ranking students in schools will be encouraged to take up infocomm as a career.

Innovative and personalized services will be developed to achieve sectoral transformation in financial services, manufacturing and logistics, tourism, hospitality and retail, e-government, digital media entertainment, education, healthcare and biomedical sciences (IDA 2006j).

Interactive and digital media

The Singapore Research, Innovation and Enterprise Council (RIEC), which is chaired by the Prime Minister, has approved plans in three strategic research sectors, including Interactive and Digital Media (IDM). The overall goal is to build up core research and development (R&D) capabilities in selected areas and to attract and develop talent to sustain advanced research activities for the long term. The approved IDM programme will build on Singapore's multicultural, multilingual identity with a strong infocomm infrastructure to create new niches in this area, including games and edutainment. This is expected to increase jobs and value-added in the sectors, resulting in broader economic benefits.

The fund allocation for the three strategic research programmes over the next five years is SGD 1.4 billion; SGD 500 million of the total is earmarked for IDM. The funds will be held in the National Research Fund and administered by the National Research Foundation (NRF).

To foster R&D, innovation and creativity, RIEC also approved the creation of a Campus for Research Excellence and Technological Enterprise (CREATE), with the Singapore-MIT Alliance for Research and Technology (SMART) Centre as the first centre within the Campus. These initiatives will develop the ability to build linkages with global institutions, to enhance relationships with other centres of research in Europe and the US, and to incubate a more inventive, innovative and entrepreneurial economy. Joint research efforts with other top universities are expected to provide the foundation to nurture a pool of new talent (NRF 2006).

Broadcasting

The Media Development Authority (MDA) is preparing the country to embrace high-definition TV (HDTV). Singapore media companies are developing new applications and solutions for HDTV, Internet Protocol TV (IPTV) and Digital Audio Broadcasting (DAB). Local service providers and content developers are being encouraged to develop and deliver compelling high-definition content and services. Trials are ongoing to bring content to the country while allowing local service providers to explore different business models.

Singapore aims to serve as a test bed for global companies seeking to test new concepts. MDA, through its Digital Technology Development Scheme, provides funding support to the broadcasting industry for the development of original and innovative products or processes that will help bring about value-added services, products and technology. Singapore-made applications are also actively promoted to the international broadcasting industry. The Scheme champions the development of innovative and experimental work to encourage more media talent to aim for greater international exposure and commercial success (MDA 2006a).

Bridging the digital divide

Recognizing the need to ensure access to technology at all levels of society, the government is aiming to improve access to infocomm for the elderly, low-income students and people with disabilities. This is in recognition of the fact that even as Singapore becomes increasingly digitally-enabled, there are some pockets of society where access to technology remains unaffordable. As part of the plan, the government will ensure that no student is denied computer and Internet access, and that the less tech-savvy are able to get connected through 'infocomm bridges'.

The 'infocomm bridges' include the NEU PC Plus programme, which seeks to provide highly-subsidized computer ownership and Internet access to 10,000 low-income households. Industry members have already committed SGD 31 million to assist in this programme. The Infocomm Accessibility Centre, another 'infocomm bridge', aims to provide people with disabilities infocomm training and employment opportunities. Specialized training and course materials at different levels of competency can lead to industry certification. Assistive technologies will be used to aid trainees with different disabilities and workshops and infocomm industry-relevant apprenticeship programmes seek to enhance the employability of disabled citizens. For the elderly, the Silver Infocomm Initiative provides workshops on the use of mobile devices, promoting an enhanced digital lifestyle and helping other senior members learn and adapt to technology.

The government's digital divide efforts have drawn the support of many industry players, and would be an essential complement to other national initiatives to achieve an all-inclusive digital society (IDA 2006o).

Multilingual support

In line with Singapore's aim to be a regional infocomm hub and building on Singapore's multiracial and multilingual society, the Singapore Network Information Centre (SGNIC) has conducted a trial to provide users the option of adopting multilingual domain names or Internationalized Domain Names (IDNs). This allows Chinese and Tamil characters to be used as part of the domain name, for the benefit of Chinese and Tamil-speaking audiences. For Malay-speaking audiences, this is not an issue because Malay is based on Roman characters. The IDN approach allows a non-English speaking person to more easily access domain names, while also offering service providers the opportunity to brand themselves using local languages. The IDN test bed was intended to gauge public response to IDN; allow the domain name registry and registrars to gain operational experience; and address technical, operational and policy issues concerning IDN. The implementation was based on established international IDN standards. Although the trial has been concluded, further plans have yet to be announced (IDA 2005; SGNIC 2006).

Enabling policies

Cyber security

The Singapore government has announced a series of new initiatives to help secure the national infocomm environment. The move is considered an important one for a country that is heavily dependent on infocomm technologies for managing its day-to-day activities.

The Infocomm Security Masterplan unveiled in March 2005 was developed after a series of consultations with the public and private sectors. This multi-agency effort led by IDA identified the key areas that need to be enhanced and proposed strategies to raise the level of awareness and preparedness against cyber attacks. Deficiencies such as lack of experienced professionals in infocomm security, difficulties faced by businesses in formulating and complying with IT security and best practices, and absence of awareness among employees were flagged for remedial action.

The Masterplan includes six integrated strategies: securing the people sector, securing the private sector, securing the public sector, developing national capability, cultivating technology and R&D and securing the national infrastructure. The approach adopted for implementing these strategies is based on current best practices adopted by the public and private sectors. The plan will focus on building capabilities, resources and skills in:

- information protection assurance and risk mitigation measures (including risk assessment, vulnerability analysis and reduction, authentication and technology assessment);
- enhanced situational awareness and contingency planning assurance (including round-the-clock vigilance and business continuity preparedness); and
- human and intellectual capital development (including cyber security awareness of Internet users, development of professional skills, and promotion of R&D in infocomm security).

Among the projects being planned are a National Authentication Infrastructure for online transactions, a Business Continuity Readiness Assessment Framework for public sector agencies, a national Cyber-Watch Centre (CWC) to monitor cyber-threats to government networks and provide early warning of impending cyber-threats round-the-clock, and an Infocomm Vulnerability Study of National Critical Infrastructure to assess the security readiness of key national infrastructures (IDA 2005b, 2006m).

Human resources development

An annual survey conducted in 2005 found that infocomm personnel in Singapore grew by 3.1 per cent to reach 111,400 in 2005. This number was higher by 5.5 per cent than during the Internet boom period of 2000. Half of infocomm personnel worked in infocomm organizations, while the other half are in end-user organizations. Job vacancies more than doubled from 2,100 in 2004 to 5,700 in 2005, in particular for skills in software development, infocomm security, database management, IT project management and Web services. Demand for high-end jobs and qualified infocomm personnel is expected to grow.

The survey results indicate that Singapore's infocomm workforce is highly educated: 83 per cent had tertiary education, 44 per cent possessed a Bachelor's degree, 24 per cent had a diploma and 15 per cent had postgraduate qualifications (postgraduate diplomas, Master's and doctorate).

IDA has a SGD 120-million Infocomm Manpower Development Roadmap designed to develop a globally competitive and infocomm-savvy workforce. The effort includes establishing a Student Infocomm Outreach programme for school children through Infocomm Clubs. Since January 2006, some 30 schools offer such clubs as part of their co-curricular activities and students earn points for joining them. Credit exemption may also be offered to students who subsequently pursue IT diplomas. The aim is to have more than 150 schools with Infocomm Clubs by 2008, and about 50 per cent of their members joining the infocomm profession eventually (IDA 2006k). The National Infocomm Scholarships will be expanded under the Outreach programme to include overseas studies with top universities. The scheme includes work opportunities with leading infocomm multinational companies who are partners with IDA under the programme.

For the general workforce, a series of activities named 'Infocomm Skills@Work' or 'InSkills@Work' targets the development of infocomm competencies in key economic sectors. Existing competency training programmes intended for infocomm professionals will also be expanded to include non-infocomm professionals.

Infocomm professionals will also be able to refer to a competency framework that sets the competency requirements and step-by-step certification process of an infocomm professional's career. The Graduate Career Development Programme equips local infocomm undergraduates with certifications for critical and essential skills needed by the industry to complement their academic qualifications (IDA 2005c).

Regulatory environment

Spam Control Bill

The Singapore government has conducted two consultations concerning a proposed Spam Control Bill that will include both email and mobile phone-based spam. The inclusion of mobile spam reflects a need resulting from the high penetration and usage of mobile messaging. The Bill contains a number of important features:

- It focuses on the real problem of egregious bulk spammers and prohibits the use of dictionary attacks or address harvesting software to indiscriminately send unsolicited e-mail.
- It requires bulk spammers to clearly label their unsolicited commercial electronic messages by including <ADV>, subject and header information that is not misleading and efficient contact information (e-mail, telephone).
- It adopts a pragmatic 'opt-out' regime through quick and workable unsubscribe requirements.
- It establishes a solid basis for enforcement by providing for a private right of action for either actual or statutory damages.

The Bill protects civil rights and provides for remedies for anyone who suffers loss or damage from spam, including statutory damages of SGD 25 per spam message subject to a cap of SGD 1 million. An opt-out approach has been recommended under the Bill, both for e-mail and mobile spam. Unsolicited fax and telemarketing are excluded. Improvements suggested by the industry during the consultation include:

- clarifying the definition of an 'unsolicited' commercial electronic message, so as to take into account pre-existing business relationships;
- encouraging the deployment of anti-spam technologies; and
- ensuring a tougher deterrent effect by allowing for increased enforcement (IDA 2005d).

ETA review

The Singapore government has undertaken a three-step exercise to review the Electronic Transactions Act (ETA) and its regulations, and to identify areas that need to be improved. The ETA was first enacted in 1998 to facilitate e-commerce transactions and to provide legal recognition of electronic signatures and records. With newer technology and solutions available since the initial enactment of the law, the consultation seeks to obtain public feedback on areas where ETA amendments are needed. Views are also sought on approaches to the regulation of certification authorities (IDA 2005e).

VoIP framework

The Singapore government announced a new policy framework governing Internet Protocol (IP) Telephony, a form of VoIP service in which a user can potentially use any broadband Internet access connection to make and receive local or international voice, data and video calls (regardless of location), with a phone number. Such solutions allow consumers to make local and international voice calls at a rate typically cheaper than rates for traditional fixed-line telephone calls. Under the framework, IDA will issue licenses and phone numbers for IP telephony services. The growth in IP telephony is expected to reduce the cost of providing telephone services, reduce prices and provide more choices for businesses and consumers in the long term. Facilities-based operators and services-based operators in Singapore can be licensed to provide the service. IDA's framework is intended to include minimal regulatory obligations so as to encourage the adoption of the emerging technology (IDA 2005f).

Telecommunication Competition Code

IDA released the amended Telecom Competition Code in February 2005. The amendment is part of a review every three years and takes into consideration feedback from the public and the industry on the state of competition in the Singapore market.

The basic principles of telecommunication competition remain unchanged. They are: reliance on market forces and proportionate regulation; regulation for effective and sustainable competition; minimum rules for consumer protection; technology neutrality; and efficient, transparent and reasoned decision-making. In the area of transparency, IDA will undertake public consultations in the review of regulatory frameworks and issue preliminary decisions on material policy or regulatory issues for comment by the public or interested parties. Greater clarity of procedures and standards will also be achieved through the publication of guidelines that stipulate the IDA processes.

The definition of dominance has been changed to one based on economic consequences, compared to the previous definition based on control over bottleneck facilities and ability to restrict output or raise prices above competitive levels. The new definition considers a player as dominant if it has operational control over facilities that are costly or difficult to replicate, or if it can exercise significant market power in providing telecom services in Singapore. Dominant licensees are now required to publish their prices on their websites, including all prices that have been approved by the regulator, such as discount structures, service availability, eligibility and termination clauses. IDA will also identify circumstances where dominant licensees have to offer services at wholesale prices and consult the public before mandating such prices. In segments where there is effective competition, wholesale services provided by licensees need to be provided at fair, reasonable and non-discriminatory prices.

IDA has also issued guidelines on dispute resolution and consolidation (that is, mergers and acquisitions) which set out in detail the framework and procedures that IDA will undertake in these circumstances, including how licensees may approach IDA to reconsider regulatory decisions (IDA 2005g).

Number portability

Following a public consultation in September 2005, IDA announced in August 2006 the implementation of a true number portability regime in Singapore so that consumers can benefit from greater choice and flexibility in mobile and fixed-line services. From the fourth quarter of 2007, consumers can switch between telecom service providers easily while maintaining full use of their existing number, rather than having to update family members, friends and business contacts about a new one. Beyond this, the new solution will result in greater competition among telecom service providers and more business opportunities in Singapore's telecom market.

The three key changes to the present regime, in place since April 1997, include:

- A common centralized database approach for operators to provide number portability for fixed-line services and post-paid and prepaid mobile services. The centralized database will be independently run and have open access.
- One number is all consumers will have when they switch between mobile service providers. Switching between mobile operators should also take about one day, compared with five days at present.
- Fixed-line operators must stop recurring monthly porting service charges. Consumers who want to port their fixed-line telephone number to a new fixed-line service provider will pay only a one-time administrative charge. Mobile service providers stopped such recurrent charges in June 2003 (IDA 2006l).

Internet regulation

As a multiracial society, Singapore places significant emphasis on maintaining a harmonious environment for different races and religions even on the Internet. Thus, MDA and its predecessor, Singapore Broadcasting Authority, have put in place since 1997, a class licensing scheme to address content issues in cyberspace. With rapid technology evolution, in 2005–06 the National Internet Advisory Committee (NIAC) also undertook a comprehensive review of the licensing scheme to take technological and market trends into account. An area given consideration was the application of the licensing regime to blogs. Declaring that blogs are no different from other types of content, NIAC applies the licensing scheme to blogs as to all other websites. Peer pressure and community norms would continue to play an important role in dealing with private blogs that may carry irresponsible or extreme views. NIAC also felt that the licensing framework should not apply to mass e-mail, so as not to stifle the *bona fide* use of e-mail as a form of communication.

Recognizing the growing pervasiveness of mobile phone usage and the availability of Internet access on mobile phones, NIAC also commended a 'Voluntary Code for Self-Regulation of Mobile Content' developed by the major mobile phone operators in March 2006 as a step forward to protect users from undesirable Internet and new media content (MDA 2006b).

Open source movement

Singapore continues to adopt a technology-neutral and pro-competition approach to the open source movement. Both open source and commercial software solutions are deployed in the industry and the government in a complementary manner. Software choices and selections are made by individual purchasers based on individual needs, merits and requirements. Purchasers recognize the benefits and strengths of both open source and commercial software. While many installations continue to use commercial software, a notable user of open source software is the Ministry of Defence, which is increasingly using OpenOffice in place of other commercial office suites. The National Library has also conducted trials of Linux desktops.

The Singapore Open Source Alliance (SOSA) has also been established. It is a consortium of global and local IT vendors such as Apple, Hewlett-Packard, IBM, Intel, Novell, Oracle, Red Hat, Resolvo Systems and Sun Microsystems, whose goals include maintaining an updated list of locally supported open source software and a hardware compatibility matrix, inviting CIOs for regular discussions to provide industry feedback on macro policies, and highlighting applications of open source systems (Computerworld 2005).

The Linux User Group of Singapore continues to be a resource for users regarding open source trends and issues.

Research and development

The Singapore government has announced that it will commit SGD 7.5 billion from 2006 to 2010 to sustain innovation-driven growth through economic-oriented R&D supporting the key industry clusters. The plan is set out in the Ministry of Trade and Industry's (MTI) Science & Technology Plan 2010 (STP2010). The broad programmes under the Plan include:

- developing the research talent in Singapore;
- strengthening and deepening our research capabilities;
- promoting private sector R&D; and
- providing infrastructure support.

The initiative is part of the broader strategy to make more substantial investments in R&D in order to increase the national R&D spending to 3 per cent of GDP. The National Research Foundation will coordinate the different research areas under the larger national framework and develop policies and plans to implement the strategic thrusts for the national R&D agenda. The promotion will be done through the Agency for Science, Technology and Research (A*STAR) and the Economic Development Board. A*STAR will develop and sustain a substantial pipeline of research talents to meet industry needs. It will also build concentrations of R&D in areas that are most relevant to developing key industry sectors. A*STAR will also encourage the integrated management and commercialization of the intellectual property arising from research at the research institutes to optimize their economic impact (MTI 2006).

Future trends

IDA has unveiled its Fifth Infocomm Technology Roadmap (ITR5) which sets out the infocomm landscape over a 10-year period. The Roadmap, which was developed through research involving visionaries, industry players, academics and government leaders, sets out how major technologies can be deployed to help Singapore address the key challenges of economic growth, national security and population demographics in the next decade. It identifies innovations in nanotechnology and biotechnology, key areas that will enable people-centric technologies. Context-aware sensors and intelligent agents will automate, analyze, synthesize and present personalized information. The result will be the creation of a world of things that think, transforming the way we live, learn, work and play. Sentient technologies are expected to make a dramatic difference and improve key sectors such as health care. They would also create smart homes and entertainment applications.

The Roadmap highlights the coming of three waves:

- *The Computing Wave*—PCs would disappear by 2015, and mainframes would be accessible through nanotechnology.
- *The Communications Wave*—The world would be covered by fibre with almost unlimited capacity. Broadband would be available everywhere.
- *The Sentient Wave*—The first two waves combined will create intelligent devices that can sense and interact with each other (IDA 2005h).

References

Computerworld. (2005). *Dear SOSA*. Retrieved 22 January 2007 from http://computerworld.com.sg/ShowPage.aspx?pagetype=2&articleid=1893&pubid=3&issueid=56

Computerworld. (2006). *Singapore to set up e-Government Leadership Centre*. Retrieved 22 January 2007 from http://www.computerworld.com.sg/ShowPage.aspx?pagetype=2&articleid=3844&pubid=3&tab=Home&issueid=92

Infocomm Development Authority of Singapore. (2004). *World cyber games 2004*. Retrieved 22 January 2007 from http://www.ida.gov.sg/News%20and%20Events/20050718123855.aspx?getPagetype=21

Infocomm Development Authority of Singapore. (2005a). *Annual survey on Infocomm industry for 2005*. Retrieved 22 January 2007 from http://www.ida.gov.sg/Publications/20061205102839.aspx

Infocomm Development Authority of Singapore. (2005b). *Three-year Infocomm Security Masterplan unveiled*. Retrieved 22 January 2007 from http://www.ida.gov.sg/News%20and%20Events/20050712110643.aspx?getPagetype=20

Infocomm Development Authority of Singapore. (2005c). *Grooming talent to make it happen*. Retrieved 22 January 2007 from http://www.ida.gov.sg/News%20and%20Events/20050705154150.aspx?getPagetype=20

Infocomm Development Authority of Singapore. (2005d). *IDA & AGC seek second round views on proposed Spam Control Bill for Singapore*. Retrieved 22 January 2007 from http://www.ida.gov.sg/News%20and%20Events/20050705172846.aspx?getPagetype=20

Infocomm Development Authority of Singapore. (2005e). *IDA and AGC review Electronic Transactions Act*. Retrieved 22 January 2007 from http://www.ida.gov.sg/News%20and%20Events/20050706162941.aspx?getPagetype=20

Infocomm Development Authority of Singapore. (2005f). *IDA launches new policy framework for Internet Protocol (IP) telephony*. Retrieved 22 January 2007 from http://www.ida.gov.sg/News%20and%20Events/20050706170936.aspx?getPagetype=20

Infocomm Development Authority of Singapore. (2005g). *Telecoms competition code offers greater regulatory transparency and clarity*. Retrieved 22 January 2007 from http://www.ida.gov.sg/News%20and%20Events/20050712112923.aspx?getPagetype=20

Infocomm Development Authority of Singapore. (2005h). *IDA shares vision of infocomm landscape with inauguration of 10 Year Infocomm Technology Roadmap*. Retrieved 22 January 2007 from http://www.ida.gov.sg/News%20and%20Events/20050711175041.aspx?getPagetype=20

Infocomm Development Authority of Singapore. (2005i). *Multilingual domain names now available for six-month trial!* Retrieved 22 January 2007 from http://www.ida.gov.sg/News%20and%20Events/20050706153720.aspx?getPagetype=20

Infocomm Development Authority of Singapore. (2006a). *Annual survey on infocomm usage in households and by individuals for 2005*. Retrieved 22 January 2007 from http://www.ida.gov.sg/Publications/20061207182001.aspx

Infocomm Development Authority of Singapore. (2006b). *Statistics on telecom services for 2006*. Retrieved 22 January 2007 from http://www.ida.gov.sg/Publications/20061205181639.aspx

Infocomm Development Authority of Singapore. (2006c). *Singapore iN2015 Masterplan offers a digital future for everyone*. Retrieved 22 January 2007 from http://www.ida.gov.sg/News%20and%20Events/20050703161451.aspx?getPagetype=20

Infocomm Development Authority of Singapore. (2006d). *IDA invites private sector to propose concepts for Singapore's next generation national broadband network*. Retrieved 22 January 2007 from http://www.ida.gov.sg/News%20and%20Events/20060417173050.aspx?getPagetype=20

Infocomm Development Authority of Singapore. (2006e). *Two years of free Wi-Fi for Singapore*. Retrieved 22 January 2007 from http://www.ida.gov.sg/News%20and%20Events/20061013120532.aspx?getPagetype=20

Infocomm Development Authority of Singapore. (2006f). *Singaporeans to enjoy the convenience of one card for electronic payments—IDA, SPRING Singapore, LTA, and Industry launch world's first nationwide interoperable micro-payment platform using Singapore Standard SS 518 CEPAS*. Retrieved 22 January 2007 from http://www.ida.gov.sg/News%20and%20Events/20050829100306.aspx?getPagetype=20

Infocomm Development Authority of Singapore. (2006g). *Singapore's e-government on the next lap with iGov2010*. Retrieved 22 January 2007 from http://www.ida.gov.sg/News%20and%20Events/20050704121154.aspx?getPagetype=20

Infocomm Development Authority of Singapore. (2006h). *Singapore delivers games to Asia's $14billion video game market*. Retrieved 22 January 2007 from http://www.ida.gov.sg/News%20and%20Events/20050704144250.aspx?getPagetype=20

Infocomm Development Authority of Singapore. (2006i). *IDA survey also shows Singaporeans are becoming more sophisticated in Internet usage*. Retrieved 22 January 2007 from http://www.ida.gov.sg/News%20and%20Events/20050705120931.aspx?getPagetype=20

Infocomm Development Authority of Singapore. (2006j). *iN2015*. Retrieved 22 January 2007 from http://www.in2015.sg

Infocomm Development Authority of Singapore. (2006k). *Number of Infocomm professionals reached new high in 2005*. Retrieved 22 January 2007 from http://www.ida.gov.sg/News%20and%20Events/20050705104523.aspx?getPagetype=20

Infocomm Development Authority of Singapore. (2006l). *IDA enhances number portability regime to benefit consumers*. Retrieved 22 January 2007 from http://www.ida.gov.sg/News%20and%20Events/20050829134538.aspx?getPagetype=20

Infocomm Development Authority of Singapore. (2006m). *Real-time response to cyber-threats by government*. Retrieved 22 January 2007 from http://www.ida.gov.sg/News%20and%20Events/20050906111323.aspx?getPagetype=20

Infocomm Development Authority of Singapore. (2006n). *Government continues to invest in technology-enabled work environment*. Retrieved 22 January 2007 from http://www.ida.gov.sg/News%20and%20Events/20050829143504.aspx?getPagetype=20

Infocomm Development Authority of Singapore. (2006o). *Bridging digital divide*. Retrieved 22 January 2007 from http://www.ida.gov.sg/News%20and%20Events/20061212162944.aspx?getPagetype=21

Media Development Authority. (2006a). *Singapore pavilion at Broadcast Asia presents the latest in digital technology*. Retrieved 12 August 2006 from http://www.mda.gov.sg/wms.www/thenewsdesk.aspx?sid=719

Media Development Authority. (2006b). *NIAC emphasis co-ordinated and multi-pronged approach in making Internet safe for all*. Retrieved 26 October 2006 from http://www.mda.gov.sg/wms.www/thenewsdesk.aspx?sid=721

Ministry of Information, Communications and the Arts. (2006). *Opening address by Dr Vivian Balakrishnan at the 10th Infocomm Commerce Conference*. Retrieved 5 September 2006 from http://www.mica.gov.sg/pressroom/press_060817.htm

Ministry of Trade and Industry. (2006). *MTI announces Science & Technology 2010 Plan*. Retrieved 12 August 2006 from http://app.mti.gov.sg/default.asp?id=148&articleID=2461&intViewCat=1&intCategory=4&txtKeyword=&txtStart=&txtEnd=&intOrderBy=1&intYear=&intQuarter=0

National Research Foundation. (2006). *The biopolis of Asia*. Retrieved 12 August 2006 from http://www.biomed-singapore.com/bms/sg/en_uk/index/newsroom/pressrelease/year_2006/7_jul_-_research_.html

Singapore Network Information Centre. (2006). *Internationalised domain names (IDA) test bed*. Retrieved 26 October 2006 from http://www.idn.sg

Statistics Singapore. (2006). *Key stats—annual statistics*. Retrieved 12 August 2006 from http://www.singstat.gov.sg/keystats/annual/indicators.html

.kr
South Korea

Jong Sung **Hwang** and Jihyun **Jun**

Total population	48,846,823 (CIA July 2006)
GDP per capita	USD 16,291
Key economic sectors	Services
Computers per 100 inhabitants	53.2
Fixed-line telephones per 100 inhabitants	57 (2004)
Mobile phone subscribers per 100 inhabitants	79.4 (ITU 2005)
Internet users per 100 inhabitants	74.8 (NIDA December 2006)
Domain names registered under '.kr'.	784,199 (NIDA March 2007)
Broadband subscribers per 100 inhabitants	26.4 (OECD June 2006)

Technology infrastructure

In the early 1980s, South Korea (referred to as Korea in this chapter) was a developing nation at best. For example, the wire phone penetration ratio was a mere 7.2 per cent in 1980 (MIC 2003). To upgrade its telecommunication infrastructure, the Korean government imported electronic switches from overseas as a transitional alternative, but embarked on a local electronic switch (TDX-1) development project in 1981. Five years later, Korea became the 10th electronic switch-producing nation and finally succeeded in improving the phone penetration ratio to an average of one phone per household in 1987.

As Internet connection service debuted in 1994 and broadband network service became commercially available in earnest in 1998, the number of Internet users increased exponentially from 3 million persons in 1998 to 33.01 million persons by the end of 2005, corresponding to 72.8 per cent of the population.

All primary, middle and high schools now have access to broadband connection and all cities in Korea are connected by fibre optic cable. Broadband network has now penetrated even to rural villages. More than KRW 8,000 billion (around USD 8.6 billion) of public funds has been invested in the nationwide broadband implementation project that went on for 11 years (1995–2005). As a result, broadband communication infrastructure now connects 144 coverage areas across the country with up-to-date fibre optic cable. There are five Internet eXchanges (IX) in Korea and KIX (www.kix.ne.kr) of the National Information Society Agency (NIA) serves as a gateway for non-commercial networks of government authorities and public institutions. In addition, 79 service providers offer commercial Internet access services.

International submarine fibre optic cable connects different countries across national boundaries to enable mass data transfer over a long distance at dozens of Tbps. Submarine cable switching stations are located in seven areas (Busan, Geoje, Goheung, Namhae, Hosan, Wooleung and Taeahn) and the country is now connected to 10 international submarine fibre optic cables (APCN, APCN-2, CUCN, C2C, EAC, FEA, etc.) whose aggregate capacity equals around 19 Tbps.

Thanks to a series of advances in the evolution and penetration of IT network services, Korea is now at a critical juncture toward active utilization of ubiquitous IT technologies. The Korean government is giving priority to the implementation of a Broadband Convergence Network (BcN), which is a service integration network that enables customers to use multimedia services converging voice, wire, wireless communication and broadcasting services anywhere and anytime based on an open API platform that allows service providers to deliver diverse services with ease. The government has invested around KRW 500 billion (about USD 539.34 million) in pilot BcN projects, core technology development initiatives and standardization efforts in Phase 1 (2004–05). Phase 2 of the programme will focus on developing such service models as BCS (Broadband Convergence Service), u-Work, u-Learning and u-City, and support commercialization of the services identified in Phase 1 in an effort to promote BcN service utilization from the customer's perspective.

An address shortage problem is foreseen with the current reliance on Internet protocol 4 (IPv4). Although the number of Internet users is increasing at a much slower pace since it exceeded 30 million, demands for IPTV addresses are increasing continuously with projects involving new technologies such as BcN, home network, portable Internet and RFID. The Ministry of Information and Telecommunication is trying to resolve the Internet address shortage by launching a next-generation Internet programme focusing on the promotion of IPv6, which provides almost limitless Internet address resources.

Key institutions dealing with ICT

The Korean government overhauled the Ministry of Posts and Telecommunications into the Ministry of Information and Telecommunication (MIC) in 1994, bringing together government functions relating to ICT and the broadcasting sectors in the Ministry of Commerce, Industry and Energy, the Ministry of Science and Technology, and the Public Information Agency.

Government policies relating to information and communication are developed and implemented through deliberation and coordination by and between the Informatization Promotion Committee (IPC), the Special Committee on e-Government (under the Presidential Committee on Government Innovation and Decentralization), the Ministry of Information and Communication (MIC) and other central government authorities, local municipalities and public organizations, including the NIA.

To implement and integrate ICT policies across different government authorities, the IPC was formed in June 1996 under the Basic Act on Informatization Promotion (BAIP). The committee is a top-level decision-making authority headed by the Prime Minister. It has 24 Ministerial-level representatives, the Minister of Finance and Economy as Vice-Chairman and the Chief Assistant to the Prime Minister as Secretary. It is mandated to develop, coordinate and assess policies, plans and programmes to promote the penetration of ICT within the nation. It is also tasked to foster the implementation and utilization of broadband ICT platforms and to determine operational guidelines for funds for facilitating informatization.

Each central government authority has an internal function responsible for planning and implementing information and communication initiatives within the bounds of its competence. In most central government authorities, this function belongs to a department (or an individual officer) of the planning and management office. However, some authorities, such as the Ministry of Education and Human Resources Development, the Ministry of National Defence, the Ministry of Government Administration and Home Affairs and the National Tax Service, operate permanent IT organizations headed by directors-general.

In addition, to effectively oversee and coordinate information and communication initiatives within each government authority and to promote utilization of ICT in government administration and public service, the government in 1997 required each government authority to appoint an IT professional as a CIO (chief information officer).

The following agencies render in-depth assistance in policy planning and implementation to the MIC:

- The NIA drafts government policies on information and communication and provides support for the development and enforcement of the informatization promotion master plan and implementation plans thereto.
- The Korea Agency for Digital Opportunity and Promotion (KADO), which was established under Article 6 of the Act on Closing the Digital Divide, performs various research and assists with policy development to close the digital divide within the nation.
- The Korea Information Security Agency (KISA) is responsible for improving information security, promoting information security awareness, responding to information security violations, safeguarding the IT infrastructure, assessing information security product quality and developing information security technologies.

Digital content initiatives

Access to online information is almost universal in both the public and private sectors and a wide variety of digital content on various subjects, including economy, society, welfare and culture, is now available in Korea. The following are just a few of the most representative examples of content digitization in the Korean public sector.

National knowledge contents digitization strategy

To foster the digitization of national knowledge content in a systematic manner, the Korean government enacted the Knowledge and Information Resource Management Act in 2002 and kicked off the Public Knowledge and Information Resource Management Project. Knowledge databases storing information resources in five key strategic sectors—science and technology, education and academics, culture, history, information and communications—as well as content in such domains as industrial economy, construction technology, and maritime affairs and fisheries, have been built for use by the general public.

Because of the high cost of comprehensive digitization programmes, the Korean government has developed an automatic knowledge and information resource registration system to turn knowledge and information into online digital format immediately upon creation. Two hundred and one automatic knowledge and information resource registration system sites were deployed in 2004. A year later, an additional 100 sites were in place.

Public service information database

As e-government services expanded and government information publication services began to take root, public demand for government information increased dramatically. Thus, in his administrative policy address to the National Assembly (on 25 October 2004), President Roh Moo-hyun stressed the need for a major investment in IT infrastructure and underlined the importance of 'migrating national databases in the public sector onto a next-generation network platform'. In November 2004, the IT Sector Investment Enhancement Strategy was announced during the joint policy workshop involving the Presidential House, the government and the ruling party; it mandated MIC and the Ministry of Government Administration and Home Affairs (MOGAHA) to launch in 2005 the Year 1 projects envisioned by the strategy.

The government information database implementation project aims to migrate public service information from administrative authorities to databases. The project is designed to prevent duplication in IT infrastructure investment by the government and to maximize returns on investment by incorporating interfaces with other IT implementation projects already launched by other government agencies and the e-government initiatives.

In 2005, eight projects for improving administrative efficiency, 14 projects for improving the quality of public service, and 11 projects for preserving government documents and artefacts were implemented. A total of 1,383,402 person hours was spent on digitizing 91,024,674 information items throughout the year, which also contributed to easing unemployment among the youth. In March 2006, the Database Standardization Guideline was issued to replace the different database standards among government agencies, to encourage them to leverage their databases more and to ensure interoperability among different government databases.

Korea knowledge portal

In November 2001, KADO opened a national knowledge portal (Korea Knowledge Portal, www.knowledge.go.kr) that as of late 2005 consisted of more than 25 million records stored in databases owned by 908 public organizations. The portal offers a comprehensive information search service encompassing a variety of fields integrating five comprehensive information centres and more than 900 in-depth information repositories.

To complement the national knowledge portal service, key government ministries, including the Ministry of Science and Technology, the Ministry of Education and Human Resources Development, and the Ministry of Culture and Tourism, have joined hands to designate and operate comprehensive information centres that would deliver in-depth knowledge content in the five strategic areas of science and technology, education and academics, culture, history, and information and communication.

Online job information services

Several hundreds of public and private online recruitment websites connecting potential employers and job seekers are now operating in Korea. The most representative online employment information service portal in the public sector is Work-Net run by the Korea Employment Information Service (KEIS). KEIS recently launched a mobile version of Work-Net to provide through a mobile platform (for example, a mobile phone or PDA) information on job openings, tips for finding jobs and employment-related news articles.

Art, culture and history content online

With increased interest in arts and culture in cyber space among the general public, the government needed to promote opportunities to appreciate art and culture online and develop an efficient management framework for knowledge and information resources relating to art and culture. Thus, the Ministry of Culture and Tourism runs an art and culture information portal (Art Way, www.art.go.kr) that classifies and presents a wide variety of digitized information from art and culture institutions and organizations. Art Way provides access to more than 1.76 million records of art and culture-specific information, 430,000 art and culture images and 3,562 video clips showing performances at the National Theater of Korea and the National Centre for Korean Traditional Performing Arts (Korea Culture Information Service 2005).

A comprehensive cultural properties information service portal (Korean National Heritage Online, www.heritage.go.kr) provides a single point of integration and management for information relating to cultural properties, national treasures and historical sites across the country. Managed by the Korean National Heritage Online, the portal provides access to 630,000 pieces of information on Korean cultural heritage from 105 institutions, including the Cultural Heritage Administration, the National Museum of Korea and the War Memorial. The in-formation was collected through six project phases from 2000 to 2005. The e-Museum (www.emuseum.go.kr) service has added a search and theme function to the comprehensive museum information service. The online museum portal also offers an opportunity for a 3D immersion experience with its cyber museum service that shows more than 4,500 major cultural artefacts on exhibit in 72 museums across Korea.

Other cultural content services online are: tourism information (www.visitkorea.or.kr), KOLIS-NET or the Korea Library Information System Network (www.nl.go.kr/kolisnet), and Q-NET, which provides job certification and license information.

ICT and ICT-related industries

The backbone of the Korean economy in the 1960s was the light industry. In the 1970s, it was the heavy and chemical industry, and in the 1980s to 1990s the electronic appliance and automobile industries. In the 21st century this distinction belongs to the ICT industry. In 2004, the ICT industry accounted for 48.15 per cent of real GDP in Korea (MIC 2005).

Recently, the aggregate growth has slowed down a little as the market has matured. However, because some sectors, such as the mobile data service sector, are still growing by leaps and bounds and new services are debuting one after another, the ICT industry will remain a key growth engine for the national economy in the future.

The volume of aggregate exports by industrial sector in Korea reached USD 284.4 billion in 2005. But the export growth rate slipped somewhat to 12 per cent. On the other hand, the aggregate industry import volume rose by 16.4 per cent from the previous year to USD 261.2 billion, which resulted in an international trade surplus of USD 23.2 billion.

The export growth rate of the ICT industry fell to 9.25 per cent and the industry posted USD 48.4 billion in trade surplus. However, this figure accounted for 209 per cent or more than twice the aggregate trade surplus of USD 23.2 billion in the industrial sector as a whole in 2005, which indicates that the ICT industry makes up for the trade deficits suffered by the other industries.

The ICT appliance industry posted USD 43.88 billion in trade surplus in 2005, up by 10.1 per cent from USD 43.93 billion in 2004 and showing a steady stream of trade surplus. Notably, mobile communication devices accounted for a sizable portion of the positive trade balance. The trade surplus from the sale of digital TV, semiconductors and display panes is also increasing.

Enabling policies and programmes

The Korean government embarked on building the information infrastructure in 1987 through the National Backbone Information Network Project, which sought to enhance public sector efficiency and lay the ground for the advancement of the information and communication industry in five key strategic areas, including government administration, finance, education, national defence and law enforcement. The government then kicked off the Government Administration Information Service Project which implemented an online network connecting all government administration authorities.

In the mid-1990s, with a strong commitment to staying ahead of others in the information revolution, the Korean government launched a government-wide information and communication programme. In March 1999, to get over the Asian financial crisis and prepare for the emergence of the knowledge-based economy, the government developed a policy blueprint called Cyber Korea 21. In 2002, e-Korea Vision 2006 was launched with the goal of positioning Korea as a global IT leader.

In 2003, new changes and challenges, including the inauguration of the new administration, major online security violation scandals and an economic slump, made it necessary for the Korean government to formulate a new vision for information and communication in Korea. Accordingly, the government announced Broadband IT Korea Vision 2007.

The successor to Vision 2007 that now underpins government projects is the u-Korea Master Plan, which seeks to build the first ubiquitous society (u-society) where people can access information anywhere, anytime using the best ubiquitous communication infrastructure (u-Infrastructure). Under the Plan, the government intends to achieve the five key visions of Friendly Government, Intelligent Land, Regenerative Economy, Secure and Safe Social Environment and Tailored u-Life Services. It also intends to optimize four major engines: Balanced Global Leadership, Ecological Industrial Infrastructure, Streamlined Social Infrastructure and Transparent Technological Infrastructure.

By leveraging information and telecommunication technologies to satisfy emerging socio-economic demands, and by repositioning Korea as the leading IT superpower in the world, the u-Korea Master Plan (2006–10) provides a vision to 'Build the 1st u-Society on the BEST u-Infrastructure'. Specifically, the government is targeting advancements in five domains—public administration, national land, economy, society, and personal life—and the optimization of four engines encompassing globalization, industrial infrastructure, social institution and framework, and technology development.

As the IT industry evolves further, conventional barriers between different industries and products are fading away and a new momentum for new industries is in the making. To turn such a momentum into a new national growth engine, MIC is pressing ahead with the 'IT839 Strategy' to introduce innovative services and premium infrastructure that are the first and best of their kinds in the world, and to develop top-notch products to contribute to the growth of the national economy. The IT839 Strategy, which was released in February 2005, refers to a strategy to stimulate the advancement of the IT industry by promoting eight new services to attract investment in three key wire/wireless communication and broadcasting infrastructure and to induce the growth of nine state-of-the art device, terminal and software content industries.

The eight services are: WiBro, DMB, Home NW, telematics, RFID-based, W-CDMA, terrestrial DTV and Internet telephony. The three key infrastructures are broadband convergence network, ubiquitous sensor network and next-generation Internet protocol (IPv6). The nine industries are: mobile telecommunications handset and equipment, digital TV and broadcasting devices, home network devices, IT system-on-chip, next-generation PC, embedded software, digital content and software solutions, telematics devices and intelligent service robots.

As the IT839 Strategy progressed from the initial market formation phase to the commercialization phase, it evolved into the u-IT839 strategy and the strategic items were readjusted to strengthen the interfaces between the eight services, three infrastructures and nine new growth engines, and to enhance the focus on software policies.

Legal and regulatory environment for ICTs

The legal framework for information and communication in Korea is founded on the Framework Act on Informatization Promotion, which consists of legislation pertaining to three areas: promotion of informatization across society, development of the ICT industry and advancement of ICT infrastructure. The Framework Act enabled the government to integrate or coordinate functions relating to information and communication across different government organizations to form a basis for more efficiency and consistency in information and communication projects at the national level.

More recently, refinements in Korea's legal and regulatory environment for information and communication are expected to create more value and enhance efficiency and productivity. For example, in line with the penetration of e-banking services and the advent of new payment instruments such as e-cash, the government is planning to enact the 'e-Financial Transaction Act' to clarify the rights and obligations of the parties to online financial transactions in consideration of the characteristics of such transactions. In addition, as violations of personal privacy become a serious social issue in the wake of rapid development of IT, the government intends to enact the 'Act on Security of Personal Information Processed by Computers in IT Networks'. The objective is to protect the privacy of individuals from potential violations resulting from the increase in online use of personal information, and to overhaul relevant legal and regulatory arrangements to prepare for privacy violations that may occur as convergence and advancement of new technologies such as RFID proceeds and 'the intelligence-based society emerges' (MIC 2005).

Education and capacity building

The government tried to improve the information and communication capability of the general public in 2000 through the 'IT Education and Training Plan for 10 Million Persons'. Ten government authorities, including MIC, jointly launched the training programme. By 2002, 13.8 million had joined the IT education and training courses offered. Later, the government sought to transform the public into creative e-Koreans and 12 government organizations, including MIC, kicked off Phase 2 of the National IT Education and Training Programme.

However, in spite of the programme, the digital divide between the majority and segments of the national population with little access to information and communication services, including the elderly, the physically challenged, the poor, farmers and fisher folk, became a social issue (MIC 2005). In response, the government developed the 'Mid- to Long-Term Plan for Reducing the Digital Divide' in 2004. Its goals were to ensure equal access to information and communication services and to educate and train 5 million persons deprived of information and communication services by 2008. A total of 1.2 million and 1.3 million elderly, persons with disabilities, low-income earners, and farmers and fisher folk were trained in 2004 and 2005, respectively.

Open source initiatives

The government is spearheading the distribution and promotion of open source software (OSS) by developing the Korean OSS standard platform named Booyo, launching open code software pilot projects, and operating technical assistance centres. It also recommends a preview process prior to adoption of OSS.

In the case of embedded software, corporate buyers are favouring foreign vendors due to the small number of Korean embedded software vendors and concerns about engineering support. In addition, the embedded software vendors are having difficulty recruiting new engineers. MIC is trying to develop and distribute standard embedded software platforms that are applicable universally to a variety of ICT devices by focusing on the promotion of the embedded software industry and operating the Embedded Software Technology Support Centre. The latter aims to provide an integrated test environment for small embedded software vendors and help them meet technological challenges.

Research and development

The government will invest approximately KRW 960 billion (about USD 1 billion) to improve the competitive edge of the Korean IT industry in 2007. The allotment is 12 per cent more than the 2006 allotment of KRW 860.4 billion. Specifically, investments in technology development will increase from KRW 633.7 billion to 707.6 billion, in human capital development from KRW 107.8 billion to 114.5 billion, in infrastructure deployment from KRW 88.7 billion to 107.8 billion, and in standardization from KRW 30.2 billion to KRW 33.5 billion. Technology development will focus on convergence among information technology, broadband technology, network technology, critical hardware parts, materials and software. Human capital development will underscore better IT education infrastructure in colleges and education programmes that focus more on practical and field experience. In terms of standardization, the government will enhance partnerships among Korea, China and Japan, and beef up the clout of Korea as a global IT standard setter. In infrastructure, the government will focus on establishing a u-IT hub and expanding common services for small and medium IT venture firms.

The IT Industry Competitiveness Promotion Programme was launched by MIC in 1999 to sponsor industrial technology development proposals of small and medium IT venture firms in recognition of their technological value and industrial innovations. Until 2005, the policy initiative made a tremendous contribution to promoting technology-innovating IT venture firms by extending a total of KRW 240 billion (about USD 258.7 million) in support of more than 1,200 technology development programmes. In 2006, MIC approved 59 out of 138 proposals by IT venture firms, including the 'Mobile Wi-MAX Terminal Chipset Development Programme', and extended KRW 15 billion in sponsorship funds.

Challenges

From the early days of the ICT revolution in Korea, the government established a government-wide implementation apparatus involving the legislature and the judiciary, with IPC as the steering body. The government also built a consensus on the need for ICT services across the entire spectrum of government agencies. In addition, the enforcement of the Framework Act on Informatization Promotion in 1995, the establishment of the Informatization Implementation Committee and Informatization Promotion in 1996, and the organization of the Informatization Strategy Forum in 1998 were significant milestones towards a single nationwide apparatus for ICT promotion. Notably, the establishment of MIC in December 1994 as a single government agency responsible for planning ICT promotion strategy, developing the ICT industry and regulating communications service industry had no precedence in any other country.

The ability to put forward a vision for ICT in society in response to various changes was also one of the factors that ensured the success of the ICT revolution and national development in Korea. By providing a comprehensive and systematic long-term vision for the nation from the early days of nationwide ICT transition, the government created demand for ICT adoption and generated ICT investment in the private sector as a pioneer of transition to ICT in Korea.

ICT development in Korea is the product of a virtuous cycle involving the ICT promotion drive of the government, the ICT industry and the communications service infrastructure. The government deserves credit for facilitating the transition to ICT adoption by creating demand for ICT and leveraging this demand to encourage growth in the ICT manufacturing sector.

Lastly, aggressive initiatives in advancing ICT adoption in consideration of the national and cultural disposition seem to have worked. The unusually high receptiveness of Koreans to the latest technologies and the 'education fever' has turned Korea into a global ICT superpower in just 20 years. The Internet cafés all over Korea have had a tremendous influence on the prompt penetration of the broadband network. In addition, government commitment to ICT training programmes such as the 'ICT Training Programme for 10 Million Koreans' prepared the public to be able to access and utilize digital knowledge and information in various ways.

However, in spite of the successful track record of ICT promotion to date, there are issues awaiting resolution if Korea is to sustain its ICT-leveraged development in the future. First, while Korea has succeeded in implementing a world-class network infrastructure ahead of other nations, it lacks a policy and strategies for protecting IT users and reducing the digital divide. This policy gap must be filled especially given the rise of ubiquitous technologies and the campaign for u-Korea (MIC 2006).

Second, while the IT industry in Korea has achieved breakthroughs in hardware (devices), large businesses and systems, the imbalance between hardware and software, between large corporations and small businesses, and between systems and parts persists, and the global competition with China and other competitors is intensifying as the growth of the global IT market slows down.

Third, a proactive policy focus on market competition has resulted in an investment boom in the telecommunications industry and in IT industry development. However, the market outlook remains unclear due to a falling growth rate in the maturing market and uncertainty in the legal framework for convergence between communication and broadcasting services.

Fourth, the IT industry has emerged as the key export driver of Korea, thanks to aggressive global marketing. However, advances into promising emerging markets have been relatively insufficient.

Moving beyond the previous milestones of communication infrastructure development and ICT adoption and utilization, Korea is now bracing itself for the advent of the era of ubiquitous ICT. As the dispersal of the broadband network brought about tremendous opportunities several years ago, the arrival of ubiquitous ICT is likely to create the momentum for further growth and development. The Korean government needs to define strategic responses to sustain industrial growth and expand the market, and to resolve outstanding social issues, in the ubiquitous ICT environment.

References

Korea Culture Information Service. (2005). *Trends in cultural informatisation in major countries*. December.

MIC. (2003). *IT policies of Korea*.

MIC. (2005). *Informatisation annual report 2005*.

Sri Lanka

Nalaka **Gunawardene**

Introduction

'Smart people, smart island' was the promotional tag-line for the ambitious e-Sri Lanka programme that Sri Lanka launched in 2003 as a major ICT development initiative. It provided a vision and a roadmap based on the premise that 'ICT is a foundation medium for the equitable distribution of opportunity and knowledge....' Proponents talked of Sri Lanka being on the threshold of a new wave of social and economic development built on peace, equity and social harmony (ICTA 2003).

When 2006 ended, however, that promise lay unfulfilled as the 2002 ceasefire was no longer able to prevent the two decade long ethnic conflict from erupting again, and the peace process was on hold. In fact, this became a 'double whammy' because the country suffered substantial loss of life and property damage from the Indian Ocean Tsunami of December 2004, the recovery from which is prolonged and incomplete. Social and political instability, and the over-stretching of the state to cope with multiple emergencies, has forced many development efforts to be scaled down or deferred.

Thus, Sri Lanka is once again pursuing development under duress, a condition it has known well during the past 30 years. It is within this broad context that ICT development is taking place.

Technology infrastructure

Telephony

Since the telecommunications sector was liberalized and deregulated in the early 1990s, both the telecommunication and information infrastructure of Sri Lanka have recorded a steady and healthy growth. The Central Bank of Sri Lanka recognizes the telecom sector as one of the most liberal, competitive and fast growing sectors in the economy. Its 2005 annual report notes that vigorous competition prevailed among the three fixed phone operators, four mobile phone operators, 32 external gateway operators, 29 data communication and Internet service providers, four paging operators, two payphone operators and two trunk radio operators (CBSL 2006).

The total telephone subscriber base of around 120,000 land phones when reforms started in 1990 has now increased by more than tenfold. The number of land phone lines (both fixed and WLL) stood at 1,509,000 in June 2006 (TRCSL 2006). However, the growth rate has not been consistent during this period. A steady increase that was recorded for a few years following initial reforms climaxed with a growth rate of 53 per cent in 1998. It gradually decreased to single digit growth rates in the period 2001–04, when the overall economic growth rate itself slowed down. Telecom growth has picked up again from 2005, when land phones using the CDMA technology were introduced.

The subscriber network expanded by 45 per cent in 2005, with the fixed access network growing by 26 per cent and the number of mobile telephone subscribers increasing by 54 per cent. The vast expansion and stiff competition has driven operators to offer more affordable user rates, enabling more low income groups and rural residents to access these services. Meanwhile, the introduction of CDMA has reduced the urban–rural disparity in telephone service penetration (CBSL 2006).

It is mobile telephones that have transformed the Sri Lankan telecommunications sector beyond recognition. Mobile services have experienced exponential growth, from 71,000 subscribers in 1996 to 4,284,256 subscribers in June 2006 (TRCSL 2006). These growth rates did not dip even when the overall economy went through periods of lean or zero growth. The number of mobile phones surpassed the number of land phones in 2002 and by end 2005, three out of every four phone connections in Sri Lanka were mobile.

By June 2006, the teledensity (number of phones per 100 inhabitants) rates were 7.5 for fixed phones and 21.5 for mobile phones. Overall teledensity was 29.1 (TRCSL 2006). While this is impressive for a South Asian country with a per capita GNP of USD 1,189, it should not distract from many gaps that remain in telecom coverage, service quality and dependability.

Meanwhile in August 2006, Sri Lanka became the first South Asian country to introduce commercial 3G services for mobile telephony—another first by Dialog Telekom, the largest mobile operator in the country.

Internet

Sri Lanka was the first South Asian country to introduce unrestricted, commercial Internet connectivity in April 1995. Despite this head start, penetration has been slow and uneven in the 11 years since. The Telecommunications Regulatory Commission estimated the number of Internet and e-mail subscribers to be 125,800 as of June 2006. The actual figure could well be 2–3 times that, because this estimate does not take into account the shared corporate accounts. The International Telecommunications Union cited a figure of 1.3 Internet users per 100 inhabitants (ITU 2006a).[1]

Broadband in Sri Lanka is largely limited to the capital Colombo and a few key cities. Both ADSL and ISDN facilities are available in urban areas and one operator was testing WiMAX services in late 2006, with plans to introduce it in main cities in 2007. In areas not currently served by broadband, the only available option is to use a direct satellite link through a VSAT, but this is too expensive for most users including small and medium enterprises. It seems unlikely that broadband services will be rolled out to semi-urban and rural areas in the near future as the market potential is limited.

Limitations in international bandwidth has been a concern for a decade, but improvements are finally being made. The construction of a submarine cable system between India and Sri Lanka was initiated in September 2005. When commissioned, this system will enable Sri Lanka to secure fast telecommunications connectivity at cheaper rates with India and other South Asian countries. Meanwhile, the Sea-Me-We-4 (Southeast Asia–Middle East–West Europe-4) submarine cable project, commissioned in 2005 with 15 international partners, is set to enhance the quality and bandwidth of Sri Lanka's telecommunication services (CBSL 2006). Submarine optical fibre cables also link Sri Lanka with India and from 2007 with neighbouring Maldives (LBO 2006a).

A main characteristic of the telecommunication and Internet infrastructure of Sri Lanka is that both are concentrated in the Greater Colombo area and a few other cities. For example, the Western Province (where Colombo is located) had 54 per cent of all fixed phones in mid-2006 (TRCSL 2006). Mobile phones have penetrated more widely, but there is no comprehensive signal coverage by a single operator. This imbalance was to be addressed under the e-Sri Lanka programme through a Regional Telecommunication Network covering five provinces, but this has not happened due to legal and other complications.

Key institutions dealing with ICT

The Information and Communication Technology Agency (ICTA) (http://www.icta.lk), which replaced the Computer and Information Technology Council (CINTEC), is the single apex body involved in ICT policy and direction. Established under ICTA Act No. 27 of 2003, the Agency is the implementing organization of the e-Sri Lanka Initiative. It was originally conceived as a 'sunset agency' with a fixed term of five years that runs out in 2008. However, at the time of writing, there were discussions whether it should continue beyond this timeframe.

The Sri Lanka Telecommunication Regulatory Commission (SLTRC) (http://www.trc.gov.lk), established under the Sri Lanka Telecommunication (Amendment) Act No. 27 of 1996, is the national regulatory agency for the telecommunications sector with the mandate to 'ensure that competition in the market is open, fair and effective'.

The Computer Society of Sri Lanka (CSSL) (http://www.cssl.lk) is the only professional body for ICT professionals in the country. Founded in 1976, CSSL works towards improving the professional status of its members.

The Sri Lanka Information and Communication Technology Association (SLICTA) (http://slicta.lk) is an association of ICT-related trade associations and professional bodies.

Legal and regulatory environment for ICTs

The Government ICT Policy 2005–08 is to be implemented 'to promote a formal framework that focuses on good governance, interoperability and standard'. It is intended for ministries, government departments, provincial councils, other levels of public administration and the local government authorities

(ICTA 2006d). However, ICTA sources said in November 2006 that the policy had not been adopted by the Cabinet of Ministers, and the agency had been asked to review and resubmit it.

Meanwhile, the telecom regulator is struggling with bigger issues that have remained unresolved for years, among them introducing an interconnection arrangement acceptable to all operators, and rationalizing the electro-magnetic spectrum for optimum use. The latter problem stems from both the shortage of spectrum and the difficulty in shifting existing users to clear the way for telecom operators' expansion plans. In June 2006, the cabinet of ministers agreed to clear spectrum in the 450 MHz, 800–900 MHz, 1800 MHz and 2 GHz radio frequencies, and distribute them among public telecommunication networks. Many government agencies had been given generous allocations of frequency in the past. Spectrum is yet to be set aside for the new fifth mobile phone operator (LBO 2006c).

As of November 2006, the regulator had not publicly addressed the transition from analog to digital broadcasting.

Digital content initiatives

In spite of the gradual diffusion of ICTs in Sri Lanka in the past 15 years, there is still a considerable dearth of digital and online content that is specifically related to Sri Lanka. In particular, the number of websites in the two local languages of Sinhala and Tamil are limited. It is unclear whether this is a cause or effect of the poor Internet penetration. The content that does exist has resulted largely from the initiative of individuals or small groups rather than through any concerted action on the part of government or national institutions.

The content on most Sri Lankan websites is provided entirely in English, in which no more than 10 per cent of the population is conversant. This includes many government websites (including that of the National Parliament) even though the Official Languages Policy stipulates that the government must communicate in all of the three official languages: English, Sinhala and Tamil.

One bottleneck preventing the development of local content is the lack of standardized fonts. Even though the Sinhala Unicode has been identified as the standard font, as of October 2006 none of the major operating systems—except for a few versions of Linux—supported it. Users will need to download and install Unicode compatible fonts before being able to read Sinhala content online. As of October 2006, none of the government websites could be read without going through this cumbersome process. As many non-technical users are unlikely to do this, even the limited Sinhala content online remains inaccessible to most people.

The situation is somewhat different with Tamil, as the latest versions of operating systems already support the Tamil Unicode. Tamil speakers also have the option of referring to content generated in India and Singapore, where Tamil is an official language.[2]

The absence of standardized Sinhala font has also inhibited the development of local language metadata (search engines, yellow pages, Web portals, etc.) and other applications such as digital dictionaries, SMS, databases and optical character recognition (OCR). Compared with other Asian languages with unique alphabets of their own, Sinhala metadata content is negligible.[3]

Struggling within these limitations, some Web developers have improvised ways to display Sinhala fonts without the users having to download and install various fonts. Newspaper websites such as www.lankadeepa.lk and www.rivira.lk, and political or news sites such as http://www.lankaenews.com/English/index.php, employ such alternative techniques. However, this is not a satisfactory solution as it can only be used for display purposes, and the content cannot be imported into any other application like Microsoft Word or Powerpoint. Some other sites (such as www.silumina.lk, www.lakehouse.lk/budusarana) make their content available in PDF format, which also has its own limitations.

Lanka Library (www.lankalibrary.com) is an English language Web portal that collects Sri Lanka-related material online. The portal categorizes the links under titles like history, archaeology, heritage, traditions and rituals, myths and mysteries, wildlife, natural resources, Buddhism, language and literature, travel and tourism, foods, education, and the like. There are also sections dedicated to the ethnic issue and the Asian Tsunami. Although it does not have much information and mostly provides links to external resources, it is a useful indexing system.

Similarly, www.kottu.org is an effort to link many blogs related to Sri Lanka. As of October 2006, it has syndicated more than 100 blogs and photoblogs. With no restrictions imposed by the compilers, it showcases a cross-section of blogs covering the full spectrum of political, religious and social opinions. Some popular personal blogs receive 300–500 hits per day.

Another interesting trend is the recent emergence of a large number of websites specially aimed at young readers. They are essentially bilingual; do not observe strict grammatical rules; and their content is mostly non-political and aimed at casual readers. They also offer audio and video content. Some examples are: www.kaputa.com, www.ananmanan.com, www.emanuka.com, www.sinhalaya.net, www.lankasri.tv, www.naagayaa.com, www.funlk.com, www.elakiri.com, www.anangaya.com, www.123srilanka.com, www.clublk.us and http://tharunaya.com/index.htm. These websites, all individual efforts, reflect Sri Lanka-related non-political content development.

In recent years, there has been a proliferation of websites that present political news and views related to the Sri Lankan conflict which has raged for a quarter of a century. As can be expected, the views are highly divergent and sometimes provocative. Most are updated several times a day. Some examples: www.tamilnet.com, www.lankaweb.com, www.lankatruth.com and www.nitharsanam.com

When it comes to development-related digital content, the gaps are wider. A commendable initiative is the set of CD-ROMs containing agricultural information produced by the Department of Agriculture. Available in local languages, these provide information on cultivating 10 types of common vegetables and fruits.

Online services

Search engines

The absence of widely indexed and frequently updated local search engines has been felt for some years. This gap would be partly filled with the launch of Sri Lanka-themed search that Google introduced in early 2006 from www.google.lk. With its launch, all queries from within Sri Lanka to access the traditional www.google.com site are automatically directed to the new .lk site, from where two options are available: search the entire Web, or search pages from Sri Lanka.

e-Commerce

Eleven years of commercial Internet connectivity have not catalyzed much e-commerce activity in Sri Lanka and the number of operators and users—as well as the volume of transactions—still remains small.

One exception is Internet banking. Most commercial banks now offer Internet banking facilities to their customers, although with varying levels of interactivity and service. The total number of commercial bank customers registered for Internet banking was 24,650 by end 2003 (Central Bank 2004). An independent researcher estimated the number of active Internet banking accounts in 2003 to be in the range of 7,500 (Kasturiratna 2003). No later data are available.

In 2006, the Central Bank introduced an image-based cheque clearing system with LankaClear, the private operator entrusted with the task. This Cheque Imaging and Truncation (CIT) system helps reduce delays in clearing systems which earlier took up to nine days depending on the location of the drawer's and payee's banks. The new system does not move physical cheques to the central clearing facility; instead, a digital image of the cheque is transmitted. While expediting clearing, it also reduces the risk of losing checks during clearing (LankaClear 2006).

Probably the most widely used e-commerce service is the fixing of appointments with medical specialists through the site http://www.echannelling.com/. This trilingual service in some commercial banks and leading pharmacies which can be accessed online has reduced commuting by patients and their families who earlier had to make two visits for one consultation: first to make the appointment and then to see the doctor.

Apart from the infrastructure limitations and the lack of broadband services, online security concerns have also inhibited the growth of Internet banking and e-commerce in Sri Lanka. Lack of confidence in safeguards prevents many customers from transacting with their banks online. On the other hand, some complain of excessive security measures that make transactions tedious.

Meanwhile, commercial banks have complained that new stamp duties introduced since April 2006 on all credit card-based payments—as a new revenue source for the government – would discourage their customers from using the banking system and credit cards (LBO 2006b). This directly affects the over 600,000 credits cards in use.

e-Government

Placing government online has been slow and difficult in Sri Lanka. As of November 2006, there was not a single full-fledged citizen service where one could proceed from enquiry to the completion of a transaction entirely online. While many e-government initiatives have matured from simply providing information to allowing some level of interaction, none had reached transaction stage yet. According to ICTA, the first complete online e-Service would be the e-Motor Revenue License, due sometime in 2007.

Several factors have contributed to this situation. Lack of ICT literacy among public officials and an attitude that has long relegated IT functions to computer technicians have held back ICT integration. At citizen level, the low levels of Internet use and the absence of standardized local fonts limit the numbers who can transact with government online.

ICTA has a programme for re-engineering government with a vision 'to provide citizen services in the most efficient manner by improving the way government works by re-engineering and technologically empowering government business processes'. It promotes the strategic use of ICT in the public sector, and aims to implement ICT-enabled administrative policies that would, among other things, share electronic data across agencies, increase transparency in government operations and have an 'always-on, user-friendly, distance-neutral information and service facilities to citizens and businesses' (ICTA 2006a).

In fact, many e-government measures amount to complete administrative reform that would streamline government and improve efficiency. Sri Lanka has one of the largest public services for any Asian country, and reorienting the formidable bureaucracy takes time, effort and investment.[4] Many ways of reducing paperwork and a multiplicity of applications and approvals have been identified in planning for e-government. For example, the Department of Pensions has found that a retiring public servant may submit three instead of eight applications for this pension, signing only in eight places instead of the current 22.

Meanwhile, efforts to deliver citizen services through ICTs are beginning to show early results. Many government forms and circulars are now available online. In 2005, the governmental Web portal was revived at www.gov.lk as a gateway to all such services, offering links to many ministries, departments and statutory bodies. A remaining challenge is to rein in some arms of government that continue to host websites outside of a systematic structure, sometimes using generic domain names such as .org and .net instead of the proper .gov.lk domain.

The Government Information Centre (GIC) launched in August 2006 is primarily a call centre with the hotline 1919, but the services are to be extended to its website, www.gic.gov.lk. GIC provides information on public services in three official languages for 12 hours a day (8 am to 8 pm), every day of the year. Calls to the number are not toll-free, although there are plans to make it so later.

ICT and ICT-related industry

Unlike India, Sri Lanka does not have a well developed ICT industry. Except for the local assembly of PCs, there is little hardware. Locally assembled PCs cater to the home market, while branded machines are preferred in the corporate sector. According to marketing research sources, the number of PCs/servers sold in 2005 was 175,950. Assuming an average lifetime of four years for a machine, the present PC/server population can be estimated to be around 700,000[5] (LBO 2006d). This market is too small to interest many PC and peripheral manufacturers.

The software export market was estimated to be USD 82 million in 2005 (CBSL 2006). The software exporters association projects this figure to exceed USD 1 billion by 2012. But given current capacity, this seems a tall order. On the other hand, this goal is modest when compared to the burgeoning Indian market. Few Sri Lankan companies have established a globally recognizable brand for software or other ICT services.

Sri Lanka was a late entrant to business process outsourcing (BPO). Although the first company to engage in this was set up in 1983, it was not until after 2000 that BPO gained momentum. The aggregate market size is not known, but a recent survey involving 21 BPO companies among 25 identified ones showed a combined investment of over USD 13 million. Between them, they employed 3,700 persons and this is expected to grow by 30 per cent in 2006. The average salary in 2006 was USD 270 per month.

In terms of services provided, 43 per cent were engaged in accounting services, 19 per cent in call centre services and 14 per cent in medical insurance services. The companies cited three major obstacles to growth: civil and political instability, transport difficulties and poor telecommunication infrastructure (LIRNE*asia* 2006).

With three fixed phone operators and four mobile operators fuelling the telecom boom, the country imports a significant volume of telecom equipment: an estimated USD 95 million in 2003, up from the previous year's USD 85 million. The major vendors are Alcatel, Ericsson, Fujitsu, NEC and Nokia (Zita and Kapur 2004).

Enabling policies and programmes

The e-Sri Lanka project was launched in 2003 'to use ICTs to develop the economy of Sri Lanka, reduce poverty and improve the quality of life of the people'. This vision is to be realized through a five-programme strategy which covers building the implementation capacity, building information infrastructure and an enabling environment, developing ICT human resources, modernizing government and delivering citizen services, and leveraging ICT for economic and social development through public–private partnerships.

This project was started with an initial credit of USD 53 million from the World Bank, which has since been supplemented by other donors (World Bank 2004).

The government that launched e-Sri Lanka is no longer in office, and the project has undergone various changes of leadership, focus and emphasis. As of November 2006, the impression created by the ICTA website was that the agency had subsumed the project. The website www.esrilanka.lk now just provides free email to anyone interested and the official e-Sri Lanka roadmap (2003–07) is no longer available on the ICTA website.

One initiative launched under e-Sri Lanka is a network of telecentres named *Nenasala* (Sinhala for 'knowledge centre') (http://www.nanasala.lk). ICTA aims to set up 1,000 telecentres by end 2008. A majority of *Nenasalas* is to follow a community-based model where the centres are established in a central place such as a temple, public library or community hall. They are to provide services including high-speed Internet access, e-mail, telephone, computer training classes and other ICT-related facilities. Content relevant to rural people is to be made available in Sinhala and Tamil.

As of November 2006, 246 *Nenasalas* had been set up in 22 administrative districts (ICTA 2006c). Though no comprehensive independent studies have been done to monitor ICT usage at these telecentres and their contribution to communities, online discussion forums (the most popular being the www.lirneasia.net blog) suggest that *Nenasalas* have yet to achieve their objectives. Criticisms have centred around the high capital and recurrent cost of telecentres, and their complete resource dependence on ICTA-provided external funds. This contrasts with an earlier scheme, called *Vishva Gnana Kendras* (VGKs), which planned to set up telecentres on a semi-franchize model where operators had to co-invest, giving them incentive to expand services.

Many *Nenasalas* are located at Buddhist temples where women traditionally do not visit without being accompanied by a male. Ignoring such cultural norms can inadvertently restrict women's access to telecentre facilities. Location in a Buddhist temple might also indirectly discourage users from other religious faiths. Buddhist priests have no incentive to sustain *Nenasalas* or introduce more services. A related concern is that these rural telecentres might be used for promoting ruling party political agendas over and above the provision of ICT facilities and information. The content mix and sourcing for the telecentres remain unclear.

Education and capacity building

Creating an ICT-literate society and ICT-skilled workforce has been recognized as a key factor for the development of IT-enabled services (ITES) and ICT in Sri Lanka.

ICTA has embarked on several ICT human resources development projects, all of which are at early stages of implementation and too premature to report progress or assess impact. These cover skills development and training at school, university, government office and industry levels, with aims to standardize and require quality controls for various types of training currently available from privately-run training centres. There is also a National e-Literacy Initiative to raise ICT literacy among Sri Lankans. Through training, awareness raising and lowering of entry barriers, the initiative hopes to make 1 million people e-literate by 2009 (ICTA 2006b).

Open source initiatives

The free and open source software (FOSS) movement is gaining popularity and momentum in Sri Lanka, but awareness levels are still low. Only a handful of advanced users have opted for FOSS. A majority of users have no incentive or need to migrate to FOSS as unlicensed versions of most types of proprietary software can be purchased in the open market for as little as USD 2. Unlicensed software use is prevalent not only among home users, but also in some corporate and public sector institutions.

The Lanka Software Foundation (http://www.opensource.lk) was established in 2003 to promote FOSS and to enable Sri Lankans to become world class open source software developers. The Lanka Linux User Group (http://www.lug.lk), founded in 1998, actively promotes GNU/Linux software, handles Linux installations and lends distribution kits. It also maintains a library of Linux resources.

The best known Sri Lankan FOSS initiative is Sahana Disaster Management software (http://www.sahana.lk/) developed by LSF in response to the Asian Tsunami disaster. Its applications include finding missing people, coordinating relief organizations, reporting on aid disbursement, matching donations with requests, tracking temporary shelters and on the whole, improving information management—and thereby, transparency—in post-disaster situations. Since the tsunami, Sahana has been deployed after the earthquake in northern Pakistan, the Guinsaugon landslide in the Philippines and the earthquake in Yogjakarta, Indonesia (LSF 2006).

Research and development

Three institutions dominate ICT-related research and development in Sri Lanka.

Learning Initiatives on Reforms for Networked Economies (LIRNE*asia*) is a regional research and advocacy organization based in Sri Lanka and operating across Asia. Since 2004, it has been carrying out action research on telecom policy and regulation, ICT diffusion and markets. Working with industry professionals, regulators and researchers, LIRNE*asia* has produced an impressive array of research products that offer new insights and policy advocacy tools in South and Southeast Asia. Their preferred mode is open source research with multiple, evolving drafts published online for wide-ranging comment and consultation (*see boxed article next page*).

An example is the LIRNE*asia* study on the telecom use of the poorer segments of society carried out in India and Sri Lanka in 2005. Titled 'Telecom Use on a Shoestring', the study revealed

> **'Open source research' on ICTs in Asia**
>
> Speed and accuracy are critical in policy-relevant research necessary to remove constraints affecting ICTs. One solution, adopted by LIRNE*asia*, is open-source research. The Colombo-based regional organization does not claim to know all of the answers. Its researchers work with multiple drafts that are published on the Web. In some cases—such as a national early warning system for Sri Lanka in the months following the tsunami—they went further, holding expert forums and public meetings and using the mass media to draw attention to the online drafts. The end result of this online and offline feedback is extensive revision of initial drafts.
>
> LIRNE*asia* says open source research converts readers into reviewers and helps improve the quality of the final product. A by-product is the faster and easier acceptance of research findings and recommendations by regulators, policymakers and other key stakeholders.
>
> 'Open source research is not the norm in universities, where peer review is the defining characteristic,' says Rohan Samarajiva, Executive Director of LIRNE*asia*. 'But it [the peer review system] has come to serve as a break on the early release of ideas, lest they be thought of as half-baked. But in the new Internet-mediated world open source is a better model for research.'
>
> For LIRNE*asia*, the circle of people interested in a given research effort or set of ideas are the true peers who will donate their time and effort to refine the work in progress. Some blog threads on the LIRNE*asia* website have received over 50 useful, substantive comments. 'If the author is willing to revise and revise again, the end result will be superior in quality and will be produced in a shorter time,' Samarajiva adds.
>
> Source: www.lirneasia.net

important insights: more then two-thirds of users do not own the phone they use; phones are used overwhelmingly for maintaining relationships than for business; and nearly half of all users found no difficulties in obtaining a connection. It also found that SMS usage was low in this user group (LIRNE*asia* 2005).

Both the University of Colombo and University of Moratuwa are engaged in ICT-related research and development. The Language Technology Research Laboratory of the University of Colombo School of Computing (UCSC) is involved in the development of Sinhala standards, tools and content as part of a wider project aimed at ICT localization (UCSC 2006).

A significant innovation is the Disaster and Emergency Warning (DEWN) system which combines the inherent strengths of GSM mobile phone technology and the widespread access provided by GSM networks. The system was developed in 2005 at the Dialog-University of Moratuwa Centre for Mobile Communication Research with the leading telecom service provider Dialog Telekom and mobile applications company Microimage, and was first field tested in 2005 (Microimage 2006).

Challenges

Political instability, protracted civil war, regulatory uncertainties, policy gaps, poor infrastructure and income poverty combine to inhibit progress in ICT-related industries as well as ICT-for-development initiatives. These present formidable challenges to government, industry and civil society as they try to find ICT-enabled solutions to deep-rooted economic and social problems.

Sri Lanka scored 0.33 in the 2005 Digital Opportunity Index. The index was developed by the International Telecommunications Union to measure and compare the levels of ICT development in countries. Ranked 106 among 180 economies assessed, Sri Lanka was ahead of India (ranked 119), Pakistan (128) and Bangladesh (139) but was behind Thailand (80), the Philippines (94) and Indonesia (105) (ITU 2006b). However, national statistics often mask the stark disparities that exist within the country due to geographical, economic, social and cultural factors.

Since 1990, Sri Lanka has pioneered telecom reforms and technology adoption and then allowed momentum to be lost, enabling late entrant neighbouring countries to roll-out services on a more sustained basis. The bigger challenge is not to look for 'firsts' but to develop 'staying power' to see reforms and programmes through to their logical end, even when some aspects might be politically or bureaucratically unpopular. It is this bold leadership in ICT-enabled growth and development that Sri Lanka needs—but currently lacks—the most.

Notes

1. Many publications incorrectly cite the number of subscribers as the number of users, which is misleading. The actual number might vary from 300,000 to 500,000. However, no survey has been done to ascertain the number.
2. Tamil is estimated to be spoken by at least 74 million people worldwide. See http://en.wikipedia.org/wiki/Tamil_language.
3. For further discussions on this subject, see http://www.akuru.org and http://www.fonts.lk/
4. According to one World Bank economist, Sri Lanka maintains 3.9 civil servants for every 100 people compared to the Asian regional average of 2.6. The rate is 1.2 in India, 1.5 in Pakistan and 0.6 in Bangladesh. Sri Lanka's public service is disproportionately large even compared to East Asian countries: China has 2.8 government servants for 100 people while Indonesia has 2.1 and Korea has 2.2.
5. There is a significant second-hand market in PCs. Used PCs are imported to be sold for a fraction of their original cost, almost exclusively for home use. Trade numbers are not available as no one tracks such data. Such PCs cost USD 50–200; a brand new entry level PC costs at least USD 700.

References

Central Bank of Sri Lanka. (2006). *Annual report 2005*. Retrieved 3 November 2006 from http://www.cbsl.lk/cbsl/AR2005data/Chap3.pdf

Information and Communications Technology Agency (ICTA), Sri Lanka. (2003). *e-Sri Lanka, 2003*. An ICT Development Roadmap.

ICTA. (2006a). *Re-engineering government: Program concept*. Retrieved 10 November 2006 from http://www.icta.lk/Insidepages/ReGov/ProgrammeConcept.asp

ICTA. (2006b). *ICT human resources capacity building*. Retrieved 10 November 2006 from http://www.icta.lk/insidepages/programmes/ICT_Human_Resources_Capacity_Building.asp

ICTA. (2006c). *Nenasala project*. Retrieved 20 November 2006 from http://www.nanasala.lk/

ICTA. (2006d). *Government ICT policy: 2005–2008*. Retrieved 20 November 2006 from http://www.icta.lk/Insidepages/ReGov/ICTPolicy/ICT_PolicyForGovernment.asp

ITU. (2006a). *World telecommunications indicators—Sri Lanka*. Retrieved 5 November 2006 from http://www.itu.int/partners/flash/index.asp?id=LKA

ITU. (2006b). *World information society report 2006*. Retrieved 20 November 2006 from http://www.itu.int/osg/spu/publications/worldinformationsociety/2006/wisr-summary.pdf

LankaClear. (2006). *CITS: Overview of cheque imaging and truncation (CIT)*. Retrieved 8 November 2006 from http://www.lankaclear.com/data/services/cits_pa.htm

LBO. (2006a). Maldives-Lanka undersea-link could be first leg of trans-Indian Ocean cable to Africa. *Lanka business online*, Colombo. Retrieved 15 November 2006 from http://www.lankabusinessonline.com/fullstory.php?newsID=193995836&no_view=1&SEARCH_TERM=5

LBO. (2006b). Tax ceiling: Sri Lankan bankers suggest lower tax rate for credit card transactions. *Lanka business online*, Colombo. Retrieved on 8 November 2006 from http://www.lankabusinessonline.com/fullstory.php?newsID=485711489&no_view=1&SEARCH_TERM=2

LBO. (2006c). Sri Lanka's mobile penetration to peak in 2008: New study. *Lanka business online*, Colombo. Retrieved on 20 November 2006 from http://www.lankabusinessonline.com/fullstoryphp?newsID=781764373&no_view=1&SEARCH_TERM=5

LBO. (2006d). Locally assembled units dominate Sri Lanka's computer sales. *Lanka business online*, Colombo. Retrieved on 20 October 2006 from http://www.lankabusinessonline.com/fullstoryphp?newsID=27047728&no_view=1&SEARCH_TERM=5

LIRNE*asia*. (2005). *Telecom on a shoestring (2005)*. Retrieved 10 November from http://www.lirneasia.net/projects/completed-projects/strategies-of-the-poor-telephone-usage/

LIRNE*asia*. (2006). *A baseline sector analysis of the business process outsourcing industry in Sri Lanka*. 29 September 2006. Retrieved 8 November 2006 from http://www.lirneasia.net/wp-content/uploads/2006/09/BPO_Report_ver3-5-Final.pdf

LSF. (2006). *Sahana: Free and open source disaster management system*. Lanka Software Foundation, Colombo. Retrieved 20 November 2006 from http://www.sahana.lk/

Microimage. (2006). *Dialog Telekom, Microimage and University of Moratuwa Research Lab wins the overall gold*. Retrieved 20 November 2006 from http://www.microimage.com/press/MicroimageDirectOctober2006.htm

TRCSL. (2006). *Statistical overview of telecommunications sector as of end of 2nd quarter 2006*. Retrieved 20 November 2006 from http://www.trc.gov.lk/pdf/statover1.pdf

UCSC. (2006). *Background to Sinhala language computerization work*. University of Colombo School of Computing. Retrieved on 20 November 2006 from http://www.ucsc.cmb.ac.lk/research/ltrl/index.html

World Bank. (2004). *Project appraisal document for the e-Sri Lanka Programme*. Retrieved 20 November 2006 from http://www-wds.worldbank.org/external/default/WDSContentServer/WDSP/IB/2004/09/10/000009486_20040910112659/Rendered/PDF/28979a.pdf

Zita, K. and A. Kapur. (2004). *Sri Lanka telecom brief*. Retrieved on 15 November 2006 from http://topics.developmentgateway.org/ict/rc/filedownload.do~itemId=1006080

Taiwan

Yu-li **Liu** and Eunice Hsiao-hui **Wang**

Total population	22,879,132 (as of February 2007)[1]
GDP per capita	USD 16,098 (NTD to USD: 32.167) (2006)[2]
Computers per 100 inhabitants	52.78 (2005)[3]
Main fixed-line telephones per 100 inhabitants	59.79 per cent (2005)[4]
Mobile phone subscribers per 100 inhabitants	100 per cent (as of June 2006)[4]
Internet users per 100 inhabitants	58.1 per cent (by 2005)[3]
Domain names registered under '.tw'	4.32 million (as of July 2006)[5]
Broadband subscribers per 100 inhabitants	20.21 per cent (2005)[4]
Internet domestic bandwidth	1,740 Gbps (as of June 2005)[4]
Internet international bandwidth	148 Gbps (as of September 2006)[4]

Sources: 1. Ministry of the Interior 2007
2. Ministry of Economic Affairs 2007
3. International Telecommunications Union 2005
4. National Communications Commission 2005, 2006
5. Internet Systems Consortium 2006

Overview

Taiwan is a major manufacturing centre for global information and communication technology (ICT) companies (Huang 2006). In the *Global Competitiveness Report 2005–2006*, Taiwan ranked fifth among 117 countries in the Growth Competitiveness Index and third in the technology index (FIND 2005). In the 2005 Digital Opportunity Index Ratings of the ITU, Taiwan's overall rating was 0.64, ranking seventh in the world and fourth in Asia Pacific. In March 2006, the *Global Information Technology Report* of the World Economic Forum also showed that Taiwan ranks seventh in the world and fourth in Asia Pacific in terms of network readiness. With respect to e-government performance, Taiwan ranked first among 198 countries surveyed by Brown University in 2002, 2004 and 2005. The Point Topic Survey in March 2006 ranked Taiwan number five worldwide in terms of broadband household penetration and number six in terms of DSL lines.

To meet the science and technology standards of a developed nation by 2010, Taiwan has been working to increase R&D funding to 3 per cent of GDP and the number of research personnel with college or advanced degrees to 32 persons per 10,000 by 2007.

Meanwhile, R&D work on IPv6 is ongoing in line with the implementation of the e-Taiwan policy.

Technology infrastructure

The ongoing integration of the fixed network and mobile communications networks aims to create a ubiquitous broadband network service environment in Taiwan. The Broadband Development Project for the establishment of next-generation fibre-to-the-home (FTTH) broadband networks was initiated in June 2004, supporting integrated connectivity of wired and wireless access, as well as voice and data services.

Broadband network

The total bandwidth of Taiwan's main broadband backbone network (running the length of the island) reached 1,740 Gbps in June 2005. The bandwidth of the international submarine cable network reached 629 Gbps, of which 87.77 Gbps was allocated for Internet connection (FIND 2005).

In accordance with the 'Broadband Conduit Deployment' plan, more and more apartment complexes in Taiwan are being wired with fibre optic networks, leased lines or high-speed ADSL connections. The total number of broadband Internet accounts in Taiwan, including ADSL, cable modem, leased line and PWLAN (Public Wireless Local Area Network), reached 4.19 million in August 2005. Less than a year later, by June 2006, the number of broadband subscribers was close to 4.3 million (FIND 2005). The Taiwanese government is aiming for six million broadband subscribers by 2008.

IPv6

Academic networks (TANet and TWAREN) and ISPs in Taiwan have been upgraded to IPv6, and have thus acquired 20 IPv6 network segments supporting 114/32 addresses. In 2005, Taiwan ranked eighth in the world in IPv6 network segment volume. Domestic IPv6 links were connected to 23 firms with a total of 11.13 Gbps, and international IPv6 links were connected to

93 firms with a total of 11.5 Gbps. By 2005, Taiwan had obtained 23 IPv6 Ready Logo Phase I international accreditation logos and two IPv6 Ready Logo Phase II international accreditation logos, which made Taiwan fourth and third worldwide, respectively. By the end of June 2006, the number of IPv6 addresses issued in Taiwan had reached 2,243 (Taiwan Network Information Center).

Mobile communications

There were 22.17 million mobile phone numbers (including GSM, PHS and 3G) in Taiwan as of the fourth quarter of 2005. The mobile phone penetration rate remained unchanged at 97 per cent. However, the number of mobile Internet subscribers has grown rapidly with a compound annual growth rate (CGAR) of 126.6 per cent from 2001 to 2005. The number of mobile Internet users in August 2005 was 7.5 million, representing 33.6 per cent of all mobile phone subscribers (National Communications Commission).

Dual-network technology integrating conventional GSM and 3G networks with WLAN or WiMAX is being developed. Seamless broadband mobile communications with dual-network handsets is viewed as the next-generation mobile communication networking.

Digital radio and digital TV broadcasting

Taiwan's five terrestrial TV stations currently operate a total of 14 digital TV channels. The projection is for Taiwan to evolve to a new era of digital radio and TV broadcasting by 2010. In 2005, the cable TV penetration rate reached 85.1 per cent (ACNielsen). A forward-looking Multimedia Home Platform (MHP) has been discussed to integrate the emerging hand-held Digital Video Broadcasting (DVB-H) and Internet Protocol Datacasting (IP-DC) service system, and to connect the ongoing IP-TV and IP-DC.

In 2005, a Digital Video Service Platform Systems and Digital Rights Management (DRM) Research Project was initiated, with the aim of establishing DRM technology standards and management methods conforming to international practice integrating interactive TV broadcasting and Internet networking.

Key institutions dealing with ICTs

One of the key government agencies dealing with ICTs in Taiwan is the National Communications Commission (NCC) established on 22 February 2006. The NCC is an independent regulator governing the telecommunications, media and information sectors. Article 2 of the NCC Organization Act provides that all laws and regulations dealing with communications, including the Telecommunications Act, Radio and Television Act, Cable Radio and Television Act and Satellite Broadcasting Act, are under the official responsibility of the NCC. Corresponding powers originally under the Ministry of Transportation and Communications (MOTC), the Government Information Office (GIO) and the Directorate General of Telecommunications (DGT) have been transferred to NCC.

The NCC's key functions include licensing, enforcement, spectrum assignment and management, technical standards, regulation of information and communication security and consumer protection. In keeping with its regulatory role, NCC is not a member of the Cabinet and is not even under the supervision of the Executive Yuan (the highest administrative organ of the State). However, the MOTC and GIO still play a role in providing guidance and incentives to the telecommunications and broadcasting industries since, according to Article 3 of the Communications Basic Law, national communications resources planning and provision of guidance and incentives for industry development are to be performed by subordinate organizations of the Executive Yuan.

The National Information and Communications Initiative Committee (NICI), which combines three ICT-related task forces under the Executive Yuan, was established in April 2001. This cabinet-level committee is responsible for accelerating the development of the IT industry, e-commerce and related businesses, improving the efficiency of government services, promoting Internet usage and related applications, and increasing the competitiveness of the IT industry.

Four government agencies are also involved with the development of the ICT industry in Taiwan. The Council for Economic Planning and Development (CEPD) is a ministerial-level agency that is responsible for drafting plans for national economic development, evaluating development projects submitted to the Executive Yuan, coordinating economic policymaking activities, monitoring the implementation of development projects, and im-plementing the goals set out in Challenge 2008, the Six-Year National Development Plan. The Industrial Development Bureau of the Ministry of Economic Affairs (MOEA) provides assistance to all industries and businesses, and has divisions dealing with industry policy, industry development, knowledge services and electronic information.

The Research, Development and Evaluation Commission (RDEC) runs administrative programmes for the Executive Yuan. Its functions include research and development, overall planning, policy innovation, government restructuring, supervision and evaluation, and information management. It also contracts academics and experts to conduct relevant surveys and studies on e-government.

The National Science Council (NSC) is the highest government agency responsible for promoting the development of science and technology (S&T). It oversees all S&T programmes of the government. It also plays a major role in developing science parks. It convenes the Science and Technology Strategic Planning Sessions and holds National Science and Technology Conferences regularly (Dahl and Lopez-Claros 2005). In May 1998, it established the National Science and Technology Programme for Telecommunications (NTP) focusing on wireless communications, broadband Internet, applications and services, and telecommunication industry promotion and development.

Four non-profit organizations—the Industrial Technology Research Institute (ITRI), Institute for Information Industry (III), Telecommunications Technology Center (TTC) and Taiwan Network Information Center (TWNIC)—are vital to the development of Taiwan's ICT industry. ITRI was founded in 1973 by the MOEA to meet the technological requirements of industrial development through applied research and technical services. It serves as the technical centre for industry and makes recommendations regarding industrial policy to the government.

III, which was founded in 1979, serves as a joint government-private sector think tank to promote the development of the ICT industry. III provides the private sector with market analysis, ICT training, interoperability standards, services and technology transfers. To the government, III proposes ICT policies (Dahl and Lopez-Claros 2005). It assisted the Science and Technology Advisory Group (STAG) of the Executive Yuan to implement the e-Taiwan Programme which is part of Taiwan's Six-Year National Development Plan. III also plays an important role in promoting digital content: it helped the government set up the Digital Contents Industry Promotion Office and the Digital Contents Institute Digital Contents Assets Appraisal and Investment Service Center to attract investment.

TTC is a telecommunications testing and certification centre founded in February 2004. It was chosen by the Association of Telecom Service Providers to be in charge of the construction and operation of the Number Portability Administration Center (NPAC) in its first five years. TTC also provides testing laboratories various services for IT security, digital TV and WiMAX.

Under NCC supervision, TWNIC oversees domain name registration and IP address allocation in Taiwan. It collaborates with international network information organizations such as ICANN and APNIC, as well as other national Internet organizations such as JPNIC, CNNIC and KRNIC. TWNIC also contracts academics and experts to conduct telephone surveys about broadband use and wireless and mobile communication use twice a year.

Digital content initiatives

Along with the spread of broadband connections and wireless networks in Taiwan, the demand for digital content on the Information Superhighway has grown exponentially. The Taiwanese government pays very close attention to the development of the digital content industries. In June 2002, the government initiated the Two Trillion and Twin Star programme, which envisions the digital content industry as one of the strategic (or rising-star) industries that would contribute an annual production value of over NTD 1 trillion (about USD 30.287 billion). In line with the vision, several digital content programmes have been formulated and designated agencies are working closely with the private sector.

National Digital Archives Program

The National Digital Archives Program (NDAP) is one of the nine National S&T programmes sponsored by the National Science Council. The first phase of this five-year initiative began on 1 January 2002. It aims to digitize cultural collections stored in Taiwan's major museums, libraries and universities. Sixteen content themes are covered: Anthropology, Archaeology, Architecture, Archives, Artefacts, Botany, Calligraphy and Painting, Chinese Classics Full-text Database, Geology, Journalism and Mass Media, Linguistics, Maps and Remote images, Rare Books, Stone and Bronze Rubbings, Video and Zoology.

One of the more noteworthy digital museum projects involves the National Palace Museum which is world-famous for preserving the finest artefacts collected by the Chinese emperors and royal families. The museum's collections include a rich variety of jade pieces, ceramics, porcelain, rare books, tapestries, embroidery, ritual bronzes, ancient calligraphic works and paintings from the Stone Age to the present. The goal of the digital museum project is to establish a metadata database of the National Palace Museum's priceless collection and to digitize the 60,000 artefacts, calligraphic works and paintings, and the 190,000 Ching archival documents housed in the museum. The digitized cultural contents would be more accessible to the public, more easily available for academic research and commercial value-added services, and very useful for promoting the museum's collections internationally.

Digital Content Promotion Office

Under the Two Trillion and Twin Star programme, the government aims to build both real and virtual industrial parks to promote the digital content industry, establish digital content colleges to introduce the newest digital technology, and train new digital talents and professionals. The Digital Content Promotion Office under the MOEA is the governing agency for all of these activities. Its mission is to promote mobile content/services, online games, 2D/3D animation, software and streaming video products. The industry's digital content production value is projected to reach USD 10.75 billion by 2006 and international sales of Taiwan's digital content industry are expected to grow by 30 per cent. The digital content industry is also expected to create new job opportunities and employ over 70,000 people.

Cultural and Creative Industries Promotion Office

As part of the government's Challenge 2008 National Development Plan, incentives are being given to industries engaged in innovation, cultural heritage and intellectual property development. It is recognized that the output of such industries will contribute to employment and to the economy as a whole, as well as help improve the overall living environment. Thirteen cultural and creative industries have been designated by the Cultural and Creative Industries Promotion Office under MOEA: visual arts, music and performance arts, cultural exhibitions and performance facilities, handicrafts, motion pictures, radio and television, publishing, advertising, design, digital games and entertainment, brand and fashion designs, innovation lifestyle and architectural design industries.

In addition, the Executive Yuan has launched the Development Fund Investment Plan for Digital Content, Software and Cultural Creative Industries in accordance with the Promotion of Industrial Upgrading Act to financially support and encourage any strategic businesses or projects in digital content, software and culture, as well as the creative industries.

Online services

With the growing accessibility of broadband connections in Taiwan, a booming online services market for video entertainment services, online music, online gaming and online learning is expected. Taiwan is ranked the third lowest in broadband pricing around the world, with a premium of USD 0.18 per 100 Kbit, which is only slightly higher than Japan's USD 0.07 and South Korea's USD 0.08 and much lower than the USA's USD 0.49 and Hong Kong's USD 0.83 (ITU 2005).

Online video entertainment

Hichannel, which is owned by Taiwan's largest ISP, has the largest subscriber base on the ARO (access rating online) index with 150,000 paid subscribers as of June 2004. It is among the four most popular video streaming sites (Insightexplorer 2005). Microsoft's Windomedia, which allows users free access to their online videos, has the broadest reach among the three most popular streaming media sites. On the other hand, subscribers who log on to Webs-tv.net, which is partially owned by shareholders from Hong Kong, seem to spend much more time watching online videos per visit. In September 2006, Webs-tv.net was officially merged with the third largest portal site, Yam.com. This merger is expected to capture 54.23 per cent of users in Taiwan, which means that Webs-tv.net will top the online video market with a total of 8.2 million users.

The incumbent fixed network, Chunghwa Telecom, also supports online entertainment services Multimedia on Demand (MOD) through its Hinet broadband ADSL. It has a very competitive basic monthly rate of NTD 100–150.

Online music

Kuro.com.tw, which is popular among student groups but controversial with its ongoing P2P copyright infringement lawsuit, ranked No.1 in paid usage (21.7 per cent of the market) in a survey conducted in March 2006 (Insightxplorer 2006). Kuro, which charges a monthly rate of NTD 99 (about USD 3), now has over 400,000 paid members.

KKBOX, a local brand supporting licensed online music, reached 16.6 per cent of paid usage. Yahoo! Music (tw.music.yahoo.com) began exploring the local online music market in early 2006 and now enjoys huge popularity with its strong international brand. Yahoo! Music has 9.9 per cent paid market share, which is higher than Ezpeer's 6.4 per cent.

Online games

With the booming Internet market, a new segment in the gaming industry has emerged—online games. The total revenues from online game services in Taiwan reached NTD 9.81 billion (about USD 297 million) in 2005, representing a growth rate of 17.8 per cent from the previous year (Market Intelligence Center 2006).

Some online game companies, such as Chinesegamer Co., also joined the m-commerce market with mobile games and other mobile value-added services. In addition, many online game companies in Taiwan, such as Gamania Digital Entertainment Co. Ltd., Wayi International Digital Entertainment Co. Ltd. and Soft World International Corp., are competing to be the local distributor of international brands and online game developers.

Online business

Total revenues from Taiwan's e-commerce market, including online sales and online purchasing, increased by 20.7 per cent from NTD 177.86 billion (about USD 5.386 billion) in 2004 to NTD 214.7 billion (about USD 445.218 million) in 2005 (FIND 2005).

In a survey (FIND 2005) of online businesses in Taiwan, 8.7 per cent of Taiwanese enterprises were found to be engaged in online selling either through their own website or through an electronic marketplace or EDI (Electronic Data Interchange). The e-selling market increased 1.1 per cent compared to 2004. Among enterprises in online sales, 45.4 per cent of companies employed a secure transaction mechanism, and nearly 30 per cent offered customers online payment services.

The online procurement penetration rate in Taiwanese industry as a whole reached 11.8 per cent in 2005, roughly the same as in 2004. The growth in e-procurement was most significant among large enterprises, where the penetration rate increased from 17.4 per cent in 2004 to 21.9 per cent in 2005. A dramatic growth was also found in the transportation, warehousing and communications industry, where the online purchasing penetration rate rose from 7.4 per cent in 2004 to 13.1 per cent in 2005.

Among Internet services applications, enterprise resource planning (ERP) was the most popular, with 12.9 per cent of enterprises surveyed adopting it and a growth rate of 2.1 per cent from 2004. The next most popular application was customer relationship management (CRM) at 8.6 per cent. In total, however, the penetration rates of e-business applications demonstrated a slightly slow growth in 2005, with an average growth rate of around 2 per cent.

ICT industries

Taiwan's semiconductor industry plays a leading role in the world. It is strong in its value chain, from IC (integrated circuit) design and manufacturing to packaging and testing. Taiwan is the world's second largest supplier of IC designs (next only to the US) and accounts for 28 per cent of the world's output.

Taiwan's flat-panel display industry grew significantly, with 110 companies creating a value of USD 8.95 billion (47 per cent exports and 53 per cent domestic sales). It is expected to achieve a value of USD 40.53 billion by 2006, which would make it the world's largest TFT-LCD supplier (*Taiwan Yearbook 2005*).

Taiwan is also the fourth largest producer of PCs.

The total production value of Taiwan's ICT industry in 2005 includes USD 5.2 billion for software and services, USD 13.3 billion for communications production, USD 22.6 billion for LCD panels, USD 34.8 billion for semiconductors and USD 81 billion for information hardware (MIC, IEK April 2006).

Taiwan's revenue from telecom services in 2005 was USD 12 billion (or 3.4 per cent of the GDP), of which 58 per cent was due to mobile networks, 22 per cent was due to fixed networks, 13 per cent was for value-added services and 7 per cent was for other services (Li 2006). Chunghwa Telecom (CHT), the state-owned dominant carrier, was privatized in August 2005, although government still owns 41.37 per cent of its shares. As of March 2006, its market share in terms of revenues was 54.47 per cent for international services and 76.18 per cent for domestic long-distance services.

Taiwan's 2G operators and their market share in 2005 are: Chunghwa 39 per cent, Far Eastone 31 per cent, and Taiwan Cellular 30 per cent. Five mobile phone companies provide 3G services.

CHT is the dominant fixed network in Taiwan. Its branch company, HiNet, also dominates 84.27 per cent of the ADSL market (TWNIC 2006.7). Asia Pacific Broadband Telecom (APBT) is a new fixed network that used to have family affiliation with the cable company MSO Eastern Multimedia Company (EMC). It provides both ADSL and cable modem services. It has acquired an ISP named Asia Pacific Online (APOL) and is now the second largest provider in the Taiwanese broadband market. Other new fixed networks include Taiwan Fixed Network (TFN) and New Century InfoCom Tech Co. Ltd. Since they are new entrants and still have problems regarding access to the last mile, they prefer to promote broadband to small and medium businesses and buildings that have access to high-speed Internet. ISPs such as Seednet and SoNet also actively promote ADSL.

EMC and Hoshin GigaMedia Center Inc. also provide broadband service via a cable modem. However, cable modem penetration is only 5.6 per cent of the broadband market. GigaMedia provides both cable modem and ADSL services.

Enabling policies and programmes

The key role played by the Taiwan government in promoting ICT production and IT use is widely recognized. The Taiwan government continues to initiate critical national programmes for strengthening indigenous production of advanced ICT technologies and upgrading the people's quality of life.

Challenge 2008 National Development Plan

Encouraged by the two decades of success of the Hsinchu Science-Based Industrial Park, the government of Taiwan in May 2002 aggressively launched the Challenge 2008 National Development Plan, which emphasizes strengthening international competitiveness, upgrading the people's quality of life, and promoting sustainable development. Under the Plan, the government and the private sector have been working closely together to transform Taiwan into a 'green silicon island' in the first decade of the new century.

To speed up the Plan's implementation, the government in November 2003 identified a number of new major construction projects, including:

- Education: Develop top-notch universities and research centres
- Digital Infrastructure: Implement the Mobile Taiwan (m-Taiwan) project
- Transportation: Build up a rapid railway transit network, that is, Taiwan High Speed Rail

e-Taiwan

In May 2002, the NICI initiated implementation of the e-Taiwan Program as a key component of the Challenge 2008 Program. ICT infrastructure development and e-government initiatives were emphasized during the first stage of the e-Taiwan Programme (implemented in 2002–04). The second stage (2005–07) focused on enhancing the use of ICT applications in Taiwan.

In June 2006, the Intelligent Community Forum (ICF) honoured Taipei as the 'Intelligent Community of the Year' for demonstrating 'sustainable competitiveness' in its use of broadband technology for economic development. Taipei, Taiwan's capital city, was also praised by ICF for its continuous demonstration of 'digital democracy' through its broadband policies. Indeed, e-government supported by broadband infrastructure is making significant contributions to the achievement of a more participatory democracy in Taipei. For example, by logging on to the Taipei e-government portal site (http://www.taipei.gov.tw), Taipei citizens can obtain government administrative planning data, project implementation status reports, on-site inspection reports and the results of public opinion surveys. They can also use a wide variety of e-services, from e-education to filing complaints with any government agency. The Taipei city government enables the public to play the role of 'cyber citizens'.

m-Taiwan for a Ubiquitous Network Society

The NICI in coordination with the Ministry of the Interior (MOI) and the MOEA, proposed the m-Taiwan Program, a key initiative to transform Taiwan into a 'Ubiquitous Network Society' (UNS) in which the public can feel the benefits of 'e-enablement'. The vision of the m-Taiwan initiative, which has a budget of NTD 37 billion (about USD 1.12 billion) from 2005 to 2009, is to develop the infrastructure for wireless Internet access for 8 million subscribers and to provide broadband internet coverage to 80 per cent of the population in urban areas. Taipei and Taichung are the pioneering 'mobile cities' established in 2005.

The m-Taiwan Program aims to solve the Last Mile problem by further building up seamless wireless networks (including WiMAX), integrating mobile phone networks, implementing the optical fibre backbones and implementing the Integrated Beyond 3rd Generation (iB3G) Double Network Integration Plan (that is, integrating WLAN and cellular communications). The government will soon be issuing licenses for WiMAX, the fourth generation of telecommunications.

Thus, Taiwan has become a WiMAX test bed for global companies like Intel and Nortel. In 2006, Nortel signed an agreement with Chunghwa Telecom to deploy a WiMAX solution in the operator's experimental park to create an environment for testing WiMAX, and integrating the technology with the Nortel Wireless Mesh Network solution deployed in Taipei and Kaohsiung as part of the government's m-Taiwan initiative. In addition, Nortel deployed a WiMAX trial system at the National Taiwan University campus, and established a Center of Excellence for Devices in Taipei. In early December 2006, Chunghwa Telecom also announced its intention to build the first fully integrated broadband wireless network driven by the Yilan local government under the m-Taiwan project. Chunghwa Telem envisions Yilan, a northern county in Taiwan, as a national showcase for ubiquitous wireless broadband services in which WiMAX infrastructure will enable broadband wireless access to services like m-learning, m-commerce, m-tour, as well as video surveillance and IPTV services.

Legal and regulatory environment for ICTs

In order to liberalize the telecommunications industry, two major Telecommunications Liberalization White Papers were propagated. The first, which was released in 1987, aimed to provide a fair and competitive environment, promoting technological progress, improving industrial efficiency, providing high-quality, diverse and cheaper telecommunications services, and triggering economic development. The second Telecommunications Liberalization White Paper, which was released in 2002, was aimed at constructing an international telecommunications environment, promoting universal service, narrowing the digital divide, promoting competition, providing creative and high-quality telecommunications services and fostering the development of the telecommunications industry (Liu 2003).

The Telecommunications Act is the main law regulating telecommunications enterprises. It was first introduced in 1958, and has been revised several times. The 1996 revised version was significant in that it separated DGT from CHT, meaning that DGT would become a governing agency and would no longer manage CHT. The Telecommunications Liberalization White Paper and the revised Telecommunications Law emphasized competition by opening up the telecommunications market and applying asymmetrical regulations, such as network interconnections and unbundling, on the dominant player CHT (Liu 2003).

There are three electronic media laws (Radio and Television Act, Cable Radio and Television Act, and Satellite Broadcasting Act) and one Telecommunications Law. With the convergence of telecommunications and broadcasting, many laws and regulations have become outdated. When the former broadcast and telecommunication regulators were merged, there were discussions about whether the telecommunications law and the electronic media related laws should be integrated into one law. The Communications Basic Act, promulgated in January 2004, was enacted to accommodate the convergence of technologies and encourage the sound development of communications. Article 6 of the Act states that the government shall encourage innovation in communication technologies and services and that it shall not impose any restriction on innovation without proper cause. Article 7 states that the government shall avoid any discriminatory administration of different transmission technologies, but this requirement does not apply to the allocation of scarce resources.

To protect the people's right to know and to further people's understanding of public affairs, the Freedom of Government Information Law was promulgated in December 2005. It was enacted to ensure that administrative measures directly related to people's rights and interests as well as other relevant government information shall be made available to the public in a timely way.

The Electronic Signature Act was enacted in April 2002 to ensure the security of electronic transactions, and to facilitate the development of e-government and e-commerce.

Education and capacity building

The National Science and Technology Program for e-Learning was initiated under the Challenge 2008 National Development Plan with a five-year budget of USD 120 million. The programme aims to:

- create a better e-learning environment to prepare *everyone* for the e-life, including the workforce, the unemployed, schoolteachers and students, government agency staff, retired senior citizens and on-duty army forces;
- bridge the digital divide between information haves and information have-nots;
- promote and develop local e-learning industries; and
- upgrade Taiwan into one of the leading e-learning and knowledge-based economies.

The project transforms established centres and institutions, such as libraries, schools, job training centres, cybercafés and citizen clubs, into e-learning classrooms. It also investigates countrywide digital divide indicators, analyses the factors contributing to the digital divide and formulates strategies for bridging the digital divide, such as e-learning programmes for the less privileged. The project supports advanced mobile learning devices such as multifunctional e-schoolbags, which are mobile learning devices that provide access to e-learning environments regardless of time and place. The project also provides for the establishment of the Network Science Park for e-Learning (e-Park), which supports e-process industries (technologies, markets, platforms and quality control) and links to physical service centres. The e-Park is expected to play an important role in the Chinese e-learning world. Indeed, one of the aims of the e-learning project as a whole is the enhancement of the global competitiveness of the local e-learning industry.

In addition, over 1,000 elementary schools and junior and senior high schools are connected via the Taiwan Academic Network (TANet), one of Taiwan's three major ISPs (the other two are HiNet and Seednet). TANet was built as a collaborative project of the Ministry of Education's Computer Centre (MOECC) and Taiwan's leading universities in July 1990. Today it supports the following Internet services along with Academia Sinica: e-library and e-periodicals, Website services, e-mail, server hosting, over 4,000 English domain names (edu.tw), IPv6, e-learning, academic e-document exchange, videoconferencing and information security for schools. The Taiwan Advanced Research and Education Network (TWAREN), TANet's second generation (TANet2), which began as an initiative under Challenge 2008, has more than 49 research and academic institutions connected to it.

For off-campus ICT education and capacity building programmes, the Digital Education Institute (DEI) under III is a professional training, e-learning and corporate training solution provider with a client portfolio that includes CEOs, CIOs, government officers, consultants, managers and staff. The training programmes offered help a wide variety of individuals improve their computer and ICT skills to meet corporate and government requirements.

The Council of Labor Affairs under the Executive Yuan also provides short-term intensive training programmes in ICT applications for the unemployed. It offers vocational lessons online through the e-learning platform.

The Ministry of Education under the e-Taiwan project expects to initiate 400 Digital Opportunity Centers (DOC) in 180 rural areas from 2005 to 2008, to provide aboriginal tribes and the elderly with computer literacy courses that introduce Internet access, e-learning fundamentals and e-commerce essentials.

Open source and open content initiatives

In June 2002, the NICI decided that the Industrial Development Bureau (IDB) under the MOEA would be the agency responsible for initiating free software promotion initiatives, and the Free Software Steering Committee would be formed to promote open source software. Thus the Free and Open Source Software (FOSS) Promotion Initiative was launched. This four-year project from 2004 to 2007 aims to facilitate the development of the Taiwanese free software industry; encourage the use of free software in Taiwan, including Linux, Chinese-language applications or other free software; and further contribute to information sharing and exchange.

The Open Source Software Foundry (OSSF) is a sub-project of the FOSS Promotion Initiative. It is partially funded by the MOEA/IDB, NSC and Academia Sinica to help Taiwan become a hub of the FOSS industry and a major contributor of FOSS worldwide. The OSSF's website (http://www.iis.sinica.edu.tw/page/research/en_OSSF.html) is a repository of numerous resources in all aspects of free software. Currently it contains:

- OpenFoundry, a FOSS repository and management system where one can access many of Taiwan's FOSS projects as well as create one's own FOSS projects;
- A database on Taiwan's FOSS talents and experts;
- Documents on intellectual property issues, and LicenseWizard, a Web-based tool to assist in the selection of FOSS license agreements;
- Resource catalogues in selected areas, including embedded software, enterprise computing and Chinese language processing;
- Announcements of FOSS-related conferences and events;
- *Open Source Newsletter*, a biweekly electronic publication of the OSSF.

The OSSF also works with government, academic and research institutes, and industry in the release of software source code under FOSS licenses. An emphasis of OSSF work is the analyses of FOSS licenses and national policies worldwide.

It provides training courses for users and helps project owners choose the appropriate FOSS licenses. The OSSF conducts surveys of local FOSS communities, designs training courses, and holds workshops. The OSSF also studies well-known FOSS projects and highlights their successful experiences in OSSF publications in order to encourage Taiwan's software developers to join FOSS.

R&D initiatives

Taiwan has been working to consolidate its S&T foundation in order to achieve the S&T standards of a developed nation by 2010. To this end, R&D input and output targets include raising R&D funding to 3 per cent of GDP and increasing the number of research personnel with college or advanced degrees to 32 persons per 10,000 by 2007. Local inventors and companies are encouraged to apply for US patents to help Taiwan reach its targeted 3.5 per cent of all US patents granted (not including new design patents) by 2007. Other goals include increasing the number of broadband users to more than six million by 2007.

The IT-led development strategy adopted by the Taiwan government has been generally recognized as one of the critical factors in Taiwan's economic development (Wang 2003). For example, the Hsinchu Science-Based Industrial Park (HSIP) was established in 1980 by the National Science Council to ignite economic development in this island country. The park's high-tech talents collaborating with academic and research institutes in close proximity has clearly created 'cluster effects' and has contributed to significant economic outputs for the last two decades.

Also in line with the IT-led development strategies, the Southern Taiwan Science Parks (STSP) were established in Tainan in 1996 and in Kaoshiung in 2001. These two new science parks have been widely recognized as critical IT R&D incubators serving the same policy function as HSIP in northern Taiwan.

Challenges and opportunities

Despite its important role in supplying the world with ICT products, the Taiwanese ICT industry is faced with several challenges. First, the revenues for original equipment manufacturing (OEM) and original design manufacturing (ODM) ICT industries have been shrinking. To achieve economies of scale, more companies are merging. The government needs to provide a sound environment for small and medium-sized companies.

Second, Taiwan's ICT industry used to make products for the world's major brands. It needs to cultivate its own brand by engaging in more marketing and promotion.

A third challenge is that although Taiwan's ICT technology is advanced, Taiwan has not done enough to improve industry standards and technology innovation. More R&D needs to be devoted to innovative technology. Also, high technology experts abroad should be attracted into Taiwan, which means that a better working environment—better living standards, good medical care, and an international-standard living environment—should be provided.

A fourth challenge relates to the fact that the Taiwanese ICT industry has built many firms in China where labour is cheap and land prices are low. These firms produce 70 per cent of China's IT output (Lin 2005). Taiwan needs to make the best use of its technology talent and disciplined labour force to contribute more to the higher ends of the value-chain.

In the area of telecommunications, a level playing field for telecom operators and fair competition must be assured. The NCC needs to ensure that the CHT will open its last mile and provide reasonable wholesale prices to competing operators. As for digitization, the Digital Opportunity Centres should provide a good environment for the companies involved in digital entertainment to grow.

The e-Taiwan, m-Taiwan and u-Taiwan initiatives provide a clear direction for the growth of the ICT industry. But determination to implement the above mentioned policies and collaboration among the relevant government agencies, are also crucial to maintain Taiwan's competitive edge in the global ICT industry.

References

Dahl, A.L. and Lopez-Claros, A. (2006). The impact of information and communication technologies on the economic competitiveness and social development of Taiwan. In Dutta et al. (eds), *The global information technology report 2005–2006: Leveraging ICT for development*. World Economic Forum.

FIND. (2005). Measurable performance: Taiwan as an ICT giant. *e-Taiwan*. Taipei: FIND.

Global competitiveness report 2005–2006. (2006). Retrieved from http://www.weforum.org/en/initiatives/gcp/Global%20Competitiveness%20Report/index.htm

Huang, J. (2006). *Taiwan ICT hardware industry 2006*. US Commercial Service, Department of Commerce, USA.

ITU Reports. Retrieved from http://www.itu.int/osg/spu/publications/internetofthings/

Li, F. (2006). *Taiwan telecom service overview*. US Commercial Service, Department of Commerce, USA.

Lin, F.C. (2005). Case study: Rearing Taiwan's ICT industry. In Dutta, S. and Lopez-Claros, A. (eds), *The global information technology report 2004–2005: Efficiency in an increasingly connected world*. World Economic Forum.

Liu, Y.L. (2003). Broadband use, competition and relevant policy in Taiwan. *Journal of interactive advertising*. Available online at http://jiad.org

Point Topic. (2006). *World broadband maps Q1 2006*. Retrieved from http://www.point-topic.com/

Taiwan yearbook 2005. (2005). Government Information Office (GIO), Taiwan.

Wang, E.H.H. (2003). Technolopolis development in Taiwan: An IT-capabilities-enhancing approach. In Meheroo, J. and Taylor, R. (eds), *IT parks in Asia and the digital divide*. NY: M.E.Sharpe.

.th
Thailand

Thaweesak **Koanantakool** and Kalaya **Udomvitid**

Total population	62.42 million (2005)
GDP per capita	USD 2,567 (USD 1 = THB 38)
Key economic sectors	Agriculture, Mining, Manufacturing, Construction, Public utilities, Transportation and communication, Trading, Banking, Insurance, Public service and Defence
Computer per 100 inhabitants	13.7
Computers (on Internet) per 100 inhabitants	6.2
Fixed-line telephones per 100 inhabitants	11.7
Mobile phone subscribers per 100 inhabitants	47.9
Internet users per 100 inhabitants	13
Domain names registered under '.th'	21,976 (in July 2006)
Broadband subscribers per 100 inhabitants	0.9
Broadband users per 100 inhabitants	5.1
Internet domestic bandwidth	37.60 Gbps (in July 2006)
Internet international bandwidth	9.32 Gbps (in July 2006)

Sources: NSO 2005; NECTEC 2006.

ICT and ICT-related industries

The total value of the Thai ICT market, which is composed of four sectors—hardware, software, computer services and communications—was approximately USD 11,090 million in 2005.[1] It is expected that by the end of 2008, the Thai ICT market would be worth USD 19,329 million, which means a 20 per cent annual growth rate in the next three years.

The communications market, which includes both services and equipment, is the largest sector in the Thai ICT industry, accounting for 74 per cent of the total market. Its growth is a direct result of government policy to introduce competition from the private sector in 1993, with TelecomAsia Company (now True Corporation) being granted permission to participate in speeding up the two million-line telephone expansion project. Later, one million more telephone lines were added by TT&T. The entry of mobile service providers resulted in the number of mobile phone subscribers exceeding the number of fixed-line subscribers in 2001.

The hardware market also improved throughout 2005, due mainly to the growth in the notebook computer, personal digital assistant (PDA) and digital camera segments. Notebook sales totalled 280,000 units, up 41 per cent from 2004 levels, while desktop PC sale levels increased only slightly. The growth of the notebook market was driven by declining prices, as well as the notebook's portability and improved features and performance. In the PDA market, the market in 2005 was valued at USD 82.28 million, representing a 186 per cent increase from 2004 figures. The digital camera market also increased significantly, earning USD 184.2 million in 2005, or 56 per cent higher from the previous year.

The software market was valued at USD 1,090 million, representing a 27 per cent growth from 2004. Outsourced software took a 48 per cent share of the market, and the rest was taken by packaged software.

Along with the growth of the ICT market, the hard disk drive (HDD) production was recognized as a driver of national economic growth. The strong manufacturing industry has attracted key global HDD players to Thailand, and the inflows of foreign investment in HDD manufacturing made Thailand the world leader in the production of HDD and HDD components in 2005. The industry generated more than USD 263 billion in export value in 2005, accounting for 9.4 per cent of total Thai exports and 42 per cent of the worldwide HDD market. HDD production volume has more than doubled, from 54.1 million units in 2003 to 119.8 million units in 2006. More than 100,000 workers are currently employed by the industry.

Technology infrastructure

Thailand is making a significant investment in ICT infrastructure. Compared to developed countries which have put in high speed Internet infrastructure typically by adding value to fixed-line telephones and the cable TV business, Thailand had a later and slower start in broadband services. However, since 2005 increased investments in infrastructure have resulted in a big growth in the number of broadband users.

Broadband technology

In 2003, there were 12,700 broadband subscribers in Thailand. But only two years later, the number has jumped to 570,000 subscribers, according to the figures of ADSL service providers. In a survey by the National Electronics and Computer Technology Center (NECTEC) in 2005, about 39 per cent of respondents said they are using broadband services, from which it can be inferred that there are about 3,183,000 broadband users. The increase is due to improved competition among service providers.

IPv6: The Next Generation Internet (NGI)

For developing countries, IP technology promises to make voice and other services available at cheaper rates, compared to traditional networks. However, the rapid migration to IP technology is resulting in the exhaustion of IP numbers in IPv4. IPv6 technology is the solution to this problem. NECTEC runs an NGI laboratory to promote and develop applications over IPv6. The Thailand IPv6 Forum, consisting of local universities and Internet service providers, was established in 2006 and new applications are being developed.

Key institutions dealing with ICTs

The main coordinating organization for ICT development in Thailand is the Ministry of ICT (MICT). Other institutions are tasked with deploying ICT in various sectors.

Ministry of Information and Communication Technology (MICT)

The MICT is responsible for formulating implementation strategies and action plans in accordance with the national IT policy (IT 2010) and ICT Master Plan (2002–06). The vision of the ICT Master Plan is to use ICT for economic and social development and to improve the quality of life of Thai people and transform Thailand into a knowledge-based society. ICT is considered the key to the development of 'e-Thailand', consisting of e-Government, e-Industry, e-Commerce, e-Education and e-Society. Government agencies, led by the MICT, are expected to focus on e-Government, which is the core mechanism to mobilize the other e-Thailand components.

Software Industry Promotion Agent (SIPA)

SIPA is a public organization tasked with promoting the Thai software industry through rapid enhancement of software worker skills, rapid increase in employment and rapid growth both in the domestic and international markets. SIPA has established development guidelines for four main areas: enterprise software, animation and multimedia, mobile application and embedded software. The four areas are in high demand in the domestic and international markets. The guidelines help software developers to focus more on these strategic areas that build up Thailand's competitive edge in the software development outsourcing industry.

Electronic Transactions Commission (ETC)

The Electronic Transactions Act of 2002 established the ETC of which the Minister of ICT is the chairman and 12 members are appointed by the Cabinet from experts in finance, electronic commerce, law, computer science, science or engineering and social science. NECTEC acts as the Commission's secretariat. The functions of the ETC are to:

1. Recommend to the Cabinet policies for the promotion and development of electronic transactions;
2. Monitor and supervise the operation of businesses relating to electronic transactions;
3. Make recommendations or give advice to the Prime Minister to issue Royal Decrees pursuant to the Electronic Transactions Act of 2002;
4. Issue rules or notifications relating to electronic signatures in compliance with this Act or with the Royal Decrees issued pursuant to this Act; and
5. Perform any other act in compliance with this Act or with other laws.

National Telecommunications Commission (NTC)

The NTC was established on 1 October 2004 as the first independent state telecommunications regulator under the Telecommunication Business Act. Its duties and responsibilities are to regulate all telecommunications services in the country by formulating a Master Plan on Telecommunications Activities, set criteria and categories of telecommunications services, issue permits for and regulate the use of the spectrum for telecommunications services, and grant licenses to telecommunications operators.

National Electronics and Computer Technology Center (NECTEC)

NECTEC is a statutory government organization under the National Science and Technology Development Agency (NSTDA) of the Ministry of Science and Technology. Its main responsibilities are to undertake, support and promote the development of electronic, computing, telecommunication and information technologies through research and development activities. NECTEC also disseminates and transfers such technologies to stimulate economic growth and social development in the country, following the National Economic and Social Development Plan.

NECTEC was a forerunner in developing the national IT policy and IT promotion as mandated by the National IT Committee in 1992. It handed over policy tasks to the MICT after the completion of the first National ICT Master Plan of 2002.

Software Park Thailand

Software Park Thailand (http://www.swpark.or.th) is a software industry community development arm of the NSTDA with strong links to the private sector. Its mission is to be the region's premier agency supporting entrepreneurs and to help create a strong world-class software industry that will enhance the strength and competitiveness of the Thai economy. As of 2005, four more regional software parks were being developed in Thailand following the model of the first software park located in Bangkok, the central city.

Digital content initiatives

Many local websites with local language content have been created. The following are examples.

Search engines in Thai

SiamGuru and SanSarn are two main search engines using the Thai language. SiamGuru (www.siamguru.com) is a free directory and search engine for anyone who needs to find information relating to or about Thailand. It offers basic search, advanced search and multimedia search services. SiamGuru also provides news and Webboard search.

SanSarn (www.sansarn.com) was developed by NECTEC as a search engine for Thai/English documents. It aims to offer faster search and more precise results, particularly from local websites. Due to the non-segmenting characteristic of written Thai, words are written continuously without the explicit use of word delimiters such as spaces. To index Thai texts, Sansarn performs word segmentation to obtain the token terms which are then used as indexes for the search. To help users search more efficiently and effectively, Sansarn has features like word prediction and word approximation. According to Truehits.net statistics (as of 2005), Sansarn is the fifth most popular search engine in Thailand, after Google, Yahoo, MSN and SiamGuru.

ParSit: An online language translator

To help reduce language barriers and bridge the digital divide, NECTEC has initiated a machine translator (MT) R&D project which aims to utilize MT capabilities to translate English to Thai. ParSit (http://come.to/parsit) is a Web-based service through which people can browse Web pages in English and translate short passages of text into Thai. Although the system does not provide a perfect translation, it does provide the gist of the text translated and therefore allows many Thai users to 'read' Web pages in English in translation.

Online services

In recent years, a number of government and private agencies have begun to offer online services and information through their websites, some of which are described below.

The Thailand Central Registration for Missing Persons in Disasters website (www.missingpersons.or.th/) was created for relatives of persons who went missing as a result of the Tsunami disaster in southern Thailand on 26 December 2004. It provides information on missing persons, survivors, the injured and those found dead. The website is fully supported by the Thai Red Cross Society and NECTEC.

The Stock Exchange of Thailand (SET) website (www.set.or.th) runs an online stock trading system (www.settrade.com) to serve and support investors who have an ac-count with qualified brokers. After opening an account with a broker, investors can trade online directly through the broker's website anywhere and anytime using any PC connected to the Internet.

As the country's central bank, the Bank of Thailand (BOT) (www.bot.go.th) has established key financial risk control mechanisms for all financial institutions operating in Thailand. It has carried out two major innovations in electronic payments/settlements and established the Inter-bank Transaction Management Exchange (ITMX). In cooperation with ETC, BOT has completed for the Cabinet's consideration a draft Royal Decree for electronic payments. There are two online settlement systems in place—BAHTNET (Bank of Thailand Automated High-value Transfer Network) and SMART (Interbank Retail Funds Transfer Clearing System). With the operation of ITMX by the Thai Bankers' Association, it is expected that retail financial transactions can be handled online by early 2007.

The website of the Office of the Board of Investment (BOI) (www.boi.go.th) offers a large volume of information, including the ASEAN Supporting Industry Database, information from the Thailand Provincial Investment Gateway and Economic Warning Indices.

e-Commerce

In 2004, over 2,500 websites operated by approximately 1,860 e-commerce entrepreneurs operating in Thailand were registered with the Department of Business Development of the Ministry of Commerce. According to the e-transaction survey conducted annually by NECTEC, the transaction value of 61 e-commerce key players in B2B and B2C businesses in 2003 was about USD 1.52 billion (THB 58.5 billion), of which 99 per cent was classified as B2B. For B2G, a report from the Comptroller General's Department states the value of government e-auctions in 2003 as USD 129 million (THB 4907 million). Approximately 80 per cent of this amount comes from expenditures of state-owned enterprises. The survey also showed that the size of enterprises, classified by number of employees, is related to the extent to which enterprises absorb new technology. Larger enterprises are more likely to adopt advanced technology. The Thai e-Commerce Association was established in 2006 to help expand the market value of e-commerce in Thailand.

Enabling policies and programmes

Policies promoting investments in ICT and related industries

The BOI is committed to promoting investments in the ICT sector. In the software industry, for example, the BOI announced that software development projects will be classified as a priority activity and will receive exemptions from machinery import duties and corporate income tax, without an upper limit, for a period of eight years. This incentive also supports HDD, wafer fabrication and integrated circuits (IC) manufacturing.

In addition to the extended tax holiday, the BOI has amended its 'Skills, Technology and Innovation' or STI incentive package for projects with components in research, design, development of Thai staff, or support for educational or research institutes. The amendments allow a company with these STI expenditures to get an extension period for corporate income tax exemptions in addition to other privileges.

Industrial cluster development programmes

To strengthen local industries and respond to the demand for R&D, the NSTDA started clustering R&D projects in 2005. Examples of the industry clusters are the food industry, automotive industry and software-microchip-electronic industry. For each industry cluster, the NSTDA aims to improve local producer technology skills to increase local value added and maintain global competitiveness.

The HDD programme, which is part of the software-microchip-electronic cluster, is a successful collaboration between private companies and NECTEC. The latter has established an HDD training institute to develop the skills of HDD suppliers in Thailand. In 2005–06, 2,000 personnel from HDD industries joined the institute's training courses. The institute also intends to stimulate collaboration in R&D and technology transfer among local universities, state agencies and industries.

National Smart ID Card

The National Smart ID Card project is part of Thailand's plan to develop a modern e-government system. The concept is to make use of smart card technology for traditional citizen IDs to enable government agencies to share the machine-readable data from each card. The Thai government aims to improve public services, enable electronic transactions and boost the counter-terrorism effort through the national smart ID system.

Legal and regulatory environment

ICT laws

ICT laws have been put in place to nurture e-commerce activities, as well as to increase the confidence of multinational investors in Thailand. Six ICT laws have been developed since 1998. Two have been approved by the Parliament and merged into one Act, while the remaining drafts are in various stages of progress in the parliamentary process.

Enacted in April 2002, the Electronic Transactions Act of B.E.2544 (2001) recognizes the legality of data messages by treating them as the functional equivalent of writing or evidence in writing, to promote the reliability of electronic transactions. The Act combines two original drafts—the Electronic Commerce Law and the Electronic Signatures Law. The Act was drafted following the Model Law on Electronic Commerce 1996 and the Model Law on Electronic Signatures 2001 of the United Nations Commission on International Trade Law (UNCITRAL).

The National Information Infrastructure (NII) Law approved by the Cabinet in 2003 provides for an equitable information infrastructure and for universal access by promoting the right to affordable access to information and communications services. The law aims to reduce Thailand's digital divide, in accordance with Section 78 of the Thai Constitution.

Under consideration by the Parliament since September 2003 are the Computer Crime Law, which criminalizes new types of offences committed in cyberspace, and the Data Protection Law, which protects privacy rights by safeguarding the personal data of individuals.

The Electronic Fund Transfers Law, which aims to facilitate electronic fund transfers, was drafted in 2005. Through the efforts of the ETC, it is now a draft Royal Decree for Electronic Payment that is expected to be enacted in 2007.

> ## The National Smart ID Card: A not-so-smart move
>
> The National Smart ID Card project was approved by the Thai Cabinet in January 2004. The objectives of the project are to improve public service, increase efficiency, reduce wasteful public investment by using one smart card to access all public services and facilitate electronic transactions. The scheme is divided into three phases. The first phase aims to produce 12 million smart cards with a budget of THB 1,670 million (about USD 50.799 million) in 2004. The second phase, with a budget of THB 3.120 billion (about USD 94.9 million), will distribute 26 million cards in 2005. The third phase, also with a budget of THB 3.120 billion, covers the production of the remaining 26 million smart cards. The MICT is in charge of the procurement and management of the multi-purpose identity cards or smart cards, as well as securing cooperation among other government bodies concerning the use of the smart cards.
>
> The first phase was to have been completed in September 2006 by a consortium of ST Microelectronics, InCard Consortium and Cisco Engineering which won the bid to produce 12 million smart cards for the MICT for THB 888 million (about USD 27 million) (*The Nation* June 2004). The MICT is processing the second bid to provide 13 million smart cards within January 2007.
>
> However, many criticisms have been levelled at the Thai smart card project. First, the project was done in a hurry, without any standard interoperability framework. There was insufficient architecture planning and consultation with qualified experts. Second, the procured goods did not meet JavaCard 2.1.1 security specifications. Third, the cards were accepted and used without the critical functions, such as PKI and fingerprint matching, stipulated in the project terms of reference. The recommendations of the experts for the MICT to conduct technical acceptance tests were ignored. Subsequently, the cards delivered were found to have duplicated chip serial numbers and more than 100,000 cards were rejected.
>
> Lesson learned: Skipping key steps in the IT development process does not speed up a project. Although there is an impression of progress in the beginning, the whole project fails in the long term.

Information security standard

The Information Security Subcommittee under the ETC has proposed the adoption of the international Information Security Standard ISO/IEC 17799 and ISO/IEC 27001. The draft Information Security Standard for Thailand is widely accepted in the industry, and key business regulators, including the SEC and BOT, plan to enforce compliance to the ISO/IEC 27001 for all businesses under their control.

In addition, the Information Security Alliance (ISA) was set up recently to help develop and promote knowledge and awareness of information security. ISA was initiated from the collaboration between the major critical infrastructure organizations in Thailand. It is expected to raise public awareness of the International Information Security Standard.

Spectrum allocation for Radio Frequency Identification (RFID)

Radio Frequency Identification (RFID) technology is becoming increasingly widespread in most countries, including Thailand. Four main frequency bands are used for RFID applications: low frequency (LF), high frequency (HF), ultra high frequency (UHF) and microwave. The use of UHF is growing in whole sale, retail sale, transportation and logistics due to the adoption of RFID by large organizations such as Wal-Mart and members of GS1.[2] Due to growing usage of RFID at UHF, the NTC approved in December 2005 NECTEC's recommendation to use the RF spectrum in the band 920–924 MHz for RFID, in addition to LF and HF which were approved by the MICT in 2004.[3] The NTC and the industry expect this to promote the adoption of the RFID system at the international level.

Education and capacity building programmes

Lack of skilled labour in ICT and related industries, such as IC, HDD and consumer electronics, is a serious problem facing high-tech firms in Thailand. A 2002 NECTEC study of demand for IT personnel showed that the ICT industry would need 40,000 personnel in 2002–06. Education and training are very important tools for developing human resources with the skills to make effective use of ICTs. The following are key human capacity building programmes in Thailand.

Thailand Knowledge Park (TK Park)

The Thailand Knowledge (TK) Park was set up as a 'dynamic living library, a provider of good books that is constantly being updated and improved'. With its wide array of books, information and multimedia learning tools, TK Park has the major role of encouraging Thais to read and to learn. Covering 4,000 sq m at the Central World Plaza, the Park also functions as a cultural hub for young people and as a recreational space for a variety of activities. Approximately 500–800 persons visit TK Park daily. It was opened on 24 January 2005.

National Software Contest (NSC)

Held annually since 1999, the competition aims to stimulate students' interest in developing computer software. Participating students receive financial support during the project development stage. Outstanding projects are showcased at the national competition, which is also a venue for junior developers to meet professional developers and researchers. Winners receive cash rewards, with the top prize being the royal trophy awarded by H.R.H. Princess Maha Chakri Sirindhorn. The NSC project is organized by NECTEC and is sponsored by a number of organizations.

Outstanding NSC projects have been nominated for and have won recognition at the Asia Pacific ICT Awards (APICTA), an international awards programme initiated by the Multimedia Development Corporation of Malaysia to increase ICT awareness in the community and help bridge the digital divide. At the 2005 APICTA, 'Mech Tournament', a game software from Chulalongkorn University Demonstration School, won the Merit Award in the category of Secondary Student Projects, while 'Battle Crossword RPG Game' from Chiangmai University won the Merit Award for Tertiary Student Projects.

Young Scientist Competition in Computer Science and Engineering Project (YSC.CS and YSC.EN)

Organized by NECTEC, these competitions aim to build S&T capacity and standards among Thai secondary students. They encourage and support students to demonstrate their scientific capabilities and research skills in computer science and engineering. Participants submit their creative projects to the competition for a research grant. Qualified students are selected as representatives of Thailand in the Intel International Science and Engineering Fair (Intel ISEF), an international competition organized in the United States. YSC.CS and YSC.EN winners have captured prizes at the Intel ISEF. 'Visible Surface Determination Technique for Moving Camera in First Person 3-D Software VSD' won the Second Award in the Computer Science Category at the Intel ISEF 2000 in Detroit, Michigan while 'Statistical-based Adaptive Binarization for Document Imaging' won the Fourth Award in the Computer Science Category at the Intel ISEF 2006 held in Indianapolis, Indiana.

Embedded Systems Training Alliance for Thai Engineers Project (ESTATE)

ESTATE was launched in May 2006 by the Japan External Trade Organization (JETRO), Association for Overseas Technical Scholarships (AOTS), Technology Promotion Alliance (TPA) and the Thai Embedded System Association (TESA) to prepare Thai Embedded System developers to compete in the world market. The Embedded Systems Training Alliance for Thai Engineers was set up with funding from the AOTS and the Ministry of International Trade and Industry (METI) of Japan. This 12-month training programme in Japan aims to prepare Thai developers to work in the Embedded Systems industry through internships in Japanese companies. The scholarship includes studying the Japanese language in the morning and studying Embedded Systems in the afternoon for six months in Thailand before the one-year training in Japan. The first 15 scholars are scheduled to leave for Japan in January 2007.

Open source software initiatives

Many government and public organizations in Thailand are moving away from complete reliance on proprietary solutions towards partial use of open source software (OSS). The Electricity Generating Authority of Thailand (EGAT) is a success case. EGAT has installed OSS in half of its servers and one-fifth of its desktops, achieving savings in software investment of up to THB 300 million (approximately USD 7.9 million) from 1994 to 2004 (*The Nation* June 2004). EGAT's Internet department is also developing an open source software application, called *EGAT Linux*, for the organization's own use. Other government organizations, such as the Revenue Department, Royal Thai Air Force, Public Health Ministry and NECTEC, are also developing applications on open source software platforms for their online services.

For users who are familiar with the Windows Operating System, SIPA and NECTEC jointly introduced in 2004 an OpenOffice.Org 2.0 package with full support for the Thai language. The software runs well on both Microsoft Windows and Linux platforms. OpenOffice.Org 2.0 is more compatible with the Thai language and is easier to use than the previous versions. In 2005, OpenOffice V2.01 was launched and it is now widely accepted in Thailand.

NECTEC also participated in the development of the Linux Operating System. Initially NECTEC provided technical support to SchoolNet members to install Linux SIS (School Internet Server) on their servers. Linux SIS for the SchoolNet project was so successful that many small businesses also adopted the software. To promote the free-license solution, NECTEC teams up with and introduces Linux SIS to system integrators and small and medium-sized businesses. Under the project, companies working with NECTEC would offer free consulting and training on OSS and Linux SIS to the first 10 system integrators, enabling them to adopt Linux SIS for future system implementation for their customers.

Campaign for open standards

In addition to OSS development, NECTEC and leading Chief Information Officers (CIOs) in the public sector and a number of companies have initiated the 'Campaign for Open Standards in Thailand'. They organized a regional conference on Open Standards in May 2006.

Thailand has also initiated OpenCARE, an open exchange system for collaborative activities in response to emergencies. OpenCARE is information middleware that enables incompatible systems to work together. It is also an information/alert dissemination system. OpenCARE supports multiple input and output incident/progress reports. It works across borders and supports multiple languages by means of ISO 10646/Unicode. OpenCARE does not replace any of the existing relief systems but simply extends their reach to wider audiences and creates an open ICT ecology.

Research and development

Lack of technology capacity, especially in small and medium-sized industries, is a problem facing the ICT industry in Thailand. Thus, both the Science and Technology Strategic Development Plan for Thailand (2004) and the National Research Strategies are targeting R&D in key technologies. As a focal point of current and emerging technology development, NECTEC has initiated 10 strategic programmes:

1. RFID technology
2. HDD technology
3. Information and Mobile Applications
4. Embedded Technology
5. Wireless Technology
6. Intelligent Transportation System
7. Biomedical Engineering

Emergency and Educational Communication Vehicle: Cutting-edge technology

From the tsunami disaster in December 2004, Asians learned that an efficient and reliable emergency communication system is a necessity no less critical than food and potable water. In a typical disaster situation, while most basic communication systems cease to function, the need to communicate increases tenfold. To address this problem, the Emergency and Educational Communication Vehicle (EECV) was designed by NECTEC in collaboration with the Interior Ministry (Disaster Prevention and Mitigation Department), Asian Disaster Preparedness Centre (ADPC) and Cisco Systems. The EECV contains 25 notebook computers, 25 WiFi mobile phones, an electricity generator that can provide electricity for three days, and a network connection to high-speed Internet that would allow people in a disaster area to communicate with people outside the area using voice and Web.

The EECV was successfully tested in communities in the north of Thailand that were hit by a big flood in May 2006. Aside from providing basic voice communication for citizens and relief workers, the EECV also sped up the registration of and search for missing persons as people could search online for their relatives instead of making lengthy phone calls. Another Web-based EECV service is matching 'needs' and 'donations'.

The EECV's broadband Internet and mobile phone facilities and its 18-metre tall retractable antenna mast make it a natural choice for a C^3I (command, control, communication and information) centre for any kind of field operation. The antenna can also host older types of analogue communication systems, as well as serve as a sturdy platform from which to view a disaster area.

Aside from addressing urgent problems in disaster areas, the EECV also aims to facilitate distance learning for people in remote areas by providing Rural Wireless Broadband Access (RWBA) to schools and communities when it is not being used in disaster mitigation.

The EECV is a cutting-edge, low-cost and field-proven technology developed by local researchers. Its success highlights the value not only of technology but also of cooperation among the responsible agencies. The initial concept is now being developed for more versatile solutions and for military use.

8. Sensor Technology
9. Security Technology
10. Intelligent/Knowledge Engineering

In 2006, NECTEC completed a number of research projects with significant economic and social impact, such as the smart agriculture system, RFID readers, intelligent transport monitoring system and the Emergency and Educational Communication Vehicle (EECV).

The EECV is a vehicle designed to meet the need for vital communication services during emergency situations. A collaborative project of several government agencies and the private sector, the EECV provides people who are living in disaster areas and remote communities with a long distance communication link via satellite, terrestrial microwave, Wi-Fi or WiMAX, as well as local community services using Wi-Fi.

Challenges

The major obstacles to universal ICT use in Thai society are barriers to Internet access and lack of productive content, including educational materials. Barriers to access include the absence of basic telecommunication facilities and electricity in the rural areas. Moreover, computers are still not affordable for low-income families. Where access is not a problem, lack of useful content for Internet users and students has been identified as a critical challenge for effective ICT use. Although computer density and Internet penetration in Thailand has been increasing over the years, the development of relevant content has been slow. Furthermore, although the Thai government has tried to create a liberal and enabling environment for the growth of ICTs, there are inadequacies that still need to be addressed, such as the lack of transparency in large projects and the shortage of skilled labour in ICT fields.

Thailand needs to become a knowledge-based society and economy through the promotion of innovation and investments in key ICT infrastructure. For example, the government must provide support to ICT projects that seek to place computers within everyone's reach, such as NECTEC's economy-class PC, Computer ICT Project (Ministry of ICT and NECTEC) and the One Laptop per Child (Ministry of Education, MIT and NECTEC) programme. The cluster approach of the NSTDA and the Ministry of Science can accelerate industrial growth to a certain extent. Thailand must likewise seek new ICT business opportunities such as business process and software outsourcing.

Thailand must address these challenging issues now in order not to get left behind other countries in the region.

Notes

1. The ICT market survey was conducted by the Software Industry Promotion Agency (SIPA), National Statistics Office (NSO) and National Electronic and Computer Technology (NECTEC) in September and December 2005.
2. GS1 is a global organization dedicated to the design and implementation of global standards and solutions to improve the efficiency and visibility of supply and demand chains globally and across sectors. The GS1 system of standards is the most widely used supply chain standards system in the world.
3. Before NTC was established.

References

Bangkok Post. (20 July 2005). NECTEC targets SMEs.
Electronic Transactions Commission Website. Available at http://www.etcommission.go.th/
Koanantakool, T. (2005). Thailand. In Chin, S.Y. (ed.), *Digital review of Asia Pacific 2005/2006*. Published by ORBICOM, PAN IDRC, and UNDP-APDIP in association with Southbound.
Lenard, T.M. and Britton, D.B. (2006). *The digital economy fact book 8th edition*. The Progress and Freedom Foundation. Retrieved from http://www.pff.org/issues-pubs/books/factbook_2006.pdf
National Statistical Office (NSO) Website. Available at http://www.nso.go.th/
SIPA and ATCI. (February 2006). *ICT market survey 2005 and Outlook 2005*.
The Nation. (24 April 2006). Emergency vehicle now fully equipped for any eventuality. Retrieved November 2006 from http://www.nationmultimedia.com/2006/04/24/byteline/Byteline_30002356.php

.tl/.tp

Timor-Leste

João Câncio **Freitas**

Total mobile phone subscribers	
Prepaid	42,608
Postpaid	1,770
Number of Internet subscribers	
Dial-up	680
SDSL	17
Telecentres in the communities	8
Internet bandwidth	
Up Link	5 Mbps
Down Link	7 Mbps

Note: All figures are as of 2006.

Background

After gaining independence in May 2002, Timor-Leste became the newest country in the world and the 191st member of the United Nations. The official domain name for Timor-Leste is '.tl' although '.tp' is also still being used. Portugal ruled the territory for almost five centuries. This colonial era was followed by a controversial period of Indonesian occupation for more than two decades since late 1975. The occupation culminated in a UN-sponsored referendum held in August 1999 where, the majority of the people rejected the Indonesian proposal for autonomy, opening the way to independence. After the referendum, a huge military campaign of killing and destruction took place, backed by pro-Indonesia militiamen. The rampage forced more than half of the population to leave their homes, destroyed about 80 per cent of the schools and clinics and damaged the country's economic and social infrastructures.

In 2004, census results indicated that the population of Timor-Leste had reached nearly 925,000. With a population growth rate that is one of the highest in the world, Timor-Leste's population is expected to double by 2022, placing increasing pressure on public services, food security and the labour market. Service delivery has steadily improved, yet services are reaching a smaller proportion of the population than had previously been thought. Unemployment rates are high and growing in the face of a rapidly expanding youth population. In this context, enhancing the provision of sustainable services and creating productive employment, especially in the rural areas, within a framework of good governance will be critical to ensuring economic growth and poverty reduction in the years to come.

The national human development report of 2006 still places Timor-Leste as one the world's least developed and poorest countries and far behind the other countries in the region. While the national human development report of 2002 placed the per capita GDP at only USD 478, the latest reports show that per capita income is even lower at only USD 370 per year. Life expectancy is short, education levels are low and a high proportion of the population live below the poverty line.

The country is still suffering from the destruction and trauma that followed the referendum in August 1999. Seven years of UN presence (since late 1999) and four years of full sovereignty have not translated into concrete material gains for the people of Timor-Leste. In fact, per capita income and infrastructure have not rebounded to the levels present during the Indonesian occupation. Poverty and disenfranchisement, as well as disillusionment with the government in some sectors of the population, are factors in the social and political crises in 2006 that led to the dismissal of the first government and the installation of the second constitutional government.

Telecommunications infrastructure

Under Indonesian administration, the telecommunications sector was large, with the government as the biggest user. The sector provided telephone services to almost all parts of Timor-Leste using digital exchanges. In 1998, there were approximately 6,750 phones connected, of which 4,800 were in Dili. An exchange had been constructed in Dili with a capacity of 15,000 customers, well in excess of the local connections at the time. Mobile phone coverage was limited to Dili, although land line access was available in all districts. At that time, this was equivalent to about 13 phone connections per thousand population in Dili and the district capitals served. On a national basis, the telephone density was about eight phones per thousand. Timor-Leste had one of the lowest teledensities among countries with roughly comparable levels of GNP per capita.

Practically the entire ICT infrastructure of Timor-Leste was destroyed in 1999. This has given rise to considerable improvization in the past four years.

The communications sector plays a key supporting role to the other sectors of the economy. ICT such as radio, television, telephones, computers and the Internet can provide access to knowledge in areas such as agriculture, micro-enterprise, education and human rights, offering a range of choices that enable the poor to improve their quality of life. Thus, the Timor-Leste government's objective for the communications sector is universally available communications that brings telephone, Internet and broadcast services to communities throughout the country.

Much remains to be done to accomplish this objective. However, the time is propitious for well-placed, relatively small investments to leverage far reaching changes in the way people communicate, learn and do business. Public telecommunications services are being rebuilt under a Build-Operate-Transfer (BOT) arrangement with Timor Telecom. Public broadcasting has been restored and, with the help of several donors, community broadcasting services have been started in all of the Districts. There is one public broadcaster operating a radio and television service (RTL and TVTL). RTL has repeater facilities in each district, enabling broadcasts to almost all Timorese households. The television service is limited to Dili and Baucau, with the latter having the facilities to broadcast only taped programmes.

The business community and other segments of the private sector have developed their own communications network based on satellite phones. Government ministries have also developed their own separate infrastructure to deliver at least some of the services they need to function effectively. In key areas such as health, the development of these networks has been supported with donor funding to ensure that the Ministry has the capacity for health-related communications, including medical emergencies. The government also operates an internal computer communications network and rents Internet access from the UNDP.

With the award of a 15-year franchise to Timor Telecom in 2002, there has been significant progress in developing a national telecommunications network. Effective December 2003, the Timor Telecom network provided nationwide voice and Internet services covering Dili and the 12 district capitals. Under the conditions of the BOT contract, there are no long distance charges; uniform tariffs apply across the entire country. Private telecommunications services are not subject to the monopoly. Both public and private institutions can build and operate their own network. However, such services cannot be resold or offered publicly.

Currently, the number of fixed connections is about 3,000, while use of mobile phones has increased appreciably and is estimated to be about 44,400. These estimates imply about 40 telephones per thousand inhabitants—still low by international standards but an appreciable increase from 1998.

As mentioned, in 1999 practically the entire communications infrastructure of the country was destroyed. Initial efforts to restore telecommunications services were led by the International Force for East Timor (INTERFET), after which the Australian telecommunications company, Telstra, provided public services on a rolling, short-term contract basis. For the most part, this service was mobile, although a small portion of the landline network was restored. In these circumstances, the United Nations Transitional Administration in East Timor (UNTAET) developed and maintained its own stand-alone telecommunications system. The business community and other segments of the private sector developed their own communications networks (satellite phones, for example), especially those located outside Dili.

In the first phase of the telecommunications development programme, the objectives were as follows:

- To complete ongoing restoration of the telecommunications infrastructure and related training and development programmes;
- After rehabilitation of the infrastructure, to establish the legislative framework for regulating the telecommunications industry; and
- To prepare a tender for an experienced international company to develop the public telecommunications system on a BOT basis.

There has been considerable progress towards these objectives. The National Development Plan (NDP) recognizes clearly that telecommunications is a capital-intensive industry and that the broad objectives of the Plan could be achieved through a BOT contract with an experienced major private international company. The objective was to support the development of the industry with 'state-of-the-art telecommunications...[and] [f]rom the start of the contract, consistent and thorough training of local technicians and administrative staff and a full transparent...system with rapid deployment of capital.' The BOT contractor would be expected to build local capacities for transfer at the end of the BOT contract.

In 2002, the principal regulations and laws pertaining to public telecommunications were incorporated into an organizational BOT tender. Following a public BOT tender, a contract was awarded to Portugal Telecom International (PTI) and Timor Telecom (TT) in July 2002 to rebuild and operate public telecommunications in Timor-Leste. The contract extends a 15-year franchise to TT. Service provision began in the middle of 2003. By that time, TT had spent a total of about USD 12 million on the development of the national network. In December 2003, the network became fully operational and TT began to provide nationwide voice and Internet services covering Dili and the 12 district capitals. The BOT contract with the government specified that TT must supply 'universal' service at the same cost anywhere in the country.

Internet in Timor-Leste

Use of the Internet in Timor-Leste is encouraging. There are around 700 Internet subscribers; of these 680 subscribe to dial-up services and the remaining use SDSL (Symmetric Digital Subscriber Line). The level of Internet access, which today is considered to be one of the more obvious indicators of whether a country belongs to the Global Information Society, is still on the low side. Currently there are two operational Internet Service Providers (ISPs) and both run their main operations in Dili. An Internet subscription is usually shared by several people, especially accounts belonging to government, non-government institutions and companies. Only a few residents have Internet access in their own private residences. The main limitations are said to be economic—that is, the relatively high costs of computers, telephone lines and ISP fees. It is likely that at least 80 per cent of all users are in Dili. Even if reliable telecommunication facilities and ISP POPs now exist particularly in Dili, unreliable electricity, high Internet costs and lack of computers and skills limit the use of the Internet outside of the capital.

There are some Internet cafés in Dili, and a few of the better hotels in Dili offer Internet access to their guests. TT has been developing eight telecentres in the communities (TT shops). These are cybercafés that offer telephone services and Internet access. Internet cafés in Dili charge about USD 3 an hour.

Broadcasting services

Radio is a major channel for communication of information in Timor-Leste. According to a survey conducted in 2002, 69 per cent of the population listen to radio; of this total 42 per cent listen to radio everyday, 13 per cent listen between three to four days a week, and 8 per cent listen from one to two days a week, bringing the overall total to 63 per cent (The Asia Foundation 2002). For example, the government uses radio broadcasts to inform the population about the National Development Plan. Radio broadcasting services are provided by the Public Broadcasting Services (PBS) and by local, community-based stations.

Public broadcasting services

The creation of the PBS is mandated by the Constitution of Timor-Leste. It was established by law in March 2002. The PBS was created as an independent service whose budget is to be guaranteed by government, but government is not to interfere with editorial decision-making. The PBS is a juridical citizen—that is, it can own property and be used independently of government. It is allowed to receive contributions from various sources and to generate revenues to cover its operating expenses.

There is one PBS radio station in Dili that is re-broadcast on 12 transmitters throughout the country, primarily to district capitals. Dili has two transmitters, one FM and one AM. These radio programmes are distributed via satellite link to the 12 district capitals at a total cost of USD 1,500 per month. The PBS plays a major role in the development of democracy, in national dialogue and reconciliation and in providing information about health, education, agriculture and other sectors that save lives and improve the quality of life of the people of Timor-Leste.

There are currently two public broadcasting television transmitters, one in Dili and one in Baucau. The former is very unreliable and is also badly in need of replacement since parts needed for the repairs are no longer available in the market. The Dili transmitter currently operates at 10 per cent of its rated power. The television transmitter in Baucau is not yet linked to the network and can only play tapes. Given the severe limitations of the domestic service, a number of individuals have installed satellite television for their personal use in order to receive international programming.

Community radio services

There are 18 locally run not-for-profit community radio stations broadcasting to local communities throughout Timor-Leste. Of the existing community broadcasters, only one station—Radio Timor Kmanek (RTK) in Dili—was on air prior to 1999. Since 1999, another 16 locally owned and operated community radio stations have started operations. They were set up by local communities with help from a range of aid agencies, including the United States Agency for International Development (USAID), the World Bank, UNESCO and Internews.

There is one community radio station based in each district outside of Dili and six stations established in Dili, of which two are not currently broadcasting. In addition, the international network of Christian broadcasters established a radio station called Voice FM in Dili in October 2003, while the Diocese of Baucau established an FM radio station in Baucau in late 2005. There are also five stations that repeat the signal of international broadcasters, including Radio Australia and BBC (British Broadcasting Corporation) World Service.

The legal and regulatory framework

In November 2003, the government promulgated the basic legal framework for telecommunications. The relevant laws are:

1. Law 11/2003, which set the basic framework for telecommunications, postal services and BOT franchise (Timor Telecom)
2. Law 12/2003, which created the Communications Regulatory Authority—*Autoridade Reguladora de Comunicações* (ARCOM)—to regulate Timor Telecom and other areas of telecommunications and postal services

In addition, UNTAET Public Broadcast regulation 5/2002 provided for the creation of an independent Public Broadcasting Service on 20 May 2002.

Although some broadcasting laws and regulations exist, as mentioned earlier, laws relating to community and private broadcasters have not yet been enacted. In effect, community broadcasting is self-regulated. In the absence of an enacted regulatory framework, the Association of Community Radios in Timor-Leste has developed a Code of Practice for all community broadcasters with support from Internews. The main issue in broadcasting is the sustainability of the various institutions.

ARCOM is intended to consist of a Council and an operational organization. However, the Council has yet to be formed and only a handful of staff is currently employed to perform some of ARCOM's regulatory function. The International Telecommunication Union (ITU) is currently providing assistance to the Government of Timor-Leste in establishing the operational structure for ARCOM. Included in this initiative will be significant support to enable ARCOM to manage and regulate the telecommunications sector in the country. Among the major outputs will be a Telecommunication Act, a Radio Communications Act and a Broadcast Act.

ITU will also assist in the elaboration of a National Frequency Plan, National Radio Regulation and a guide for spectrum monitoring. Likewise, ITU will assist in the introduction of WinBASMS as the computerized Spectrum Management System for the country. Assistance will be provided for the design of a coverage map for MW, FM and TV broadcasting, and the enhancement of local competence in all of the areas mentioned. The project will include an integrated capacity building process for upgrading the skills of ARCOM staff. It is expected that at the end of three years, Timor-Leste will have established a robust regulatory organization, acquired sufficient know-how and formulated some of the most significant regulatory policies towards the development of a healthy telecommunications sector.

National ICT initiatives

Aside from building up the telecoms regulatory organization and framework, the Government of Timor-Leste is currently undertaking a convergence study to lay the foundations for integrated voice and data services within the government in a cost-effective manner. This study will be implemented in conjunction with two other new projects to improve government telephone services, and to improve government Internet services. These initiatives could save the government over USD 1 million a year in communications costs.

Another initiative to reduce poverty and increase growth through the improvement of rural infrastructure is the Internet for Isolated Villages Pilot project. The project is designed to provide Internet service in remote and/or isolated village schools and communities. Introducing the Internet into village schools can not only transform education in rural Timor-Leste but also help modernize communications for agriculture and rural commerce.

Reference

The Asia Foundation. (2002). *National survey of citizen knowledge.*

.vn Vietnam

Tran Ngoc Ca and Tran Ba Thai

Total population	83.22 million (in 2005)
GDP per capita	USD 650
Key economic sectors	Agriculture (rice, coffee, cashew nut), Mining, Oil and Gas, Manufacturing (garment, textile, footwear, electronics etc.)
Computers per 100 inhabitants	2.4
Fixed-line telephones per 100 inhabitants	18.73
Mobile phone subscribers per 100 inhabitants	10.68
Internet users per 100 inhabitants	16.14
Broadband subscribers	227,000 (ADSL)
Domain names registered under '.vn'	21,481
Internet international bandwidth	5,795 Mbps (as of 2005)

Technology infrastructure

Over the past few years, Vietnam has both strengthened existing ICT infrastructure and set up new ICT infrastructure. The International Gateway Switch is being operated alongside the National Transit Switch. There are six national switching systems in Hanoi, Danang and Ho Chi Minh City. The local tandem switch is located in Hanoi and Ho Chi Minh City, and local switches are operating in many provinces. PSTN has been replaced by new technology such as Next Generation Network (NGN). ADSL services and broadband access are also being used to improve the quality of Internet connections. There are currently three international telecom centres linked to each other by optical fibre cable. The provincial network system relies on a north-south optical fibre cable of 20 Gbps and an electricity network with a 500 KV transmission. The backup system is a digital transmission system of 140 Mbps. The access network consists of a local cable network (coaxial and optical fibre), a wireless subscriber network and a VSAT system. Data transmission relies on X25 technology.

However, the use of telecom services in the rural areas remains poor. Although more than 92 per cent of communes in Vietnam have access to telephone services, less than 2 per cent of rural households have a telephone. Likewise, Internet use is not equally spread among regions, with 86 per cent of Internet users living in Hanoi and Ho Chi Minh City. Thus, Internet use in the rural areas is still negligible.

Fixed telephone infrastructure

According to ITU statistics, the rate of growth of fixed-line telephones in Vietnam in 2000–2005 was 44.1 per cent, the highest in the world. The Asian average for the same period was 11.9 per cent and the world average was 5.3 per cent. In 2005, there were 18.73 fixed-line telephones per 100 people in Vietnam.

The providers of fixed-line telephones services are Vietnam Post and Telecom Co. (VNPT), Sai Gon Post and Telecom Share Holding (Saigon Postel), Electricity Vietnam Telecom (EVN Telcom), Hanoi Telecom and Military Telecom Co. (Viettel). International telecom services are provided by VNPT, EVN Telcom and Viettel.

Mobile telephone infrastructure

The mobile telephone market experienced a 62.7 per cent annual growth rate between 2000 and 2005. However, the ratio of mobile phones to every 100 inhabitants is still only 10.68, lower than that of Asia and the world. Counting both fixed and mobile phones, the ratio is 29.42 per 100 inhabitants.

VNPT with VinaPhone and MobiPhone provides mobile phone services to 4.4 million subscribers using GSM technology. They have begun to provide GPRS/MMS services to prepare for 3G services. Saigon Postel uses CDMA1x technology and has 1.4 million subscribers. EVN Telcom also uses CDMA450. Hanoi Telecom began with CDMA2000 technology. Viettel uses GSM technology and has 1.5 million subscribers.

Internet infrastructure

Since it began operations on 19 November 1997, the Vietnam Internet Network has increasingly become an important tool in many socio-economic, cultural and State management areas. Internet costs continue to decrease and favourable rates are charged during off-peak times. Software parks receive discounted rates. From April 2003, prices of 12 telecommunication services, including the Internet, mobile phone connections and international calls, were reduced by 20–38 per cent.

Between 2005 and 2006, the Internet in Vietnam continued to grow rapidly and the subscriber base grew by 53 per cent from 1.8 million to 3.6 million. However, users far outnumber subscribers at 13.4 million in 2006, representing 16.14 per cent of the total population of Vietnam. The annual growth rate of Internet subscribers and users is 1.5 per cent.

The total bandwidth linking Vietnam internationally is 5,795 Mbps. There are 21,481 registered domain names (.vn) and 764,672 IP addresses. There are 985,364 Kbps leased-line subscribers among 339,734 Internet subscribers, 13 licensed ISPs and six companies licensed for IXP (Internet eXchange Point). VNPT has the largest share of the market (43.13 per cent in 2006), followed by FPT Tel (24.08 per cent) and Viettel (18.61 per cent).

In 2005, the number of ADSL subscribers increased nearly 300 per cent, reaching a total of 227,000. VNPT, FPT Telecom and Viettel have a combined ADSL market share of 98 per cent.

As of late 2006, all 64 provinces and cities have completed the Internet development programme in universities, colleges and schools. In addition, most government agencies at the central and provincial levels have access to the Internet, which enables them to provide e-government services.

Organizations and companies like Intel, the US Agency for International Development (USAID) and Vietnam Datacommunication Company (VDC) have experimented with new technologies such as WiMAX. For example, Lao Cai, a mountainous province near Chinese border, is the test location for Intel's Asian Broadband Campaign.

The Master Plan for Internet and telecom development in Vietnam sets the following targets for 2010:

- A telephone penetration rate of 32–42 per 100 inhabitants (with a fixed-line telephone penetration of 14–16 per 100)
- An Internet subscriber rate of 8–12 per 100 (of which 30 per cent are broadband subscribers)
- 25–35 per cent of the population as Internet users
- Access to the Internet for majority of teachers, students, doctors and school children
- An Internet connection in all communes, and public Internet access points in 70 per cent of communes
- Broadband Internet services in 100 per cent of districts and key economic zones
- Broadband Internet connection and links to the government WAN for all government organizations and agencies, down to the district level
- Broadband Internet access points for all research institutes, universities, colleges and secondary schools
- Internet connection for 90 per cent of primary schools and hospitals

Digital content initiatives

The Vietnamese language has been standardized to Unicode UTF 8. There are many websites in Vietnamese. Several operating systems, especially open source systems like Linux, have been localized by some local companies. The domain name system is also supported in Vietnamese script. Other Vietnamese languages like Thai, Cham, Jarai, Bah'nar, Êđê, M'nông, Sê đăng and K'hor are also being standardized in Unicode.

Among local groups, the Vietkey Group is the most active (http://www.vietkey.net) in developing local language content. The Group is working on the localization of foreign fonts to support Vietnamese in Windows, Linux and PDA (Windows CE, Palm) environments. The Group also provides free Vietnamese language solutions. The EVietnam Group (http://evietnamese. net) provides e-learning training in the local language for foreigners. Its website is in five languages: Vietnamese, English, French, Japanese and Korean. Two other local language websites are Vietnameseonline (http://vietnameseonline.net) and eVietnamese (http://evietnamese.org).

The Ministry of Post and Telematics (MPT) initiated in 1997 the Commune's Cultural Post Office. This digital content initiative is based on traditional village cultural and social life and involves the setting up of a public access point equipped with newspapers, telephones and an Internet connection, as well as development of Web content in Vietnamese. The content is suitable to the conditions and interests of poor farmers living in mountainous and remote areas. Currently there are 8,000 public access points in 10,800 communes. Most have dial-up Internet access while a few have ADSL connections. Informal assessments indicate that farmers have benefitted from the use of these public access points. On the downside, some of the currently available content is not suitable for rural life and agricultural activities and the Internet connection is not always reliable.

Online services

Online gaming

A range of online services has been developed, including mobile content, e-news and online newspapers, online games and online databases. The most popular are online games and value-added services over mobile phones. In 2005, the leading company in the online game business (VinaGames) reached a turnover of USD 5 million while the Center for Mobile Business (VASC), another company, earned USD 9 million. Six locally developed online games were featured in a competition in 2005. Most of the games were built using 3D technology while some were written using Direct X 9.C technology.

Currently, there are about 10 companies with 300 employees working on the production of local online games. At a projected growth rate of 70 per cent per year, there should be at least 83 online gaming companies in 2010, with a turnover of USD 1 million each. It is also expected that by then, locally made games would have a share of 25 per cent of the gaming software market. The most active player in promoting online games is VASC, a company under the Vietnamnet Media Share Holding Group associated with VNPT (www.vietnamnetgroup.com), FPT Communication and VDC, another firm under VNPT. VASC is also collaborating with several foreign suppliers to bring online games into the domestic market.

e-Government

Efforts to computerize government offices under the previous National IT Programme have achieved some success. The State Administrative Management Computerization Project (known as Project 112) between 2001 and 2005 sought to build and put into operation an electronic information system to improve the effectiveness and efficiency of government administration. Some successful cases include online business registration and licensing in Ho Chi Minh City, customs declaration in the Dong Nai province and electronic forestry monitoring in Dak Lak. However, the Project will only fully succeed with more rigorous reforms in traditional public administration. Key reforms must include a reorganization of the public administration system in terms of reporting procedures, information control and information sharing over the Internet, as well as improved inter-agency cooperation. Furthermore, provincial and central administration agencies must find a way to harmonize their reporting systems to eliminate difficulties in exchanging files. Currently, different users have different Vietnamese fonts for Microsoft Word files. Users in the north use the VnTime font while users in the south prefer VNI. The use of Unicode as a common platform has not yet been implemented.

The first experiences of some government agencies in using ICT as a means of providing information and timely services to citizens are encouraging. In Dong Nai province, the Customs Office has put custom declaration forms online. In Ho Chi Minh City, e-registration for business licenses has been implemented. Other attempts to introduce a 'one-stop shop' for public and administrative services have been undertaken in Hanoi, Haiphong, Quang Ninh, Thanh hoa, Ba ria-Vung tau and Da Nang, with some degree of success. Other central government organizations, such as the Ministry of Trade and the Ministry of Planning and Investment, are to follow suit in providing legal documents, regulations and other information on their websites. The websites are updated regularly to make sure visitors can get the latest information about new regulations and relevant business matters. Nevertheless, to make ICT a really useful e-government tool, a more comprehensive legal framework is required, as well as staff training and more vigorous changes in public administration.

Recently, the government put up its own website (www.gov.vn) to create an electronic database on government activities to serve as the official source of information about government policies and to provide an exchange forum for government agencies, government officials, corporations and the general public. The overall aim is to provide citizens with better access to public services.

e-Commerce

According to several surveys of the status of e-commerce in Vietnam (MOT 2005), 82.9 per cent of commercial companies have Internet access and 25.32 per cent have their own website. Seventy percent (70.14 per cent) use ADSL, 16.29 per cent use leased lines, and 13.75 per cent use dial-up. Seventy-three percent (73.91 per cent) are engaged in B2B transactions and 56.09 per cent are engaged in B2C transactions. In 32.9 per cent of the companies surveyed, there is a person designated to work on their online presence. However, the investment structure for ICT is still unbalanced: 62 per cent is for hardware, 29 per cent for software and only 12 per cent for training (MOT 2005).

Overall, the last few years have seen rapid development of various kinds of online services, such as information supply, online marketing, Internet-Intranet, software and e-commerce solutions, tourism, online learning and consultancies. The Vietnam e-Commerce Portal (www.ecvn.gov.vn) was created in 2003 to support companies interested in e-commerce (MOT 2005).

e-Banking

One of the most popular online services is e-banking, which includes Internet banking, mobile banking, phone banking and e-payment services. There are 6,400 ATMs in Ho Chi Minh City alone, and approximately three million ATM cards in the whole country. In 2006, banks experimented with new services, such as home banking and payment of salary and pension funds over the Internet.

ICT industries

In 2005, the Vietnam ICT market (that is, ICT spending, including production for domestic use and import) reached USD 828 million, which represents a 20.9 per cent growth since 2004. Of this total, the hardware market accounts for USD 630 million (76 per cent) and the software market accounts for USD 198 million (24 per cent).

The total industry production value (for both domestic use and export) in 2005 was USD 1.4 billion, which represents a 49.6 per cent increase since 2004. Domestic use accounts for USD 288 million (USD 108 million for hardware and USD 180 million for software), while export value accounts for USD 1,112 billion (USD 1,042 billion for hardware and USD 70 million for software).

The hardware industry includes production of PCs, telecom equipment and electronic products and components. Several Vietnamese companies manufacture and assemble PCs from imported semi-knocked down (SKD) components. Some companies cooperate with foreign companies in assembling computers with a Vietnamese brand name, such as CMC, SingPC, Mekong Green, VINACom, T&H, Robo and Elead. These cooperative efforts have resulted in lower prices for local consumers. A similar approach is being followed for the production of telecom equipment, such as switching systems.

In 2005, the Vietnamese hardware industry reached a turnover of more than USD 1 billion for the first time. Of this amount, USD 1,042 billion was for export and USD 108 million was for the domestic market. ICT became one of the top seven export sectors, after crude oil, garment, seafood, footwear, furniture and rice. The leading exporters were 100 per cent foreign-owned firms like Fujitsu (printed circuit board, USD 515 million) and Canon Vietnam (printers, USD 450 million). Other multinational companies began to increase their investment in the Vietnamese ICT market. For example, Intel invested more than USD 300 million in Ho Chi Minh City and recently upgraded this to a USD 1 billion investment.

The software industry in Vietnam consists mostly of small companies with an average of 20–30 staff members. These companies focus on Vietnam-specific requirements and on providing software services. There are currently about 6,000 software companies with 15,000 staff and a productivity of USD 10 million in 2005. Overall, the number of software and service companies as well as software engineers increased 23 per cent in 2005.

Subcontracting software for export reached USD 70 million in 2005, a 55.5 per cent growth over 2004 figures. The software and services industry grew 47 per cent and achieved a turnover of USD 250 million. Of this, USD 180 million (61.1 per cent) came from the domestic market and USD 70 million (38.9 per cent) came from exports.

The targets for software production by 2010 are: an average annual growth rate of 35–40 per cent; a turnover of USD 1 billion with an export share of 40 per cent; training of 150,000 engineers and IT experts, of which 40–50 per cent should be professional software specialists; becoming one of the top 15 software exporters; reducing the rate of IPR violations to the region's average; and mastering some technologies in key products.

There are nine software parks in Vietnam, including Saigon Software Park (SSP), QuangTrung Software Park (QTSP) and E-Town.

Needing more attention is the content industry which remains underdeveloped in terms of producers and technical facilities despite recent efforts to promote its growth.

Key institutions dealing with ICTs

To ensure the development of a cohesive regulatory framework, the government set up a new Ministry of Post and Telematics (MPT) in 2002. Most functions and responsibilities concerning ICT development have been transferred to MPT from the Ministry of Science and Technology (MOST) and the Ministry of Industry. However, several other ICT-related activities are under the supervision of different ministries. For example, MOST is responsible for R&D management and provides funding for high-technology development, among which ICT is a key priority. R&D activities are undertaken in several institutions, the most notable being the Institute for Information Technology under the Vietnam Academy of Science and Technology (VAST). The Ministry of Trade is responsible for commercial activities, especially those related to e-commerce development. The Ministry of Education and Training (MOET) is in charge of ICT training and education, while the Ministry of Industry has jurisdiction over industrial production activities.

MPT also hosts the Secretariat of Steering Committee for 58 Directive, which aims to facilitate the application and development of information technology in industrialization and modernization for the 2001–2005 period. In 2004, a network of Departments of Post and Telematics (DPT) under MPT was set up in provinces and cities to deal with ICT-related matters on behalf of MPT. In 2005, MPT also created the Department for ICT Application to oversee broader ICT opportunities in all areas of the economy and society, such as those for improving productivity in the manufacturing sector and agricultural information. To oversee Internet promotion, management and development, MPT set up the Vietnam Internet Center. The Centre is also responsible for the allocation of Internet addresses and domain names (under .vn). In 2006, MPT set up the VTC-Multimedia Corporation as the basis for a television development company.

ICT is also being promoted by non-governmental organizations. Within the Central Committee of the Communist Party, there is a board tasked with promoting ICT applications among Party units and organizations. Within the National Assembly structure, the Committee for Science, Technology and Environment proposes various initiatives related to technology development, especially ICT-related legislation. A notable example is the e-Transaction Law enacted by the National Assembly.

Private sector organizations also play an important role in ICT development. The Vietnam Association of Information Processing (VAIP), Hanoi Association for ICT, Ho Chi Minh City Computer Association (HCA), Vietnam Association of Electronics Enterprises, Vietnam Software Association (VINASA) and Vietnam Chamber of Commerce and Industry (VCCI) are very active and work closely with the government to promote the growth of ICT. For example, these associations can involve ICT companies and organizations in the discussion and preparation of policy recommendations to the government. Their role has become important enough for the government to assign to VCCI the task of organizing implementation activities for ICT applications in SMEs. Other important association activities are the publication of an ICT yearbook, the organization of an ICT Olympiad competition for students, annual provincial meetings on ICT-related issues, and coming up with an ICT index (ranking) of cities and provinces in cooperation with foreign actors like the International Data Group and Asia-Oceanian Computing Industry Organization.

Enabling policies and programmes

Several important documents on ICT strategy and master planning were promulgated in 2005–06, such as:

- Decision No. 191/2005/QĐ-TTg promulgated on 29 July 2005, and establishing a programme to assist small and medium-scale enterprises in using ICT to increase productivity and performance for the period 2005–10.
- Decision No. 222/2005/QĐ-TTg promulgated on 15 September 2005, and putting in place a master plan to develop e-commerce in 2006–10.
- Decision No. 246/2005/QĐ-TTg promulgated on 10 June 2005 to develop ICTs and ICT-supported development in Vietnam.
- Decision No. 32/2006/QĐ-TTg promulgated on 7 February 2006, mandating development planning to promote the development of telecommunications and the Internet in Vietnam until 2010.

In general, these strategies or plans would provide a platform and roadmap for the development of e-commerce, telecoms and the Internet up to 2010 and, in some cases, beyond 2010. More important, they provide for a tentative financial allocation for implementing the proposed activities.

ICT laws and regulation

The National Assembly enacted the law on electronic transactions on 29 November 2005 and the intellectual property rights law, which has a special section on software development, on 12 December 2005. In June 2006, a joint document (No. 60/2006) to regulate the production, supply and use of online games in Vietnam was issued jointly by the Ministry of Culture and Information, MPT and Ministry of Police. Also issued in June 2006 were the Decree on e-Commerce and the Information Technology Law. The Decree was issued to promote all e-commerce-related activities, while the IT Law sets the legal framework for infrastructure development, ICT application and industrial development.

There are others laws that are related to ICT, such as the Law on Commerce. In addition, a new law on telecommunication is being drafted. Since joining the World Trade Organization (WTO) in 2006, Vietnam also had to prepare for the Information Technology Agreement clauses, in particular the exemption of import duty on ICT products among WTO member economies.

All of these policy documents, laws, decrees and regulations constitute the legal framework for ICT development in Vietnam. As such, they clarify the 'rules of the game' for ICT development in the country. However, their implementation and enforcement is a problem as it usually takes some time before laws are transformed into concrete regulations. Moreover, different ministries and implementing bodies sometimes have different interpretations of policy documents.

Security issues

To deal with emerging security problems arising from inappropriate use of the Internet, the Ministry of Police issued in 2004 Resolution No. 71/2004 regulating Internet businesses in Vietnam.

To address potential emergency situations in the computing environment, MPT also set up in 2006 the Vietnam Centre for Computer Emergency Response Team (VNCERT). The Centre is tasked to deal with Internet safety, security and technical standards.

The most important security organization in Vietnam is the Bach Khoa Center for Internetwork Security (BKIS) under the Hanoi Technology University (www.bkis.org.vn). This R&D organization, which was created in 2001, provides services addressing virus attacks and hacking problems. BKIS supplies BKAV (Bach Khoa AntiVirus), the most popular antivirus software in Vietnam. It helps the government solve security problems, and works with foreign organizations and members of the Asia Pacific Computer Emergency Response Teams (APCERT). In 2004, the government upgraded it into a national emergency centre with 55 members of staff. In general, BKIS works as a core element within the VNCERT community.

The Center for Network Solutions and Computer Rehabilitation (http://www.sinhviennghco.biz) and Vietnam Security (http://www.security.com.vn) are two other organizations working on ICT-related security issues in Vietnam. Within the Ministry of Police, there is a taskforce for computer-related or high-tech crimes.

Education and capacity building

Many colleges and universities have set up faculties teaching ICT subjects. By 2000, there were about 20 IT faculties in Vietnamese universities and more than 20,000 students had completed undergraduate (Bachelor's) degrees in ICT-related fields. This translated to 10,000 people working in professional IT units (that is, IT companies, R&D, and IT education and training).

The increase in the number of trained IT personnel is the result of Decision No. 331/QĐ-TTg (6/4/2004), which set up a programme to develop human resources for IT-related fields until 2010. In 2005, MPT was assigned to develop a master plan for ICT human resource development and many other strategies were formulated, including Resolution No. 05/2005/NQ-CP which emphasizes the involvement of private sector organizations in ICT education and training.

On 14 June 2005, the National Assembly passed the Law on Education which provides for the development of private universities. A network of private universities has been set up, with several institutions working on ICT training, such as the ICT University in Ho Chi Minh City, FPT University (run by FPT, an ICT company) and the Vietnam-Korea College for ICT. Currently, there are approximately 80 universities providing university degrees in ICT-related fields, with an intake of more than 10,000 students per year. Counting the 103 colleges (which are middle-level educational institutions between high schools and universities) offering ICT degree programmes, the annual ICT student intake goes up to 20,000. There are also 60 diploma and certificate programmes (non-university degrees) in ICT offered by many training centres and units. A notable cooperative initiative in ICT education is being pursued by Microsoft Vietnam and Partners in Learning to provide ICT training and support to 50,000 schoolteachers and two million school children.

Research and development

The Ministry of Science and Technology (MOST) oversees the state's research programme in ICT, coded KC-01. One of the main goals of R&D is to establish a close link between the market, research and education, including a special orientation toward a Vietnamese-machine interface such as voice recognition.

The leading institute working in IT is the Institute of Information Technology (IOIT), which is under VAST. There are other institutes, such as on telecommunication, microelectronics and information systems, working with several ministries.

Every year, MOST selects key R&D programmes in ICT. In 2006, the projects selected included: development of key products in voice recognition and Vietnamese text, grid computing to support complex tasks, development of a buffer system and Internet security, and production of audio process and receiver using digital technology. The institutes involved in these projects are IOIT, the Department for ICT, the General Department of Technology and Voice of Vietnam Radio.

Open source initiatives

Linux was introduced in Vietnam in the mid-1990s through research institutes such as Francophone IT University and technology universities in Ho Chi Minh City and Hanoi. Both Linux Office and Microsoft Office are now available in Vietnamese. Open source software (OSS) is a main research thrust of the VietKey group, which has been using a Linux base to develop Vietnamese fonts.

To promote OSS, the Vietnamese government issued in 2004 Decision No. 235/QDD-TTG, which outlines a Master Plan for Applying and Developing Open Source Software in Vietnam for the period 2004–08. According to the plan, OSS will be tested in selected organizations. A range of incentive policies and programmes have been designed, such as staff training, technical assistance and standards creation. The Master Plan's road map identfes the creation of some key software products by 2007. MOST has formed a special National Steering group to work on OSS development. The total investment for the period 2003–07 is about USD 20 million. This amount is divided among nine projects conducted by various ministries, such as MOST, MPT, MOET, the Ministry of Home Affairs, the Ministry of Labour and Social Affairs and the Ho Chi Minh City government.

There are no official statistics on the use of OSS in Vietnam, but MOST is conducting a survey with the support of the Organization Internationale de la Francophonie. It is safe to say that there are many OSS users, such as organizations under the e-government programme, Communist Party organizations and some small firms that provide Internet solutions and services.

Challenges

ICT development in Vietnam is still dominated by an industry that mostly seeks to generate profits. The notion of ICT as an enabler of economic development has not yet received adequate attention from key stakeholders in society. A more radical change of attitude at all levels and layers of society is needed to develop Vietnam into a more open, transparent and impartial ICT-friendly society.

For one, in spite of various efforts to encourage more competition, the mighty monopoly of VNPT remains almost intact. As it joins the WTO, Vietnam will have to confront the challenge of foreign players demanding the opening of the domestic market for telecom services.

Recent changes in state management of ICT also need some attention. Although creating a separate government body for ICT makes sense and is in line with trends in many economies around the world, it is not clear how this body, MPT, can effectively take over all ICT-related activities. MPT is responsible for regulating the telecommunications sector in Vietnam. It undertakes the dual role of policymaker and regulatory authority (Global Internet Policy Initiative 2003). This combination of functions is not common. Moreover, important areas of concern are still under the management of the Central Government Office (Project 112), the Ministry of Trade (e-commerce), the Ministry of Industry (PC production), MOST (R&D and other technology development programmes), MOET (education and training in ICT) and so on. On the other hand, no organization has overall responsibility for developing an e-economy and e-society. A balance must be struck for MPT to perform its organizational functions in coordination with other agencies. At the same time, MPT must assume leadership in the development of an e-society as one of its main functions in addition to telecom infrastructure and ICT industry development.

There is a need for impartial treatment of all telecoms players. MPT has a close relationship with business entities such as VNPT, the biggest player in the telecom and Internet market. Officially, the running of VNPT has been separated from state management. However, the previous relationship between the government and VNPT makes impartial treatment of this company difficult.

Despite the movement toward more liberalization and competition, vestiges of monopoly by some actors remain, which hampers the achievement of ICT4D goals. There are several points to note in this regard. At the macro policy level, there is no clear separation between the perception of ICT as enabler and the role of industry. At the meso and micro level of using ICT for development, the cost of access to ICTs for poor people and for rural areas remains high. The issue of affordability to achieve equity in access to information needs to be addressed as soon as possible. Deregulation and opening up the market is a step in this direction.

References

Economist Intelligence Unit. *Annual report.* 2002–2004.
Gartner Dataquest. (2005). *IT spending report.*
Global Internet Policy Initiative. (2003). *Draft report on telecommunications and Internet policy of Vietnam.* Hanoi.
Ho Chi Minh City Computer Association. *Vietnam ICT panorama* 2001, 2002, 2003, 2004, 2005.
ITU. *Statistics data.* 2001–2006.
Ministry of Trade. (2005). *E-commerce report 2005.* Hanoi.
PCWorld Vietnam. *Annual review* 2001–2006.
UNDP. *Achieving MDGs through ICT: Experiences and challenges in Vietnam.* Hanoi.
World Ecomomic Forum. *Annual reports* 2002–2004.
World Bank. *World development report* 2002–2004.

Review of sub-regional associations

Association of Southeast Asian Nations

South Asian Association for Regional Cooperation

Association of Southeast Asian Nations

Lorraine Carlos **Salazar** and Shelah **Lardizabal-Vallarino**

Introduction

The Association of Southeast Asian Nations (ASEAN), which was established in August 1967, has grown in the past three decades from five members—Indonesia, Malaysia, Philippines, Singapore and Thailand—to include Brunei Darussalam, Vietnam, Lao PDR, Myanmar and Cambodia. ASEAN encompasses a vast region with a total population of about 530 million and a land area of 4.5 million square kilometres of archipelagic nations as well as countries located in Continental Asia. In 2005, ASEAN's combined gross domestic product (GDP) reached almost USD 700 billion while its total trade amounted to USD 850 billion.

ASEAN's foundational aim was to 'accelerate economic growth, social progress and cultural development' and 'promote regional peace and stability'. In particular, ASEAN leaders concurred on the importance of integrating their economies in order to remain competitive and to reduce poverty and socio-economic disparities within and among the member states.

Acknowledging the need to narrow the developmental divide among member countries, ASEAN leaders launched the Initiative for ASEAN Integration (IAI) in December 2000, which focuses on bridging the gap between the older and the newer members, in particular Cambodia, Laos, Myanmar and Vietnam (collectively known as CLMV). The IAI includes infrastructure development, human resource development, information and communication technology and promoting regional economic integration.

Moreover, in October 2003, ASEAN leaders reached an agreement to create an ASEAN Community by 2020. The plan, also known as the Bali Concorde II, is made up of three components: the ASEAN Security Community, the ASEAN Socio-Cultural Community and the ASEAN Economic Community (AEC). At present, the establishment of an ASEAN Economic Community (AEC) is spearheading ASEAN regional integration. The AEC is envisioned to create a single market and production base characterized by a free flow of goods, services, investment, capital and skilled labour by 2020.[1]

There is a consensus within ASEAN that information and communication technologies (ICTs) have a significant role to play in efforts to realize the goal of regional economic integration and the vision of an ASEAN Community. This chapter reviews the efforts of ASEAN as a regional body in enhancing the region's ICT capacity and utilization. First, the chapter considers the key institutions in ASEAN that deal with ICTs. Next, key ASEAN ICT policy statements and programmes are reviewed, with emphasis on initiatives under the Vientiane Action Plan 2004 until the present.[2] The article then examines ASEAN's digital content, online services, as well as initiatives and programmes in education and capacity building, open source and research and development. The chapter concludes with the issues and challenges faced by ASEAN in utilizing ICTs to better the lives of its citizens and to foster the spirit of ASEAN community.

Key ICT Institutions

This section reviews the institutional mechanisms and bodies within ASEAN that are in charge of ICT policies and their implementation.

Telecommunications and IT Ministers' Meeting (TELMIN)[3]

In recognition of the increasingly important role of ICTs and the challenges and opportunities they offer, ASEAN established the Telecommunications and IT Ministers' Meeting (TELMIN) in 2001. Its aim is to strengthen and institutionalize regional cooperation on telecommunications and ICT issues. Before the creation of the TELMIN, the ASEAN Economic Ministers Meeting (AEM) was in charge of telecommunications and IT matters, along with a long list of economic issues.

The TELMIN meets annually. Some of the meetings are sessions with ASEAN Dialogue partners, such as China, Japan and Korea, as well as India and the European Union.[4] Following ASEAN custom, the TELMIN chair is rotated annually among all members, with the incoming chair hosting the next annual meeting. In its policy planning for the ICT sector, the TELMIN engages telecommunications and ICT industry players, especially through the e-ASEAN Business Council which is composed of private sector representatives from all of the ASEAN member countries. In addition, since its third meeting in 2003, the TELMIN has included a session in its meetings to engage ASEAN youth, seeking their views and discussing their thoughts on how ICTs affect their lives and the future of ASEAN.

At its 6th Meeting, held in Brunei on 18–19 September 2006, the TELMIN reviewed the progress and achievements in the implementation of the ASEAN ICT Focus 2005–10, the 2005 Hanoi ICT Action Agenda, and in particular the e-ASEAN Integration Roadmap and the Vientiane Action Programme's (VAP) telecommunications and IT action agenda. They discussed common issues in the creation of a conducive, competitive and sustainable ICT environment, digital content development, network security and important initiatives for strengthening human and institutional capacities in the ASEAN ICT sector. They highlighted the importance of building capacity in areas such as ICT literacy, full integration of ICTs in education and training at all levels, and creation of a critical mass of qualified and skilled ICT professionals and experts. The Ministers stressed that enhancing ASEAN's competitiveness in these areas is crucial in ensuring its relevance in the borderless world.

The Ministers announced that the ASEAN ICT Fund, created to finance priority ICT projects, is now in place. The Fund consists of contributions from ASEAN member countries as well as commitments from ASEAN Dialogue partners Japan, China, India and Korea. The Ministers agreed that the Fund will finance priority initiatives in the implementation of the Brunei Action Plan in 2007, in particular the development of ICT masterplans in CLMV, and capacity building, including: (a) training for home workers in ASEAN countries, (b) workshops on public domain and content development, (c) e-learning, e-culture and e-heritage training for ASEAN youth, and (d) ASEAN ICT skills standard development (Sonia 2006).

In addition, the 6th TELMIN endorsed the creation in 2007 of an ASEAN ICT Centre (AICTC) to strengthen management and coordination mechanisms in the implementation of ASEAN ICT Focus and Work Programmes. The Centre's Programme Director and the two Deputy Directors will be seconded from Malaysia, and Indonesia and Vietnam, respectively. The Ministers also welcomed the regular dialogues and collaborative projects with dialogue partners (China, the European Union, India, Japan and Korea) and key international organizations like the International Telecommunication Union (ITU) and the Asia-Pacific Telecommunity (APT). The Ministers said they look forward to the implementation of the cooperation work plans with China, Korea and Japan namely, the Plan of Action to Implement the Beijing Declaration on ASEAN-China ICT Cooperative Partnership for Common Development, ICT Co-operation for Co-Prosperity in East Asia 2007–11, and Japan's Asia Broadband Programme: ICT Cooperation with ASEAN. An ASEAN-India ICT Ministerial and Industry Forum is also being organized for 2007.

Finally, the Ministers agreed to revitalize the e-ASEAN Youth Forum and the e-ASEAN Business Council by ensuring that activities and priorities are relevant to and supportive of the ASEAN ICT Focus/Work Programmes.

Telecommunications & IT Senior Officials' Meeting (TELSOM)

The ASEAN Telecommunications & IT Senior Officials' Meeting (TELSOM) was created to serve as the coordinating arm of the TELMIN. Specifically, TELSOM's mandate is to 'supervise, co-ordinate and implement policies, programmes and activities for telecommunications and Information Technology cooperation in ASEAN, in line with the directions and priorities set by TELMIN'. To facilitate coordination, the TELMIN country chair also chairs TELSOM.

The impetus to create TELSOM grew out of a May 2000 meeting among APEC Ministers of Communication and Information Technology where the need to create a regional platform for dealing with ICT issues was raised. Thus, in October 2000, the first ASEAN TELSOM meeting was convened in Brunei. TELSOM is composed of senior telecommunications officials designated by each of the 10 ASEAN member countries. They meet at least once a year and special meetings are held from time to time upon the TELMIN's request or direction. TELSOM works with the ASEAN Telecommunications Regulators'

Council (ATRC), the e-ASEAN Business Council and ASEAN's Dialogue Partners, in particular to:

- identify, implement and monitor cooperation programmes and activities to meet the telecommunications and IT requirements of the ASEAN region;
- serve as a forum for information exchange, discussion and consultation on major regional or international issues and developments in telecommunications and IT that are of common interest to Member Countries;
- provide the mechanism to promote participation of the private sector, regional/international organizations and non-governmental organizations in the development and implementation of its programmes and activities;
- establish, whenever necessary, working groups/expert groups to assist in the development and implementation of its cooperative programmes and activities;
- report its progress to the ASEAN TELMIN; and
- carry out other activities that may be mandated by the TELMIN and/or requested by other relevant ASEAN bodies.

In the First TELSOM Meeting in Brunei, TELMIN and TELSOM were designated as the key institutions responsible for the ICT aspects of ASEAN Vision 2020. Their tasks are articulated in the Hanoi Plan of Action as follows:

- To achieve interoperability and interconnectivity of National Information Infrastructures (NIIs) of Member States by 2010;
- To develop and implement an ASEAN Plan of Action on Regional Broadband Interconnectivity; and
- To intensify cooperation in ensuring seamless roaming of telecommunications services (that is, wireless communications) within the region, as well as in facilitating intra-ASEAN trade in telecommunications equipment and services.

Currently, TELSOM is composed of four working groups (WG) and chaired by a designated lead or shepherd country for two years. The chairmanship is rotated in alphabetical order among the shepherd countries. The WG which meet at least twice a year, are:

1. ASEAN Information Infrastructure (AII)—Convened in 2002, the AII Working Group (WG-AII) was formed to facilitate the establishment of an ASEAN Information Infrastructure by enhancing the design and standards of the National Information Infrastructure (NII) of member countries and ensuring their interoperability and interconnectivity; to work towards establishing high-speed direct connection between the NII with a view to transforming this interconnection into an ASEAN Information Infrastructure backbone; to facilitate the setting up of national and regional Internet exchanges and Internet gateways, including regional caching and mirroring; and to promote the security and integrity of ASEAN Information Infrastructure.

   In line with its mandate, WG-AII has undertaken several initiatives such as the formation of national Computer Emergency Response Teams (CERT), guidelines for information sharing among CERTs, a convergence policy framework and an NII database.

2. e-Commerce and ICT Trade Facilitation (EC & ITF)—This WG is tasked to establish dialogue mechanisms and cooperative arrangements among ASEAN TELSOM administrations to discuss regulatory and policy issues pertaining to trade and investments in the telecommunications and IT sector and technological aspects of e-commerce. Its second task is to encourage and facilitate the adoption of e-commerce regulatory and legislative frameworks that will build confidence among consumers and facilitate the transformation of ASEAN businesses to become e-enabled. Third, it is to establish dialogue mechanisms with the private sector in ASEAN and with ASEAN Dialogue Partners on technological issues related to e-commerce and ICT trade facilitation.

   Aside from formulating guidelines on e-commerce legal infrastructure in ASEAN member countries, the WG on EC and ITF has undertaken the creation of an ASEAN trade and investment policy and regulatory database. The WG is currently considering the following projects for implementation: interoperability of data exchange for ASEAN economic development (proposed by Thailand), setting up regional cooperation to curb e-mail spam, establishing an ASEAN e-Commerce Centre of Excellence where an exchange of experts in the field of training and legal framework is undertaken (proposed by Malaysia), and creating an ASEAN Frequent Traveller Card to facilitate the entry/exit of ASEAN nationals.

3. e-Society and ICT Capacity Building—This WG was created as a platform to develop an e-society in ASEAN and enhance capacity building by promoting awareness, general knowledge and appreciation of ICT, particularly the Internet. It is tasked to promote positive use of the Internet through seminars, workshops, common ASEAN guidelines and website development; and to recommend incentives and policy guidance in collaboration with the private sector for the development of local and ASEAN content and

other applications to maximize use of existing and planned infrastructures for ICT, as well as generate opportunities for business, livelihood and employment.

4. Universal Access, Digital Divide and e-Government[5]—This WG was created to enhance access, adoption and usage through universally accessible ICT networks and competitive and affordable ICT products and services, and to help reduce the digital divide among ASEAN member countries. It also aims to encourage and facilitate the provision of a wide range of government services online, facilitate linkages and/or consultations between the public and private sectors, and enhance inter-governmental cooperation by promoting electronic procurement of goods and services and by facilitating the freer flow of goods, information and people within ASEAN.

ASEAN Telecommunications Regulators' Council (ATRC)

The ATRC was established in 1995 at the height of revolutionary technological changes in the industry. Its aim is to facilitate the development of the telecommunications industry in the region in line with the establishment of an ASEAN Free Trade Area (AFTA).

In 2001, the ATRC formally became an adviser to the TELMIN. One of its tasks is to discuss and coordinate policy, as well as strategic and regulatory issues in telecommunications that are of mutual interest to the telecommunications administrations of ASEAN nations, such as radio frequency coordination, standards, regulatory trends and issues, strategies for telecommunications development and international affairs. It also aims to identify and promote areas of potential cooperation in telecommunications among ASEAN members and facilitate the exchange of information in these areas through seminars, training programmes and workshops.

The chairmanship of the ATRC rotates annually among its members. The following are the key activities undertaken by the ATRC:

- Harmonization of frequency spectrum allocation and use within the ASEAN region, such as satellite services, mobile cellular services and radio paging services to enhance mobility;
- Harmonization of technical compliance processes and standards for telecommunications equipment to support the effort of ASEAN manufacturers and suppliers to market their products in one another's market and to enhance the range and availability of this equipment for users. One concrete output of this is the formulation of the ATRC Mutual Recognition Arrangement (MRA) for Telecommunications Equipment.
- Promotion of the development, interconnectivity and interoperability of the NII of ASEAN countries through harmonization of broadband multimedia networks standards; and
- Development of human resource expertise to support the growing needs of the region's telecommunication sector through the establishment of linkages among members to facilitate skills training and enhancement in telecommunications fields.

In 2006, the ATRC focused on network and cyber security, next-generation networks, radio frequency identification, and MRA for trade in telecom equipment. CERT Incident Drills were held in July 2006. The ATRC plans to expand the CERT Drills to include ASEAN's dialogue partners, test responses to simulated computer attacks and further strengthen cooperation among national CERTs. In 2007, the ATRC will focus on new initiatives in international mobile roaming charges, mobile number portability, prepaid SIM registration and information sharing on consumer protection, among others.

Enabling policies and programmes

This section reviews the ICT policy pronouncements and programmes of ASEAN that have been released since 2004.

Brunei Action Plan (BAP) 2006

The BAP, which was issued on 19 September 2006 after the Sixth TELMIN in Brunei, is ASEAN's newest ICT plan of action. It has four goals: (a) creating a conductive, competitive and sustainable ICT environment; (b) developing digital content; (c) ensuring network security; and (d) strengthening human and institutional capacities in the ASEAN ICT sector. It has eight points of action, as follows:

1. Build ASEAN ICT capacity by: (a) intensifying capacity building and training programmes to facilitate cross-border electronic transactions and the use of electronic signatures; (b) enhancing capacity building activities to support the development of the e-readiness and ICT master plans of ASEAN member countries by 2008; (c) pursuing the development of ASEAN professional ICT skills standards; (d) intensifying capacity building and training programmes for national CERTs; (e) supporting local content development to increase ICT usage at all levels of society; and (f) promoting the widespread use of ICT as a tool to

boost capacity building across all genders and segments of the society, including home workers.
2. Develop the ASEAN Information Infrastructure as the foundation for the sustainable development of an information society by: (a) strengthening the region's cyber-security network through expansion of the ASEAN CERT Incident Drills to include ASEAN's Dialogue Partners and (b) deepening policy and regulatory cooperation to deal with opportunities and challenges in the area of next-generation networks, including convergence issues and Voice over Internet Protocol (VoIP).
3. Achieve broader economic and social benefits through wider access to ICT by: (a) exploring open standards and open source technologies to increase ICT access and interoperability; (b) facilitating affordable universal access and connectivity through the use of new technology, such as wireless broadband, particularly for rural and remote communities; (c) promoting ASEAN-wide online services through the development of local and ASEAN content and the delivery of government, social and commercial services through the Internet; and (d) facilitating the development of ASEAN research and education networks.
4. Facilitate ICT trade and electronic commerce by addressing non-tariff barriers to trade and laying the policy and legal infrastructure for electronic commerce by: (a) welcoming increased participation in the ASEAN MRA on telecom equipment, (b) sharing best practices in implementing telecommunications competition policies, (c) fostering the preparation of domestic legislation to enable e-commerce transactions in all ASEAN member countries, (d) facilitating mutual recognition of digital signatures to enable cross-border transactions in ASEAN, and (e) establishing a networking forum among businesses in ASEAN and Dialogue Partners as a platform to promote trade and investment.
5. Exchange information on and harmonize, where it is appropriate, policies and regulations to increase ASEAN's ICT competitiveness—for example, the ATRC 2006–07 work plan, particularly in the areas of: (a) mobile number portability and international mobile roaming charges; (b) wireless broadband deployment, including in rural areas; and (c) prepaid SIM card registration.
6. Engage the private sector and the youth by: (a) strengthening cooperation with the e-ASEAN Business Council (e-ABC) through high-level strategic dialogues between ASEAN ICT communities and TELMIN, (b) encouraging e-ABC to play a leading role in promoting ASEAN as an attractive region for ICT investment, and (c) working with institutes of higher learning to establish a dedicated forum for ASEAN youth to address their ICT capacity building needs in areas such as applications software development, digital content development and e-learning beginning 2007.
7. Forge links with strategic partners and key ICT international organizations to pool resources and expertise by: (a) deepening cooperative activities with Dialogue Partners, in particular China, the European Commission, India, Japan and Korea, in areas like broadband, RFID, content development, standards, Internet exchanges, emerging new technologies, telecom network security and ICT infrastructure; and (b) continuing engagement with international organizations like the ITU and APT through regular dialogues and collaborative projects particularly in the area of capacity building, especially in human resource development.
8. Strengthen institutional foundations to achieve the programmes elaborated above by: (a) ensuring effective and efficient utilization of the ASEAN ICT Fund for projects of mutual benefit to all ASEAN countries, (b) setting up the ASEAN ICT Centre (AICTC) by 31 March 2007 to manage and administer the ASEAN ICT Work Programme, and (c) encouraging the establishment of Centres of ICT Excellence in ASEAN member countries that could serve as test beds or proofs of concept for new ICT technologies, applications and services beginning 2007.

The BAP cites 11 ICT priority projects for implementation in 2007 under the four TELSOM Working Groups, namely:

- Engaging ASEAN Dialogue Partners in ASEAN CERT Incidence Drills (ACID II)
- Training workshop on Open Document Format (ODF)
- Capacity building on the establishment of CERT for Lao PDR and Cambodia
- ASEAN ICT e-mall to facilitate ICT trading
- Use of ICTs to empower home workers in ASEAN countries
- ASEAN workshop on public domain and content development
- e-Learning and e-Culture/e-Heritage for the Youth
- ASEAN ICT skills standards development
- Engagement of ASEAN Dialogue Partners in IPV6 capacity building initiatives
- Framework for a Research and Education Network (REN)
- Free/Libre and Open Source Software (FLOSS) distribution kiosks

ASEAN ICT Focus 2005–10

Issued in September 2005 during the Fifth TELMIN, this work plan is described as 'the overall guiding document in the medium term, for concerted and collective cooperation in building the information society and in enhancing the region's connectivity and competitiveness' (Joint Media Statement of the Fifth TELMIN). The activities and projects included in the work plan are organized under the four main goals of (a) facilitating the establishment of a pervasive, interconnected and secure ASEAN Information Infrastructure; (b) facilitating the growth of electronic commerce and ICT trade in the ASEAN region; (c) developing an e-society in ASEAN and capacity building to reduce the digital divide within and among individual ASEAN members; and (d) striving for universal access to ICT infrastructure and services and creating digital opportunities and applications in the delivery of government services (e-government). Thus, the work plan integrates the goals and concrete action points contained in the e-ASEAN Framework Agreement, the Vientiane Action Programme 2004–2010, the Roadmap for Integration of the e-ASEAN Sector, and the terms of reference of the TELSOM Working Groups.

Hanoi Agenda on promoting online services and applications to realize e-ASEAN (2005)

This agreement, which was also released in September 2005 after the Fifth TELMIN, emphasizes cooperation in five areas: (1) enabling policy and regulatory environment for online services and applications, (2) interconnectivity and interoperability of ICT networks, (3) creation of digital content and online services, (4) improving network security, and (5) ICT capacity building initiatives focusing on the youth and the underprivileged. TELSOM and ATRC officials have been tasked to carry out the necessary measures to implement the Hanoi Agenda.

Roadmap for integration of e-ASEAN sector, 2004–10

ASEAN Leaders launched the Roadmap in November 2004 as the ICT component of the *ASEAN Framework Agreement for the Integration of Priority Sectors* in which ICT is identified as one of 11 key sectors. The Roadmap concretizes four objectives, namely: (a) to liberalize trade in ICT products, services and investments; (b) to develop, strengthen and enhance the competitiveness of the ICT sector in ASEAN; (c) to reduce the digital divide within individual ASEAN member states and among ASEAN member states; and (d) to promote cooperation between the public and private sectors in realizing e-ASEAN. The Roadmap contains specific measures such as the elimination of tariffs on about 683 ICT products;[6] elimination of non-tariff measures; establishment of rules of origin; improvement of customs procedures, standards and conformance; development of integrated transport logistics services within ASEAN; outsourcing and industrial complementation; reduction of restrictive investment measures; and many more.[7]

Vientiane Action Programme (VAP) on telecommunications and the IT sector

The leaders of ASEAN issued the VAP in November 2004 as a guiding plan for the establishment of an ASEAN Community. The ASEAN Economic Community includes a strategy statement on key sectors, including telecommunications and IT. In the VAP, the leaders of ASEAN recognize the importance of leveraging ICTs via public–private sector partnerships and strong external linkages to build a connected, vibrant and secure ASEAN community by:

- Targeting universal access to ICT infrastructure and services;
- Encouraging the development of a pervasive, interconnected and secure ASEAN information infrastructure;
- Strengthening cooperation and assistance on regulatory policy and strategy issues;
- Creating digital opportunities through e-government, e-commerce and e-society initiatives;
- Enhancing the competitiveness and dynamism of the ASEAN ICT sector by promoting and facilitating trade and investment in ICT services; and
- Developing highly skilled ICT human resources.

Digital content, online services, capacity building, R&D and other ICT initiatives

The ASEAN Secretariat's website, www.ASEANsec.org, serves as the repository of information about the organization—its members, meetings and activities. The website also serves as an archive of ASEAN documents, declarations, papers and speeches. In addition, http://www.ASEANsec.org/99.htm provides a link to all ASEAN-related information and links.

In September 2003, ASEANconnect (http://www.ASEANconnect.gov.my/index.php) was set up to serve as the Association's ICT portal. ASEANconnect is an online database containing information and indicators on the telecommunications and IT industry in ASEAN as well as existing initiatives and programmes for bridging the digital divide and facilitating trade undertaken by the TELMIN, TELSOM and ATRC. The website was developed by the TELSOM WG on Universal Access and Digital Divide within ASEAN under Malaysia's leadership and funding. However, the information on the website, especially the ICT indicators database, needs updating.

From 1999 to 2005, 36 e-ASEAN projects were recorded in ASEAN's Projects List by year and sector. These can be accessed at http://www.ASEANsec.org/14490.htm.

One of the most recent ICT projects in ASEAN is the ICT4D ASEAN Collaboratory (http://www.ict4dASEAN.org) (see boxed article below).

In line with the Association's efforts to bridge the developmental divide among its members, the Initiative for ASEAN Integration (IAI) was launched in 2000. ICT is identified as one of the four key areas for narrowing the developmental gap in ASEAN.[8] As of September 2006, 22 of the 132 IAI projects, including 14 that have been completed, are on ICT.[9]

In the 2005–06 period, the following activities and initiatives were undertaken by TELMIN:

- MRA for telecommunications equipment by Brunei, Indonesia, Malaysia and Singapore in 2004–05, and Malaysia-Indonesia and Malaysia-Brunei to be concluded in 2006.
- Harmonization of legal infrastructure for e-commerce in ASEAN, especially on issues such as e-transactions and e-signatures.
- ASEAN cyberlaw survey.
- ASEAN connect Web portal.
- Seminar on e-learning.
- 2nd ASEAN-China ICT Week.
- Cooperation with India on systems security, e-learning technology for the visually impaired, the tele-education and tele-medicine network, and the IT industry forum.
- Framework for cooperation on network security and its action plan especially on spam.
- ASEAN Chief Information Officers' Workshop.

Also in 2005, Malaysia offered an additional grant of up to USD 500,000 for 2006–08 to support the Smart Schools Project for CLMV. The Project aims to provide these countries with ICT facilities and a human resource development programme package.

With regard to research on telecommunications liberalization and harmonization of policies in the ASEAN region, the Australian Aid-funded Regional Economic Policy Support Facility (REPSF) has funded three studies: (a) ASEAN Telecommunications and IT Sectors—Towards Closer ASEAN Integration; (b) Regulatory Models for ASEAN Telecoms; and (c) Liberalization and Harmonization of ASEAN Telecommunications.[10]

The ICT4D ASEAN Collaboratory

The ICT4D ASEAN Collaboratory was established by the ASEAN Foundation in partnership and with funding support from Canada's International Development Research Centre (IDRC) to serve as: (a) a regional ICT incubator and training facility to cater to the needs of individuals and organizations in ASEAN countries, as well as Asian countries outside ASEAN; (b) an Internet-based technology bed to support experimentation and testing of ICT applications; and (c) a one-stop showcase of hard, soft and process technologies and tools used by progressive developing countries in ASEAN as well as Asian countries outside ASEAN, in support of their funded digital development programmes and projects.

The Collaboratory targets development practitioners in government and non-government institutions as well as ICT specialists and end-users of ICTs in the developing countries of Asia. It focuses on such themes as Women and ICTs, ICTs in Small and Medium-Scale Enterprises, and ICTs for Rural Communities. Customized and regularly scheduled ICT training courses for individuals and groups are conducted, such as face-to-face training or individualized hands-on training in Jakarta and distance-training courses and incubatory trials and pilots in e-commerce, database-building and website development. In the incubatory trials, learners are given step-by-step Web-based instructions on how to set up their applications and to practice in a closed site before going live with their applications on the ICT4D Collaboratory server to service their clientele.

The Collaboratory also offers Web hosting, video hosting and e-commerce (http://www.panASEANemall.org). The target for the next three years is to serve as a regional digital incubator and training facility in e-commerce, educational technologies, e-government, expert systems, geo-information system, bibliographic and textual database system, system-generated discussion and mailing list, electronic conferencing, multimedia applications and other Web-based technologies.

> ### ICT education projects for CLMV
>
> Under the Association's Initiative for ASEAN Integration (IAI) programme, older ASEAN members (also known as the ASEAN-6: Brunei, Indonesia, Malaysia, Philippines, Singapore and Thailand) have agreed to help the newer members, Cambodia, Laos, Myanmar and Vietnam (collectively known as CLMV) bridge the developmental gap through a series of projects focusing on four areas: infrastructure, human resources development, ICT and regional economic integration. Since the Association has limited financial resources to extend to its newer members, it has relied on funding from international financial institutions or developed countries to finance infrastructure projects. In areas where the ASEAN-6 feel that they are capable of helping the CLMV countries, domestic resources are mobilized to extend help. Two examples of ICT initiatives under the IAI are Malaysia's Smart Schools Project and ICT Certification Training for the CLMV countries.
>
> The Malaysian government's Economic Planning Unit's Technical Cooperation Programme manages the Smart Schools Project, which aims to extend assistance in developing ICT competencies in CLMV. The first phase of the project was implemented in 2002–03 in three schools in Myanmar and two schools in Lao PDR. The Malaysian government has set aside USD 500,000 to fund the project's second phase from 2006 to 2008. The second phase will include the expansion and enhancement of the project in Myanmar and Lao PDR, and introduction of the Smart Schools initiative in Cambodia and Vietnam.
>
> Moreover, since 2004, Malaysia has been offering CLMV participants ICT Certification Training covering basic, intermediate and advanced courses, such as computing and the evolution of networking, advanced computing and mobile satellite communications systems, network management and mastering technology in creating a digital environment. In addition, Malaysia is offering two scholarships to each of the CLMV countries for Postgraduate Research Studies in ICT at the Multimedia University of Malaysia.

Challenges

ASEAN as a regional body has been actively undertaking various cooperative actions to advance the region's utilization of ICTs toward regional integration and development. The Association has established key institutions that meet regularly to thresh out ideas and issues, and to develop and implement plans and programmes.

The creation of the ASEANconnect website as a repository of documents and information on the work of the TELMIN, TELSOM and ATRC, separate from the ASEAN Secretariat's website, is a commendable effort. However, the website needs to be maintained and regularly updated so that it can serve its function as the central portal for ICT information in the region.

Another laudable development is ASEAN's move to engage its dialogue partners in ICT-related cooperation. This is a step towards institutionalizing official mechanisms and dialogue for addressing ICT issues and challenges in a collaborative way.

Funding for pilot projects and for continuing good projects is perhaps the main challenge facing ASEAN. The ICT Fund is a concrete step forward. It is not enough, however. Bilateral and aid financing from international funding institutions, as well as support from developed countries, will continue to play an important role in developing ICT capabilities in the region, especially for CLMV.

Finally, as this review indicates, ASEAN is not lacking in policy agreements and initiatives to bridge the developmental divide and to ensure regional cooperation in ICT4D. Perhaps, rather than making a long list of projects, it is time for the Association to concentrate on a few key projects each year and to undertake a full assessment of each before moving on to other areas.

Notes

1. At the 11th ASEAN Leaders' Summit in December 2005, it was agreed to put forward the goal of realizing an ASEAN Economic Community by 2015.
2. Earlier ASEAN policy statements like the e-ASEAN Framework Agreement, e-ASEAN initiative, and the Hanoi Plan of Action were discussed in the *Digital Review of Asia Pacific 2005/2006* and will not be discussed here.
3. See www.ASEANconnect.gov.my/telmin/telmin.php
4. See for instance http://www.ASEANsec.org/18490.htm for a discussion of Euro-Southeast Asia Forum on Information and

Communication Technologies 2006 and http://www.ASEANsec.org/18368.htm on the Joint Ministerial Statement: ASEAN-China ICT Ministerial Forum on 26 April 2006 in Penang, Malaysia.
5. See http://www.ASEANconnect.gov.my/telsom/wgUADDAbout.php?groupId=4
6. For the complete list, see http://www.ASEANsec.org/Roadmaps%20Annex/e-ASEAN.pdf
7. See http://www.ASEANsec.org/16690.htm
8. See http://www.aseansec.org/934.htm. Aside from ICTs, the other three key areas identified in the Hanoi Declaration on Narrowing the Development Gap for Closer ASEAN are: (a) integration of infrastructure, particularly transportation and energy; (b) human resource development, particularly the strengthening of training institutes and programmes, English proficiency, skills for the knowledge-based economy and the information age, and civil service training; and (c) regional economic integration, with emphasis on raising the capacity of the newer members to integrate their economies to the regional economy.
9. See http://www.aseansec.org/iai_update.doc for the full report.
10. For a complete list of REPSF publications, see http://www.ASEANsec.org/16945.htm

References

ASEANconnect Website. (2003–05). Available at www.ASEANconnect.gov.my

ASEAN Secretariat Website. (2003). Available at www.aseansec.org

ICT4D ASEAN Collaboratory Website. (2003). Available at www.ict4dASEAN.org

Progress of IAI Work Plan: Status Update. (September 2006). Retrieved from www.aseansec.org/iai_update.doc

Sonia, K. (2006). Brunei Action Plan for ICT savvy ASEAN. *1-BN Online*. Retrieved 20 October 2006 from http://www.onebrunei.com/news_ict_item.php?newsid=464

South Asian Association for Regional Cooperation

Atanu **Garai** and Kapil **Chawla**

Introduction

Since its founding in 1985, the South Asian Association for Regional Cooperation (SAARC) has been acting as the nodal intergovernmental agency to collaborate across the seven countries in South Asia—Bangladesh, Bhutan, India, Maldives, Nepal, Pakistan and Sri Lanka. Member states have developed cooperative agreements and linkages in the areas of agriculture and rural development; health and population activities; women, youth and children; environment and forestry; science, technology and meteorology; human resources development and transport. In these areas of development cooperation, SAARC has formed several Integrated Programmes of Action (IPA) to provide policy recommendations and coordination support to the initiatives and activities being undertaken by the member countries in cooperation with the SAARC Secretariat. Several high-level working groups have been formed in areas like biotechnology, intellectual property rights, tourism and energy.

In the field of ICT, South Asia as a region is witnessing a phenomenal growth in ICT infrastructure and ICT usage among the population. During the past five years (2000–05), countries have made tremendous progress in all sectors of information society development—connectivity, content, community, commerce, culture, capacity, cooperation and capital (Rao 2005). Overall, member countries have achieved significant progress in telecom connectivity, especially in the mobile phone segment. Against this backdrop of a burgeoning ICT penetration across the region, the role of SAARC as the leading coordinating mechanism among the governments of South Asian nations has been unremarkable. More specifically, SAARC has limited itself to initiating several dialogues among the ministers of communication, telecommunications and IT, on the subject of developing a common ICT framework.

Regional cooperation in the ICT sector

SAARC started building development cooperation in the ICT sector during the Ninth SAARC Meeting held in May 1997 in Malé, the capital of Maldives. The Ninth SAARC Summit also identified the inadequate and poor communication infrastructure prevalent in member countries as one of the significant challenges to development cooperation across the region. The Summit recommended simplifying complex documentation procedure and use of transactional software to increase economic cooperation among the member nations.

The Malé Summit led to the First SAARC Communications Ministers' Conference in Colombo in May 1998. In that conference, the Ministers adopted a Plan of Action on Telecommunications to give impetus to the telecommunications sector in the region. The Plan of Action, which was adopted during the Second Meeting of the Communications Ministers held in Islamabad in June 2004, has the following goals and objectives:

1. to promote cooperation in the enhancement of telecommunication links and utilization of information technologies within the region;
2. to minimize disparities within and among Member States in the telecommunications field;

3. to harness telecommunication technology for the social and economic uplift of the region through infrastructure development by optimal sharing of available resources and enhanced cooperation in technology transfer, standardization and human resource development; and
4. to evolve a coordinated approach on issues of common concern in international telecommunications fora.

The Second Conference also adopted a common position on issues of concern to the region in the telecommunications sector to be presented at the World Summit on the Information Society (WSIS) in Tunis in November 2005. The conference also recognized the need to develop a framework of knowledge sharing on ICT development across the region. The Working Group on Telecommunications and ICT was given the responsibility of developing ICT performance indicators. SAARC also identified the need to partner with two institutions—the SAARC Human Resource Development Centre, Islamabad and the Asia Pacific Telecommunity—to implement the proposed plan of action.

SAARC institutions working on ICT for development

SAARC works primarily through the participating government line ministries, such as the Ministry of Information and Broadcasting, Ministry of Information Technology or Ministry of Telecommunications. Apart from the SAARC Secretariat and relevant line Ministries in the national governments, several specialized organs within the SAARC Secretariat have been created to develop technical expertise, exchange human resources and foster research and learning in various areas of development in SAARC member countries. In 2004, SAARC reconstituted its High-Level Working Groups and Technical Committees. Thus the Working Group on ICT was formed in the SAARC Council of Ministers' Meeting in January 2004 in Pakistan.

In the last two decades SAARC has established several centres of excellence in the priority areas of development cooperation:

1. SAARC Agricultural Information Centre (SAIC), Dhaka
2. SAARC Meteorological Research Centre (SMRC), Dhaka
3. SAARC Tuberculosis Centre (STC), Kathmandu
4. SAARC Documentation Centre (SDC), New Delhi
5. SAARC Human Resources Development Centre (SHRDC), Islamabad
6. SAARC Coastal Zone Management Centre, Maldives
7. SAARC Information Centre, Nepal
8. SAARC Energy Centre, Pakistan

SAIC, which was set up in Dhaka in 1988, was the first regional institution established. It links several national agricultural research institutions in South Asian countries to facilitate the exchange of technical information to strengthen agricultural research and development activities in the region. The main activities of SAIC include providing information services through publication of the Directory of Agricultural Institutions, Directory of Agricultural Scientists and Technologists and Agricultural Periodicals of the SAARC Countries. It completed a Bibliography of Women in Agriculture and a Bibliography of Agro-forestry in the SAARC Region. It also maintains databases on Fish Diseases and on Potato Cultivation in the SAARC Region. Recognizing the value of ICT use in its work and in agricultural development, SAIC organized a three-day workshop on 'Attempts and Successes of ICT Roadmap to Villages in the SAARC Countries' on 6–8 October 2004 (SAIC 2004).

Media content exchange

SAARC has developed several initiatives to promote exchanges of media content among member countries. Successive meetings by the Information and Communication Ministers of SAARC member countries led to the development of a common position for the development and exchange of media and content throughout the region, as reflected in the 'Dhaka Communiqué: A SAARC Plan of Action on Media and Information.' Adopted during the First Meeting of Information Ministers of SAARC countries held in Dhaka on 25–26 April 1998, the Plan of Action articulates a Comprehensive Strategy for Cooperation in the field of Media and Information with the following goals and objectives:

1. To actively encourage greater flow of information in the SAARC region on all issues of common concern to member countries for the promotion of peace and harmony in South Asia as well as sustained development of all peoples of the region;
2. To generate, disseminate and exchange information materials in support of SAARC and all SAARC initiatives in important areas, with special emphasis on trade and investment, social and cultural development, functional cooperation, environmental protection and human resource development;
3. To promote the optimal utilization of available resources and facilities in the SAARC region to strengthen cooperation

in the field of media and information and upgrade the professionalism of media persons through human resource development programmes and regional exchanges;
4. To initiate collective regional actions to enable member countries to fully benefit from the use of new technologies to ensure greater flow of information within the region and between South Asia and the outside world; and
5. To consistently work to project and promote a positive image of SAARC abroad as well as provide regular information on specific SAARC initiatives.

In order to achieve these objectives, Information Ministers agreed to implement the following:

1. Ensure the free flow of information, newspapers, periodicals, books and other publications.
2. Reduce postal and telecommunication rates for media transmission and information materials.
3. Increase cooperation among news agencies of SAARC countries.
4. Facilitate easier travel for media persons within the region.
5. Work towards the evolution of a SAARC-recognized Regional Media Forum.
6. Hold an annual conference of editors and working journalists from SAARC countries.
7. Create a Web page for exchange of news among news agencies of SAARC countries.
8. Enhance exchange of data through e-mail and the Internet.
9. Arrange regular exchange of TV and radio programmes.
10. Organize regular exchange and joint production of documentaries and films and hold periodic SAARC film festivals.
11. Arrange training for media persons of SAARC countries.
12. Include SAARC orientation modules in the syllabi of national media training institutes.
13. Improve the programmes under the SAARC Audio Visual Exchange (SAVE) Programme by making them more attractive and popular and increasing their frequency.
14. Hold annual meetings of heads of national TV and radio organizations to review the SAVE Programmes.
15. Evolve model guidelines on transnational satellite broadcasting in the region.
16. Examine the financial and technical feasibility of establishing a SAARC satellite.
17. Explore the feasibility of setting up a SAARC Information Centre with Media Production, Research and Training units, as well as a SAARC Media Development Fund.
18. Discourage negative projection of member countries by media in SAARC countries.

The Ministers agreed that an appropriate mechanism should be set up to oversee the implementation of the SAARC Plan of Action on Media and Information. For one, the SAARC Audio Visual Exchange (SAVE) Committee has been entrusted with the task of producing and implementing the SAVE Programme. The Committee conceptualized and promoted the SAARC Radio and TV Music Festival to promote regional culture. It has also agreed to air Radio and TV Music Festivals on a regular basis.

Conclusion and recommendations

Given its status as the nodal intergovernmental coordination institution for South Asian countries, SAARC's role as a catalyst for using ICTs in regional development cannot be overemphasized. Through the SAARC Centres of Excellence and a better coordination mechanism through the governments of member countries, SAARC can help deepen the uses of ICTs in specific contexts. SAARC regional institutions, especially those dealing directly with issues related to the Millennium Development Goals, should increase their efforts to harness innovative applications of ICTs in their work, and encourage rapid exchange of knowledge and expertise within the region. SAARC should be able to work more closely with the private sector and civil society organizations that have excelled in this field to promote knowledge sharing and exchange of experts across South Asia.

References

Rao, M. (2005). India. In *From the digital divide to digital opportunities: Measuring infostates for development*. Montreal: Orbicom.

SAARC. *The Dhaka communiqué: A SAARC plan of action on media and information*. Retrieved from http://www.saarc-sec.org/main.php?t=2.3.8

SAARC Homepage. Available online at www.saarc-sec.org/main.php

SAARC News: Newsletter of the South Asian Association for the Regional Cooperation. Retrieved from http://www.saarc-sec.org/saarcnews.php?id=45

SAIC. (2004). *SAIC annual report 2004*. Retrieved from http://saic-dhaka.org/AnnualReport2004.htm

About the contributing authors

Frederick John Abo

As Technical Manager of the Public Health in Emergencies Team (PHE) of the Asian Disaster Preparedness Center (ADPC), John is responsible for providing inputs for the health and emergency response capacity building programmes of ADPC. He joined ADPC as a Training Manager under the Programme for Enhancement of Emergency Response (PEER) under which he conducted training courses on Medical First Responder (MFR) and Collapsed Structure Search and Rescue (CSSR) in South and Southeast Asian countries. A registered nurse and a paramedic by profession, John acquired extensive experience in pre-hospital emergency care in Manila, Philippines while working as a mobile intensive care unit nurse in the private EMS company Lifeline Arrows Medical Specialist and as the head of the Emergency Preparedness and Response Division of the Metro Manila Development Authority (MMDA)—Directorate for Special Operations Metro Rescue.

Musa Abu Hassan

Dr Musa Abu Hassan is Professor of Communication at the Department of Communication, Faculty of Modern Languages and Communication, Universiti Putra Malaysia. His research interests are the use and impact of media and communication technology on society and communication in the workplace. He has headed and completed two researches funded by the Malaysian Ministry of Science, Technology and Innovation (MoSTI), namely, 'The readiness of Malaysian society to accept IT' and 'Rural society and ICT projects: Development of a k-society'. Dr Musa handles graduate courses in Trends and Challenges of Mass Communication, Advance Communication Theory, Communication Research Methods, Statistics for Communication Research, and Presentation Media. His book *Roles and Utilization of ICT among Members of Society* (2002) has become a course textbook.

Ilyas Ahmed

Ilyas Ahmed is Director at the Telecommunications Authority of Maldives (TAM), where he looks after Regulatory Services. Upon completion of his MSc in Operational Telecommunications at Coventry University, United Kingdom (UK) in 2001, he joined TAM in 2002. Since then, he has been heavily involved in activities to restructure and open up the telecommunication sector of the Maldives. Prior to this, he worked for the national telecommunications company Dhiraagu as the Product Manager Mobile. He was a key member of the team that introduced GSM mobile services to the Maldives. He has extensive knowledge of ICT development in the Maldives and has worked on various development initiatives. He has also attended and actively participated in many regional and international events and conferences on ICT.

Zorayda Ruth Andam

Atty. Zorayda Ruth B. Andam is a practicing lawyer and Associate of the ANGARA ABELLO CONCEPCION REGALA and CRUZ Law Offices (ACCRALAW). She has a Bachelor of Laws from the University of the Philippines College of Law and a bachelor's degree in Business Economics, also from the University of the Philippines. As a law student, she co-authored a legal paper titled 'Regulating Communications in a Converging Environment: Technology, Markets and Dilemmas', which won the Professor Maria Clara Lopez Campos Award for Best Paper in Commercial Law in 2004. She is also co-author of *An Introduction to Electronic Commerce* (2000) and *SMEs and e-Commerce in Three Philippine Cities* (April 2003). Atty. Andam was also part of the USAID team that provided technical assistance to the Philippine government in the development and passage of the country's e-Commerce Law.

Lkhagvasuren Ariunaa

Lkhagvasuren Ariunaa is CEO of Information Technology Consulting (InTeC), based in Ulaanbaatar, Mongolia. Her company is currently involved in the implementation of a number of projects funded by the Asian Development Bank, World Bank, USAID and International Commission on Technology and Accessibility. Ms Ariunaa has authored a number of studies related to ICT development in Mongolia, including the Mongolia chapter in the 2005/2006 edition of the *Digital Review of Asia Pacific*. She is a member of the Mongolian Australian Association.

Batpurev **Batchuluun**

Batpurev Batchuluun is director of Infocon at the Information Technology Consulting (InTeC) company of Mongolia. He has been involved in a number of projects funded by the International Development Research Centre (IDRC), including ICT for help services in rural Mongolia.

Axel **Bruns**

Dr Axel Bruns lectures in the Creative Industries Faculty at Queensland University of Technology in Brisbane, Australia. He is the author of *Gatewatching: Collaborative Online News Production* (2005). With Joanne Jacobs, he is also editor of *Uses of Blogs*. In 1997 Dr Bruns co-founded *M/C—Media and Culture*, and he continues to serve as *M/C*'s General Editor.

Danny **Butt**

Danny Butt (http://www.dannybutt.net) is a consultant in new media, culture and development, and Partner at Suma Media Consulting (http://www.sumamedia.com). He was founding director of the Creative Industries Research Centre at Wintec, New Zealand. He is on the Editorial Board of the *Digital Review of Asia Pacific* and is an associate member of The ORBICOM International Network of UNESCO Chairs in Communications. His research interests centre on the social impact of new media technologies, colonization and settler culture, and the development of the creative sector in the Asia-Pacific region.

Donny **B.U.**

Donny B.U. is a senior researcher at the Indonesia-based Center for ICT Studies Foundation (known as ICT Watch) which he co-founded in 2001. He is also a faculty member and e-Business Discipline Coordinator at the Faculty of Economics of the Bina Nusantara University, Indonesia (www.binus.ac.id). He is also Managing Editor of *detikINET*, an online publication (www.detikinet.com), and a member of the Indonesian Infocom Society (www.mastel.or.id) and Indonesian Communication Scholars Post-Graduate Association. Donny has a Bachelor's degree in Computer Science and a Master's degree in Communication Management. Born in 1974 in Indonesia, he often travels to many cities in Indonesia to promote ICT and to encourage people to use it correctly and wisely. E-mail: donnybu@ictwatch.com

Elizabeth V. **Cardoza**

Elizabeth V. Cardoza holds a Bachelor of Laws (Hons.) degree (1976) from the University of Singapore, and is a member of both the California Bar and the Singapore Bar. Her Intellectual Property (IP) career began as an Examining Attorney with the Registry of Trademarks and Patents, Singapore. She later became a California Attorney-at-Law and in-house IP Counsel to VERITAS Software Corporation in the Silicon Valley. In 2002, she joined the Intellectual Property Office of Singapore (IPOS) to head its newly-established Legal Policy and International Affairs Division, with focus on domestic and international IP issues, including IP negotiations in bilateral and multilateral agreements. She speaks frequently on a range of IP issues at regional and international events, such as the International Trademark Association (INTA) and the IP Academy, Singapore, and she has presented papers (2005, 2006, 2007) at the Asia IP Law Day of the Fordham University School of Law, New York and the Fordham Annual Conference on International Intellectual Law and Policy. Ms Cardoza is now Director of CARDOZA IP, an entrepreneurial venture focused on the Asia Pacific region. She is also Adjunct Faculty at the National University of Singapore Communications and New Media programme and at the Singapore Management University School of Law. She was recently appointed to sit on the Free Trade Agreement Panel of Advisors, an initiative of International Enterprise Singapore.

Claude-Yves **Charron**

Dr Claude-Yves Charron is Secretary General of Orbicom and Vice Rector of Academic and Technological Development at the University of Quebec in Montreal, Canada. A former Canadian and UN diplomat, he has been a Professor of International Communication and Development for over 30 years. He has conducted 69 missions within the Asia-Pacific region.

Kapil **Chawla**

Kapil Chawla has more than 10 years working experience in the IT industry at various levels. His current position at the Asia Pacific Network Information Center (APNIC) as Liaison Officer for South Asia involves gathering, compiling and reporting information, to generate advice and to liaise with South Asian members of APNIC, including members from India, Pakistan, Sri Lanka,

Nepal, Bangladesh, Afghanistan, Maldives and Bhutan. He has presented on behalf of APNIC at various trainings and conferences, and recently organized a successful webcast of APNIC training. Kapil also assisted the United Nations Development Programme (UNDP) Asia-Pacific Development Information Programme (APDIP) in its Internet Governance survey. Between 2002 and 2004, he worked with the Internet Service Providers Association of India as Technical Coordinator of the National Internet Exchange of India (NIXI) project. Kapil has a postgraduate diploma in Advanced Software Design & Development and a Bachelor of Computer Science from the Jain Vishva Bharati Institute University, India.

Masoud Davarinejad

Masoud Davarinejad joined the Planning and Budget Organization of Iran in 1984 where he directed several offices for 16 years. He played a major role in promoting satellite remote sensing and was the Executive Manager of the Iran Remote Sensing Center from 1986 to 1989. In 1988, he was appointed deputy to the vice-president and as such, served as head of the High Council of Informatics (HCI) for eight years and as chair of the Board of Directors of Data Processing Iran (DPI) and Iran Argham Company (IAC). He also chaired an office known as SITTDEC Tehran which collaborated with SITTDEC (South Investment, Technology, Trade Data Exchange), an initiative of G-15 countries with headquarters in Kuala Lumpur, Malaysia. In June 2000, he was appointed the deputy mayor of Tehran in charge of planning and development of IT and GIS and he implemented a number of e-government projects. In September 2002, he joined the Ministry of PTT as deputy minister for telecommunications. With the restructuring of the telecommunications sector including the MPTT, he became president of the first telecommunications regulatory authority in Iran until June 2005. Davarinejad is now a freelance consultant on telecommunications and IT.

Deng Jianguo

Deng Jianguo is a candidate for a PhD in Communication Studies at the School of Journalism, Fudan University, China. An ex-journalist with significant media experience, he now researches new media, media convergence and international communication. He is the author of a number of journal articles in these fields and has been active in international exchanges in communication studies.

Hj Abd Rahim Derus

Haji Abdul Rahim Derus is the Assistant Director of Information and Communication Technology (ICT) Department of the Ministry of Education, Brunei Darussalam. He is actively involved in many IT projects at the national level, including e-Education projects. He is also actively involved in national and international ICT awards, namely, INFORAMA, Brunei ICT Awards (BICTA) and Asia Pacific ICT Award (APICTA). His area of interest is ICT in education and he has been involved in various researches on educational computing technology.

João Câncio Freitas

Dr João Câncio Freitas is a Timorese and is currently Executive Director of the Dili Institute of Technology. He received his PhD in government and business management from Victoria University, Australia. Since 2003, Dr Freitas has been teaching, conducting research and providing consultancy services to the Timor-Leste Government, World Bank, Asian Development Bank, UNDP and other international and national organizations. His research focuses on public and social policy development in developing countries, particularly those in post-conflict situations such as East Timor.

John Fung

Dr John Yat Chu Fung is a hybrid of social service and ICT. He has been practicing as a social worker since 1985 and, with his computer science qualification and experience, is a forerunner in ICT application in social services. Dr Fung is the Director of the Information Technology Resource Centre of the Hong Kong Council of Social Service, a unit he founded to develop the ICT capability of civil society organizations. He also founded a Digital Solidarity Fund to finance local digital inclusion initiatives. With support from the Hong Kong government, Dr Fung has organized the e-inclusion category of the first ICT Award. He is also founder of a popular senior person portal (www.cybersenior.org.hk) and founding member of the Internet Society, Hong Kong Chapter. In addition, Dr Fung is Honorary Research Associate to a range of digital inclusion research projects at the University of Hong Kong and he has lectured in a number of overseas and local universities about ICT and Society. His doctoral thesis is about empowerment of senior persons via exposure to ICT. He was a member of the Advisory Committee of the UN ITU Telecom World 2006 Forum.

Atanu Garai

Atanu Garai is an Online Networking Specialist at Globethics.net. He specializes in theoretical and practical aspects of knowledge management in development contexts, with particular emphasis on information-enabled development. His research and practice interests include electronic government, participation and democratization in information society, and information management in a networked environment. Before joining Globethics.net, Garai worked with OneWorld as a Knowledge Coordinator for e-Governance, with primary responsibility for editing the Digital Opportunity Channel, a global portal carrying research knowledge on ICT-enabled development. He is co-author of the book *Taking ICT to Every Indian Village: Opportunities and Challenges*, a historiographical account of policy and implementation of ICT-enabled development in India since the late 1990s.

Goh Seow Hiong

Goh Seow Hiong is Director, Software Policy for Asia at the Business Software Alliance (BSA). At BSA, he is responsible for representing the views of BSA member companies before governments and the marketplace in the Asia Pacific region, covering China, Japan, South Korea, Taiwan, Hong Kong, the Philippines, Thailand, Malaysia, Indonesia, Singapore, Vietnam, India and Australia. Prior to joining BSA, he was in legal practice at the Singapore law firm Rajah and Tann, advising local and foreign clients on technological issues. Goh also served in the Singapore public service for 10 years—as Deputy Director of Infocomm Development Policy of the Infocomm Development Authority of Singapore (IDA), Special Assistant to the Chief Executive Officer of IDA, and Deputy Director for Infocomm Security. Prior to the creation of IDA in 1999, he led a team responsible for the e-commerce initiatives at the then Singapore National Computer Board (NCB) and was principal staff to the chief executive and chairman of NCB. Goh helped draft the Singapore Electronic Transactions Act, its Regulations and Security Guidelines, the Evidence (Computer Output) Regulations and the Amendments to the Computer Misuse Act. He continues to be actively involved in various inter-Ministry working committees on legal and policy matters relating to ICT and e-commerce.

Lelia Green

Dr Lelia Green is Professor of Communications at Edith Cowan University in Perth, Western Australia. Author of *Technoculture* (2002), Lelia is also on the Editorial Boards of *Media International Australia* and the *Australian Journal of Communication*. Her research interests include the social and cultural dimensions of new communication technologies.

Nalaka Gunawardene

Nalaka Gunawardene counts 20 years of experience in science writing and development communication. Trained as a science journalist, he worked for several years with Sri Lanka's print and broadcast media, covering science, technology and development issues, and winning three national awards. From 1993, he has held various positions as a UN consultant, media researcher, television commentator and columnist. He has authored four books and edited several volumes on media and communications issues. He is Director and CEO of TVE Asia Pacific (www.tveap.org), a regionally operating non-profit media organization that uses television, video and new media to communicate sustainable development and social justice issues. He is also a Trustee of the Science and Development Network (www.scidev.net) and the Director of Panos South Asia (www.panossouthasia.org). E-mail: alien@nalaka.org

Shah M. Ahsan Habib

Shah M. Ahsan Habib is an Associate Professor at the Bangladesh Institute of Bank Management (BIBM), a premier national education, training, research and consultancy institute in banking and finance in Bangladesh. He is also part of the Research Division of the Development Research Network (D.Net), Dhaka, Bangladesh. He completed his MA in Economics in 1995 and a PhD in Economics (specialization: International Finance) in 2004, at the Banaras Hindu University, India. He has been visiting professor in a number of universities, including the Institute of Business Administration at the University of Dhaka. He has diversified research interests, including international trade payment and finance, international financial management, corporate social responsibility of banks and ICT and poverty reduction.

Mohd Safar Hasim

Dr Mohd Safar Hasim is a Professor of Journalism at the School of Media and Communication Studies, Universiti Kebangsaan Malaysia. His research work in ICT includes the recently completed research on the Malaysian Network Society funded under the Intensified Research in Prioritiy Areas (IRPA) of the Malaysian Ministry of Science, Technology and Innovation (MoSTI). His work in journalism includes a book, *The Press and Power: Development of the Press System in Malaysia since 1806*, published by the University of Malaya Press in 1996 and which has been reprinted four times. He has been cited in the latest issue of *Marquis' Who's Who in the World 2007*.

Sarmad **Hussain**

Sarmad Hussain is Professor and head of the Center for Research in Urdu Language Processing (www.crulp.org) at the National University of Computer and Emerging Sciences (NUCES) in Lahore, Pakistan. He is actively involved in the development of local language computing and coordinates the Script, Speech and Language Processing programme at NUCES. He also serves on committees formed by the Ministry of IT and National Language Authority to develop national IT standards for Pakistan. He has represented the Pakistan National Body at ISO's JTC1 SC2 WG2 and is also a member of Bidi-List of Unicode, which looks into issues on the standardization of bidirectional languages. Professor Hussain has completed many script and language processing projects, including the Nafees Nastaleeq and Naskh fonts, Urdu Spell-Checker, and searchable Sindhi online dictionary. He is currently leading the development of English-Urdu Machine Translation, Urdu Text-To-Speech and Urdu Computational Lexicon systems through a project funded by the e-Government Directorate of the Government of Pakistan. He is also leading the IDRC-funded PAN Localization project focusing on developing local language support for Asian languages. Professor Hussain has been awarded the Dr M.N. Azam Prize (2002) for Computer Science by the Pakistan Academy of Sciences, Government of Pakistan, for his work in language computing in Pakistan.

Jong Sung **Hwang**

Dr Jong Sung **Hwang** is the Executive Vice President of the National Information Society Agency (NIA), heading the IT Policy Division. Since he joined NIA in 1995, he has played a leading role in developing Korea's national IT strategies, including the Informatization Promotion Basic Plan in 1996, Cyber Korea 21 in 1999 and the u-Korea Master Plan in 2006. Dr Hwang also contributed to the development of IT evaluation policy by setting up evaluation methods, framework and process for public IT projects in 2001. Since 2005, he has been in charge of promoting ubiquitous technology and services, including RFID, USN and u-City. As head of the Korea USN Center in 2005 and 2006, he coordinated RFID/USN projects in the public centre. He is now the acting chairman of the steering committee of the u-Korea Forum. Dr Hwang received his PhD in Political Science from Yonsei University, Korea in 1994. He is also an adjunct professor of the Graduate School of Information and Telecommunications, Konkuk University, Korea.

Malika **Ibrahim**

Malika Ibrahim currently heads the Telecom and ICT Development Division of the Telecommunications Authority of Maldives. Prior to this, she worked in the Ministry of Communication, Science and Technology. After completing her postgraduate studies in Information Systems Management at the London South Bank University, UK in 2001, she worked mainly on ICT development projects in the Maldives, particularly with respect to ICT policy and community-based ICT. She was a key participant in the formulation of the e-Government Project. Malika received the National Award for Special Achievement from the President of the Maldives in 2003. She also lectures part-time at the Faculty of Management and Computing of the Maldives College of Higher Education and is presently a member of the Faculty's Advisory Council.

Seungkwon **Jang**

Seungkwon Jang, PhD (Lancaster University Management School, UK) is Associate Professor at the Department of Logistics and Information Systems at Sungkonghoe University, Seoul, Korea. His interests are organization theory and ICTs. His current research is concerned with the application of complexity science to organizational and technological innovations such as open source software.

Jihyun **Jun**

Jihyun Jun is a senior researcher at the National Information Society Agency (NIA), IT Policy Division. She is the principal publisher of *NIA IT Issues Weekly* which serves over 3,500 subscribers online in Korea. Her major research interest is national ICT visions and programmes of major countries and she has written on various aspects of ICT and future ICT strategies. Jihyun has a master's degree in e-Business from the Yonsei University, Seoul, Korea and a BA from Michigan State University, USA. She received the Ministerial Award from the Ministry of Information and Communication for her contribution to ICT policy research in 2004.

Keisuke Kamimura

Keisuke Kamimura is a senior research fellow at the Center for Global Communications (GLOCOM), International University of Japan. His research focus is the interaction between technology and society. He has been engaged in studies of Australia's telecommunications market and national broadband strategy and *ex post* evaluation surveys of Japanese-funded ICT projects commissioned by the Japan Bank for International Cooperation (JBIC). Keisuke has also been consulting for the Civil Society Session of the Tokyo Ubiquitous Network Conference and has been commissioned by Japan's Ministry of Internal Affairs and Communications to coordinate a fellowship programme for representatives of civil society groups to participate in the Conference. More recently, he has been engaged in research on the digital divide, multilingualism on the Internet, and network and information security policy in the US and major European countries and the European Union. He was one of the panellists for the Diversity Session at the inaugural meeting of the Internet Governance Forum in Athens in 2006. Keisuke is also a visiting senior research scientist at the Center of the International Cooperation for Computerization. He has a Master of Arts from Osaka University, Japan.

Kyungmin Ko

Kyungmin Ko received his PhD from Konkuk University in Korea, and is a Research Professor of Brain Korea 21 at the Department of Political Science and Diplomacy of the Cheju National University in Jeju, Korea. He has authored books and research papers on the political process, Internet and democracy, and ICT and development in North Korea. His main research interests are the political economy of Internet usage in non-democratic states and Internet politics in Asian (developmental) states.

Thaweesak Koanantakool

Thaweesak 'Hugh' Koanantakool received his Bachelor and PhD degrees in Electrical Engineering from Imperial College of Science and Technology, University of London, UK in 1975 and 1980, respectively. He taught at Prince of Songkla University and Thammasat University prior to 1994, when he was appointed Deputy Director of the National Electronics and Computer Technology Center (NECTEC), a national centre under the National Science and Technology Development Agency (NSTDA) of Thailand. His mission was to lead the Network/Software Technology laboratories. In 1995, Thaweesak co-founded the first Internet service provider in Thailand, Internet Thailand Company Limited (INET). During 1996–97, he led the SchoolNet Thailand Programme and the Information Superhighway testbed project at NECTEC. From 1998 until 2006, he served as Director of NECTEC. He also served as the Executive Secretary of the country's IT planning body, the National Information Technology Committee (NITC). Now Thaweesak is a Vice President of the NSTDA. He has published more than 12 books and 170 papers and articles in Thai and English.

Shelah Lardizabal-Vallarino

Until the 1 March 2007, Shelah Lardizabal-Vallarino was the Programme Officer of the Asia Pacific Economic Cooperation (APEC) Business Advisory Council International Secretariat. Prior to joining the ABAC team, she was part of the e-ASEAN Task Force Secretariat. Shelah contributed to the work of the Task Force in paving the way for the signing of the ASEAN Framework Agreement on ICT Products, Services and Investments by ASEAN Heads in 2000, a landmark step in ASEAN's efforts to create a coherent strategy for ICT development in the region. Shelah has a Master's degree in Asian Studies from the University of the Philippines.

Heejin Lee

Heejin Lee is Associate Professor at the Graduate School of International Studies (GSIS), Yonsei University, Korea. Before joining GSIS, he worked for Brunel University, UK and the University of Melbourne, Australia. He obtained his PhD at the London School of Economics. His research interests include IT/IS (information systems) in developing countries, ICT standards and electronic commerce. He is currently involved in an international research project on the digital divide in Malaysian rural communities with Monash University Malaysia.

Lawrence Liang

Lawrence Liang is an Indian legal researcher and lawyer of Chinese descent, who is based in the city of Bangalore. He is a founder of the Alternative Law Forum and as of 2006, he has emerged as a prominent spokesperson against concepts like 'intellectual property'. Lawrence graduated from the National Law School of India and subsequently pursued his Master's degree in Warwick, England on a Chevening Scholarship. His key areas of interest are law, technology and culture, and the politics of copyright. He has been working closely with Sarai, New Delhi on a joint research project on 'Intellectual Property and the Knowledge/Culture Commons'. A keen follower of the open source movement in software, Lawrence has been working on ways of translating open source ideas into the cultural domain. E-mail: lawrence@altlawforum.org

Yu-li Liu

Since February 2006, Dr Yu-li Liu has been serving as one of the Commissioners of the National Communications Commission (NCC) of Taiwan. She is also a Professor at the Department of Radio and TV at National Chengchi University, Taiwan. She has published a number of books, including *Multi-channel TV and Audience*, *Cable TV Management and Programming Strategy*, *Cable TV Programming and Policy in China*, *Radio and TV* and *Telecommunications*. Dr Liu earned her PhD in telecommunications at Indiana University, USA in 1992. She was a Fulbright visiting scholar of the Graduate Telecommunications Programme of George Washington University, USA from August 2002 to February 2003.

Geoff Long

Based in Bangkok, Geoff Long has been covering ICT developments in Asia Pacific for more than a decade. He previously edited IDRC's *Pan Asia Networking Yearbook* and has contributed chapters to UNESCO's *World Communication and Information Report* and the book *Asian Cyberactivism: Freedom of Expression and Media Censorship*. He is a former staff writer on technology for *The Australian* newspaper and has contributed articles to most of the major technology magazines in Asia Pacific. Currently he writes a popular weekly technology column in the *Bangkok Post* and a monthly column for *Network World Asia*, contributes to the *CommsDay* industry newsletter and is editor of a new weekly e-publication called *BroadBand Communities*.

Salman Malik

Salman Malik has been working for the ICT sector development in Pakistan since 2000. Initially associated with COMSATS, an intergovernmental organization of 21 member countries, and subsequently with the Ministry of IT, Pakistan, Malik has been involved in various key initiatives aimed at ICT proliferation and growth. In COMSATS, he was involved in various projects, including the creation of the Industrial Information Network for small and medium enterprises and the establishment of an IT Center in Syria. At the Ministry of IT, Malik has been contributing to the smooth implementation of deregulation policies and providing input on policy measures. He has been a key member of the teams responsible for formulating the policies for National ICT R&D Fund and Universal Service Fund. Malik is an engineering graduate from the National University of Sciences and Technology (NUST), Pakistan. He has a Master of Business Administration from Pakistan and a Master of Science in Development Finance from the UK.

Muhammad Aimal Marjan

Muhammad Aimal Marjan is Director General of the ICT Directorate of the Ministry of Communications and IT (MoC) of Afghanistan. Previously, he was National ICT Advisor to the MoC, where he advised the Minister in areas such as ICT policy and strategies, ISP licensing and regulations, Internet exchanges, cctlds, e-government and general capacity building issues. Prior to joining the ministry, Marjan served as an IT Advisor to the Director of the Afghan Assistance Coordination Authority (AACA). In addition, he has been project manager of GDLN, a World Bank project which provided eight government offices with Internet and Intranet facilities. In 1999, he founded and became President of the Afghan Computer Science Association. Marjan was also the Technical and Academic Advisor of the Computer Science Department of the Islamic University for Science and Technology (IUST), Jalalabad, Afghanistan. He remains very close to the academe, having been a visiting faculty member of the Economics Department of the International al Islamic University of Islamabad (IIUI), where he completed his Bachelor's degree in Computer Science and his Master's degree in Computer Networks, and a visiting faculty member of the Computer Science Department, Kabul University.

Jamshed Masood

Jamshed Masood has been working in the telecommunication sector of Pakistan since 1995, focusing on telecommunication technologies and regulatory and policy issues. He was part of the team that set up the Pakistan Telecommunication Authority in 1995 and brought into operation the basic telecom regulatory framework. In 1998, he joined the Privatization Commission as a consultant for telecommunication sector policy and regulatory issues arising from the privatization of PTCL, the dominant telecom operator in Pakistan. In 2000, he joined the newly established IT and

Telecommunication Ministry and led the development and implementation of the sector deregulation policy in 2004. Since then, Jamshed has worked in the private sector, setting up two long distance and international telecommunication business operations with overall responsibility for developing their business plans, long-term strategy, organizational structure, product development, sales and marketing activities, and profit and loss accounts of the Strategic Business Unit. He has also been a member of the management team of the largest private sector integrated telecom service provider in Pakistan and is credited with introducing the concept of Virtual Network Operations and wholesale services in the industry. Jamshed is a graduate of the University of Engineering and Technology, Lahore (EME College), Pakistan; University of Pennsylvania, USA and Institute of Business Administration (IBA), Karachi, Pakistan.

Ram **Mohan**

Ram Mohan is the Vice President of Business Operations and Chief Technology Officer at Afilias Limited, the world's largest TLD (Top Level Domains) registry services provider. He helped write the IDN Guidelines that has been widely adopted by registries worldwide, and is an active participant in internationalization efforts, including the ICANN President's Advisory Committee (PAC) on IDNs. He is also the Chairman of the GNSO IDN Working Group, an international multi-stakeholder group focused on policy issues surrounding IDNs. A member of the ICANN Security and Stability Committee (SSAC), Ram has served on three ICANN Nominating Committees (NomCom). He was the technical leader on the winning .ORG bid that led to the creation of the Public Interest Registry (pir.org), and he helped create the business, technology and policy foundations behind the successful .mobi and .asia TLDs. In 2004, Ram and his company were appointed as technology advisor to the Indian government to liberalize the .IN country code domain name. Working in close coordination with government, industry and civil society, Ram led the team responsible for the transformation of the .IN domain from 7,000 names registered at the start of 2005 to 200,000 names in 16 months. Ram also sits on the boards of various educational and leadership organizations in the Philadelphia, USA area. He can be reached at rmohan@afilias.info

Charles **Mok**

Charles Mok is founding chairman of Internet Society Hong Kong (ISOC HK), and immediate past president of the Hong Kong Information Technology Federation (HKITF), as well as a past chairman and a co-founder of the Hong Kong Internet Service Providers Association (HKISPA). Mok is also a director of Computancy Limited, an Internet and mobile media technology and consulting services provider, chairman of Snappa International Limited, and a director of Globe Technology Development Limited. Previously, he was Deputy Managing Director and a co-founder of HKNet Company Limited, one of Hong Kong's first Internet service providers. Mok is a member of many key Hong Kong government statutory bodies or advisory committees, including the Committee on Economic Development and Economic Cooperation with the Mainland of the Commission on Strategic Development, Consumer Council, Digital 21 Strategy Advisory Committee, Hospital Authority, Transport Advisory Committee and Non-Local Higher and Professional Education Appeal Board. Internationally, Mok sits on the Policy Advisory Board of dotMobi, the operator of '.mobi', the sponsored Top Level Domain operator dedicated to mobile Internet. Mok graduated from Purdue University in the US with Bachelor and Master's degrees in Electrical Engineering. In 1999, he was recognized as one of Hong Kong's 'Ten Outstanding Young Digi Persons'. He has been writing extensively on technology and management, currently appearing regularly in the *Hong Kong Economic Journal*, the *Sun* and *CUP magazine*.

Rapin **Mudiardjo**

Born in 1976 in Indonesia, Rapin Mudiardjo is an advocate and solicitor with the Indonesian law firm Karim Mudiardjo and Partners Law Office, which he founded. His areas of specialization are intellectual property, information technology, media and commercial litigation. Former chairman of the ASEAN Law Students Association (ALSA), Rapin received his Bachelor of Laws from and established the Research Institute of Law and Technology in the University of Indonesia. He also has a Bachelor's degree in Systems Information, major in cyber crime. He co-founded ICT Watch in 2001 and is active as a researcher and writer of articles on ICT law and policy. E-mail: rapin@ictwatch.com

Maria **Ng** Lee Hoon

Maria Ng Lee Hoon has represented IDRC's interests in the Asia and Pacific region, guiding its program of work and budget to reflect the concerns of the region, for almost three decades. She has contributed substantially to developing IDRC's Information and Communications

Technologies for Development/Information Systems/ Information Technology (IDT4D/IS/IT) programs in Asia, leading it from the disciplines of librarianship, information sciences, development communication sciences, to the Internet and networking sciences, in tune with the changing technologies. Before joining IDRC in 1976, she spent 10 years working at the Singapore Ministry of Defence and the National University of Singapore. Maria holds a Bachelor of Arts degree in English Language and Literature from the Open University, UK, and a Master of Distance Education (MDE) and Advanced Graduate Diploma in Distance Education (Technology) from Athabasca University, the Open University of Canada. She is a Chartered Member of The Chartered Institute of Library and Information Professionals in London, UK.

Frederick **Noronha**

Frederick Noronha is a journalist who focuses on ICT and ICT for development issues in India. He co-founded BytesForAll in 1999, has been the unofficial chronicler of the Free/Libre and Open Source Software movement in South Asia and is a member of the UN Global Alliance on ICT and Development Champion's Network.

Thein **Oo**

Thein Oo is president of the Myanmar Computer Federation and chairman of the ICT Standardization Steering Committee of the e-National Task Force of Myanmar, and the ASEAN Working Group for Universal Access, Digital Divide and e-Government. He is also chairman of the Myanmar Info-Tech Co-operation, and of the Working Group for Natural Language Processing Standard and Applications, as well as Secretary of the Myanmar Natural Language Implementation Committee.

Sushil **Pandey**

Sushil Pandey has a Master's degree in computer science from the Asian Institute of Technology, Bangkok, Thailand. He worked as a systems analyst at the Regional Computer Centre of the Institute prior to joining the International Centre for Integrated Mountain Development as Systems Officer. He has several years of experience on ICT technologies and has been following its development in the region. E-mail: spandey@icimod.org.np

Adam **Peake**

Adam Peake is a senior researcher at GLOCOM, Tokyo. He works on projects related to telecommunications, Internet and broadband policy, network and information security policy and trends, and follow-up activities for the World Summit on the Information Society (WSIS). Adam has been working on Internet policy related projects in the Asia Pacific region since the mid-1990s, and he has been active in policymaking activities for Internet resource allocation since then. He was a participant in the G8 Dot Force, which made recommendations to the Group of Eight nations for action to address the digital divide. Recently he was appointed as a member of the UN Secretary-General's Advisory Group on the Internet Governance Forum (IGF). He is also the Associate Chair of the ICANN (Internet Corporation for Assigned Names and Numbers) 2007 Nominating Committee. Before coming to Japan in 1989, Adam was employed at British Telecom as a project manager working on the interconnection of Other Licensed Operators (cellular radio, radio paging and competitive telephony carriers.)

Phonpasit **Phissamay**

Phonpasit Phissamay is an engineer-economist specializing in management information systems. He has been with the Lao Science, Technology and Environment Agency since 1995. Phonpasit was the manager of the PAN-Laos project, which launched the first e-mail service in the country. In 2000, he was appointed director of the Information Technology Centre and is now responsible for ICT R&D, training and services throughout the country. He is also the national project director of the Lao Localization Project and leader of the Natural Language Processing Group conducting research into ways of improving information processing in the Lao language. Phonpasit has also been appointed director of the National Internet Service Centre of the Lao National Internet Committee which operates the National Internet Exchange Point and the ISP serving all government agencies.

Gopi **Pradhan**

Gopi Pradhan of Bhutan has worked in the field of ICT and human development for more than 15 years. He took part in the implementation of the domestic telecom network project funded by the Government of Japan, and was the national coordinator of the first telecentre pilot project in Bhutan in 1996. He has worked with the UNDP's Asia-Pacific Development Information Programme on ICT and knowledge management and provided ICT advisory services to UNDP programmes in the Cook Islands, Lao PDR, Cambodia and the Maldives. In early 2005, he served as the ICT Project Manager in

Afghanistan and subsequently as research and programme consultant for the Regional Human Development Reports Unit of UNDP's Regional Centre in Colombo, Sri Lanka. He recently completed two assignments with the United Nations Economic and Social Commission for Asia and the Pacific (UNESCAP) Asian and Pacific Centre for Information and Communication Technology for Development (APCICT) and the Asian Development Bank. Gopi holds an MBA from the International University of Japan.

Ananya **Raihan**

Ananya Raihan is the Executive Director of D.Net, a premier development research institution in Bangladesh. He is also a member of the National Advisory Committee on People's Forum on MDG. Dr Raihan completed his MS in Economics (specialization: Economic Cybernetics) in 1990 and was awarded a PhD in Economics in 1994 by the V.M. Glushkov Institute of Cybernetics, National Academy of Science, Ukraine. Dr Raihan started his professional career at Kharkov State University, Ukraine as an Assistant Professor in 1993. He also served as a Senior Research Fellow at the Centre for Policy Dialogue (CPD), a civil society think tank in Bangladesh. He was Associate Professor at the Bangladesh Institute of Bank Management (BIBM) and consultant at the Bangladesh Institute of Development Studies (BIDS). Dr Raihan has done work for a number of international and regional organizations, including the UN Conference on Trade and Development (UNCTAD), World Trade Organization International Trade Centre (WTO-ITC), SAARCFINANCE, OXFAM, Centre for Trade and Development (CENTAD), GTZ, the World Bank and the International Centre for Trade and Sustainable Development (ICTSD). His diverse research interests include access to information, international trade, financial sector reform, corporate social responsibility, and SME development. In 2004, Dr Raihan was awarded the Ashoka Fellowship in recognition of his contribution as a social innovator in the area of ICT for the rural community.

Naomi **Robinson**

Naomi Robinson developed an interest in computer technology through osmosis from her father, Brian Unger. After working in the Canadian film and television industry for four years, she moved to Cambodia in 2006 to enhance her career in film and to pursue her interests in visual arts and literature. She is currently working as a journalist for Phnom Penh's *Asia Life Magazine*, and is also a freelance filmmaker.

Massood **Saffari**

Massood Saffari joined the Iran Radio and Television Organization in 1981 as an engineer and served as laboratory administrator of TV and FM radio transmitters until 1984, when he joined a senior research team at Iran Telecommunication Research Center working on the digital PABX project. For almost six years at the Planning and Budget Office, he was responsible for coordination and consolidation of provincial budgets. In 1991, he was appointed secretary of the High Council of Informatics (HCI) where, for five years, he played a key role in policymaking, drafting regulations on intellectual property and copyright, the legal and criminal aspects of the emerging cyber life, and settlement of claims. Saffari implemented the first high-speed Internet link in Iran at the Institute for Studies in Theoretical Physics and Mathematics-IPM (1996), and then the high-speed Internet link at the Data Processing Iran Co. (1997). He was a member of the Electronics Council Secretariat (1999–2001), the Board of the Electronics Support Fund for Research and Development (since 1999), the Board of Injazat Technology Fund (2001–06), and the Council of ICT Planning for Iran's Fourth Five-Year National Development Plan. He was Commissioner of the Communication Regulatory Commission in 2005–06.

Lorraine Carlos **Salazar**

Lorraine Carlos Salazar is a visiting research fellow at the Institute of Southeast Asian Studies (ISEAS), Singapore and an Assistant Professor of Political Science at the University of the Philippines in Diliman. She earned her PhD from the Australian National University in 2004, and a BA-MA Political Science (Honours) degree from the University of the Philippines in 1996. Her research and teaching interest is comparative political economy issues in Southeast Asia, focusing on the politics of market reform, political dynamics and contemporary developments in the region. She has also done work on telecommunications, e-commerce, and information and communication technology issues and policies. Her book, *Getting a Dial Tone: Telecommunications Liberalisation in Malaysia and the Philippines*, was published by ISEAS in 2007.

George **Sciadas**

Dr George Sciadas is Chief of Information Society Research and Analysis in the Innovation and Electronic Information Division of Statistics Canada, and the editor of the Connectedness Series. He is also collaborating with the OECD Secretariat for the Working Party on Indicators for the Information Society and is the Scientific Director for Orbicom's Monitoring the Digital Divide project. Dr Sciadas has worked extensively and for many years at the national and international levels on conceptual, measurement and analytical issues concerning the information society. He obtained his PhD from McGill University and has taught economics at McGill, Concordia and Carleton universities. He is a frequent participant in international events and has published numerous studies, which are quoted widely.

Basanta **Shrestha**

Basanta Shrestha has a Master's degree in computer science from the Asian Institute of Technology, Bangkok, Thailand and a bachelor's degree in electrical and electronic engineering from Madras University, India. He has extensive experience in IT and has published numerous reports and articles on various aspects of ICTs and their applications. He currently heads the Mountain Environment and Natural Information Systems Programme at the International Centre for Integrated Mountain Development. E-mail: bshrestha@icimod.org.np

Abhishek **Singh**

Abhishek Singh is an Associate at the Singapore-based Rajah and Tann's iTec Practice Group (intellectual property, Technology, entertainment and communications). He earned a BEc/LLB (Hons) from the University of Sydney, Australia in 2004. Abhishek's present practice with the iTec Group at Rajah & Tann focuses on telecommunications and technology law matters, with an emphasis on advising on regulatory and competition issues in mergers and acquisitions (M&As) and large multi-year outsourcing projects in the telecommunications sector.

Rajesh **Sreenivasan**

Rajesh Sreenivasan is a Partner at Rajah and Tann's iTec Practice Group. He has been advising clients—which include state governments, multinational corporations, government linked companies and statutory boards—on matters relating to telecommunications, electronic commerce, IT contracts, digital forensics and digital media for over 10 years. On the regional front, Rajesh has been engaged by the ASEAN Secretariat to facilitate a pan-ASEAN forum on legislative and regulatory reforms to collectively address convergence of IT, telecoms and broadcasting across all 10 member countries. Rajesh also assisted the telecoms regulator in Brunei in formulating Brunei's telecoms licensing regime. On the international front, Rajesh has been engaged by the World Bank to review IT legislation in Mongolia, to lead and conduct an electronic commerce legislative benchmarking exercise involving over 20 East Asian and Pacific nations, and in 2006, to assist in eASEAN advisory matters. Rajesh has also been engaged by the Ministry overseeing ICT in the Kingdom of Lesotho, Africa to draft its ICT legislation. In addition, he has advised state authorities in Fiji and Canada on regulatory compliance and statutory interpretation matters and leading organizations in Malaysia and Indonesia in drafting and negotiating complex IT procurement contracts and outsourcing arrangements. Rajesh has also written articles and book chapters highlighting the interplay between IT and the law in areas such as IT law, harmonization of ICT legislation, and digital forensics. An international panel of lawyers selected Rajesh to be listed in the *Guide to the World's Leading Technology, Media and Telecoms Lawyers* 2002 Edition, a Euromoney Legal expert guides publication. He has been listed as a leading lawyer in telecoms by *Global Counsel* and *International Who's Who of Telecommunications Lawyers*, noted by *Asia Pacific Legal 500*, *AsiaLaw Profiles 2001* and the *International Financial Law Review IT and Telecoms Survey*, and recognized by the *Legal Who's Who Singapore* 2003 as a leading practitioner in IT, telecoms and biotechnology law.

Krishnamurthy **Sriramesh**

Dr Krishnamurthy Sriramesh is Associate Professor at the School of Communication and Information, Nanyang Technological University, Singapore. Prior to relocating to Singapore, he had worked as Assistant Professor in the Department of Communication at Purdue University, USA and as Associate Professor in the Department of Public Relations at the University of Florida, USA. He won the Charles W. Redding Award for Teaching Excellence at Purdue University, Teacher of the Year award at the University of Florida, and the Faculty Award for Research and Golden Gator Award for research also at the University of Florida. In 2004, he was awarded the prestigious Pathfinder Award from the Institute for Public Relations (USA) for 'original scholarly research contributing to the public relations body of knowledge'. Dr Sriramesh co-edited *The Handbook of Global Public*

Relations: Theory, Research, and Practice, which won the PRIDE award presented by the National Communication Association in the United States. He has also edited *Public Relations in Asia: An anthology*. In addition, he has published over 50 journal articles and book chapters and has presented over 60 research papers, seminars and invited talks in over 20 countries. He serves as the Associate Editor of the *Journal of Communication Management* and is a member of the editorial board of the *Journal of Public Relations Research*, *Public Relations Review*, *Digital Review of Asia Pacific*, *Journal of Communication Studies* and the *Journal of Information and Knowledge Management*. Dr Sriramesh currently serves as Recognized Supervisor for PhD and Doctor of Business Administration students at Henley Management College in London, UK. He has served as a Visiting Professor at the Annenberg School for Communication of the University of Southern California (Los Angeles, USA), School of Management of Waikato University (Hamilton, New Zealand) and School of Communication of Charles Sturt University (Bathurst, Australia).

Tan Geok Leng

Dr Tan Geok Leng is the Chief Technology Officer and Senior Director (Technology and Planning) at the Infocomm Development Authority of Singapore (IDA). He is responsible for IDA's strategic planning and his team provides technology direction and consultancy to others within the organization. Dr Tan headed the team responsible for the recently launched Intelligent Nation (iN2015) Infocomm Masterplan which is Singapore's infocomm roadmap for the next 10 years. He graduated from Birmingham University, UK with a BSc (First Class Honours) in Electronics and Communications, and from Cambridge University, UK with a PhD in Engineering.

Suranart **Tanvejsilp**

Suranart Tanvejsilp recently worked for the National Electronic and Computer Technology Center (NECTEC) as Advisor to the Director and was in charge of about 40 projects relating to mobile devices application research and development. He worked for Microsoft Thailand as a Managing Consultant, for SVOA Thailand as Deputy Executive Director and for IBM (US) as Design and Test Engineer for PC. Suranart graduated from Pratt Institute, New York, USA with a BS in Electrical Engineering, from Sasin Chulalonkorn, Thailand with a Master of Management and recently from Ramkeamhang University, Thailand with a BS in Law.

Myint Myint **Than**

Dr Myint Myint Than has been Director of the Myanmar Computer Federation since 2003. She is also a member of the Security Working Group of the Myanmar Natural Language Processing Standard and Applications. Dr Than has a Bachelor of Science degree in Mathematics and a PhD in Engineering Mathematics.

Tran Ba Thai

Dr Tran Ba Thai is Director of Netnam, a company under the Institute of Information Technology (IOIT), Vietnam Academy of Science and Technology (VAST). He is one of the ICT pioneers in Vietnam, introducing and promoting Internet development in the country.

Tran Ngoc Ca

Dr Tran Ngoc Ca is Deputy Director of the National Institute for S&T Policy and Strategy (NISTPASS) and Director of the Secretariat for the National Council for S&T Policy (NCSTP) of Vietnam. He has a PhD from the University of Edinburgh, UK. He has written extensively on ICT development in developing countries and Vietnam and has published and co-edited several books and articles, such as 'Impact of policy on development of e-commerce in Vietnam' in *E-commerce in the Asian Context: Selected Case Studies* edited by Lafond and Sinha and published by IDRC and the Institute of Southeast Asian Studies in 2005, 'Harnessing ICT for development: Facing the challenges, closing the divide, grasping the opportunities' in *Globalisation and ICT: The Role of Government, Private Sector and Civil society in an Information Society for All* published by IKED, Ministry of Foreign Affairs of Sweden in 2004, *ICT for Achieving the Millennium Development Goals* published by UNDP in 2004 and *ICT Handbook for Enterprises* under the Vietnam Competitiveness Initiative funded by the USAID in 2005.

Kalaya **Udomvitid**

Dr Kalaya Udomvitid has been working as a policy researcher at Thailand's National Electronics and Computer Technology Center (NECTEC) for almost 10 years. She earned her PhD in Economics at the Colorado State University, USA. Her research centres on technology policy and its implementation particularly in electronics, telecommunication and information technology. In recent years, she has also participated in technology foresight and road mapping studies, such as the Thailand Hard Disk Drive technology and industry road mapping, and Embedded System Technology road mapping.

Brian Unger

Dr Brian Unger has lived and worked in Cambodia since 2005. He is a 'Special Advisor' for an informatics for rural development project of the Cambodian Ministry of Commerce and the Canadian International Development Research Centre (IDRC). Dr Unger founded and co-led several initiatives in Canada, including the Grid Research Centre (http://grid.ucalgary.ca/), the Netera Alliance (http://www.netera.ca/), WestGrid (http://www.westgrid.ca/) and the informatics Circle of Research Excellence (iCORE, http://www.icore.ca/). As President and CEO of iCORE, he led the investment of USD 43 million in Alberta University research chairs that now support over 500 faculty members, graduate students and research staff. Dr Unger was named a Canadian Pioneer of Computing in 2005 and received the IWAY Public Leadership award for outstanding contributions to Canada's information society in 2004. He has published over 150 refereed research journal and conference papers. He earned his PhD in Computer Science at the University of California, San Diego, and is currently a Professor Emeritus at the University of Calgary, Alberta, Canada.

Sajan Venniyoor

Sajan Venniyoor is with the UN initiative in India Solution Exchange as resource person and moderator for ICT for Development Community. He is a professional broadcaster and was until end of 2006, programme executive for DTH and narrowcasting at India's first TV broadcaster, Doordarshan (Prasar Bharati). His involvement with the community radio campaign drew him to South Asian ICT for development issues.

Eunice Hsiao-hui Wang

Eunice Hsiao-hui Wang is Associate Professor of Information and Communication at the Yuan Ze University in Taiwan. She has taught, researched and published on ICT and economic development, ICT policy reforms and convergence, competitiveness analysis of Chinese Internet content in Great China market, streaming media development in Taiwan, and the business value of e-commerce. Dr Wang is a former core member of the Cable TV Premium Review Committee appointed by the Tao Yuan County government in Taiwan. She has a PhD in Communication and Information Sciences from the University of Hawaii at Manoa.

Sangay Wangchuk

Sangay Wangchuk has over 17 years of experience in ICTs in government. For the first 10 years in government service, he served as systems analyst and developed many government applications. Currently, he is Chief ICT Officer in Bhutan's Department of Information of Technology and is responsible for formulating ICT policy, strategy and programmes. He is also a project leader of many ICT projects in Bhutan, including e-governance, Dzongkha localization, and rural access connectivity and community information centres.

Chanuka Wattegama

Chanuka Wattegama, an electronics engineer by training and with an MBA from the University of Colombo, Sri Lanka, counts more than 13 years experience at specialist and senior management level. His specialization is the use of ICT for development (ICT4D). He is also one of the key ICT4D researchers in Sri Lanka. Chanuka now works for the UNDP Asia-Pacific Development Information Programme, which promotes, designs and implements national and regional level ICT4D projects in Asia Pacific. He has also worked as a consultant for the Information and Communication Technology Agency (ICTA), the apex body for ICT-related activities in Sri Lanka, leading one of its largest projects. He also works as a senior researcher for LIRNEasia, a regional telecommunication research organization based in Sri Lanka.

Esther Batiri Williams

Dr Esther Batiri Williams is Acting Vice Chancellor at the University of the South Pacific (USP), Fiji. She has served in various capacities including Pro-Vice Chancellor, University Librarian and Director of Planning and Development at the USP and currently chairs a number of university committees. She is on the PANAsia ICT Research and Development Committee and has served on the International Federation of Library Associations and Institutions subcommittees and a number of international professional organizations. Dr Williams is on the Board of the Reserve Bank of Fiji and is a commissioner for the Fiji Commerce Commission and the Fiji Audio Visual Commission. She was a member of the Fiji Education Commission in 2000 and project manager of the European Union's project development team for the Fiji Rural Education Project in 2003. Dr Williams obtained her PhD from the University of Queensland, Australia. She has undertaken recent research in ICT curriculum and education, national political elections and telecommunications policy. She has published two books and numerous articles in the areas of information, communication, libraries and elections.

Andy Williamson

Andy Williamson is Managing Director of the New Zealand-based Wairua Consulting. Focusing on the strategic and policy aspects of ICT, Andy has undertaken research and consultancy projects relating to community informatics, educational technology and broadband. His background in strategy, policy development and evaluation, particularly relating to ICTs, is underpinned by expertise in knowledge management and information architecture. Andy is recognized as one of New Zealand's leading community ICT researchers. He is a member of the New Zealand Government's Digital Strategy Advisory Group and chairs the WaitakereOnline Portal Editorial Board. He is an associate of AUT University's Tourism Research Institute and the Centre for Community Networking Research at Monash University, Australia.

Yong Chee Tuan

Dr Yong Chee Tuan is the Director of the ICT Centre of Universiti Brunei Darussalam. He is also actively involved in lecturing, research and consultancy work related to e-government and managing IT projects. He has consulted for a number of government agencies and corporations on strategies and IT development. He was involved in the drafting of the 2000 and 2001 APEC Business Forums held at Bandar Seri Begawan and Shanghai, respectively.

Dr Yong currently leads two of the national e-education projects, the knowledge management system and the e-learning system. He is author of the Brunei Chapter in the 2003/2004 and 2005/2006 editions of the *Digital Review of Asia Pacific*.

Zhang Guoliang

Zhang Guoliang is Professor and Dean of the School of Media and Design, Shanghai Jiaotong University, China. He was the first chairman of the Chinese Association of Communication and director of the Research Center for Information & Communication of Fudan University, which is a national key base for communication studies in China. He has headed over 30 national-level communication researches, including the project on mass communication and national development. He is the author of many academic works and articles.

Zhang Xinhua

Zhang Xinhua is a researcher at the Information Research Institute of the Shanghai Academy of Social Sciences (SASS). He is also the director of the Academy's Policy and Strategy Research Center and the director of the Information Studies Center of SASS. He publishes widely on information studies, international strategy, social and enterprise changes and development strategy.

Index

ADSNet, 31
advanced communication services
　issues of literacy and language, 25
　native language capabilities, 25
Advanced language computing
　requirements for, 45
Afghanistan
　ICT(s):
　　and ICT related industries in, 90
　　challenges for growth of, 90
　　digital content initiatives, 90
　　growth in, 89
　　growth of teledensity, 89
　　education and capacity building in, 90
　　Internet penetration, 89
　　key institutions dealing with, 89
　　legal and regulatory environment for, 90
　　online services and government efforts, 90
　　penetration of telecom infrastructure, 89
AlertNet, 30
AP, 22
APEC, 67, 68
APEC Business Advisory Council (ABAC), 68
ASEAN, 67
　about, 329
　ICT education projects for CLMV, 342
　ICT(s):
　　challenges in development of, 342
　　digital content, online services, capacity building, R&D, other initiatives in, 340
　　enabling policies and programmes, 338:
　　　ASEAN ICT Focus 2005–2010, 339–340
　　　Brunei Action Plan (BAP) 2006, 338
　　　Hanoi agenda on e-ASEAN (2005), 340
　　　Roadmap for integration of e-ASEAN sector, 2004–10, 340
　　　Vientiane Action Programme (VAP) on telecom and IT sector, 340
　　key institutions, 335:
　　　ASEAN Telecommunications Regulators' Council (ATRC), 338
　　　Telecommunication and IT Ministers' Meeting (TELMIN), 336
　　　Telecommunications Senior Officials' Meeting (TELSOM), 336–338

ICT4D, ASEAN Collaboratory, 341
ASEAN-Disease-Surveillance.Net (ADSNet), 31
Asian Disaster Preparedness Center (ADPC), 31
Association of Southeast Asian Nations (ASEAN), 67
audio-visual Internet communication, 5
Australia
　and its relationship with neighbours, 99
　Copyright law in, 96
　digital content initiatives in, 93
　Education and R&D, 97
　effect of USFTA and TPMs, 96–97
　ICT(s):
　　and ICT related industries, 94
　　enabling policies and programmes, 95
　　funding schemes for research, 93–94
　　history of engagement with, 94
　　infrastructure, penetration in, 92
　　key institutions dealing with, 92
　　legal and regulatory environment for, 95
　　online services in, 93
　　open source, open content initiatives, 97
　　policy on teaching and learning of, 98
　　related research and Research Quality Framework, 97
　ICT4D, involvement with, 99
　opposition to Creative commons, 97
　privatisation of Telstra and conflicts, 96
　Research Quality Framework (RQF), 97
automatic machine language translators, 25

Babel Fish, 26
Bangladesh
　ICT(s):
　　and ICT related industries, 106
　　challenges to penetration of, 107–108
　　D.net and content delivery on demand, 104
　　D.net and telecentres, 104
　　digital content initiatives, 103
　　education and capacity building programmes, 107
　　first introduction of computers, 102
　　for poverty alleviation, 103
　　growth in outsourcing of, 106
　　growth of companies in, 106
　　key institutions dealing with, 103
　　market size and break-up, 106
　　number of companies in, 106

online services, 104
open source, open content initiatives, 107
penetration and usage status of, 102
policies and regulatory framework for, 106
R&D intiatives related to, 107
role of private sector in industry, 103
technology infrastructure, 102
telecentres, 104
telecom infrastructure and teledensity, 102
use in developmental activities for underprivileged, 102
use in e-Governance, 105
base stations, 20
cellular arrangement of, 25
Basel Ban, 6
Basel Convention 1992, 6
basic localization
initial linguistic details necessary for, 44
Bhutan
ICT(s):
agriculture, advisory services, 113–114
and rural development, 113
challenges and opportunities ahead, 115
e-governance initiatives, 111
e-governance, examples, 112
health management services, 114
in education, 114
industries, 110
information media and portals, 112–113
key institutions for, 110
Online services, 111
open source initiatives, 115
penetration status, 109
policies and regulatory framework, 110
projects, CICs—a case study, 112
Research and development in, 115
rural telecommunication initiatives, 113
scope for MNC industries, 116
technology infrastructure in, 109
Blogging, 38
blogs, numbers within Indonesia, 10
Bluetooth, 19
BPO, 7
broadband, 4
Brunei
ICT(s):
as enablers for other industries, 119
Digital content initiatives and online services, 118
e-governance initiatives, investment and efforts in, 117
enabling policies and programmes, 119
hardware industries, factors hindering set up and growth of, 119
industries, growth prospects of, 118–119
industry, need for and hindrances to growth, 120
key institutions for development of, 118
Key institutions for e-governance, 118
open source and R&D initiatives, 120
Research, hindrances, opportunities, 120
technology infrastructure, 117
vendors, aspects to be addressed by, 119
buddy call plans, 21
Business Process Outsourcing (BPO), 7

Caller-Party-Pays (CPP), 21
Cambodia
forms of entertainment, 122–123
ICT(s):
basic indicators, 123
challenges and opportunities, 129–130
comparison with neighbours, 123
donor support, 128–129
education and capacity building, 127–128
efforts to promote, 124
in education, 124–125
infrastructure, current status and growth, 122–123
key organizations responsible for, 123–125
localization initiatives, 129
long-term development vision, 126
opportunity index, 123
penetration of, 122
policy and regulation, 126–127
services, NGOs involved in, 125
skills and gaps in graduates, programmes to overcome them, 127–129
ICT4D:
current, recent donor projects, 128–129
cellular broadcasting, 34
China
e-governance, progress achieved in, 131
e-government research report, 140
ICT(s):
achievements during 2001–05:
3G mobile phone networks, 133
CPUs made in China, 132
IPv6, next generation Internet, 132
awareness and training, 131
Basic indicators, 131
contribution of industry to GDP, 131
digital content:
agricultural information, 133
e-Commerce, 133

government affairs, 133
    IT information, 133
    portal websites, 133
    top 10 portal websites, 134
    Web 2.0 content, 135
Education and capacity building programmes, 137–138
Enabling policies and development strategies, 136
Future trends, e-governance, policies and social informatization, 139–140
industries and services, 135
innovations during 2001–05, 132
penetration:
    achieved in, 132
    disparities between growth in urban and rural areas, 132
    of, 131
Regulatory environment:
    e-signatures, 137
    Internet Copyright, 137
Research, patents, indigenous chips, 139
training for leaders, public servants and professionals, 138
information development legislation and standardization, 131
Information industry, growth rate, 131
Information sector:
    progress achieved in development of, 131
information security and safety management, 131
Information society:
    Three Golds Projects, adoption of, 131
Information technology:
    areas affected by growth, 131
Informatization development report, 140, 141
National defence and military information, 131
next generation Internet, 141
online information, 131
Online services:
    e-Commerce, 135
    e-Community, 135
    e-Government services, 135
    search services, 135
Open source movement:
    adoption and penetration strategies, 138
    telecom penetration levels, 131–133
China Mobile, 25
China Next-Generation Internet (CNGI), 132
Code of Practice for Information Security Management, 8
communication:
    technological developments in, 4
    management software, 20
convergence, 4, 5, 12

copyright:
    and its impact on access to knowledge and technology, 59, 60
    case study depicting its disadvantage to visually impaired, 62
cordless phones, 24
coverage areas, 20
    of base station, 20
Creative commons, 97, 98
    in Australia, 98
creative industries, 10
crises
    faced by Asia Pacific, 29
    types of, n2, 39

data communications channels, 25
DECT, 19
developed ICT Market Countries, 10
developing ICT Market Countries, 10
development agencies, 19
development-friendly telecom regulatory policies, 26
digital divide, 12
disasters
    ICTs in risk communication of impending, 32
Domain Name System (DNS), 9

Early Warning Outbreak Recognition System (EWORS), 31
e-Commerce, 3
e-Governance, 11
e-Government
    Leading by example, 11
    and regulatory issues, 10
EIRP, 27
Electronic health initiatives, 35
Electronic waste, 5
Emerging Market Handset Programme (EMH), 21
e-waste
    positive aspects of international trade in, 6

fatpipe, 12
First Mile Solutions, 23
fixed-line telephone(s)
    penetration of, 20
    rates of growth during 2000 to 2005, 19
    users that existed in world during 2005, 19
frequency reuse, 20

game machines, 24
General Agreement on Tariffs and Trade (GATT), 59
General IPR and ICT laws, 12
General Packet Radio Service (GPRS), 20
Global Public Health Intelligence Network (GPHIN), 32
Globalization, meaning of, 43

GSM, 20
GSM Association (GSMA), Role played by, 21
GSMA's Asia Mobile Innovation Award, 25

handset(s)
  low cost, 25
  secondhand, 21
  subsidies, 21
hazard, 29
hazardous waste, 6
High Altitude Platforms, 19
High Speed Download Packet Access (HSDPA), 25
highly directional antennas, 22
home operator, 21
Home Wi-Fi market, 23
Hong Kong
  ICT(s):
    basic indicators, 142
    Content and services:
      DigitalCopyright.hk, 147
      Electronic Health Records, 148
      ICT awards, 147
      Software, 147
    digital TV broadcasts, 144
    e-readiness, 142
    Future trends and initiatives, 148
    infrastructure, 143
    Key national initiatives, policies and strategies, 144–145
    Network readiness index, 142
    Open Source Software (OSS), use of, 148
    penetration and usage statistics, 142
    R&D in, 145
    Regulatory environment, 146
Hotspots, 24
HSDPA, 25

ICANN, 9
ICT(s)
  and disaster alleviation, 6
  and economic inequality, 6
  and poverty alleviation, 6
  as tools in long-term recovery after disasters, 38
  blogging for risk communication, 38
  challenges for societies investing in, 4
  comparison between North America/Europe and Asia Pacific, 43
  Electronic health initiatives, 35
  engagement strategies for businesses, governments and individuals with, 4
  in affecting factors that exacerbate poverty, 6
  in Asia Pacific, 2005, 76, 80–85
  in risk communication of impending disasters, 32
  internationalizing, 43
  last mile communication in disaster warning, 33
  overall effects on economies, 3
  risk communication during crises, 36
  role in social and economic development, 3
  role played in risk management, 39
  technological developments in, 4
  use of tariffs to protect local manufacturers, 7
  use of Wireless LAN technology in post disaster periods, 38
ICT adoption
  advantages for developing countries, 11
  differences in policy and regulation between developed and developing ICT market countries, 15
  high-level champions, 16
  main challenges ahead, 15
  open source software, 12
  political will and high-level support, 12
  role of open content, 12
  role of policy and regulation, 12
    Managing the convergence process, 12
    Managing the digital divide, 12
ICT industries
  clustering of, 7
  tariffs to protect local, 7
ICT opportunity index
  aggregate picture of Asia Pacific region, 74
  data for specific ICTs and individual countries, 74
  in Asia Pacific, 2005, 75
  magnitude of gaps among Asia Pacific economies, 74
  Pacific island states, 2005, 78
ICT policy and regulation
  holisitic view of national and regional landscape, 13
  managing the convergence process, 12
  managing the digital divide, 12
ICT services
  peering and exchanges affecting, 7
  security in, 8
ICT skills
  e-Business skills, 15
  Practitioner skills, 15
  user skills, 15
ICT4D, 1, 3, 4, 5, 6, 14, 15, 16, 39, 105
  most common discussion agendas of, 3
  policy drivers for, 3
  policy initiatives in Asia Pacific, 3
India
  community radio stations, 151, 159
  ICT(s):
    basic indicators, 150

digital content:
    Manthan Awards, 151
    Open Access (OA), 152
    Traditional Knowledge Digital Library (TKDL), 151
    Wikipedia, 152
FLOSS:
    contributions by Indian techies to, 157
    efforts to promote, 157
    initiatives in, 156
incubation of R&D companies in telecom infrastructure, 156
Industry initiatives:
    technology solution for rural areas, 155
Internet growth, 151
local language solutions:
    Angkur Supporting Bangla, 159
    BharateeyaOO.o: OpenOffice.org in Indian Languages, 159
    developments in, 157
Media Lab Asia, 155
Mission 2007, 153
Natural Disaster Information System (NDIS), 152
penetration of, 150–151
policies and programmes, 153
R&D in e-commerce and cyber laws, 159
Regulation and security, e-Security, 156
tools for diverse groups, 158
Industry initiatives, 155
Indonesia
    ICT(s):
        and ICT related industries, 163
        basic indicators of penetration, 161
        Bandung High Tech Valley, 165
        challenges for the development of, 169
        digital content initiatives, 166
        e-Indonesia initiatives, 165
        education, capacity building efforts, 168
        Enabling policies and programmes, 165
        Key institutions responsible for development of, 163
        National information system, 165
        online services:
            e-Pabelan telecentres, 169
            available at, 166
        Open source and open content initiatives:
            books for the blind, 168
            blogs, 167
            Computer knowledge for free, 167
            National computer camp for blind, 168
        R&D in, 168–169
        Technology infrastructure, progress and plans for
            increasing penetration, 162
            use by Tax directorate, 167
Industrial, Scientific and Medical (ISM), 22
information and communication technologies (ICTs), diffusion of, 43
information security, issues in, 8
Information Security Management
    Code of Practice for, 8
information society
    challenges for, 4
    processes that characterize, 14
Information Village Research Project, 34
intellectual property
    benefits of multilateralism, 69
    bilateral agreements and imposition of TRIPS plus standards, 66
    bilateral and regional free trade agreements (FTAs), 66
    challenges with reference to open content, 66
    copyright, fair use and DRM
        scope and implications, 64
    effects of stringent enforcement vs. benefits of lower levels of enforcement, 66
    Fair use, 63
    forms of flexibilities under TRIPS, 63
    forms that impact ICT industries, 59
    FOSS, open access, open content, 65
    key issues to be addressed by policymakers in Asia Pacific, 59, 60
    Knowledge (A2K) Treaty, 65
    need for policies with public interest approach, 69
    need for policymakers to intertwine IPRs with ICT agenda, 69
    need for policymakers to strike balance between enforcement and self-reliance, 66
    case study on Asia Pacific—Singapore approach to FTAs, 67
    promoting non-proprietary models, 63
    statutory or compulsory licenses, 63
    stronger IPR enforcement, 66
    'three-step test' of copyrights, 63
    using flexibilities under TRIPS, 63
interconnection
    between mobile operators, 21
    between mobile and landline operators, 21
Intergovernmental Coordination Group for the Indian Ocean Tsunami Warning and Mitigation System (ICG/IOTWS), 33
International Health Regulation (IHR), 32
International Telecommunications Union (ITU), 20, 73
    work of, 21
Internationalization, meaning of, 43
Internationalized Domain Names (IDNs), 9

Internet:
　comparison of penetration levels, 43
　culture and local language content, 9
　technical issues to localization, 9
Internet communication
　audio-visual, 5
Internet Governance
　issues related to, 8, 9
Internet Governance Forum (IGF), 8
Internet Service Provider (ISP), 23
Internet-based language translation tools
　Google's language translation service, 26
　Yahoo's Babel Fish, 26
Iran
　ICT(s):
　　and ICT related industries, 176
　　basic indicators, 172
　　challenges for growth of, 178
　　digital content, 175
　　education and capacity building, 178
　　enabling policies and programmes for, 176
　　Key institutions dealing with, 173–174
　　penetration of, 172, 173
　　R&D intiatives, 178
　Legal and regulatory environment, 177
　online services, 175
ISO/IEC 17799, 8
iSpeech, 25, 26

Japan
　ICTs:
　　basic indicators, 180
　　digital content and life online:
　　　online gaming, 186
　　　online music, 185
　　　Social Network Services (SNS), 185
　　education and human resource development, 186
　　enabling policies and programmes, 184
　　Intellectual property rights, 185
　　Key government institutions dealing with, 183
　　penetration of, 180
　　R&D intitiatives, 187
　　technology infrastructure:
　　　Fixed-line phones, broadband and VoIP, 181
　　　mobile broadcasting (HDTV), 183
　　　mobiles, 182
　　　NGNs, 183
　　　telecom, 180

　knowledge economy, 14

language corpus, 46
language translation engine, 25
languages, break-up of number of, 43
Lao PDR
　ICT(s):
　　basic indicators, 188
　　digital content, important websites, 190
　　education, 194–195
　　industries and services, status of, 189
　　legal and regulatory environment for, 193
　　policies and programmes:
　　　e-government action plan, 192
　　　telecommunication, 191
　　Overview, 188
　　R&D, 193
　　Technology infrastructure at present and future plans, 188
　ICT4D:
　　policy framework and digital standardization for Lao information exchange, 192
　　National Internet Exchange Point, 192
　　telecentre, 192
　open source:
　　efforts in and adoption of, 193
　Report on Internet Development in, 195
localization
　advanced application development, 44
　advanced language computing application, 45
　applications for content generation, 46
　applications to provide information access, 46
　Asia Commons, 56
　Asia Open Source Software (AOSS), 56
　Asian federation on natural language processing (AFNLP), 50
　automatic speech recognition systems, 46
　basic application localization, 45
　breadth vs. depth of, 54
　Case study, PAN localization project, 54
　computing platforms, 56
　contribution of ISO, 49
　contributions of regional and international organizations, 47
　creation of IDN TLDs, principles, 49
　Creative Commons, 47
　development of IDNs for .IN domain, 48
　Dobhase project, 9
　encoding, 44
　European Language Resource Association (ELRA), 50
　fonts and rendering, 45
　FOSS initiatives, 49
　Free and Open Source Software in Asia Pacific (FOSSAP), 56
　highly localized languages, 51

human resource training, 54
IBM International Standards for Unicode (ILU), 50
information retrieval systems, 46
interactive Voice Response (IVR), 46
International Open Source Network (IOSN), 56
internationalized domain names (IDNs), 48
ISO 10646, 49
ISO 3166 for country codes, 49
ISO 639 for language codes, 49
keyboard and input method, 44
keyboard mapping, 44
language corpus, 46
language models, 46
    requirements for creating, 46
language resources, 46
language resources and vendor initiatives, 50
language understanding systems, 46
linguistic analysis, 44
Linguistic Data Consortium (LDC), 50
local language interface, 45
locale, 45
machine translation systems, 47
majority vs. minority languages, 54
meaning of, 43
Microsoft's Language Interface Packs (LIPs), 50
moderately localized languages, 52
need for effort to convert translation of policy into projects, 57
need for liberal licensing regimes, 56
non-localized languages, 53
obstacles to, 44
of Internet, 9
participatory standardization
    problems associated and need for proactive participation, 56
policy considerations in Asia Pacific for, 53
POS tagger, 46
process of, 44
regional and international partnerships and resource sharing, 56
search engines, 46
somewhat localized languages, 53
speech interface, 46
standardization, 44
status of language technology, 50
steps in, 44
text-to-speech systems, 46
very localized languages, 51
Wikipedia, 47
location-aware advertisements, 23

Macau
  ICT(s):
    basic indicators, 217
    content and services, 220
    enabling policies, 218
    future trends, 221
    in the gaming industry, 219
    infrastructure and penetration, 218
    intellectual property rights, 221
    key national initiatives:
      e-Macao, 219
      UNeGOV.net, 220
    open source community, 221
    regulatory environment, 220
machine language translators, 25
Maharashtra Emergency Earthquake Rehabilitation Programme, 30
Malaysia
  ICT(s):
    basic indicators, 196
    challenges to development of, 202
    digital content initiatives, 201
    educational programmes and capacity building, 199
    enabling policies and programmes, 198
    government efforts in promoting, 196
    industries, 198
    Key institutions dealing with, 197
    legal and regulatory environment, 198
    online services, 200
    open source initiatives, 200
    R&D in, 201
    security issues, 199
    Technology infrastructure, penetration of, 196
Maldives
  ICT(s):
    and ICT related industries, 207
    basic indicators, 204
    capacity building and R&D, 209
    challenges for development of, 209
    digital content initiatives, 206
    enabling policies and programmes, 208
    key institutions, 206
    legal and regulatory environment for, 209
    online services, 206–207
    open source initiatives, 209
    penetration of, 204
    technology infrastructure, 204
Managing the convergence process, 12
Managing the digital divide, 12
mesh networking, what does it do?, 24, 25

meshed wireless networks, 19
mobile:
    and wireless technologies
        barriers in Asia Pacific to use of, 19
        definition of, 27
        different forms in use today, traditional approach, 19
        equipment manufacturers, 21
        handsets, secondhand, 21
        impediments to, 21
        interconnection between, 21
        number portability, 21
        phone ecosystem, 21
    phone operators:
        role played by, 21
        share of SMS in revenue of, 20
    phone providers, 20
    phone system:
        comparison with Wi-Fi systems, 23, 25
        constituents of, 20
        first generation, 20
        future evolutions of, 24
        second generation, 20
        third generation, 20
    phones:
        affordability of, 20
        authentication of subscribers, 20
        billing of subscribers, 20
        design of, 20
        factors that drive the growth of, 20
        features of, 20
        first introduction of, 20
        geographical areas with prospects of growth in future, 20
        number of network operators in world, 20
        penetration, 20
        players in, 20
        prepaid subscribers, 21
        rates of growth during 2000 to 2005, 19
        roaming across national boundaries, 20
        tools for growing price sensitive market segment, 22
        users that existed in the world during 2005, 19
mode of communications
    most common in the world, 19
Model for Public Health Management of Disasters for South Asia, 30
Mongolia
    ICT(s):
        and ICT related industries, 213
        challenges for development of, 216
        digital content initiatives, 213
        education and capacity building, 215
        enabling policies and programmes, 214
        key indicators, 211
        key institutions dealing with, 212
        Last mile initiative, involvement in, 214
        online services, 213
        open source, open content initiatives, 215
        penetration of, 211
        R&D, 216
        technology infrastructure, 211
Multilingual Internet Names Consortium (MINC), 9
multimedia applications, 20
Multiple Input Multiple Output (MIMO), 22
Myanmar
    ICT(s):
        current status and penetration of, 223–224
        education, 228–229
        e-Government, 227–228
        future outlook, 229
        key institutions dealing with industry, 225
        legal and regulatory environment for, 228
        policies, 226–227

NAMRU-2, 31
Near Field Communications (NFC), 19
NECTEC, 25
Nepal
    ICT(s):
        basic indicators, 230
        digital content initiatives, 234
        education and capacity building, 234
        enabling policies and programmes, 231
        hardware and software, 232
        industries, 232
        Internet service providers, 233
        IT enabled services, 232
        legal and regulatory environment, 231
        online services, 233
        open source and open content, 234
        R&D in, 235
        security, 232
        Technology infrastructure:
            optical fibre network and Asian super highway, 230
        telecom sector, 232
    ICT4D, key institutions, 233
Network neutrality, 7
Networks index, in Asia Pacific, 74, 75
New Zealand
    ICT(s):
        basic indicators, 236
        digital content initiatives, 236

educational programmes, 241
enabling policies and programmes, 238
example of online safety initiative, 237
industries, 238
legal and regulatory environment, 239
online services, 237
open source, 239
R&D, 240
security issues, 240
technology infrastructure, 241
ICT4D, key institutions, 240
North Korea
ICT(s):
basic indicators, 244
challenges for development of, 248
cooperation between north and south, 248
digital content and online services, 247
education, 246
industries, 245
infrastructure, 244
open source initiatives, 247
single leap strategy and key institutions, 245–246

Office Wi-Fi market, 22
off-net calls, 24
omni-directional antennas, 22
on-net telephone calls, 24
'one-seg' telephones, 5
open content, 12
outsourcing, 7

Pacific Disease Surveillance Network (PACNET), 31
Pacific Island States
ICT(s):
capacity building, 255
current development, proliferation of, 251
in education, 256
infrastructure and development, 252
local online content, 253
networks, 256
online services and industries, 253
open source movement, 256
Pacific Plan Digital Strategy, 253–254, 259
policy, 253
trends in, 258
Pacific Tsunami Warning Centre (PTWC), 33
Pakistan
ICT(s):
and ICT related industries, 263
basic indicators, 263

current proliferation of, 263
education and capacity building programmes, 265
e-Government services, 265
enabling policies and programmes, 264
key institutions dealing with, 264
open source initiatives, 266
R&D, 266
Pan Asia Networking Program, 26
PAN Localization Project, 26
Pandemic Alert and Response, 32
pay-as-you-go price plans, 21
Peering and exchanges, 7
Philippines
ICT(s):
basic indicators, 267
education and capacity building, 275
infrastructure, 268–270
laws, 270
next BPO hub?, 273
open source initiatives, 276
plans and initiatives, 272
proliferation of, current, 267
regulatory bodies, 271
web presence, 275
ICT4D projects, 274
photography website, 5
prepaid service plans, 21
prepaid subscribers, 21
Public Hotspot market, 23, 24
public–private partnerships (PPPs), 10

radio frequency spectrum
quantum and pricing of, 21
Receiving-Party-Pays, 22
regulatory policies, 26
risk, meaning of, 29
risk communication
cellular broadcasting, 34
forecasting crises, 33
ICTs in the face of impending disasters, 32
importance of ICTs in, 39
Information Dissemination Project, 35
Information Village Research Project, 34
last mile communication in disaster warning, 33
meaning of, 29
role of political commitment and government, 39
role of regional efforts in, 39
role of telecentres in, 39
role of telecommunications regulators, 39
types of media used for, 29

risk management
- alternative communication channels in the aftermath, 37
- ASEAN-Disease Surveillance Net (ADSNet), 31
- cellular broadcasting, 34
- ICT network of disaster management, Maharashtra, 30
- ICTs as tools for assisting long term recovery post disaster, 38
- last mile communication in disaster warning, 33
- Last Mile Hazard Information Dissemination Project, 35
- long-term programmes for ICT use in, 30
- Long-term recovery, 38
- Model for Public Health Management of disasters for South Asia, 30
- Reuter's AlertNet, 30
- role of ICT in, 30
- role of the Internet, 37
- use of new media such as blogging, 38
- Wireless LAN technology as ICT tool in post disaster periods, 38

SAARC
- about, 345
- exchanging media content, 346
- ICT sector, regional cooperation in, 345
- ICT4D, institutions working on, 346

satellite telephone networks, 19
secondhand handsets, 21
Security, 8
Singapore, 23
- ICT(s):
  - basic indicators, 278
  - enabling policies:
    - cyber security, 283
    - human resource development, 283
  - future trends, 286
  - industries, 280
  - key national initiatives:
    - bridging the digital divide, 282
    - Broadcasting, 282
    - interactive and digital media, 282
    - multilingual support, 283
    - Singapore iN2015 Masterplan, 281
  - online gaming, 281
  - online services:
    - e-Government, 279
    - e-Lifestyle, 280
    - e-Payment, 279
    - Web services, 280
  - open source movement, 286
  - proliferation of, current, 278
  - R&D, 286
  - regulatory environment:
    - ETA review, 284
    - Internet regulation, 281
    - number portability, 281
    - spam control bill, 284
    - telecommunication competition code, 281
    - VoIP framework, 284
  - technology infrastructure, 278–279

skill-based technological change, 14
Skype, 24
SMS (Short Message Service)
- capabilities of, 20
- heaviest and most innovative users of, 26
- in Khmer language, 25
- in Korean language (Hangul), 25
- in Tamil language, 25
- innovative uses of, 26
- numbers sent during 2006, 20
- reasons for popularity of, 26
- sent in Philippines during 2006, 26
- translating English text messages to Thai, 25

South Korea
- ICT(s):
  - and ICT related industries, 292
  - basic indicators, 289
  - challenges to the development of, 294–295
  - digital content initiatives:
    - Art, culture and history, 291
    - Korea knowledge portal, 291
    - national knowledge contents digitization strategy, 290
    - online job information services, 291
    - sharing public service information database, 291
  - education and capacity building in, 293
  - enabling policies and programmes, 292
  - key institutions dealing with, 290
  - legal and regulatory environment for, 293
  - open source initiatives, 293
  - R&D in, 294
  - technology infrastructure, current, 289

South Pacific Applied Geoscience Commission (SOPAC), 30
speech recognition program, iSpeech, 25
speech-to-speech, 26
speech-to-speech translation application, 26
Sri lanka
- ICT(s):
  - and ICT related industry, 300
  - challenges to the development of, 302
  - contribution to open source research on ICTs in Asia, 302
  - digital content initiatives, 298
  - education and capacity building, 301

enabling policies and programmes, 300
infrastructure:
    penetration of Internet, 297
    penetration of telephony, 295
key institutions dealing with, 297
legal and regulatory environment for, 297
online services:
    e-Commerce, 299
    e-Government, 299
    search engines, 299
open source Initiatives, 301
R&D, 301
standards and regulations, 21
Strengthening Animal Health Management and Biosecurity in ASEAN (SAHMBA), 32
subscriber identification number, 20
subscriber identification module (SIM), 21

tablet personal computers, 24
Taiwan
    ICT(s):
        basic indicators, 304
        challenges and opportunities for, 311
        digital content initiatives:
            Cultural and Creative Industries Promotion Office, 307
            Digital Content Promotion Office, 306
            National Digital Archives Program, 306
        education and capacity building, 310
        enabling policies and programmes:
            Challenge 2008 National Development Plan, 308
            e-Taiwan, 309
            m-Taiwan for a ubiquitous Network Society, 309
        industries, 308
        key institutions dealing with, 305–306
        legal and regulatory environment for, 309
        online services:
            online business, 307
            online games, 307
            online music, 307
            online video entertainment, 307
        open source, open content initiatives, 311
        R&D initiatives, 311
        technology infrastructure:
            Broadband network, 304
            digital radio and digital TV broadcasting, 305
            IPv6, 304
            mobile communications, 305
technological change, skill-biased, 14
technological developments:
    electronic waste, environmental impacts, 5
    in Broadband, 4

in communication, 4
in Convergence, 5
in Wireless, 5
reaction of regulators to, 5
that reshape social and economic opportunities, 4
telecommunications ducts and pipes, 21
telecommunications regulator(s)
    role in development of telecom, 21, 26
telecommunication towers and exchanges, 21
Text messages
    sending and receiving in Tamil language, 25
text-to-speech conversion software, 25
Thailand
    Emergency and educational communcation vehicle, 320
    ICT(s):
        and ICT related industries, 313
        basic indicators, 313
        campaign for open standards, 319
        challenges for development of, 320
        digital content initiatives, 315
        education and capacity building programmes, 317
        enabling policies and programmes, 316
        key institutions dealing with, 314
        legal and regulatory environment, 316
        online services, 315
        open source software initiatives, 318
        R&D, 319
        technology infrastructure, 313–314
Timor-Leste
    ICT(s):
        basic indicators, 321
        broadcasting services, 323
        Internet, 323
        legal and regulatory framework for telecommunications, 324
        National initiatives in, 324
        telecommunications infrastructure, 321
translators, 25
    speech-to-speech, 26
    English-to-Malay, 26
TRIPS, obligations under, 59

UN Working Group on Internet Governance (WGIG), 8
Universal Service Provision (USP), 27
universal telecommunication services, 19

Vaja, 25, 26
Vietnam
    ICT(s):
        basic indicators, 325
        challenges to development of, 331

    digital content initiatives, 326
    education and capacity building, 330
    enabling policies and programmes, 329
    industries, 328
    key institutions dealing with, 328
    laws and regulation, 329
    online services, 327–328
    open source initiatives, 330
    R&D, 330
    security issues, 330
    Technology infrastructure, 325–326
voice communications, 20, 25
Voice over Internet Protocol (VoIP), 4, 24

wage inequality, 14
WHO Global Alert, 31
wideband digital communications, 20
Wi-Fi
    Alliance, 22
    card, 22
    comparison with mobile phones, 25
    equipped motorcycles (Motomen), 23
    Future evolutions of, 24
    LAN systems, 19
    radio network, 23
    recent developments in, 24
    systems:
        comparison with mobile phone, 23
        setting up of, 23
        weaknesses of, 24
WiMAX, 19
    Forum, 24
    LAN technology, 24
WIPO, 60
WIPO Copyright Treaty (WCT), 63
wired, copper-based infrastructure, 19
wireless
    and mobile space, 4, 5
    current technologies, 27
    definition of, 27
    local area network (WLAN), 22
    local area networking, 22
    technologies, 19
Wireless@Sg, 23, 24
WLAN, 22, 27
World Intellectual Property Organization (WIPO), 60
world Internet population, 43
World Trade Organization (WTO), 26, 59